Waste to Profit

Waste to Profit: Environmental Concerns and Sustainable Development gives information about selecting the most suitable technology for waste treatment and energy recovery under different conditions. It contains techno-economic analysis, life cycle assessment, optimization of tools and technologies, including overview of various technologies involved in the treatment of wastes and factors influencing the involved processes. Finally, it explores the environmental, socioeconomic, and sustainability impact of different waste-to-energy systems.

Features:

- Reviews energy sources and technologies from waste, their environmental interactions, and the relevant global energy policies
- Provides overview of waste-to-energy technologies for a sustainable future
- Explores physicochemical properties involved in the pertinent process and technologies
- Gives a multidisciplinary view about energy conversion and management, planning, controlling, and monitoring processes
- Discusses information in transferring the technologies' industrial level and global level to meet the requirements of different countries

This book is aimed at researchers and graduate students in environmental engineering, energy engineering, waste management, waste to energy, and bioenergy.

Waste to Profit
Environmental Concerns and Sustainable Development

Edited by
Meera Sheriffa Begum K.M., Anand Ramanathan,
Amaro Olimpio Pereira Junior, Dmitrii O. Glushkov, and
M. Angkayarkan Vinayakaselvi

CRC Press is an imprint of the
Taylor & Francis Group, an **informa** business

Designed cover image: © Shutterstock

First edition published 2023
by CRC Press
6000 Broken Sound Parkway NW, Suite 300, Boca Raton, FL 33487–2742

and by CRC Press
4 Park Square, Milton Park, Abingdon, Oxon, OX14 4RN

CRC Press is an imprint of Taylor & Francis Group, LLC

© 2023 selection and editorial matter, Meera Sheriffa Begum K.M, Anand Ramanathan, Amaro Olimpio Pereira Junior, Dmitrii O. Glushkov, and M. Angkayarkan Vinayakaselvi; individual chapters, the contributors

Reasonable efforts have been made to publish reliable data and information, but the author and publisher cannot assume responsibility for the validity of all materials or the consequences of their use. The authors and publishers have attempted to trace the copyright holders of all material reproduced in this publication and apologize to copyright holders if permission to publish in this form has not been obtained. If any copyright material has not been acknowledged, please write and let us know so we may rectify in any future reprint.

Except as permitted under U.S. Copyright Law, no part of this book may be reprinted, reproduced, transmitted, or utilized in any form by any electronic, mechanical, or other means, now known or hereafter invented, including photocopying, microfilming, and recording, or in any information storage or retrieval system, without written permission from the publishers.

For permission to photocopy or use material electronically from this work, access www.copyright.com or contact the Copyright Clearance Center, Inc. (CCC), 222 Rosewood Drive, Danvers, MA 01923, 978–750–8400. For works that are not available on CCC please contact mpkbookspermissions@tandf.co.uk

Trademark notice: Product or corporate names may be trademarks or registered trademarks and are used only for identification and explanation without intent to infringe.

Library of Congress Cataloging-in-Publication Data
Names: Begum, K. M. Meera Sheriffa, editor.
Title: Waste to profit : environmental concerns and sustainable development / edited by Meera Sheriffa Begum K.M., Anand Ramanathan, Amaro Olimpio Pereira, Dmitrii Glushkov, Angkayarkan Vinayakaselvi.
Description: First edition. | Boca Raton, FL : CRC Press, [2023] | Includes bibliographical references and index.
Subjects: LCSH: Refuse and refuse disposal—Economic aspects. | Recycling (Waste, etc.)—Economic aspects. | Sustainable development. | Waste products as fuel. | Renewable energy sources.
Classification: LCC TD793 .W39 2023 (print) | LCC TD793 (ebook) | DDC 628.4/458—dc23/eng/20230111
LC record available at https://lccn.loc.gov/2022058801
LC ebook record available at https://lccn.loc.gov/2022058802

ISBN: 978-1-032-36906-8 (hbk)
ISBN: 978-1-032-36907-5 (pbk)
ISBN: 978-1-003-33441-5 (ebk)

DOI: 10.1201/9781003334415

Typeset in Times
by Apex CoVantage, LLC

Contents

About the Editors ix
List of Contributors xi
Preface..xv

Chapter 1 Crop Residue to Fuel, Fertilizer, and Other By-products: An Approach toward Circular Economy 1

Subarna Maiti, Himanshu Patel, and Pratyush Maiti

Chapter 2 Cost-effective Sustainable Electrodes for Bioelectrocatalysis toward Electricity Generation 15

Samsudeen Naina Mohamed, Meera Sheriffa Begum K.M., and Vigneshhwaran Ganesan

Chapter 3 Green Synthesis of Nanoparticles from Agro-waste: A Sustainable Approach 27

Muthukumar K., Jayapriya M., Arulmozhi M., Senthilkumar T., and Krithikadevi R.

Chapter 4 Conversion of Waste Plastics into Sustainable Fuel 41

Mohanraj C., Senthilkumar T., Chandrasekar M., and Arulmozhi M.

Chapter 5 Innovations in Sludge-Conversion Techniques 53

Chithra K.

Chapter 6 Scale-up of Microbial Fuel Cells: A Waste-to-Energy Option 69

Swathi S., Akanksha R., Karthick S., Sumisha A., Karnapa A., and Haribabu K.

Chapter 7 Critical Role of Catalysts in Pyrolysis Reactions 93

Anjana P. Anantharaman

Chapter 8 Engineering Perspectives on the Application of Photosynthetic Algal Microbial Fuel Cells for Simultaneous Wastewater Remediation and Bioelectricity Generation 103

Baishali Dey, Nageshwari Krishnamoorthy, Rayanee Chaudhuri, Alisha Zaffer, Sivaraman Jayaraman, and Balasubramanian Paramasivan

Chapter 9 Pyrolysis and Steam Gasification of Biomass Waste for Hydrogen Production 121

Prakash Parthasarathy, Tareq Al-Ansari, Gordon McKay, and K. Sheeba Narayanan

Chapter 10	Insight into Current Scenario of Electronic Waste to Nanomaterials Conversion........133	

Menaka Jha, Sunaina, Sapna Devi, and Nausad Khan

Chapter 11 Wastewater Treatment Using Nanoadsorbents Derived from Waste Materials........147

Menaka Jha, Sunaina, Arushi Arora, and Kritika Sood

Chapter 12 Environmental Sustainability: An Interdisciplinary Approach..............................165

M. Angkayarkan Vinayakaselvi

Chapter 13 Environmental Movements and Law in India: A Brief Introduction......................177

M. Angkayarkan Vinayakaselvi and Abinaya R.

Chapter 14 Co-pyrolysis of Biomass with Polymer Waste for the Production of High-quality Biofuel..189

Dineshkumar Muniyappan and Anand Ramanathan

Chapter 15 Thermal Degradation Behaviors and Kinetics of Pyrolysis.....................................205

Uthayakumar Azhagu and Anand Ramanathan

Chapter 16 Techniques for Biodiesel Production from Wastes...221

Gopi R. and Anand Ramanathan

Chapter 17 Environmental Impact of Municipal Waste Energy Recovery Plant......................233

Mane Yogesh G. and Anand Ramanathan

Chapter 18 A Comprehensive Review on the Modeling of Biomass Gasification Process for Hydrogen-rich Syngas Generation...241

Kalil Basha Jeelan Basha, Sathishkumar Balasubramani, and Vedharaj Sivasankaralingam

Chapter 19 A Comprehensive Review of Bio-catalyst Synthesis, Characterization, and Feedstock Selection for Biodiesel Synthesis Using Different Methods...................261

Babu Dharmalingam, Malinee Sriariyanun, Anand Ramanathan, Santhoshkumar A., Selvakumar Ramalingam, Deepakkumar R., and Kasturi Bhattacharya

Chapter 20 A Techno-economic Analysis of Green Hydrogen Production from Agricultural Residues and Municipal Solid Waste through Biomass-steam Gasification Process 271

Pon Pavithiran C.K., P. Raman, and D. Sakthivadivel

Contents vii

Chapter 21 The Energy Potential of Brazilian Organic Waste 287

Luciano Basto Oliveira, Amaro Olímpio Pereira Júnior, Ingrid Roberta de França Soares Alves, Marcelo de Miranda Reis, and Adriana Fiorotti Campos

Chapter 22 Life Cycle Sustainability Assessment of Bioenergy: Literature Review and Case Study 297

João Gabriel Lassio and Denise Ferreira de Matos

Chapter 23 Environmental, Social, and Economic Aspects of Waste-to-energy Technologies in Brazil: Gasification and Pyrolysis 311

Suani Teixeira Coelho, Luciano Infiesta, and Vanessa Pecora Garcilasso

Chapter 24 Life Cycle Assessment of Lubricant Oil Plastic Containers in Brazil: Comparing Disposal Scenarios 323

Maria Clara Brandt and Alessandra Magrini

Chapter 25 Step toward Sustainability: Fuel Production and Hybrid Vehicles 335

Manoj Eswara Vel S.B., Chandru R., Dhanalakshmi K., Joshua George Stanly, and Anand Ramanathan

Chapter 26 Microwave Pyrolysis of Composite Fuels with Biomass 349

Dmitrii O. Glushkov, Pavel A. Strizhak, Anatoly S. Shvets, and Ksenia Y. Vershinina

Chapter 27 Combustion and Pyrolysis Characteristics of Composite Fuels with Waste-derived and Low-grade Components 363

Galina S. Nyashina, Pavel A. Strizhak, and Ksenia Y. Vershinina

Chapter 28 Analysis of Gaseous Anthropogenic Emissions from Coal and Slurry Fuel Combustion and Pyrolysis 377

Mark R. Akhmetshin, Galina S. Nyashina, and Pavel A. Strizhak

Chapter 29 Allothermal Approach for Thermochemical Conversion of Coal-enrichment Waste 389

Roman I. Egorov, Roman I. Taburchinov, Zhenyu Zhao, and Xin Gao

Chapter 30 Membrane Technology in Circular Economy: Current Status and Future Projections 399

Lukka Thuyavan Yogarathinam, Ahmad Fauzi Ismail, Pei Sean Goh, and Anatharaman Narayanan

Index 411

About the Editors

Meera Sheriffa Begum K.M. graduated from Anna University, Chennai. She worked at Chennai Petroleum Corporation Ltd. (formerly Madras Refineries Ltd.) in R&D division as "MRL Research Fellow" from 1991 to 1995. She is currently Professor in the Department of Chemical Engineering at National Institute of Technology, Tiruchirappalli, India. She has received many best paper awards in international conferences, has been granted two patents, and has published more than 100 articles in renowned international journals (h-index: 28; i-index: 49; citations 2550). Her research interest is in the fields of separation processes, functional materials for water treatment, alternate fuels, and medical applications. She has authored three chemical engineering textbooks – *Process Calculations*, *Elements of Mass Transfer (Part I)*, and *Mass Transfer Theory and Practice* – and coauthored *A Thermo Economic Approach to Energy from Waste*. She has also authored seven book chapters. Under her supervision, she has guided several Ph.D. scholars, postgraduates, and undergraduates. She has contributed research, sponsored, and provided consultancy on projects on sustainable functional materials toward environment and energy applications funded by MHRD, DST-SERB, CSIR, MHRD-SPARC, and DST-BRICS. She has undertaken research collaboration and training at NUS, Singapore, through TEQIP in 2006 and has involved herself in consultancy projects on downstreaming applications in distilleries, dairy industries, Southern Railways, etc., professionally related activities, and administrative responsibilities to serve the community.

Anand Ramanathan is currently Associate Professor in the Department of Mechanical Engineering and Associate Dean (Research & Consultancy) in Industry & Outreach at National Institute of Technology, Trichy. He is a recipient of the Australian Endeavour Fellow and worked in the area of solar fuels at Australian National University, Canberra, Australia, in 2015. He also serves as Expert Member of the Centre of Excellence in Corrosion and in Corrosion and Surface Engineering. As a recognition of his contributions, he was awarded N.K. Iyengar Memorial Prize in 2009, Outstanding scientist VIRA Award in 2017, Dr. Radhakrishnan Award in 2018, and NITT's Best Innovator Award in 2022. His area of specialization is Internal Combustion Engines, and it expands to the field of alternative fuels, waste-to-energy conversion technologies, emission control, and fuel cells. He has guided eight Ph.D. and four MS research scholars and is at present guiding six Ph.D. scholars. His research-oriented scholarship has facilitated him to publish more than 60 Science Citation SCI/Scopus Indexed research journals and presented papers in several international conferences besides presenting a paper in Applied Energy (ICAE2018 – Hong Kong), ASME, USA, and SAE International conferences at Detroit, USA. He has been granted four Indian patents in the area of biocatalyst and biofuel. He has visited Germany, Hong Kong, Italy, France, the United States, and Bangladesh for academic collaborations. He has authored 3 books and 11 book chapters in renowned publications. He has received sponsored projects from IEI-India, DST-SERB, DST-YSS, DST-UKERI, DRDO-GTRE, MHRD-SPARC, and DST-BRICS and developed well-equipped Fuels Laboratory in the Department of Mechanical Engineering at NIT, Trichy. He has undertaken Consultancy for Industry and Administrative responsibilities at NITT.

Amaro Olimpio Pereira Junior is Economist. He has a Ph.D. in Energy Planning from Federal University of Rio de Janeiro. He has worked as technical advisor in the Energy and Environment Department at Energy Research Company (EPE) in Brazil. He has served as Visiting Professor at the Pierre Mendès-France University in Grenoble, France, and at the University of Texas at Austin, Texas. He has worked as Research Fellow at CIRED (Centre International de Recherche sur l'Environement et Dévélopement) in France. Currently, he is Associated Professor of the Energy Planning Program of the Institute of Graduate Studies in Engineering at Federal University of Rio de Janeiro (PPE/COPPE/UFRJ), Researcher at Centro-Clima, Director of the Institute for Strategic

Development of the Energy Sector – ILUMINA and Member of the permanent technical committee at LIFE. He has experience in energy and environmental modeling, besides working in the areas of regulation of energy sectors, integration of new technologies, and different energy sources on issues related to climate change. He is author of books, book chapters, and several papers in international journals.

Dmitrii O. Glushkov has a Ph.D. in Physical and Mathematical Sciences. He is Associate Professor at the National Research Tomsk Polytechnic University, Tomsk, Russia, and has 11 years of research and teaching experience. He is a specialist in combustion theory. His fields of research are combustion processes of composite solid propellants and gel fuels, co-combustion of solid fossil fuels and municipal solid waste, microexplosion of composite fuels, and hot spot ignition. Dr. Glushkov is the head of more than ten major research projects. The main scientific results of over the past 5 years have been published in more than 25 highly rated journals and three monographs.

M. Angkayarkan Vinayakaselvi, Associate Professor of English, Bharathidasan University, Tiruchirappalli, India, has 22 years of teaching experience, including her service at University of Madras, Chennai, and Mother Teresa Women's University, India. She has won the Travel Grant Award to present paper in North East Modern Language Association, USA. She has authored a book and several research articles and has won Best Presentation and Best Paper Awards. Currently, she is specializing in Environmental Humanities.

Contributors

Santhoshkumar A.
Kongu Engineering College, Perundurai, Tamil Nadu, India

Sumisha A.
Department of Chemical Engineering, National Institute of Technology Calicut, Kozhikode, Kerala, India

Karnapa A.
Department of Chemical Engineering, National Institute of Technology Calicut, Kozhikode, Kerala, India

Mark R. Akhmetshin
National Research Tomsk Polytechnic University, 30, Lenin Avenue, Tomsk, Russia

Tareq Al-Ansari
Division of Sustainable Development, College of Science and Engineering, Hamad Bin Khalifa University, Qatar Foundation, Doha, Qatar

Ingrid Roberta de França Soares Alves
Military Institute of Engineering – IME, Rio de Janeiro, Brazil

Anjana P. Anantharaman
Department of Chemical Engineering, National Institute of Technology, Warangal, Tamil Nadu, India

Arushi Arora
Institute of Nano Science and Technology, Sahibzada Ajit Singh Nagar, Punjab, India

Uthayakumar Azhagu
Bioenergy Laboratory, Department of Mechanical Engineering, National Institute of Technology, Tiruchirappalli, Tamil Nadu, India

Sathishkumar Balasubramani
Clean Combustion Laboratory, Department of Mechanical Engineering, National Institute of Technology, Tiruchirappalli, Tamil Nadu, India

Kalil Basha Jeelan Basha
Clean Combustion Laboratory, Department of Mechanical Engineering, National Institute of Technology, Tiruchirappalli, Tamil Nadu, India

Kasturi Bhattacharya
Vellore Institute of Technology, Vellore, Tamil Nadu, India

Maria Clara Brandt
Energy Planning Programme – COPPE/UFRJ, Rio de Janeiro, Brazil

Mohanraj C.
Department of Mechanical Engineering, M. Kumarasamy College of Engineering, Karur, Tamil Nadu, India

Adriana Fiorotti Campos
Federal University of Espírito Santo – UFES, Vitória, Espírito Santo, Brazil

Rayanee Chaudhuri
Department of Biotechnology and Medical Engineering, National Institute of Technology Rourkela, Odisha, India

Suani Teixeira Coelho
Research Group on Bioenergy (GBIO), Institute of Energy and Environment (IEE), University of São Paulo (USP), São Paulo, Brazil

Sakthivadivel D.
School of Mechanical Engineering, Vellore Institute of Technology, Vellore, Tamil Nadu, India

Sapna Devi
Institute of Nano Science and Technology, Sahibzada Ajit Singh Nagar, Punjab, India

Baishali Dey
Department of Biotechnology and Medical Engineering, National Institute of Technology Rourkela, Odisha, India

Babu Dharmalingam
King Mongkut's University of Technology, North Bangkok, Bang Sue, Bangkok, Thailand

Roman I. Egorov
National Research Tomsk Polytechnic University, Tomsk, Russia

Mane Yogesh G.
Thermal Engineering Laboratory, Department of Mechanical Engineering, National Institute of Technology, Tiruchirappalli, Tamil Nadu, India

Vigneshhwaran Ganesan
Department of Chemical Engineering, National Institute of Technology, Tiruchirappalli, Tamil Nadu, India

Xin Gao
School of Chemical Engineering and Technology, Tianjin University, Tianjin, China
National Engineering Research Center of Distillation Technology, Collaborative Innovation Center of Chemical Science and Engineering (Tianjin), Tianjin, China

Vanessa Pecora Garcilasso
Research Group on Bioenergy (GBIO), Institute of Energy and Environment (IEE), University of São Paulo (USP), São Paulo, Brazil

Dmitrii O. Glushkov
National Research Tomsk Polytechnic University, Tomsk, Russia

Pei Sean Goh
Advanced Membrane Technology Research Centre (AMTEC), School of Chemical and Energy Engineering, Universiti Teknologi Malaysia, Skudai, Johor, Malaysia

Luciano Infiesta
Carbogas LTDA, São Paulo, Brazil

Ahmad Fauzi Ismail
Advanced Membrane Technology Research Centre (AMTEC), School of Chemical and Energy Engineering, Universiti Teknologi Malaysia, Skudai, Johor, Malaysia

Sivaraman Jayaraman
Department of Biotechnology and Medical Engineering, National Institute of Technology Rourkela, Odisha, India

Menaka Jha
Institute of Nano Science and Technology, Sahibzada Ajit Singh Nagar, Punjab, India

Amaro Olímpio Pereira Júnior
Energy Planning Programme – COPPE/UFRJ, Rio de Janeiro, Brazil

Chithra K.
Department of Chemical Engineering, A.C. Tech, Anna University, Chennai, Tamil Nadu, India

Dhanalakshmi K.
Department of Mechanical Engineering, National Institute of Technology, Tiruchirappalli, Tamil Nadu, India

Haribabu K.
Department of Chemical Engineering, National Institute of Technology Calicut, Kozhikode, Kerala, India

Muthukumar K.
Department of Petrochemical Technology, University College of Engineering (BIT Campus), Anna University, Tiruchirappalli, Tamil Nadu, India

Sheeba Narayanan K.
Department of Chemical Engineering, National Institute of Technology, Tiruchirappalli, Tamil Nadu, India

Pon Pavithiran C.K.
School of Mechanical Engineering, Vellore Institute of Technology, Vellore, Tamil Nadu, India

Nausad Khan
Institute of Nano Science and Technology, Sahibzada Ajit Singh Nagar, Punjab, India

Nageshwari Krishnamoorthy
Department of Biotechnology and Medical Engineering, National Institute of Technology Rourkela, Odisha, India

Contributors

João Gabriel Lassio
Energy Planning Program, Federal University of Rio de Janeiro, Rio de Janeiro, Brazil

Angkayarkan Vinayakaselvi M.
Department of English, Bharathidasan University, Tiruchirappalli, Tamil Nadu, India

Arulmozhi M.
Department of Petrochemical Technology, University College of Engineering, BIT Campus, Anna University, Tiruchirappalli, Tamil Nadu, India

Chandrasekar M.
Department of Mechanical Engineering, University College of Engineering, BIT Campus, Anna University, Tiruchirappalli, Tamil Nadu, India

Jayapriya M.
Department of Petrochemical Technology, University College of Engineering, BIT Campus, Anna University, Tiruchirappalli, Tamil Nadu, India

Meera Sheriffa Begum K.M.
Department of Chemical Engineering, National Institute of Technology, Tiruchirappalli, Tamil Nadu, India

Alessandra Magrini
Energy Planning Programme – COPPE/UFRJ, Rio de Janeiro, Brazil

Pratyush Maiti
CSIR-Central Salt & Marine Chemicals Research Institute, Bhavnagar, Gujarat, India

Subarna Maiti
CSIR-Central Salt & Marine Chemicals Research Institute, Bhavnagar, Gujarat, India

Denise Ferreira de Matos
Brazilian Electric Power Research Center, Rio de Janeiro, Brazil

Gordon McKay
Division of Sustainable Development, College of Science and Engineering, Hamad Bin Khalifa University, Qatar Foundation, Doha, Qatar

Samsudeen Naina Mohamed
Department of Chemical Engineering, National Institute of Technology, Tiruchirappalli, Tamil Nadu, India

Dineshkumar Muniyappan
Bioenergy Laboratory, Department of Mechanical Engineering, National Institute of Technology, Tiruchirappalli, Tamil Nadu, India

Anatharaman Narayanan
Department of Chemical Engineering, National Institute of Technology, Tamil Nadu, India

Galina S. Nyashina
National Research Tomsk Polytechnic University, Tomsk, Russia

Luciano Basto Oliveira
Energy Planning Programme – COPPE/UFRJ, Rio de Janeiro, Brazil

Raman P.
Energy Efficiency and Environment P. Ltd., New Delhi, India

Balasubramanian Paramasivan
Department of Biotechnology and Medical Engineering, National Institute of Technology Rourkela, Odisha, India

Prakash Parthasarathy
Division of Sustainable Development, College of Science and Engineering, Hamad Bin Khalifa University, Qatar Foundation, Doha, Qatar

Himanshu Patel
CSIR-Central Salt & Marine Chemicals Research Institute, Bhavnagar, Gujarat, India

Abinaya R.
Department of English, Bharathidasan University, Tiruchirappalli, Tamil Nadu, India

Akanksha R.
Department of Chemical Engineering, National Institute of Technology Calicut, Kozhikode, Kerala, India

Chandru R.
Department of Mechanical Engineering, National Institute of Technology, Tiruchirappalli, Tamil Nadu, India

Deepakkumar R.
Vellore Institute of Technology, Vellore Tamil Nadu, India

Gopi R.
Fuels Laboratory, Department of Mechanical Engineering, National Institute of Technology, Tiruchirappalli, Tamil Nadu, India

Krithikadevi R.
Centre for Advanced Materials Research, University of Sharjah, UAE

Selvakumar Ramalingam
Bharath Institute of Science and Technology, Chennai, Tamil Nadu, India

Anand Ramanathan
Bioenergy Laboratory, Fuels Laboratory, Thermal Laboratory, Department of Mechanical Engineering, National Institute of Technology, Tiruchirappalli, Tamil Nadu, India

Marcelo de Miranda Reis
Military Institute of Engineering – IME, Rio de Janeiro, Brazil

Karthick S.
Department of Chemical Engineering, National Institute of Technology Calicut, Kozhikode, Kerala, India

Sunaina
Institute of Nano Science and Technology, Sahibzada Ajit Singh Nagar, Punjab, India

Swathi S.
Department of Chemical Engineering, National Institute of Technology Calicut, Kozhikode, Kerala, India

Manoj Eswara Vel S.B.
Department of Mechanical Engineering, National Institute of Technology, Tiruchirappalli, Tamil Nadu, India

Anatoly S. Shvets
National Research Tomsk Polytechnic University, Tomsk, Russia

Vedharaj Sivasankaralingam
Clean Combustion Laboratory, Department of Mechanical Engineering; Centre for Combustion and Emission Studies, National Institute of Technology, Tiruchirappalli, India

Kritika Sood
Institute of Nano Science and Technology, Sahibzada Ajit Singh Nagar, Punjab, India

Malinee Sriariyanun
King Mongkut's University of Technology North Bangkok, Bang Sue, Bangkok, Thailand

Joshua George Stanly
Department of Mechanical Engineering, National Institute of Technology, Tiruchirappalli, Tamil Nadu, India

Pavel A. Strizhak
National Research Tomsk Polytechnic University, Tomsk, Russia

Senthilkumar T.
Department of Automobile Engineering, University College of Engineering, BIT Campus, Anna University, Tiruchirappalli, Tamil Nadu, India

Roman I. Taburchinov
National Research Tomsk Polytechnic University, Tomsk, Russia

Ksenia Y. Vershinina
National Research Tomsk Polytechnic University, Tomsk, Russia

Lukka Thuyavan Yogarathinam
Advanced Membrane Technology Research Centre (AMTEC), School of Chemical and Energy Engineering, Universiti Teknologi Malaysia, Skudai, Johor, Malaysia

Alisha Zaffer
Department of Biotechnology and Medical Engineering, National Institute of Technology Rourkela, Odisha, India

Zhenyu Zhao
School of Chemical Engineering and Technology, Tianjin University, Tianjin, China
National Engineering Research Center of Distillation Technology, Collaborative Innovation Center of Chemical Science and Engineering (Tianjin), Tianjin, China

Preface

Globally, developing nations have waste management systems that tremendously need development in order to meet engineering demands as well as health and safety requirements. A waste to profit facility is one of the most difficult public projects which could be of prime importance for a community to undertake toward Sustainable Development Goals (SDGs). It requires the expertise of scientists, researchers, in-house staff, and consultants, as well as the active participation of community decision-makers and the general public for its implementation. This book gives information and tools in selecting the most suitable technology for waste treatment and energy recovery under different conditions. This book deals with techno-economic analysis, life cycle assessment, and optimization of tools and technologies toward waste to wealth. In addition, the book presents an overview of various technologies involved in the treatment of wastes and factors influencing the processes for better understanding. Finally, the book discusses the environmental, socioeconomic, and sustainability of the waste-to-energy systems covering few of the geographic locations.

The book helps to identify the various parameters influencing the treatment processes of waste resources. The technologies are needed for all types of scenarios to provide an efficient state of art in reducing and reusing the wastes, to address the problems associated with waste management, issues concerned with policies of management, and waste-to-energy initiatives. It includes the global problems and their case studies to make involvement in industrial and municipal management. It is helpful to assess the environmental degradation of various streams, production methods of biofuel, and select the suitable eco-friendly processes among different alternatives. This book provides inputs for readers to help understand the fundamental science behind the water treatment, renewable biofuel production, particularly thermochemical and biochemical conversion processes, which could lead to environmental impacts associated with more value-added products.

Editorial Team

1 Crop Residue to Fuel, Fertilizer, and Other By-products
An Approach toward Circular Economy

Subarna Maiti, Himanshu Patel, and Pratyush Maiti

CONTENTS

1.1 Introduction 1
1.2 Selection and Characterization of Ideal Crop Residue 3
1.3 Thermochemical Conversion for Energy and Value-added Products 4
    1.3.1 Solar Gasification 4
        1.3.1.1 Experimental Setup 4
        1.3.1.2 Experimental Outcome 5
    1.3.2 Pyrolysis...6
        1.3.2.1 Experimental Setup 6
        1.3.2.2 Experimental Outcome 7
    1.3.3 Synthesis of Value-added Products from Process Residue 9
        1.3.3.1 Experimental Setup 9
        1.3.3.2 Experimental Outcome 10
1.4 Techno-economic Analysis..11
1.5 Conclusion 12
Acknowledgments..13
References..13

1.1 INTRODUCTION

The global energy and food demand are rising at an unprecedented rate due to the ever-increasing human population. The demand must be satisfied by preserving ecological resources of the planet. Energy is crucial to drive global socioeconomic growth, and energy consumption is expected to grow at an average rate of 1.2% per year till 2040 (BP Energy Outlook 2019). In 2019, more than 75% of primary energy comes from fossil fuels, which makes the energy sector responsible for 72% greenhouse gases emissions (GHGs). Resultantly, the global energy sector is dealing with a dual challenge: increasing energy demand and reducing carbon emissions. These challenges can be addressed by increasing our dependence on renewable and carbon-neutral energy sources. Being the second largest country by population and fifth largest economy in the world, India's share in total global primary energy consumption is expected to double by 2040.

    Worldwide food demand is expected to rise by 59–98% by 2050. The rise in food demand will significantly increase global fertilizer demand (Tian et al. 2021). Along with nitrogen and phosphorous, potassium is one of the critical plant nutrients. Although potassium is one of the most

DOI: 10.1201/9781003334415-1

abundant elements in the earth's crust, most of it is not plant-available (Zörb, Senbayram, and Peiter 2014). Therefore, crops need to be supplied with potassium as fertilizer. No wonder more than 95% of global production of potash chemicals is utilized as fertilizer. Although fertilizer demand is almost universal, only six countries supply 85% of the total global market. Being an agricultural country, India requires significant amount of potash fertilizer. Due to the absence of commercially exploitable potash reserves, India imports the entire demand. To balance non-uniform distribution of conventional potash reserves, secondary sources of potash need to be explored.

On the other hand, in many developing countries, surplus crop residues are allowed to decay naturally or burned to clear the field for the next cultivation cycle. Apart from energy loss, uncontrolled burning adversely affects soil fertility and human health. Technologies like thermochemical conversion has the potential to harness energy from these surplus crop residues. Though biofuels derived via thermochemical conversion are more or less comparable to fossil fuels, their production as a single source of revenue remain economically non-feasible compared to their fossil rivals. Biorefinery approach, considering the recovery of numerous products from a single biomass source, can significantly improve the cost-competitiveness of biofuels (Ghosh et al. 2015).

The chapter reports the biorefinery approaches based on two thermochemical processes: (i) solar gasification and (ii) slow co-pyrolysis. The outline of the chapter is illustrated in Figure 1.1. Non-fodder crop residue (i.e., empty cotton boll) was selected as a feedstock. The solar gasification-based approach involved syngas synthesis and co-production of potash fertilizer from the residual ash. However, slow co-pyrolysis-based approach considered the synthesis of bio-oil, potash fertilizer from biochar and activated carbon from spent biochar left after potash recovery. Application of bio-oil as an alternative fuel in spark-ignition engine was tested. As a part of the potassium recovery process, water leaching of residual ash (left after solar gasification) and biochar (left after slow co-pyrolysis) was optimized. For the synthesis of potash fertilizer from the leachate, a method of selective precipitation using tartaric acid was adopted. The spent biochar left after potassium recovery was utilized as a source of activated carbon and CO_2 adsorption capacity of derived activated carbon was tested. Techno-economics of biorefinery approaches based on thermochemical conversions to energy and value-added products were evaluated and compared. However, the detailed experimental results are reported elsewhere (Müller et al. 2018; Patel, Mangukiya et al. 2020; Patel

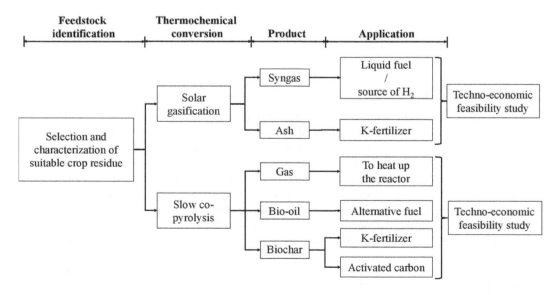

FIGURE 1.1 Outline of the chapter.

et al. 2019; Patel, Müller et al. 2020; Patel, Maiti, and Maiti 2022). The chapter highlights only the critical features of the study.

1.2 SELECTION AND CHARACTERIZATION OF IDEAL CROP RESIDUE

Based on three key criteria – (i) availability, (ii) avoidance of "food vs. fuel" conflict, and (iii) potassium content – empty cotton boll was selected as a renewable source of energy and potash from a large pool of crop residues available. In India, the total annual production of surplus cotton boll is estimated to be ~29.7 Mton, equivalent to the total energy potential of ~456 PJ/a and potash potential of ~0.32 Mton/a (in terms of K_2O). Table 1.1 describes complete characterization of empty cotton boll.

TABLE 1.1
Characterization of Empty Cotton Boll

	Unit	Empty Cotton Boll
Bulk density (unprocessed)	kg/m^3	250
Elemental analysis		
C	wt%	46.63
H	wt%	5.88
N	wt%	1.02
S	wt%	0.49
O (by difference)	wt%	45.98
Proximate analysis		
Ash content	wt%	5.46
Moisture content	wt%	9.29
Volatile matter content	wt%	64.99
Fixed carbon (by difference)	wt%	20.26
Calorific values		
HHV	MJ/kg	15.57
LHV	MJ/kg	14.29
Fiber composition		
Hemicellulose	wt%	3.00
Cellulose	wt%	53.98
Lignin	wt%	21.07
Ash composition (wt% of ash)		
Na_2O	wt%	6.17
MgO	wt%	4.87
Al_2O_3	wt%	3.66
SiO_2	wt%	6.62
K_2O	wt%	45.24
CaO	wt%	8.59
TiO_2	wt%	0.19
Fe_2O_3	wt%	1.77
Others	wt%	22.89
Ash fusion testing		
Initial deformation	°C	601
Softening	°C	609
Hemispherical	°C	627
Fluid	°C	635

1.3 THERMOCHEMICAL CONVERSION FOR ENERGY AND VALUE-ADDED PRODUCTS

1.3.1 Solar Gasification

As gasification is an endothermic process, it requires an external heat source to run the process. In solar gasification, concentrated solar radiation is used as a heat source to drive the process. Solar gasification offers several advantages over conventional gasification.

1.3.1.1 Experimental Setup

Figure 1.2 portrays a setup of solar gasification. A high-flux solar simulator with an assembly of ten Xenon arc lamps, each close-coupled to an ellipsoidal concentrator, was used for the experiments. The setup is at Solar Technology Laboratory, Paul Scherrer Institute, Switzerland, and the investigation was conducted at the Department of Mechanical and Process Engineering, ETH Zurich. In terms of radiative heat transfer characteristics, the high-flux solar simulator is equivalent to a highly concentrated solar system. Several xenon arc lamps and mechanical shutters could control solar

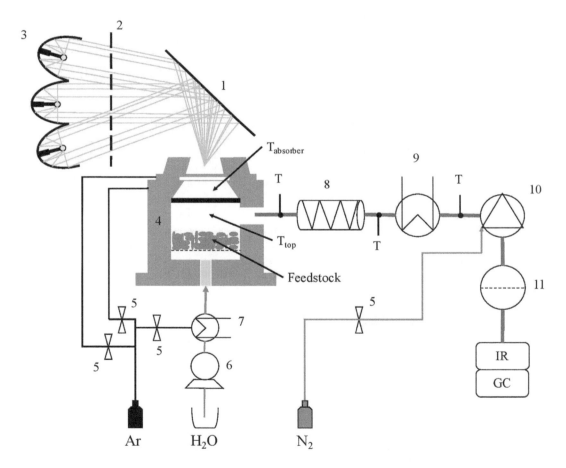

FIGURE 1.2 Solar gasification experimental setup at Solar Technology Laboratory, Paul Scherrer Institute, Switzerland. $T_{absorber}$: temperature of upper cavity, T_{top}: temperature of lower cavity above packed bed, T: thermocouple, 1: deflection mirror, 2: mechanical shutter, 3: high-flux solar simulator, 4: packed bed solar reactor, 5: mass flow controllers, 6: peristaltic pump, 7: heater, 8: tar cracker, 9: condensation tube, 10: scrubber, and 11: gas filter.

radiative power input. The incoming solar radiation was deflected down toward the top of the solar reactor using a deflection mirror. The reactor was comprised of two cavities: the upper cavity acted as an absorber of the incoming solar reactor, and the lower cavity worked as a reaction chamber. The upper cavity was purged with argon, and the lower cavity was purged with a mixture of argon and steam. Gas formed during gasification was directed toward an electric-heated tar cracker, a condensation tube to condense evaporated minerals, a wet scrubber to cool down hot gases to ambient temperature, and a solid particle filter (0.2 μm). The molar flow rate of gaseous species was determined by purging N_2 as a tracer. Infrared detectors and gas chromatography analyzed gas composition.

1.3.1.2 Experimental Outcome

Process operating conditions, experimental results, and performance indicators for solar gasification of empty cotton boll are listed in Table 1.2. Due to the lower bulk density of feedstock, only 160–347 g of CB was solar steam gasified. The steam injection was started once the T_{bottom} reached 100 °C. Gasification time denotes the time between the start of steam injection and the moment syngas production stops. Almost 2.3–8.2 g carbon-free ash was collected.

The variables which indicate the overall performance of the solar gasification process can be defined as follows.

Carbon conversion indicates the initial amount of total carbon available in feedstock that has been converted during the process:

$$\text{Carbon conversion} = \frac{\text{mass of carbon in feedstock} - \text{mass of caron in ash}}{\text{mass of carbon in feedstock}} \quad (1.1)$$

Carbon gas yield indicates the initial molar amount of carbon available in feedstock which has been converted to CH_4, CO, and CO_2:

$$\text{Carbon gas yield} = \frac{n_{CH_4} + n_{CO} + n_{CO_2}}{\text{moles of carbon in feedstock}} \quad (1.2)$$

TABLE 1.2
Operating Conditions, Results, and Performance Indicators of Solar Gasification of Empty Cotton Boll

Parameter	Experiment 1	Experiment 2	Experiment 3	Experiment 4
$T_{absorber}$, °C	1284	1276	1251	1087
T_{top}, °C	1112	1266	1239	947
Gasification time, min	73	70	115	78
Feedstock weight, g	184	166	347	160
n_{H_2O}, mol	8.73	7.24	20.46	13.35
n_{H_2}, mol	6.18	5.01	10.96	5.07
n_{CO}, mol	4.26	3.56	7.31	2.42
n_{CO_2}, mol	1.69	1.38	2.75	1.71
n_{CH_4}, mol	0.39	0.35	0.78	0.38
Ash collected, g	3.0	2.3	8.2	4.3
Performance indicators				
Carbon conversion, %	100.0	100.0	100.0	99.9
Carbon gas yield, %	99.8	96.8	95.2	85.6
$\eta_{solar-to-fuel}$, %	14.0	11.4	15.0	13.6
Energetic upgrade factor	1.07	0.99	1.02	0.92

where n_i denotes the molar amount of i gas species formed during the conversion (i = CH$_4$, CO, and CO$_2$); solar-to-fuel efficiency ($\eta_{\text{solar-to-fuel}}$) can be defined as the ratio of syngas heating value to solar radiative energy input and feedstock heating value:

$$\eta_{\text{Solar-to-fuel}} = \frac{m_{\text{syn}} \cdot LHV_{\text{syn}}}{\text{total solar radiatve energy input} + m_{\text{feedstock}} \cdot LHV_{\text{feedstock}}} \tag{1.3}$$

The energetic upgrade factor is the ratio of products' heating value (i.e., syngas) to the heating value of the feedstock:

$$\text{Energetic upgrade factor} = \frac{m_{\text{syn}} \cdot LHV_{\text{syn}}}{m_{\text{feedstock}} \cdot LHV_{\text{feedstock}}} \tag{1.4}$$

Gasification of empty cotton boll was carried out at different gasification temperatures (T_{top}). H$_2$ and CO were the leading constituents in the syngas composition, which was consistent for all four runs. Their cumulative molar composition ranged within 8.57–18.27. Molar ratio of H$_2$:CO, CH$_4$:CO, and CO$_2$:CO decreased at higher T_{top}. Carbon conversion of almost 100% was achieved for all gasification runs. Carbon gas yield improved significantly at elevated T_{top}.

In contrast, solar-to-fuel efficiency suffered at higher T_{top}, presumably because of the competitive effects of enhanced gasification rates versus higher heat loss at elevated T_{top}. The highest solar-to-fuel efficiency of 15% was achieved at T_{top} of 1239 °C. Energetic upgrade factor improved significantly at higher T_{top}. An energetic upgrade factor higher than unity indicates more heating value of the product stream than the heating value of the feedstock. For conventional autothermal gasification, this value ranges within 0.7–0.8. A significant fraction of feedstock undergoes combustion to reach the desired temperature. The energetic upgrade factor was below unity for T_{top} < 1100 °C. The net heating value of solar syngas was 10.34–10.94 MJ/m³, which was significantly higher than that of the syngas produced via autothermal gasification due to avoidance of combustion. For instance, autothermal gasification of *Jatropha* shell, legume straw, rice husk, and wood chip generated syngas with a net calorific value of 5.2, 5.9, 6.5, and 8.8 MJ/m³, respectively (Maiti et al. 2014; Cao et al. 2019).

1.3.2 Pyrolysis

1.3.2.1 Experimental Setup

Figure 1.3 indicates the process flow diagram for slow co-pyrolysis. Empty cotton boll and plastic waste (19:1 w/w) were slowly co-pyrolyzed in a fixed bed pilot-scale reactor in N$_2$ atmosphere at 500 °C, at a heating rate of 10 °C/min. Institute dry waste consisting of laboratory and packaging waste was considered plastic waste. The waste was comprised of various plastics; however, it is difficult to calculate the exact mass fraction of each plastic species available in the waste. The electrically heated furnace heated the cylindrical pyrolysis reactor until the flow of volatiles stopped (~60 min after reaching 500 °C) and the reactor cooled down naturally. Three Pt-Pt-Rh-type thermocouples were placed across the length of the reactor. The reactor temperature and heating rate were controlled using a microprocessor PID controller. Pyrolysis vapor coming out of the reactor was condensed (5 ± 5 °C) using two shell and tube condensers (connected in series) in a countercurrent manner. The composition of non-condensable gases was analyzed using non-dispersive infrared and gas chromatography. At the end of the experiment, the weight of solid residue left inside the reactor per unit amount of feedstock was considered char yield. The char was used as a source of potash fertilizer and activated carbon. The raw bio-oil yield was determined from the weight difference between the final and initial weight of the raw bio-oil-collecting vessel. Raw bio-oil contained a significant amount of water, which was removed via liquid–liquid extraction using ethyl acetate (bio-oil:ethyl acetate = 1:4 v/v). Upon mixing, aqueous and organic fractions of raw bio-oil got separated. Further,

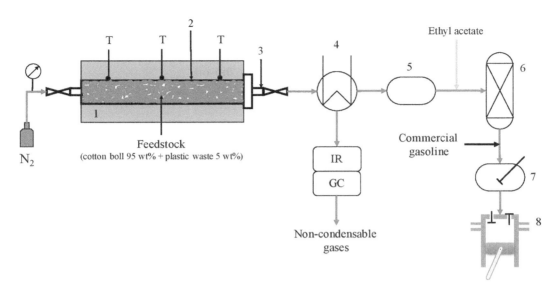

FIGURE 1.3 Process flow diagram for slow pyrolysis of empty cotton boll and plastic waste. T: thermocouple, 1: electrically heated furnace, 2: fixed bed pilot-scale reactor, 3: reactor outlet, 4: shell and tube glass condensers, 5: bio-oil-collecting vessel, 6: bio-oil upgradation unit, 7: bio-oil gasoline blending unit, and 8: four-stroke single-cylinder variable compression ratio gasoline engine.

the ethyl acetate mixture and bio-oil organic fraction were subjected to evaporation at 39 °C and 210 torr (absolute). Recovered ethyl acetate could be reused and the organic fraction was dried over anhydrous Na_2SO_4, filtered, and labeled as bio-oil. Without using any surfactant, bio-oil was blended with commercial gasoline in the range of 10–30% (v/v) bio-oil in gasoline. This bio-oil + gasoline blends were used as an alternative fuel in four-stroke single-cylinder gasoline engine. The effect of engine-operating variables (compression ratio, engine load, and blend proportion) on engine performance variables (brake thermal efficiency and brake-specific fuel consumption) and engine emissions (CO, CO_2, unburnt hydrocarbon and NO_x) was studied.

1.3.2.2 Experimental Outcome

Pyrolysis of 15 kg feedstock at 500 °C yielded 5.59 kg of raw bio-oil, 5.81 kg of biochar, and 3.12 kg of non-condensable gases. Upon water removal, 1.67 kg of water-free bio-oil was obtained. Carbon conversion yield, non-condensable gas, and bio-oil composition are presented in Table 1.3. The majority of total carbon available in feedstock was converted into biochar. Only 15.34% of total carbon in feedstock was transformed into bio-oil. CO and CO_2 were the dominating species in the non-condensable gases. GC-MS analysis indicated phenols were significantly available in bio-oil. Table 1.4 highlights essential fuel properties of gasoline and bio-oil relevant to engine performance study. Usually, bio-oils are acidic. Comparatively, present bio-oil pH was almost neutral due to relatively lower organic acids in bio-oil. Bio-oil was denser and more viscous than gasoline. However, its dynamic viscosity was well below the acceptable upper limit as fuel. Bio-oil stability study indicated that change in bio-oil dynamic viscosity was <1% after 30 days of examination. The observation was consistent for all three storage temperature conditions (0 °C, room temperature 25 °C, and open to atmosphere 30 ± 5 °C). Intensive properties like density, viscosity, and surface tension control fuel's average droplet size and atomization tendency, which ultimately controls combustion efficiency and emissions. Both fuels' surface tension was relatively close, suggesting that the existing fuel injection assembly could handle bio-oil. Ethyl acetate treatment reduced bio-oil moisture content by <0.1 wt%, which was considerably lower than the recommended upper limit. The higher

TABLE 1.3
Carbon Conversion Yield, Gas Composition, and GC-MS Analysis of Moisture-free Bio-oil

Component	Unit	Quantity
Carbon conversion yield, g C/g C fed		
Biochar	%	47.58
Bio-oil		15.34
Aqueous		2.93
Gas		26.34
Total		92.19
Gas composition, g/g feedstock		
H_2	%	0.22
CH_4		1.27
CO		10.92
CO_2		6.57
C_2-C_4		5.83
Bio-oil composition (GC-MS data)		
Aliphatic HC	Area %	26.61
Aromatic HC		28.12
Phenols		21.16
Methoxy phenols		4.51
N-containing HC		16.13
Others		3.47

TABLE 1.4
Fuel Properties of Bio-oil and Commercial Gasoline

Property	Unit	Bio-oil	Gasoline
pH	–	6.52	6.61
Density*	kg/m³	994.20	735.95
Viscosity*	cP	4.60	0.85
Surface tension*	mN/m	24.78	20.70
Refractive index*	–	1.48	1.42
Moisture content	wt%	<0.1	<0.1
Cloud point	°C	–	Below detection limit
Pour point	°C	−48.00	Below detection limit
Flash point	°C	27.00	–
HHV	MJ/kg	31.74	44.41
C	wt%	66.89	87.06
H		8.90	12.79
N		1.02	0.04
S		0.00	0.00
O (by difference)		23.19	0.11
H/C molar ratio	–	1.60	1.76
C/O molar ratio	–	3.85	1055.27
H/O molar ratio	–	6.14	1860.36
Sulfur content	ppm	769.50	386.20

* Measured at 28 °C.

heating value of bio-oil was almost double that of its precursor (i.e., empty cotton boll). It is because of higher C and H content and lower O content in bio-oil compared to empty cotton boll.

Bio-oil derived from pyrolysis of only empty cotton boll was immiscible with commercial gasoline. It was possibly due to higher O content that could have increased bio-oil polarity, making it immiscible with non-polar gasoline. Co-pyrolysis of crop residues with plastic waste is the most cost-effective technique to improve bio-oil quality. With an aim to make bio-oil miscible with gasoline, empty cotton boll was co-pyrolyzed with plastic waste. As plastic waste is not a green source, its mass fraction in feedstock was kept as low as possible. The lowest possible mass fraction of plastic waste for which the derived bio-oil was miscible with gasoline was 5 wt%. Without using any surfactant, bio-oil was blended with gasoline in the proportion of 10%, 20%, and 30% v/v.

Effects of three engine operating variables like compression ratio, engine load, and blend proportion on engine performance variables like brake thermal efficiency, brake-specific fuel consumption, and emission were assessed. Brake thermal efficiency is a ratio of brake power produced to the total energy input to the engine, whereas brake-specific fuel consumption measures the amount of fuel consumed to produce the unit brake power output. Both brake thermal efficiency and brake-specific fuel consumption were most affected by variation in engine load. While the brake thermal efficiency increased, the brake-specific fuel consumption decreased at higher engine load and compression ratio. The observation was consistent for the entire tested range of blend proportion. At elevated load and compression ratio, operating pressure and temperature of combustion chamber increase, improving combustion efficiency and enhancing brake power output. CO_2 emission increased, and unburnt hydrocarbon emission decreased at a higher compression ratio and engine load. It could be attributed to the complete combustion of fuel at an elevated compression ratio and engine load. With the increasing blend proportion, CO_2 emission declined marginally, possibly due to a drop in carbon content in fuel mixture at a higher blend proportion. CO emission was almost unaffected by the change in all three engine-operating variables. NO_x emission surged significantly at high engine load and compression ratio. Extreme engine temperature and pressure at higher compression ratio and load could have triggered thermal NO_x formation. At higher blend proportions, NO_x emissions declined considerably.

At engine-optimized conditions (i.e., compression ratio = 10 and engine load = 10 kg), engine performance was better for 30% bio-oil blend compared to engine performance for pure gasoline. For 30% blend, NO_x emission was five times lower compared to that for pure gasoline at stipulated conditions.

1.3.3 Synthesis of Value-added Products from Process Residue

Value-added products like potash fertilizer and activated carbon were synthesized from process residues left after thermochemical conversions. As residual ash left after solar gasification was free from carbon, it was used as a source of potassium. However, the carbon content in biochar left after co-pyrolysis was significantly higher, and it was utilized as a precursor for activated carbon after potassium recovery.

1.3.3.1 Experimental Setup

Mineral compositions of process residues were estimated using flame photometry and ICP-OES. The samples were prepared by single-stage water leaching of the process residues (residue:water = 1:30 w/w) for 24 h at ambient temperature. Ca^{2+} and Mg^{2+} concentrations were measured by complexometric titration with EDTA. SO_4^{2-} and Cl^- concentration was measured by gravimetric analysis and Mohr method (using $AgNO_3$), respectively.

Potash recovery from solid residues involved two significant steps: (1) water leaching of the solid residues and (2) recovery of pure potassium salts from the leachate. Batch-operated multi-stage countercurrent water leaching method was employed to leach out water-soluble components from residues. Process variables affecting water leaching were optimized using response surface methodology. Along with K⁺, leachate contained other minerals too. To derive pure potassium salts of fertilizer grade, a method of selective precipitation using tartaric acid was used.

Spent biochar left after water leaching was chemically activated using KOH as an activating agent. Impregnation ratio, i.e., spent biochar-to-KOH ratio of 1:2 w/w was maintained. In a fixed bed reactor (the same reactor used during co-pyrolysis), under an inert atmosphere (N$_2$ 120 mL/min), impregnated spent biochar was heated to 700 °C for 90 min. The reactor was allowed to cool down naturally. The activated sample was washed thoroughly using demineralized water and HCl until neutral pH. The derived activated carbon was dried and subjected to characterization and CO$_2$ adsorption.

1.3.3.2 Experimental Outcome

Process variables (like water-to-ash/biochar ratio, particle size, number of leaching stages, and leaching time) affecting multistage countercurrent leaching were optimized to leach out the maximum amount of potassium from ash/biochar at minimum energy expense. Among the tested process variables, leaching time showed the most negligible effect on potassium leaching efficiency. The water-to-ash/biochar ratio exhibited the most prominent impact on leaching efficiency, which increased considerably with a rise in water-to-ash/biochar ratio up to one point. Further increase in water-to-ash/biochar ratio had a less significant effect, the observation was consistent over the entire tested range of leaching stages. Potassium leaching efficiency increased remarkably with decreasing particle size. For smaller particle size, improved active surface area and internal diffusion of water into the solid particles ultimately increased the leaching efficiency. At optimized leaching parameters (leaching time = 0.5 h, number of leaching stages = 2, water-to-ash/biochar ratio = 3.75 mL/g, and particle size of 60 mesh), 98 wt% of potassium available in cotton boll ash and almost 93 wt% of potassium available in biochar were recovered. The mineral composition of leachate obtained after ash and biochar water leaching (only relevant to potash recovery) is represented in Table 1.5.

The mineral composition of leachate indicated that along with K⁺ several other undesired inorganic species (i.e., Na⁺) were also available in the leachate. As Na⁺ has a detrimental effect on plant growth, derived potash salts of fertilizer grade must be free from Na⁺. Therefore, K⁺ was selectively precipitated from the leachate using tartaric acid. Based on the ion selectivity of tartaric acid, selective precipitation of K⁺ from leachate can be summarized as follows:

$$K^+_{(aqs)} + MgT_{(s)} \rightarrow KHT_{(s)} + Mg^{2+}_{(aqs)} \quad (1.5)$$

Under acidic conditions (pH = 2.5–2.8), the reaction propagates in forward direction, i.e., KHT starts to precipitate. One liter of leachate was treated with a molar equivalent amount of tartaric acid and the pH of the solution was adjusted between 2.5 and 2.8 by gradual addition of Mg(OH)$_2$ slurry. The precipitated KHT was mechanically separated and KHT was decomposed into MgT via reverse reaction under alkaline conditions (pH = 7.5–8). According to Equation 1.6 (X can be NO$_3^-$, Cl⁻, or SO$_4^{2-}$), a wide range of potassium salts can be derived, upon treating KHT with an appropriate mineral acid. With an aim to derive K-N compound fertilizer (KNO$_3$), KHT was treated

TABLE 1.5
Mineral Composition of Ash and Biochar Leachate (in w/v%)

	K⁺	Na⁺	Ca^{2+}	Mg^{2+}	Cl⁻	SO$_4^{2-}$
Ash leachate	6.38	0.1	BDL*	BDL*	2.66	2.74
Biochar leachate	2.89	0.06	BDL*	BDL*	0.29	0.35

*Below detection level.

with stoichiometric amount of $Mg(OH)_2$ and HNO_3. During the conversion, pH was maintained at 7.5, and water-insoluble MgT and filtrate containing KNO_3 were obtained. Collected MgT was recycled in the next batch of KHT formation and filtrate was subjected to evaporation and KNO_3 crystallization. Via selective precipitation, more than 96 wt% and 98 wt% of potassium available in ash leachate and biochar leachate, respectively, were recovered as KNO_3.

$$KHT_{(s)} + Mg(OH)_{2(s)} + HX_{(aqs)} \rightarrow KX_{(aqs)} + MgT_{(s)} \quad (1.6)$$

Activated carbon yield with respect to spent biochar was 72.39 wt%. BET surface area and methylene blue value of 1381 m²/g and 435 mg/g, respectively, were on a par with commercially available activated carbon. Activated carbon shown higher CO_2 adsorption capacity of 5.14 mmol/g at 25 °C, at atmospheric pressure. It was significantly higher than the reported values of CO_2 adsorption capacity for many activated carbons derived from other precursors. For example, activated carbon derived via KOH activation from *Arundo donax*, date sheets, pomegranate peels, and cotton stalk having CO_2 adsorption capacity of 3.20, 3.65, 4.11, and 4.24 mmol/g, respectively, at 25 °C and 1 atm pressure (Singh et al. 2018; Li et al. 2019; Serafin et al. 2017; Pramanik et al. 2021). The higher CO_2 adsorption capacity for the activated carbon derived in the present study could be attributed to lower average pore diameter of 2.15 nm and higher micropore area of 1095 m²/g.

1.4 TECHNO-ECONOMIC ANALYSIS

During the techno-economic analysis, all the mass flow rates considered were based on lab-scale experimental results. All the monetary data provided in the study are in US$ (January 2020). Considering all the major unit operations and unit processes, process flow diagrams for both the thermochemical conversion–based plants were prepared and based on that the total equipment purchase cost was determined using literature and local suppliers' quotations. Based on total equipment purchase cost, other cost components of techno-economic models were determined. For solar gasification plant, outsourcing cost of syngas upgradation to Fischer–Tropsch diesel was estimated at $1.02/GJ of syngas produced.

Techno-economic analysis indicated that synthesis of value-added products improved overall profitability of the plant. Potash fertilizer and activated carbon synthesis had significantly dropped the levelized cost of biofuels. The analysis suggested that solar gasification as well as slow co-pyrolysis-based plants was more economical at larger plant capacities. In other words, installation of one large centralized plant would be more economically feasible than the installation of many decentralized plants, which was consistent for both the thermochemical processes.

Table 1.6 compares the economic aspects of both the thermochemical processes for synthesis of energy and value-added products from cotton boll. For the comparison, solar gasification- and slow co-pyrolysis-based plants of same capacity in terms of feedstock handling were considered. Selling price of all the products except liquid fuel derived via both the thermochemical processes were kept same. Liquid fuel produced via solar gasification followed by Fischer–Tropsch synthesis is expected to be oxygen-free, unlike the liquid fuel produced via slow co-pyrolysis. That is why selling price of liquid fuel generated via solar gasification route was more than three times higher than that of liquid fuel produced via slow co-pyrolysis. Due to higher carbon conversion yield experienced during solar gasification, liquid fuel production rate was three times higher for solar gasification than that for slow co-pyrolysis, whereas potash fertilizer production rate was ~1.6 times higher for slow co-pyrolysis-based plant than that for solar gasification plant. This could be attributed to extreme temperature conditions during solar gasification which resulted into loss of potassium. Also, this can be inferred from lower potassium recovery observed during solar gasification. Potassium recovery can be defined as a ratio of total potassium recovered as potash fertilizer to the total potassium available in the feedstock. For solar gasification-based plant, solar concentration assembly contributed more

TABLE 1.6
Economic Comparison of Solar Gasification and Co-pyrolysis for Energy and Value-added Products

	Thermochemical Conversion	
Parameter	Solar Gasification	Slow Co-pyrolysis
Feedstock	Cotton boll	Cotton boll + plastic waste (19:1)
Feedstock flow rate, ton/day	122.44	122.44
Products	Fuel + potash fertilizer	Fuel + potash fertilizer + activated carbon
Liquid fuel selling price, $/L	1.25	0.37
KNO_3 selling price, $/t	720	720
Activated carbon-selling price, $/t	–	1430
Liquid fuel production rate, m^3/day	42.97	13.71
Potash fertilizer production rate, ton/day	3.86	6.11
Potassium recovery, %	59.38	93.99
Activated carbon production rate ton/day	–	28.49
Total capital expenditure, M$	52.07	10.40
Internal rate of return, %	18.31	30.05
Payback time, y	5.43	3.30

than 50% in total equipment purchase cost. Thus, for the same plant capacity, solar gasification plant was approximately five times more expensive than the slow co-pyrolysis plant (on comparison of total capital expenditure). Higher internal rate of return and lower payback time for slow co-pyrolysis plant indicated that crop residue valorization via slow co-pyrolysis route is more lucrative than the solar gasification route.

1.5 CONCLUSION

The chapter reports on the thermochemical conversion of non-fodder crop residues to energy vectors and value-added products like potash fertilizer and activated carbon. Solar-driven gasification and slow co-pyrolysis were considered as thermochemical conversion and empty cotton boll as a feedstock. Solar gasification at different gasification temperatures was investigated. The better volume-specific heating value of syngas and energetic upgrade factor above unity suggested solar gasification is advantageous over conventional gasification. A biorefinery model based on slow co-pyrolysis of cotton boll and plastic waste resulted into bio-oil (moisture free), potash fertilizer, and activated carbon and their yield distribution was 11.13, 4.99, and 23.27 wt%, respectively. Bio-oil derived via slow co-pyrolysis was miscible with commercial gasoline without the aid of any surfactant. Bio-oil + gasoline blend was used as an alternative fuel in the spark-ignition engine, and engine performance was optimized. Compared to pure gasoline, better engine performance and lower NO_x emission were observed when the engine was fueled with bio-oil + gasoline blend. Cotton boll ash and biochar were subjected to water leaching for potash recovery, and process variables were optimized. At optimized conditions, 98 wt% of potassium available in cotton boll ash was recovered and almost 93 wt% of potassium available in biochar was recovered in the leachate. The method of selective precipitation using tartaric acid was used to derive fertilizer-grade potassium salts from the leachate. Spent biochar obtained after potash recovery was subjected to chemical activation, and activated carbon with BET surface area of 1381 m^2/g was obtained. Activated carbon exhibited CO_2 adsorption capacity of 5.14 mmol/g at 25 °C and atmospheric pressure. Techno-economic feasibility of biorefinery approaches based on both the thermochemical processes to energy and value-added chemicals was tested.

In conclusion, crop residue valorization is beneficial for local and global environment, as it co-produces a carbon-neutral synthetic fuel, natural fertilizer, and activated carbon, having lower carbon footprint, using renewable energy without competing with the food chain. Apart from advantages in terms of energy and value-added products, the commercialization of these technologies can affect the livelihood of millions of farmers associated in cotton farming in India (Sahay 2019) and worldwide (Voora, Larrea, and Bermudez 2020).

ACKNOWLEDGMENTS

Department of Science and Technology, Government of India (INT/SWISS/SNSF/P-54/2015), and CSIR network project K-TEN are gratefully acknowledged for funding. Authors also acknowledge CSIR-CSMCRI AESDCIF for analytical investigation. Prof. Aldo Steinfeld and Dr. Fabian Müller of ETH-Zurich are gratefully acknowledged for the fruitful collaboration. This is CSIR-CSMCRI 119/2022.

REFERENCES

BP Energy Outlook. 2019. *BP Energy Outlook 2019 Edition*. BP Plc.

Cao, Y., Quan Wang, J. Du, and J. Chen. 2019. "Oxygen-Enriched Air Gasification of Biomass Materials for High-Quality Syngas Production." *Energy Conversion and Management* 199: 111628.

Ghosh, Debashish, Diptarka Dasgupta, Deepti Agrawal, Savita Kaul, Dilip Kumar Adhikari, Akhilesh Kumar Kurmi, Pankaj K. Arya, Dinesh Bangwal, and Mahendra Singh Negi. 2015. "Fuels and Chemicals from Lignocellulosic Biomass: An Integrated Biorefinery Approach." *Energy & Fuels* 29 (5): 3149–57.

Li, Jiaxin, Beata Michalkiewicz, Jiakang Min, Changde Ma, Xuecheng Chen, Jiang Gong, Ewa Mijowska, and Tao Tang. 2019. "Selective Preparation of Biomass-Derived Porous Carbon with Controllable Pore Sizes toward Highly Efficient CO2 Capture." *Chemical Engineering Journal* 360: 250–9.

Maiti, Subarna, Pratap Bapat, Prasanta Das, and Pushpito K. Ghosh. 2014. "Feasibility Study of Jatropha Shell Gasification for Captive Power Generation in Biodiesel Production Process from Whole Dry Fruits." *Fuel* 121: 126–32.

Müller, F., H. Patel, D. Blumenthal, P. Poživil, P. Das, C. Wieckert, P. Maiti, S. Maiti, and A. Steinfeld. 2018. "Co-Production of Syngas and Potassium-Based Fertilizer by Solar-Driven Thermochemical Conversion of Crop Residues." *Fuel Processing Technology* 171 (May): 89–99. https://doi.org/10.1016/j.fuproc.2017.08.006.

Patel, Himanshu, Pratyush Maiti, and Subarna Maiti. 2022. "Techno-Economic Assessment of Bio-Refinery Model Based on Co-Pyrolysis of Cotton Boll Crop-Residue and Plastic Waste." *Biofuels, Bioproducts and Biorefining* 16 (1): 155–71.

Patel, Himanshu, Subarna Maiti, Fabian Müller, and Pratyush Maiti. 2019. "Sustainable Methodology for Production of Potassic Fertilizer from Agro-Residues: Case Study Using Empty Cotton Boll." *Journal of Cleaner Production* 215 (2019): 22–33. https://doi.org/10.1016/j.jclepro.2019.01.003.

Patel, Himanshu, Hiren Mangukiya, Pratyush Maiti, and Subarna Maiti. 2020. "Empty Cotton Boll Crop-Residue and Plastic Waste Valorization to Bio-Oil, Potassic Fertilizer and Activated Carbon – a Bio-Refinery Model." *Journal of Cleaner Production* 125738. https://doi.org/10.1016/j.jclepro.2020.125738.

Patel, Himanshu, Fabian Müller, Pratyush Maiti, and Subarna Maiti. 2020. "Economic Evaluation of Solar-Driven Thermochemical Conversion of Empty Cotton Boll Biomass to Syngas and Potassic Fertilizer." *Energy Conversion and Management* 209 (April): 112631. https://doi.org/10.1016/j.enconman.2020.112631.

Pramanik, P., H. Patel, S. Charola, S. Neogi, and S. Maiti. 2021. "High Surface Area Porous Carbon from Cotton Stalk Agro-Residue for CO_2 Adsorption and Study of Techno-Economic Viability of Commercial Production." *Journal of CO_2 Utilization* 45. https://doi.org/10.1016/j.jcou.2021.101450.

Sahay, Arun. 2019. "Cotton Plantations in India: The Environmental and Social Challenges." *Yuridika* 34 (3): 429–42.

Serafin, Jarosław, Urszula Narkiewicz, Antoni W. Morawski, Rafał J Wróbel, and Beata Michalkiewicz. 2017. "Highly Microporous Activated Carbons from Biomass for CO_2 Capture and Effective Micropores at Different Conditions." *Journal of CO_2 Utilization* 18: 73–9.

Singh, Gurwinder, Kripal S. Lakhi, Kavitha Ramadass, Sungho Kim, Declan Stockdale, and Ajayan Vinu. 2018. "A Combined Strategy of Acid-Assisted Polymerization and Solid State Activation to Synthesize Functionalized Nanoporous Activated Biocarbons from Biomass for CO2 Capture." *Microporous and Mesoporous Materials* 271: 23–32.

Tian, Xiaoyu, Bernie A. Engel, Haiyang Qian, En Hua, Shikun Sun, and Yubao Wang. 2021. "Will Reaching the Maximum Achievable Yield Potential Meet Future Global Food Demand?" *Journal of Cleaner Production* 294: 126285.

Voora, Vivek, Cristina Larrea, and Steffany Bermudez. 2020. *Global Market Report: Cotton*. JSTOR.

Zörb, Christian, Mehmet Senbayram, and Edgar Peiter. 2014. "Potassium in Agriculture – Status and Perspectives." *Journal of Plant Physiology* 171 (9): 656–69. https://doi.org/10.1016/j.jplph.2013.08.008.

2 Cost-effective Sustainable Electrodes for Bioelectrocatalysis toward Electricity Generation

Samsudeen Naina Mohamed, Meera Sheriffa Begum K.M., and Vigneshhwaran Ganesan

CONTENTS

2.1 Introduction 15
2.2 Electrode for Bioelectrocatalysis in BES... 16
2.3 Type of Anode Material... 17
    2.3.1 Carbon-based Anode 17
    2.3.2 Metal-based Anode 17
    2.3.3 Biomass/Waste-derived Anode 18
2.4 Surface Modification/Functionalization of Anode....................................... 19
2.5 Challenges and Outlook... 21
2.6 Conclusion 22
References... 22

2.1 INTRODUCTION

Rapid development of urbanization and industrialization causes fossil fuel depletion and global warming, which leads to energy crisis. To meet the energy demand, there is a need of alternative power generation which needs to be renewable and sustainable without affecting the environment (Hosenuzzaman et al. 2015). One of the promising solutions for energy crisis is to generate the energy from different waste utilities. Utilization of waste resources for power generation results in benefits with respect to environmental impacts and economic approach, respectively. Across the world, huge amount of wastewater is generated from domestic and industrial sources which affects the waterbodies and the living organism that lives and depends on waterbodies (Sonawane et al. 2017). So, generation of power from wastewater solves the two major crisis simultaneously. Among the treatment methods, bioelectrochemical system (BES), also known as microbial fuel cell (MFC), would be a promising technology that helps in generation of electricity from wastewater treatment through microbial metabolism (Balat 2010).

A typical BES system consists of an anode in the anodic chamber and a cathode in the cathodic chamber separated by a membrane such as proton-exchange membrane (PEM) or ion-exchange membrane (IEM). The microbes present in the anolyte (wastewater) act as a biocatalyst that consumes organic and inorganic substance and produce electrons. The produced electrons are transferred from anode to cathode through external circuit, whereas the proton produced are passed through the PEM and produces electricity along with water treatment (Mostafa Rahimnejad a et al. 2015). BES are influenced by various parameters such as design, architecture, operating pH, temperature, substrate loading, electrodes, and microbes employed (Javed et al. 2018). In aspect of

DOI: 10.1201/9781003334415-2

electrodes, both cathode and anode play a significant role in energy generation and wastewater treatment. But the anode part is highly responsible for the growth of microbes, rate of removal, generation of electrons, and the transformation to the cathode. Thus, designing an anode becomes an essential step since the anode provides a high surface area for the biocatalyst (microbes) to produce electricity (Yaqoob et al. 2020). The anode's major properties that affect the microbial adhesion and its extracellular electron transfer include anode's surface area, electrochemical activity, anode's surface hydrophilicity, and its surface, which ultimately affects the overall performance of the MFC (Y. X. Wang et al. 2019).

The material type and its surface morphology determine the energy generation and organics/waste treatment in MFC. These are the critical factors that influence the enrichment of biofilm on the anode surface, the rate of electron transfer from electrogens to the anode, and the internal resistance, respectively. In general, the essential properties that need to be considered for the selection of the anode in BES include high electrical conductivity, surface area, porosity, biocompatibility, stability, durability, accessibility, and cost of the material, respectively (Yaqoob, Ibrahim, and Rodríguez-Couto 2020). Improper adhesion of microbes on the anode, high internal resistance, and mass transfer losses limit the electron transfer efficiency between the anode and the microbe, resulting in low power output, and disturb the practical adoption of this technology.

To improve the performance of anode, various materials had been investigated to develop anode to satisfy the ideal MFC. Further, surface modifications by different strategies improve the performance of an anode that leads to the achievement of an efficient MFC. Thus, this chapter deals with the significance of anode for bioelectrocatalysis, interaction of biocatalyst with anode, various cost-effective anode material that developed from synthetic materials and natural/waste-derived sources, strategies for surface modifications/functionalization that improve anodic performance, and challenges and future directions for electrode development.

2.2 ELECTRODE FOR BIOELECTROCATALYSIS IN BES

The anode chamber has been considered as the heart of BES since it is responsible for the degradation of organic compounds, electron generation and transfer, growth of microbes, etc. The performance of anode chamber was influenced by the biocatalyst, electrode used, mediators, pH, substrate, and so on. Among these, anode plays a significant role of bacterial adhesion and transfer of electrons to the cathode. Further, electrode should be performed for overcoming the losses such as activation loss, metabolism loss, ohmic loss, and mass transfer loss, respectively (Logan 2008; Bruno, Jothinathan, and Rajkumar 2018).

Electron transfer in BES occurs either directly through direct adhesion of microbes on electrode or indirectly through self-mediators/exogenously added mediators and achieve oxidation of primary metabolites. Very few microbes such as *Geobacillus* sp. and *Shewanella* sp. have potential to transfer electrons through extracellular electron transfer (Kalathil, Patil, and Pant 2018).

Recently, it was reported that algae have also been used as the biocatalyst for the generation of electricity in BES. Green algae *Chlorella pyrenoidosa* has been studied for bioelectricity generation in dual-chambered MFC. Two models were studied: one model in which algae have been used as anode and potassium ferricyanide as cathode, and in other model algae were used to act as both anode and cathode, respectively, and it was reported that the second model showed higher power density of 6030 mW/m^2 (C. Xu et al. 2015). However, many studies have been reported the algae were suitable as cathode catalyst and anodic fed rather than biocatalyst for electron generation (Arun et al. 2020). Thus, bacteria were most preferably suited for biocatalyst in anode that helps to produce more electrons than algae. Interaction of biocatalyst with anode depends on electrode's surface roughness, porosity, surface area, surface chemical constituents, hydrophilicity, and material type (Xie, Criddle, and Cui 2015). The electrode material should be designed considering the physico-electrochemical properties and microbe–electrode interactions to achieve efficient anode performance.

2.3 TYPE OF ANODE MATERIAL

2.3.1 CARBON-BASED ANODE

Among the variable materials identified, the carbon-based material has attracted special attention due to its large surface area, low porosity, higher conductivity, better biocompatibility, easy availability, high stability, and good ability regarding electron transfer kinetics. The various carbon-based materials developed include carbon cloth, carbon paper, carbon foam, carbon mesh, carbon nanotube, carbon felt, carbon brush, and RVC (reticulated vitreous carbon) (Yaqoob, Ibrahim, and Rodríguez-Couto 2020; Dumitru and Scott 2016). Commercially available carbon cloth, carbon paper, and RVC possess favorable electrode characteristics and provide sufficient surface area for the microbial development and also possess flexible characteristics in nature. However, due to the low bacterial capacity and poor mechanical strength, these materials were not suitable for large-scale development (Yaqoob, Ibrahim, and Guerrero-Barajas 2021).

Compared to the other available carbon materials, carbon brush anode possesses ideal characteristics such as high porosity, large surface area, easy to manufacture for brush size, and ability to produce high power densities with better electrogenic performance. However, metal like titanium wire is required during the synthesis process as a joint, which leads to increased cost and hence requires alternative development in the synthesis method (Zhao et al. 2019), (Lanas and Logan 2013).

Graphite is another form of carbon-based material, consisting of a huge graphene layer (a crystalline structure with sp^2 hybridization), the most commonly used superior anode material compared to other carbon-based materials (M. Chen et al. 2016). Still, as a plain base anode, graphite shows better conductivity and stability and acts as a potential anode among all anode materials developed. They are fabricated in different forms similar to carbon-based material such as sheets, cloth, plate, granules, brush, felt, etc. Generally, graphite plates are cost-effective, easy to handle, and have a defined surface area. Similarly, graphite felt has three times more surface area than the rod form and is capable of producing large power densities. Comparative studies on effects of morphology, surface area, and porosity of graphite rod, graphite felt, and graphite foam as anodes with voltage output showed that by increasing the surface area of electrodes, the performance of MFC is increased (Chaudhuri and Lovley 2003; Kumari, Shankar, and Mondal 2018). Among the various forms, graphite fiber brushes (GFBs) are considered better for the graphite anode of MFCs because of its outstanding properties and able to improve the performance of MFC ten times more than the graphite rod as the anode. But the high cost due to the metal joint, clamping of fibers, and diffusion of substrate into the brush interior is the major limitation for large-scale development (Feng et al. 2010).

The aforementioned electrodes were not considered to be directly used in MFC as anode in recent years due to the bacterial clogging, low surface area, brittleness, and low durability (Dumitru and Scott 2016). Thus, the modified carbon-based electrodes using surface modification or functional materials were preferred to improve the properties of anode. To develop the cost-effective carbon electrode, natural biomass or waste derived materials were preferred for anode development. Graphene tends to be recent novel anode material for BES due to the chemical inertness, enhanced electrical conductivity, high electron mobility, large surface area, stability, and material strength (Olabi et al. 2020).

2.3.2 METAL-BASED ANODE

Metallic material possesses higher conductivity than the carbon-based material, but they are mostly not applied in MFC due to the non-corrosive requirements and smooth surface. Further, it affects the bacterial adhesion and growth on the electrode's surface. Many metals were avoided due to these limitations; however, some metals possess properties suitable for MFC anode, although with certain demerits (Mostafa Rahimnejad et al. 2015). SS mesh was the most commonly proposed metal-based anode in BES system for bioelectricity generation due to the high mechanical strength, corrosion resistance, and low cost (Pocaznoi et al. 2012), which may help for long-term operation and scale-up development. However, they lower the power density and hence limits its potential as anode (Dumas et al. 2007; Dumas, Basseguy, and Bergel 2008). Some noble metals like gold, silver, and platinum are able to decrease the internal resistance in MFC owing to their potential, but they cannot be implemented in large-scale studies due to poor bacterial adhesion. Some non-noble metals like titanium, nickel, cobalt, and copper were studied for their applicability in the MFC anode. Titanium is non-corrosive and has high stability and better biocompatibility nature which makes it suitable for anode, but due to the limiting current, surface modification is necessary for the titanium anode. Recently, nickel foam shows higher electrical conductivity and produces higher bioelectricity than carbon-based electrodes (Nosek, Jachimowicz, and Cydzik-Kwiatkowska 2020), but it also exhibits ohmic and activation losses (Kumari, Shankar, and Mondal 2018; Dumitru and Scott 2016). Thus, metal-based anodes were considered as conventional anodes that cannot be implemented in BES without modifications. The surface of metal-based anodes were modified with non-metal like carbon-nanotube, polymers, etc. to make its suitable for MFC (Yaqoob, Ibrahim, and Rodríguez-Couto 2020).

2.3.3 Biomass/Waste-derived Anode

Fabrication of anode from the natural sources and wastes has attracted researchers' interest in recent years due to the abundancy, stability, and recyclability of the material (Yaqoob, Ibrahim, and Rodríguez-Couto 2020). It also reduces the cost of the electrode, improves the performance of BES, and achieves sustainability and circular economy (Sonawane et al. 2017). Biomass are rich in carbon and the derived carbon through various synthesis methods possesses large surface area, porosity, and high conductivity, which improve the electrochemical performance; its applications were considered, for example, in supercapacitors (Lu and Zhao 2017). Table 2.1 presents

TABLE 2.1
Electrode Developed for Bioelectricity Generation Using Bio/Waste-derived Anode

S. No.	Bio/Waste source	Electrode Developed	Type of MFC	Power Density (mW/m^2)	Current Density (mA/m^2)	Reference
1	*Amaranthus dubius* leaf extract	FeO nanoparticle coated on carbon paper	Dual chamber	145	250	(Harshiny et al. 2017)
2	Coconut fiber	Activated carbon coated on graphite plate	Dual chamber	9500	2760	(Samsudeen et al. 2016)
3	*Vernonia amygdalina* leaf extract	α-MnO$_2$ nanoparticle forms composite with polyaniline and coated on pencil graphite	Dual chamber	426.26	2485.51	(Dessie et al. 2021)
4	Aluminum-based waste medicine wrapper	Utilized as anode	Dual chamber	21	80	(Noori et al. 2020)
5	Corn cob	H$_2$O$_2$ treated corn cob derived carbon electrode	Single chamber	89.7	500	(Sonu et al. 2022)
6	Devil fish bone waste	Char-based anode	Cubic single chamber	4.26	10	(Aguilera Flores et al. 2021)

TABLE 2.1 (Continued)
Electrode Developed for Bioelectricity Generation Using Bio/Waste-derived Anode

S. No.	Bio/Waste source	Electrode Developed	Type of MFC	Power Density (mW/m²)	Current Density (mA/m²)	Reference
7	Cocklebur fruit	Char-based anode	Single chamber	140.54 µW/g		(C. Yang et al. 2019)
8	Coconut husk	C-quantum dots	Multichamber	191	200	(Vishwanathan et al. 2016)
9	Coconut shell	Activated biochar	Dual chamber	47.48	169	(K. Senthilkumar and Naveenkumar 2021)
10	Waste paper	Waste paper–derived carbon aerogel	Dual chamber	152	1000	(N. Senthilkumar et al. 2020)
11	Almond shell	Biochar	H-type	4346	2000	(Meizhen Li et al. 2019)
12	Municipal sludge	Carbon	Single chamber	615.2	1750	(Meng Li et al. 2020)
13	Activated wood	Char	H-type	249.8	1410	(Adelaja et al. 2019)
14	Oil palm empty fruit bunch	Graphene oxide–derived anode	Dual chamber	0.1076	15.65	(Yaqoob et al. 2021)
15	Silver grass	Activated carbon	H type	963	2750	(Rethinasabapathy et al. 2020)
16	Chestnut shell	Microporous–mesoporous carbon	Cubic single chamber	23.58 W/m³	0.85 mA/cm³	(Q. Chen et al. 2018)
17	Waste paper–derived carbon aerogel	N-doped reduced graphene oxide/ CeO$_2$	Dual chamber	1486	4500	(N. Senthilkumar et al. 2020)
18	Mango wood	Carbon anode	Single chamber	589.8	~1900	(Meng et al. 2020)
19	Bread	Carbon foam anode	H type	3134	7560	(Zhang et al. 2018)

the anode developed from waste and biomass and its respective power density. From the table, it could be identified that natural/waste-derived electrode enhances the power density from 0.1 mW/m² to 10 W/m². However, some biomass reports lower power density, and this was due to the insufficient studies in development of anode. The process parameters of the carbon synthesis need to be optimized to get better performance as reported for almond shell biochar in which the carbon obtained at 800 °C shows better output than carbon obtained from other temperatures (Meizhen Li et al. 2019).

2.4 SURFACE MODIFICATION/FUNCTIONALIZATION OF ANODE

The surface of the anode need to be modified to achieve higher power production rate and to improve the properties of the electrode. Anodes were modified through surface modification and functionalization. Surface modification involves treating the electrode with various agents or route, and functionalization involves incorporating the functional materials on the anode's surface to improve the properties. Various surface modifications include ammonia treatment (for the improvement of bacterial adhesion by enhancing the positive charge of the electrode), thermal treatment (for increasing the power density), acid treatment (for enhancing the surface area and porosity), and electrochemical oxidation (improves the functional group of the electrode) (Kumar, Sarathi, and Nahm 2013).

In the recent decade, the functionalization of anode involved functional materials such as carbon nanotube, graphene, conductive polymers such as polyaniline, metal oxides, etc. for enhancing the current density and power density of the electrode (Kaur et al. 2020). Some of the functional materials reported are listed in Table 2.2. It can be observed that in many cases, functionalization of electrode increases the power density as compared to conventional anodes. Apart from power

TABLE 2.2
Functionalized Electrode for Bioelectricity Generation

S. No.	Electrode	Surface Modification/ Functionalization	Type of MFC Studied	Power Density (mW/m^2)	Current Density (mA/m^2)	Reference
1	Carbon cloth	Nickel oxide/carbon nanotube/polyaniline nanocomposite	Dual chamber	2.25	25	(Nourbakhsh, Mohsennia, and Pazouki 2017)
2	Carbon cloth	Microbe-immobilized carbon nanoparticles	Single chamber	1795	0.8	(Yuan et al. 2011)
3	Sponge	N-doped carbon nanotubes/chitosan/ polyaniline composites	Dual chamber	2816.67 mW/m^3	2.75	(H. Xu et al. 2020)
4	Carbon felt	Graphene-Fe$_2$O$_3$ nanocomposite	Dual chamber	334	1200	(Fu et al. 2020)
5	Copper wire mesh	Polyaniline	Dual chamber	1.5 mW/m^3		(Mwale, Munyati, and Nyirenda 2021)
6	Exfoliated graphite	Multiwalled carbon nanotube	Cube type	1444	1000	(Song, Kim, and Woo 2015)
7	Graphite felt	Clinoptilolite	Dual chamber with salt bridge	940.3	28	(Kardi et al. 2017)
8	Copper mesh	Tin	Dual chamber	271	1250	(Taskan and Hasar 2015)
9	Stainless steel mesh	Single-wall carbon nanohorns	Single chamber	327		(J. Yang et al. 2017)
10	Carbon cloth	Heteroatom-doped carbon nanoparticle	Air-cathode single chamber	1720	4520	(Zhu et al. 2022)
11	Graphite felt	Ethylene diamine	H type	545	1350	(Du et al. 2016)
12	Titanium oxide	Graphene/polyaniline	Air-cathode single chamber	2073	4500	(Z. L. Li et al. 2020)
13	Carbon felt	Nano-CeO$_2$	Dual chamber	2.94	0.7	(Yin et al. 2016)
14	Graphite	Iron(II) molybdate	Dual chamber	106	275	(Naina Mohamed et al. 2020)
15	Carbon paper	Biosynthesized FeO nanoparticles	Dual chamber	145	250	(Harshiny et al. 2017)
16	Pencil graphite	Biosynthesized α-MnO$_2$-based polyaniline nanocomposite	Dual chamber	426.26	2485.51	(Dessie et al. 2021)

TABLE 2.2 (Continued)
Functionalized Electrode for Bioelectricity Generation

S. No.	Electrode	Surface Modification/ Functionalization	Type of MFC Studied	Power Density (mW/m²)	Current Density (mA/m²)	Reference
17	Carbon felt	Ni$_3$Mo$_3$C	Cube MFC	2100	12.5 A/m³	(Zeng, Zhao, and Li 2015)
18	3D porous N-doped carbon nanotube sponge	Polypyrrole/ carboxymethyl cellulose composite	Dual chamber	4880	5240	(Y. Wang et al. 2020)
19	Carbon felt	MnFe$_2$O$_4$	Dual chamber	3836	~3300	(Xue et al. 2021)
20	Carbon felt	Poly(aniline-1,8-diaminonaphthalene)	H type	29.1	155.8	(C. Li et al. 2021)

density, functionalization improves the surface area and hydrophilicity and enhances the bacterial adhesion (Nosek, Jachimowicz, and Cydzik-Kwiatkowska 2020).

2.5 CHALLENGES AND OUTLOOK

From the extensive review, it was observed that the cost-effective anode developed from biomass/ waste sources were less and needs to be explored. Though the anode could achieve the cost-effective and circular economy, it lacks its practical implications due to the poor bioelectricity production as compared to existing commercial electrode. It is due to the fact that most of these cheap precursor electrodes were performed without proper optimization of the carbon synthesis. The carbonization methods and process parameters needs to be optimized to get the anode with better physico-electrochemical properties. In recent years, graphene possesses better properties such as large surface area, pore size distribution, chemical inertness, stability, and high electrical conductivity. Thus, the synthesis of graphene from waste sources may also improve the performance of anode in MFC. Apart from biomass and natural waste, other waste resources such as plastic waste and industrial waste may support the efficient anode development since it has been reported that these wastes also possess better electrochemical properties and were studied for applications in supercapacitors and capacitive deionization, respectively (Shang and Jin 2016; Utetiwabo et al. 2020). Surface functionalization of the low-cost carbon further improves the performance. Though various nanoparticles such as metal oxides, bimetallic oxides, carbon nanotube (CNT), and polymers improve the performance, low feasibility in aspects of economy and environmental impacts of the material hinders its implementations in large-scale MFC. So, apart from low-cost electrode, it is also necessary to synthesize cost-effective and eco-friendly functional materials (Brar et al. 2022). Figure 2.1 depicts the future scope of the sustainable bioelectrochemical system through sustainable electrodes. The industrial solid waste generated that suits for electrode development would be used to produce anode for BES which generates bioelectricity, water reclamation and potential for the extraction of value added products thus able to achieve zero waste emission and sustainability. Thus, for developing the cost-effective sustainable anode for bioelectrocatalysis, the research needs to be more focused on the following aspects: multidisciplinary research would be highly recommended for achieving the aforementioned aspects since BES possesses the flavor of various fields such as biology, energy, environment, chemistry, materials, sensing, etc., which helps to develop the suitable electrode that covers all aspects in terms of properties and functions.

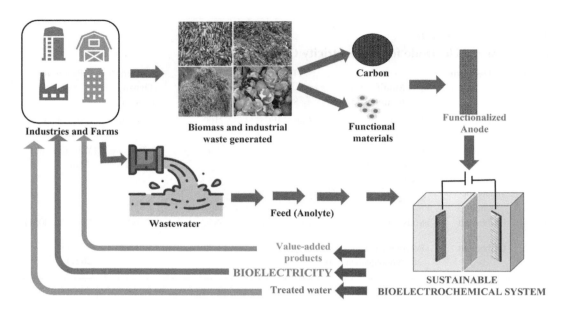

FIGURE 2.1 Future scope of the sustainable BES through sustainable electrodes.

2.6 CONCLUSION

The performance of BES for bioelectricity production highly depends on electrode since it directly influences the critical factors like bacterial adhesion, external electron transfer, and substrate oxidation. Various electrodes and their surface modifications were explored for the improvement of anode properties such as conductivity, biocompatibility, and surface area, which shows better improvement in power density and current density that indicates the effective degradation of substrate by microbes. However, the cost of the electrode, availability, and stability hinder the improvement of BES to next stage. So, design and development of anode from bio/waste-derived sources helps to achieve the cost-effective electrodes which were under progressive research. The selection of the suitable electrodes and surface modifications along with scale-up studies brings the development of sustainable technology that converts bio/solid waste to functionalized electrodes and wastewater as a source of substrate to achieve efficient bioelectricity generation.

REFERENCES

Adelaja, O.A., M.A. Ibrahim, L.A. Bello, and A.F. Aiyesanmi. 2019. "Bio-Electrochemical Treatment of Food Wastewater and Copper Recovery from Copper-Contaminated Plant with Electricity Production Using Biomaterial Anode." *International Journal of Energy and Water Resources* 3 (3): 187–201. https://doi.org/10.1007/s42108-019-00020-0.

Aguilera Flores, Miguel Mauricio, Verónica Ávila Vázquez, Nahum Andrés Medellín Castillo, Candy Carranza Álvarez, Antonio Cardona Benavides, Raul Ocampo Pérez, Gladis Judith Labrada Delgado, and Sergio Miguel Durón Torres. 2021. "Ibuprofen Degradation and Energy Generation in a Microbial Fuel Cell Using a Bioanode Fabricated from Devil Fish Bone Char." *Journal of Environmental Science and Health – Part A Toxic/Hazardous Substances and Environmental Engineering* 56 (8): 874–85. https://doi.org/10.1080/10934529.2021.1934357.

Arun, S., Arindam Sinharoy, Kannan Pakshirajan, and Piet N.L. Lens. 2020. "Algae Based Microbial Fuel Cells for Wastewater Treatment and Recovery of Value-Added Products." *Renewable and Sustainable Energy Reviews* 132. https://doi.org/10.1016/j.rser.2020.110041.

Balat, Mustafa. 2010. "Microbial Fuel Cells as an Alternative Energy Option." *Energy Sources, Part A: Recovery, Utilization and Environmental Effects* 32 (1): 26–35. https://doi.org/10.1080/15567030802466045.

Brar, Kamalpreet Kaur, Sara Magdouli, Amina Othmani, Javad Ghanei, Vivek Narisetty, Raveendran Sindhu, Parameswaran Binod, Arivalagan Pugazhendhi, Mukesh Kumar Awasthi, and Ashok Pandey. 2022. "Green Route for Recycling of Low-Cost Waste Resources for the Biosynthesis of Nanoparticles (NPs) and Nanomaterials (NMs)-A Review." *Environmental Research* 207 (September 2021): 112202. https://doi.org/10.1016/j.envres.2021.112202.

Bruno, L. Benedict, Deepika Jothinathan, and M. Rajkumar. 2018. "Microbial Fuel Cells: Fundamentals, Types, Significance and Limitations." *Microbial Fuel Cell Technology for Bioelectricity* 23–48. https://doi.org/10.1007/978-3-319-92904-0_2.

Chaudhuri, Swades K., and Derek R. Lovley. 2003. "Electricity Generation by Direct Oxidation of Glucose in Mediatorless Microbial Fuel Cells." *Nature Biotechnology* 21 (10): 1229–32. https://doi.org/10.1038/nbt867.

Chen, Meiqiong, Yinxiang Zeng, Yitong Zhao, Minghao Yu, Faliang Cheng, Xihong Lu, and Yexiang Tong. 2016. "Monolithic Three-Dimensional Graphene Frameworks Derived from Inexpensive Graphite Paper as Advanced Anodes for Microbial Fuel Cells." *Journal of Materials Chemistry A* 4 (17): 6342–9. https://doi.org/10.1039/c6ta00992a.

Chen, Qin, Wenhong Pu, Huijie Hou, Jingping Hu, Bingchuan Liu, Jianfeng Li, Kai Cheng, L. Huang, X. Yuan, C. Yang, and J. Yang. 2018. "Activated Microporous-Mesoporous Carbon Derived from Chestnut Shell as a Sustainable Anode Material for High Performance Microbial Fuel Cells." *Bioresource Technology* 249: 567–73. https://doi.org/10.1016/j.biortech.2017.09.086.

Dessie, Yilkal, Sisay Tadesse, Rajalakshmanan Eswaramoorthy, and Yeshaneh Adimasu. 2021. "Biosynthesized α-MnO2-Based Polyaniline Binary Composite as Efficient Bioanode Catalyst for High-Performance Microbial Fuel Cell." *All Life* 14 (1): 541–68. https://doi.org/10.1080/26895293.2021.1934123.

Du, Haoyue, Yunfei Bu, Yehui Shi, Qin Zhong, and Juan Wang. 2016. "Effect of an Anode Modified with Nitrogenous Compounds on the Performance of a Microbial Fuel Cell." *Energy Sources, Part A: Recovery, Utilization and Environmental Effects* 38 (4): 527–33. https://doi.org/10.1080/15567036.2013.798716.

Dumas, C., Régine Basseguy, and Alain Bergel. 2008. "Electrochemical Activity of Geobacter Sulfurreducens Biofilms on Stainless Steel Anodes." *Electrochimica Acta* 53 (16): 5235–41. https://doi.org/10.1016/j.electacta.2008.02.056.

Dumas, C., A. Mollica, D. Féron, R. Basséguy, L. Etcheverry, and A. Bergel. 2007. "Marine Microbial Fuel Cell: Use of Stainless Steel Electrodes as Anode and Cathode Materials." *Electrochimica Acta* 53 (2): 468–73. https://doi.org/10.1016/j.electacta.2007.06.069.

Dumitru, A., and Keith Scott. 2016. "Anode Materials for Microbial Fuel Cells." *Microbial Electrochemical and Fuel Cells: Fundamentals and Applications* 117–52. https://doi.org/10.1016/B978-1-78242-375-1.00004-6.

Feng, Yujie, Qiao Yang, Xin Wang, and Bruce E. Logan. 2010. "Treatment of Carbon Fiber Brush Anodes for Improving Power Generation in Air-Cathode Microbial Fuel Cells." *Journal of Power Sources* 195 (7). ACS National Meeting Book of Abstracts.

Fu, Lin, Haoqi Wang, Qiong Huang, Tian shun Song, and Jingjing Xie. 2020. "Modification of Carbon Felt Anode with Graphene/Fe2O3 Composite for Enhancing the Performance of Microbial Fuel Cell." *Bioprocess and Biosystems Engineering* 43 (3): 373–81. https://doi.org/10.1007/s00449-019-02233-3.

Harshiny, M., N. Samsudeen, Rao Jana Kameswara, and Manickam Matheswaran. 2017. "Biosynthesized FeO Nanoparticles Coated Carbon Anode for Improving the Performance of Microbial Fuel Cell." *International Journal of Hydrogen Energy* 42 (42): 26488–95. https://doi.org/10.1016/j.ijhydene.2017.07.084.

Hosenuzzaman, M., N.A. Rahim, J. Selvaraj, M. Hasanuzzaman, A. B.M.A. Malek, and A. Nahar. 2015. "Global Prospects, Progress, Policies, and Environmental Impact of Solar Photovoltaic Power Generation." *Renewable and Sustainable Energy Reviews* 41: 284–97. https://doi.org/10.1016/j.rser.2014.08.046.

Javed, Muhammad Mohsin, Muhammad Azhar Nisar, Muhammad Usman Ahmad, Nighat Yasmeen, and Sana Zahoor. 2018. "Microbial Fuel Cells as an Alternative Energy Source: Current Status." *Biotechnology and Genetic Engineering Reviews* 34 (2): 216–42. https://doi.org/10.1080/02648725.2018.1482108.

Kalathil, S., S.A. Patil, and D. Pant. 2018. "Microbial Fuel Cells: Electrode Materials." *Encyclopedia of Interfacial Chemistry: Surface Science and Electrochemistry* 309–18. https://doi.org/10.1016/B978-0-12-409547-2.13459-6.

Kardi, Seyedeh Nazanin, Norahim Ibrahim, Ghasem Najafpour Darzi, Noor Aini Abdul Rashid, and José Villaseñor. 2017. "Dye Removal of AR27 with Enhanced Degradation and Power Generation in a Microbial Fuel Cell Using Bioanode of Treated Clinoptilolite-Modified Graphite Felt." *Environmental Science and Pollution Research* 24 (23): 19444–57. https://doi.org/10.1007/s11356-017-9204-1.

Kaur, Rajnish, Aanchal Marwaha, Varun A. Chhabra, Ki Hyun Kim, and S.K. Tripathi. 2020. "Recent Developments on Functional Nanomaterial-Based Electrodes for Microbial Fuel Cells." *Renewable and Sustainable Energy Reviews* 119. https://doi.org/10.1016/j.rser.2019.109551.

Kumar, G. Gnana, V.G. Sathiya Sarathi, and Kee Suk Nahm. 2013. "Recent Advances and Challenges in the Anode Architecture and Their Modifications for the Applications of Microbial Fuel Cells." *Biosensors and Bioelectronics* 43 (1): 461–75. https://doi.org/10.1016/j.bios.2012.12.048.

Kumari, Usha, Ravi Shankar, and Prasenjit Mondal. 2018. "Electrodes for Microbial Fuel Cells." *Progress and Recent Trends in Microbial Fuel Cells* 125–41. https://doi.org/10.1016/B978-0-444-64017-8.00008-7.

Lanas, Vanessa, and Bruce E. Logan. 2013. "Evaluation of Multi-Brush Anode Systems in Microbial Fuel Cells." *Bioresource Technology* 148: 379–85. https://doi.org/10.1016/j.biortech.2013.08.154.

Li, Chao, Miaomiao Luo, Shihua Zhou, Hanyue He, Jiashun Cao, and Jingyang Luo. 2021. "Comparison Analysis on Simultaneous Decolorization of Congo Red and Electricity Generation in Microbial Fuel Cell (MFC) with l-Threonine-/Conductive Polymer-Modified Anodes." *Environmental Science and Pollution Research* 28 (4): 4262–75. https://doi.org/10.1007/s11356-020-10130-6.

Li, Meizhen, Suqin Ci, Yichun Ding, and Zhenhai Wen. 2019. "Almond Shell Derived Porous Carbon for a High-Performance Anode of Microbial Fuel Cells." *Sustainable Energy and Fuels* 3 (12): 3415–21. https://doi.org/10.1039/c9se00659a.

Li, Meng, Yan Wen Li, Xiao Long Yu, Jing Jie Guo, Lei Xiang, Bai Lin Liu, Hai Ming Zhao, M.Y. Xu, N.X. Feng, P.F. Yu, and Q.Y. Cai. 2020. "Improved Bio-Electricity Production in Bio-Electrochemical Reactor for Wastewater Treatment Using Biomass Carbon Derived from Sludge Supported Carbon Felt Anode." *Science of the Total Environment* 726. https://doi.org/10.1016/j.scitotenv.2020.138573.

Li, Zhi Liang, Sheng Ke Yang, Ya'nan Song, Hai Yang Xu, Zong Zhou Wang, Wen Ke Wang, and Ya Qian Zhao. 2020. "Performance Evaluation of Treating Oil-Containing Restaurant Wastewater in Microbial Fuel Cell Using in Situ Graphene/Polyaniline Modified Titanium Oxide Anode." *Environmental Technology (United Kingdom)* 41 (4): 420–9. https://doi.org/10.1080/09593330.2018.1499814.

Logan, Bruce E. 2008. "Microbial Fuel Cells." *Microbial Fuel Cells* 1–200. https://doi.org/10.1002/9780470258590.

Lu, Hao, and X.S. Zhao. 2017. "Biomass-Derived Carbon Electrode Materials for Supercapacitors." *Sustainable Energy and Fuels* 1 (6): 1265–81. https://doi.org/10.1039/C7SE00099E.

Meng, Li, Li YanWen, Cai QuanYing, Zhou ShaoQi, and Mo CeHui. 2020. "Spraying Carbon Powder Derived from Mango Wood Biomass as High-Performance Anode in Bio-Electrochemical System." *Bioresource Technology* 300: 122623. www.sciencedirect.com/science/article/abs/pii/S096085241931853X%0Awww.cabdirect.org/cabdirect/abstract/20203119431.

Mostafa Rahimnejad, Arash Adhami, Soheil Darvari, Alireza Zirepour, and Sang-Eun Oh. 2015. "Microbial Fuel Cell as New Technology for Bioelectricity Generation: A Review." *Alexandria Engineering Journal* 54: 745–56. http://ac.els-cdn.com/S1110016815000484/1-s2.0-S1110016815000484-main.pdf?_tid=50686514-8e74-11e7-a353-00000aab0f01&acdnat=1504201767_ace29e1e137875d50804a851b17b9bca.

Mwale, S., M.O. Munyati, and J. Nyirenda. 2021. "Preparation, Characterization, and Optimization of a Porous Polyaniline-Copper Anode Microbial Fuel Cell." *Journal of Solid State Electrochemistry* 25 (2): 639–50. https://doi.org/10.1007/s10008-020-04839-0.

Naina Mohamed, Samsudeen, Nikhil Thomas, J. Tamilmani, T. Boobalan, Manickam Matheswaran, P. Kalaichelvi, Arun Alagarsamy, and Arivalagan Pugazhendhi. 2020. "Bioelectricity Generation Using Iron(II) Molybdate Nanocatalyst Coated Anode during Treatment of Sugar Wastewater in Microbial Fuel Cell." *Fuel* 277. https://doi.org/10.1016/j.fuel.2020.118119.

Noori, Md T., G.D. Bhowmick, B.R. Tiwari, I. Das, M.M. Ghangrekar, and C.K. Mukherjee. 2020. "Utilisation of Waste Medicine Wrappers as an Efficient Low-Cost Electrode Material for Microbial Fuel Cell." *Environmental Technology (United Kingdom)* 41 (10): 1209–18. https://doi.org/10.1080/09593330.2018.1526216.

Nosek, Dawid, Piotr Jachimowicz, and Agnieszka Cydzik-Kwiatkowska. 2020. "Anode Modification as an Alternative Approach to Improve Electricity Generation in Microbial Fuel Cells." *Energies* 13 (24). https://doi.org/10.3390/en13246596.

Nourbakhsh, Fatemeh, Mohsen Mohsennia, and Mohammad Pazouki. 2017. "Nickel Oxide/Carbon Nanotube/Polyaniline Nanocomposite as Bifunctional Anode Catalyst for High-Performance Shewanella-Based Dual-Chamber Microbial Fuel Cell." *Bioprocess and Biosystems Engineering* 40 (11): 1669–77. https://doi.org/10.1007/s00449-017-1822-y.

Olabi, A.G., Tabbi Wilberforce, Enas Taha Sayed, Khaled Elsaid, Hegazy Rezk, and Mohammad Ali Abdelkareem. 2020. "Recent Progress of Graphene Based Nanomaterials in Bioelectrochemical Systems." *Science of the Total Environment* 749. https://doi.org/10.1016/j.scitotenv.2020.141225.

Pocaznoi, Diana, Benjamin Erable, Luc Etcheverry, Marie Line Delia, and Alain Bergel. 2012. "Towards an Engineering-Oriented Strategy for Building Microbial Anodes for Microbial Fuel Cells." *Physical Chemistry Chemical Physics* 14 (38): 13332–43. https://doi.org/10.1039/c2cp42571h.

Rethinasabapathy, Muruganantham, Jeong Han Lee, Kwang Chul Roh, Sung Min Kang, Seo Yeong Oh, Bumjun Park, Go Woon Lee, Young Lok Cha, and Yun Suk Huh. 2020. "Silver Grass-Derived Activated Carbon with Coexisting Micro-, Meso- and Macropores as Excellent Bioanodes for Microbial Colonization and Power Generation in Sustainable Microbial Fuel Cells." *Bioresource Technology* 300. https://doi.org/10.1016/j.biortech.2019.122646.

Samsudeen, N., Shivanand Chavan, T.K. Radhakrishnan, and Manickam Matheswaran. 2016. "Performance of Microbial Fuel Cell Using Chemically Synthesized Activated Carbon Coated Anode." *Journal of Renewable and Sustainable Energy* 8 (4). https://doi.org/10.1063/1.4955110.

Senthilkumar, K., and M. Naveenkumar. 2021. "Enhanced Performance Study of Microbial Fuel Cell Using Waste Biomass-Derived Carbon Electrode." *Biomass Conversion and Biorefinery*. https://doi.org/10.1007/s13399-021-01505-x.

Senthilkumar, Nangan, Md Abdul Aziz, Mehboobali Pannipara, A. Therasa Alphonsa, Abdullah G. Al-Sehemi, A. Balasubramani, and G. Gnana kumar. 2020. "Waste Paper Derived Three-Dimensional Carbon Aerogel Integrated with Ceria/Nitrogen-Doped Reduced Graphene Oxide as Freestanding Anode for High Performance and Durable Microbial Fuel Cells." *Bioprocess and Biosystems Engineering* 43 (1): 97–109. https://doi.org/10.1007/s00449-019-02208-4.

Shang, Tong Xin, and Xiao Juan Jin. 2016. "Waste Particleboard-Derived Nitrogen-Containing Activated Carbon Through KOH Activation for Supercapacitors." *Journal of Solid State Electrochemistry* 20 (7): 2029–36. https://doi.org/10.1007/s10008-016-3209-4.

Sonawane, Jayesh M., Abhishek Yadav, Prakash C. Ghosh, and Samuel B. Adeloju. 2017. "Recent Advances in the Development and Utilization of Modern Anode Materials for High Performance Microbial Fuel Cells." *Biosensors and Bioelectronics* 90 (October 2016): 558–76. https://doi.org/10.1016/j.bios.2016.10.014.

Song, Y.C., D.S. Kim, and Jung Hui Woo. 2015. "Effect of Multiwall Carbon Nanotube Contained in the Exfoliated Graphite Anode on the Power Production and Internal Resistance of Microbial Fuel Cells." *KSCE Journal of Civil Engineering* 19 (4): 857–63. https://doi.org/10.1007/s12205-013-0643-z.

Sonu, Kumar, Monika Sogani, Zainab Syed, and Jayana Rajvanshi. 2022. "Improved Degradation of Dye Wastewater and Enhanced Power Output in Microbial Fuel Cells with Chemically Treated Corncob Anodes." *Biomass Conversion and Biorefinery*. https://doi.org/10.1007/s13399-021-02254-7.

Taskan, Ergin, and Halil Hasar. 2015. "Comprehensive Comparison of a New Tin-Coated Copper Mesh and a Graphite Plate Electrode as an Anode Material in Microbial Fuel Cell." *Applied Biochemistry and Biotechnology* 175 (4): 2300–8. https://doi.org/10.1007/s12010-014-1439-4.

Utetiwabo, Wellars, Le Yang, Muhammad Khurram Tufail, Lei Zhou, Renjie Chen, Yimeng Lian, and Wen Yang. 2020. "Electrode Materials Derived from Plastic Wastes and Other Industrial Wastes for Supercapacitors." *Chinese Chemical Letters* 31 (6): 1474–89. https://doi.org/10.1016/j.cclet.2020.01.003.

Vishwanathan, A.S., Kartik S. Aiyer, L.A.A. Chunduri, K. Venkataramaniah, S. Siva Sankara Sai, and Govind Rao. 2016. "Carbon Quantum Dots Shuttle Electrons to the Anode of a Microbial Fuel Cell." *3 Biotech* 6 (2). https://doi.org/10.1007/s13205-016-0552-1.

Wang, Yi Xuan, Wen Qiang Li, Wen Ming Zong, Tian Yu Su, and Yang Mu. 2019. "Polyaniline-Decorated Honeycomb-like Structured Macroporous Carbon Composite as an Anode Modifier for Enhanced Bioelectricity Generation." *Science of the Total Environment* 696: 133980. https://doi.org/10.1016/j.scitotenv.2019.133980.

Wang, Yuyang, Xu Pan, Ye Chen, Qing Wen, Cunguo Lin, Jiyong Zheng, Wei Li, Haitao Xu, and Lijuan Qi. 2020. "A 3D Porous Nitrogen-Doped Carbon Nanotube Sponge Anode Modified with Polypyrrole and Carboxymethyl Cellulose for High-Performance Microbial Fuel Cells." *Journal of Applied Electrochemistry* 50 (12): 1281–90. https://doi.org/10.1007/s10800-020-01488-z.

Xie, Xing, Craig Criddle, and Yi Cui. 2015. "Design and Fabrication of Bioelectrodes for Microbial Bioelectrochemical Systems." *Energy and Environmental Science* 8 (12): 3418–41. https://doi.org/10.1039/c5ee01862e.

Xu, Chang, Karen Poon, Martin M.F. Choi, and Ruihua Wang. 2015. "Using Live Algae at the Anode of a Microbial Fuel Cell to Generate Electricity." *Environmental Science and Pollution Research* 22 (20): 15621–35. https://doi.org/10.1007/s11356-015-4744-8.

Xu, Haitao, Luguang Wang, Cunguo Lin, Jiyong Zheng, Qing Wen, Ye Chen, Yuyang Wang, and Lijuan Qi. 2020. "Improved Simultaneous Decolorization and Power Generation in a Microbial Fuel Cell with the Sponge Anode Modified by Polyaniline and Chitosan." *Applied Biochemistry and Biotechnology* 192 (2): 698–718. https://doi.org/10.1007/s12010-020-03346-2.

Xue, Ping, Shuai Jiang, Wenlong Li, Keren Shi, Lan Ma, and Peng Li. 2021. "Bimetallic Oxide MnFe2O4 Modified Carbon Felt Anode by Drip Coating: An Effective Approach Enhancing Power Generation Performance of Microbial Fuel Cell." *Bioprocess and Biosystems Engineering* 44 (6): 1119–30. https://doi.org/10.1007/s00449-021-02511-z.

Yang, Cuicui, Mengjie Chen, Yijun Qian, Lu Zhang, Min Lu, Xiaoji Xie, Ling Huang, and Wei Huang. 2019. "Packed Anode Derived from Cocklebur Fruit for Improving Long-Term Performance of Microbial Fuel Cells." *Science China Materials* 62 (5): 645–52. https://doi.org/10.1007/s40843-018-9368-y.

Yang, Jiawei, Shaoan Cheng, Yi Sun, and Chaochao Li. 2017. "Improving the Power Generation of Microbial Fuel Cells by Modifying the Anode with Single-Wall Carbon Nanohorns." *Biotechnology Letters* 39 (10): 1515–20. https://doi.org/10.1007/s10529-017-2384-4.

Yaqoob, Asim Ali, Mohamad Nasir Mohamad Ibrahim, and Claudia Guerrero-Barajas. 2021. "Modern Trend of Anodes in Microbial Fuel Cells (MFCs): An Overview." *Environmental Technology and Innovation* 23. https://doi.org/10.1016/j.eti.2021.101579.

Yaqoob, Asim Ali, Mohamad Nasir Mohamad Ibrahim, Mohd Rafatullah, Yong Shen Chua, Akil Ahmad, and Khalid Umar. 2020. "Recent Advances in Anodes for Microbial Fuel Cells: An Overview." *Materials* 13 (9): 1–27. https://doi.org/10.3390/ma13092078.

Yaqoob, Asim Ali, Mohamad Nasir Mohamad Ibrahim, and Susana Rodríguez-Couto. 2020. "Development and Modification of Materials to Build Cost-Effective Anodes for Microbial Fuel Cells (MFCs): An Overview." *Biochemical Engineering Journal* 164. https://doi.org/10.1016/j.bej.2020.107779.

Yaqoob, Asim Ali, Albert Serrà, Mohamad Nasir Mohamad Ibrahim, and Amira Suriaty Yaakop. 2021. "Self-Assembled Oil Palm Biomass-Derived Modified Graphene Oxide Anode: An Efficient Medium for Energy Transportation and Bioremediating Cd (II) via Microbial Fuel Cells." *Arabian Journal of Chemistry* 14 (5). https://doi.org/10.1016/j.arabjc.2021.103121.

Yin, Yao, Guangtuan Huang, Ningbo Zhou, Yongdi Liu, and Lehua Zhang. 2016. "Increasing Power Generation of Microbial Fuel Cells with a Nano-CeO2 Modified Anode." *Energy Sources, Part A: Recovery, Utilization and Environmental Effects* 38 (9): 1212–18. https://doi.org/10.1080/15567036.2014.898112.

Yuan, Yong, Shungui Zhou, Nan Xu, and Li Zhuang. 2011. "Microorganism-Immobilized Carbon Nanoparticle Anode for Microbial Fuel Cells Based on Direct Electron Transfer." *Applied Microbiology and Biotechnology* 89 (5): 1629–35. https://doi.org/10.1007/s00253-010-3013-5.

Zeng, Li Zhen, Shao Fei Zhao, and Wei Shan Li. 2015. "Ni3Mo3C as Anode Catalyst for High-Performance Microbial Fuel Cells." *Applied Biochemistry and Biotechnology* 175 (5): 2637–46. https://doi.org/10.1007/s12010-014-1458-1.

Zhang, Lijuan, Weihua He, Junchuan Yang, Jiqing Sun, Huidong Li, Bing Han, Shenlong Zhao, Y. Shi, Y. Feng, Z. Tang, and S. Liu. 2018. "Bread-Derived 3D Macroporous Carbon Foams as High Performance Free-Standing Anode in Microbial Fuel Cells." *Biosensors and Bioelectronics* 122: 217–23. https://doi.org/10.1016/j.bios.2018.09.005.

Zhao, Na, Zhaokun Ma, Huaihe Song, Yangen Xie, and Man Zhang. 2019. "Enhancement of Bioelectricity Generation by Synergistic Modification of Vertical Carbon Nanotubes/Polypyrrole for the Carbon Fibers Anode in Microbial Fuel Cell." *Electrochimica Acta* 296: 69–74. https://doi.org/10.1016/j.electacta.2018.11.039.

Zhu, Kaili, Shuangfei Wang, Hui Liu, Shijie Liu, Jian Zhang, Jinxia Yuan, Wencai Fu, W. Dang, Y. Xu, X. Yang, and Z. Wang. 2022. "Heteroatom-Doped Porous Carbon Nanoparticle-Decorated Carbon Cloth (HPCN/CC) as Efficient Anode Electrode for Microbial Fuel Cells (MFCs)." *Journal of Cleaner Production* 336. https://doi.org/10.1016/j.jclepro.2022.130374.

3 Green Synthesis of Nanoparticles from Agro-waste
A Sustainable Approach

Muthukumar K., Jayapriya M., Arulmozhi M., Senthilkumar T., and Krithikadevi R.

CONTENTS

3.1	Introduction	27
3.2	Agro-waste Nanomaterials	28
3.3	Properties of Nanoparticles	28
	3.3.1 Zinc Oxide Nanoparticles	29
	3.3.2 Silver Nanoparticles	30
3.4	Nanobiotechnology	31
3.5	Application of NPs/NCs	31
3.6	Future Perspectives	33
3.7	Conclusion	34
Acknowledgments		35
References		35

3.1 INTRODUCTION

Nanotechnology is an important field of modern research dealing with the synthesis, strategy, and manipulation of particle structures ranging from approximately 1 to 100 nm in size in any one direction (Naseer et al., 2018). Nanotechnology is being considered the next step logical in integrating technology-based science with other sister disciplines, including biology, chemistry, and physics (Huang and Wang, 2019; Bayda et al., 2019). Within this size range, all the properties (chemical, physical, and biological) change. Current nanotechnology is the building device of microscopic or even molecular size, which will potentially be benefiting medicine, environmental protection, energy, and space exploration (Singh and Nalwa, 2011). As we know more about nanoscience and the ability to engineer new products and services, it would not be far before the entire history can be compressed inside our pockets. In recent years, the term "nanotechnology" has been inflated and has almost become synonymous with things that are innovative and highly promising.

Intriguingly, these nanomaterials embody distinctive physicochemical and biological properties compared to their conventional counterparts, which endow them with their beneficial characteristics (Jeevanandam et al., 2018). Nanotechnology enables us to create functional materials, devices, and systems by controlling matters at the atomic and molecular scales and exploiting novel properties and phenomena. Substantially smaller size, lower weight, more modest power requirements, greater sensitivity, and better specificity are just a few of the improvements in sensor design (Ahmed et al., 2016; Patil and Chandrasekaran, 2020). Nanoscience and nanotechnology are some of the fastest-growing research and technology areas. Engineered nanomaterials are used in our daily life in the form of cosmetics, food packaging, drug delivery, therapeutics, biosensors, and so forth and, with these, unprecedented avenues for exposure of nanoparticles (NPs) and nanocomposites (NCs) to the environment and living beings are increasing (Aslam et al., 2018).

3.2 AGRO-WASTE NANOMATERIALS

Agro-waste (biomass) is generated from agricultural products such as fruits, vegetables, crops, animal waste, poultry, and dairy products (Obi et al., 2016). The management of the waste is critical and needs to be recycled to maintain a balance in the ecosystem. The agro-waste contains soluble substances like proteins and organic acids, and insoluble substances like cellulose and lignin (Martins et al., 2011). This biomass is now used to produce fuel (bioethanol, biogas, biohydrogen), biofertilizer, and animal feed. In the present scenario, the development of new innovative nanomaterials generated from biomass is indeed, and needs to be, highly appreciated. Thus, this chapter emphasizes the ongoing developments made in biomass and its wide application in environmental management.

In recent years, agro-waste like plants, sugarcane bagasse ash, rice husk, wild giant reed, sawdust, groundnut shell, oyster shell, and algae accruing studies have shown the efficacy of nanosized materials for tumor targeting, diagnosis (imaging), and therapy. Nearly about 350 million tons of agro-waste are generated every year in India alone and about 998 million tons are generated globally (Seetharaman et al., 2017). This includes crop waste, food processing waste, animal waste, and also hazardous and toxic waste like pesticides insecticides, herbicides, etc. Although there are technologies utilizing this waste to generate energy, the organic waste coming from agriculture remains a challenge. Researchers and investigators need to focus on more and more ways to reduce this waste and derive direct and indirect benefits from it for sustainable growth. Anyhow, one of the solutions to this giant pile of problems may be answered by nanotechnology.

NPs and NCs are currently being tested for molecular imaging to achieve a more precise diagnosis with high-quality images (Ponarulselvam et al., 2012). NPs are nanosized materials that can embed drugs, imaging agents, and genes (Kumar et al., 2010). NPs can deliver high doses of therapeutic factors into tumor cells while bypassing normal cells. While the scaffold structure of NPs enables the attachment of drugs and contrast agents, their surface facilitates bio-distribution and specific delivery through conjugation with ligands that bind to tumor biomarkers (Dubey et al., 2010). Green NPs and NCs are important areas in the field of nanotechnology which have more economic and eco-friendly benefits than the chemical and physical methods of synthesis. It is growing rapidly on various fronts due to its completely new and enhanced properties based on morphology and distribution. In hot research, scientists and researchers avoiding toxic chemicals and hazardous in the procedure of NPs/NCs through green technology used to develop biocompatibility and biodegradability. It has been an interesting field due to its non-requirement of high pressure, temperature, energy, toxic chemicals, viable, facile methodology, and low-cost rather than physical and chemical methods. One of the most attractive areas of NPs research is the formation of NC materials. The attention in hybrid nano-sized structures is purposed on the utilization and change of the properties of the individual elements and the development of new and enhanced functionalities (Krishnaraj et al., 2010).

3.3 PROPERTIES OF NANOPARTICLES

The particles in the bulk size will have different physical characteristics when compared with nanoparticles (Mustapha et al., 2022). An important parameter of nanoparticles is the surface-to-volume ratio property; as particles size grows bigger, the number of atoms will be decreased. Namra et al. (2022) proved that 30 nm of iron nanoparticles have 5% atoms on the surface, while 3 nm of iron nanoparticles have 50% atoms on the surface which have a significant impact on the physical characteristics. This increase in atoms will also increase the surface activity which influences the rate of chemical reactions. Also if the surface-to-volume ratio increases, it will reflect the changes in the property of melting temperatures, magnetization, and catalytic activity of the nanoparticles. Another important property of metal/metallic oxide nanoparticles is localized surface plasmon resonance (LSPR). This results in a collision of oscillation of electrons in the dispersed medium; metal NPs have strong size-dependent optical properties that exhibit spectrum in the UV-vis spectrophotometer which is not present in the bulk materials. The peak wavelength of LSPR is directly proportional to the size, shape, dielectric constant, and other factors. The other important feature of nanoparticles is magnetic property; Reiss and Hutten (Reiss and Hutten, 2005) demonstrated that when the nanoparticles are in the size range of 10–20 nm, the nanoparticles dominated with magnetic properties effectively. The uneven distribution of electrons in the nanoparticles is the reason for the magnetic property. For any application, the mechanical property of the materials is very important. The parameters such as hardness, stress and strain, elastic modulus, adhesion, and friction contribute to the mechanical nature of nanoparticles (Eastman et al., 1996). Other parameters such as surface coating, lubrication, and coagulation also aid in the mechanical nature of the nanoparticles. Thermal property is an important phenomenon in nanoparticles; generically, nanoparticles have high thermal stability in solid form compared with fluid form. For example, copper exhibits thermal properties very high at room temperature, greater by 700 than that of water (Wong et al., 2009; Khandel and Shahi, 2016). Henceforth, it is very important to understand the nanoparticle's properties before applying it to the application studies (Figure 3.1).

3.3.1 Zinc Oxide Nanoparticles

Green nanotechnologies are recently emerging as a fast-growing field in the technological applications of science with its contribution of eco-friendly nanoscale materials (Hinton et al., 1985). There is a growing urgent requirement to develop environmental technologies without using toxic compounds and replace those with the synthesis of green nanoparticles (NPs) (Hosseini et al., 2015). Nowadays, green synthetic methodologies employing green extracts drawn from metal nanoparticles provide a sustainable solution due to a green, bio-safe, biocompatible, viable, and facile methodology rather than toxic, classical, physical and chemical methods (Jeyabharathi et al., 2017). Synthesis of green zinc oxide nanoparticles (ZnO-NPs) has several economic advantages, compared to physical and chemical methods, such as lower cost and white appearance (Chang et al.,

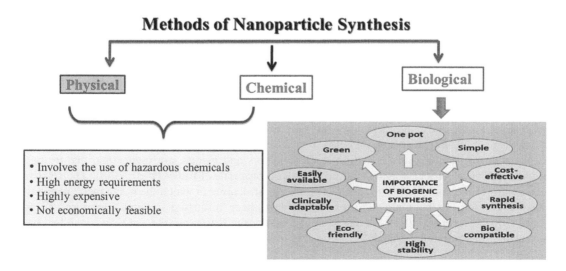

FIGURE 3.1 Methods of nanoparticles synthesis.

2016; Brayner et al., 2006). Among various NPs, ZnO-NPs are considered to be the most promising semiconductors acting as a green promising technology and providing alternative ways for antibacterial activity, which are effective in killing pathogenic and non-pathogenic bacteria. In particular, ZnO-NPs attracted attention owing to their large bandgaps and excitation binding energy, high photosensitivity, and stability (Azizi et al., 2014). ZnO-NPs were also found to be non-toxic, biosafe, and biocompatible and have been extensively used as drug carriers, in cosmetics, solar cells, automotive, and for fillings in medical applications (Harinee et al., 2019; Subramanian et al., 2022).

3.3.2 Silver Nanoparticles

In the current era, significant nanotechnological revolutions are taking place at synthetic, technological, and biological levels with the hope of solving human as well as environmental problems (Muthukumar et al., 2015). Biogenic syntheses of Ag-NPs comprising metals, their oxides, and carbonaceous materials by employing resources taken from marine sources are produced and have received immense attention due to their abundance and eco-friendly nature (Zhang and Feng, 2006). From this point of view, green synthesis of Ag-NPs by using agro-waste is better than conventional methods because it involves very few chemicals that are non-toxic to the environment and also provides an economic alternative to physical and chemical approaches. They have recently emerged as promising materials in biomedical sciences due to their antimicrobial activities toward a wide variety of pathogens and viruses (Shanmugam et al., 2013). Meanwhile, semiconductor usage by Ag-NPs was effectively documented as antibacterial materials in the past and widely applied in technological applications that were established at the molecular level (Bharathi et al., 2018).

Currently, modern green technologies provide an innovative field that demands the control of anthropogenic activities in sustainable ways. In addition, materials provided by zinc oxide (ZnO) and silver (Ag) nanoparticles (NPs) and nanocomposites (NCs) caused attention to researchers and scientists because of their unique properties as well as superior photocatalytic characteristics (Zare et al., 2019; Gao et al., 2011). Among those, ZnO-NPs and Ag-NPs are generally recognized as safe for various applications and play a key role in the field of nanotechnology, i.e., drug delivery, catalysis, antibacterial activity, photocatalytic activity, and other applications without affecting human life (Thareja and Shukla, 2017; Heiligtag and Niederberger, 2013).

Research concerned with the biosynthesis of NPs and NCs provides important areas in the field of nanotechnology with sustainable hygienic, low-cost, avoiding toxic chemicals, and eco-friendly techniques than chemical or/and physical methods (Hasnidawani et al., 2016; Li et al., 2012). Recently, green processes were progressing through the use of biological products such as crop waste (Akhayere and Kavaz, 2022), food processing waste (Vasyliev and Vorobyova, 2020), animal waste (MuhammadIdris et al., 2009), microorganisms (Ahmad et al., 2017), fungi (Senthilkumar et al., 2016), plant leaves (Inbakandan et al., 2016), microalgae (Muthusamy et al., 2017), seagrass (Harinee et al., 2019), and sponges (Nagarajan and Kuppusamy, 2013) that provide eco-friendly and viable alternatives with several potential applications. Agro-waste is easily available and has applications in traditional medicines. They contain various metabolites together with flavonoids, polyphenols, terpenoids, and alkaloids, which are responsible for the green production of metal NPs (Aragao et al., 2019; Ishwarya et al., 2018).

Nanocomposite materials, including semiconductors and metals, are interesting, due to their unique optical, electrical, biomedical, and catalytic properties (Yadav et al., 2019). The relative advantages of the core–shell arrangement over other structures in which the noble metal nanoparticles are rested on the semiconductor surface have been discussed in some literature (Xiang et al., 2016; Azizi et al., 2013). Among such nanocomposite structures, the ZnO-Ag core–shell has attracted attention not only because ZnO is less toxic and a favorable material for different applications, including UV screening applications, solar cells, biomedical, and photocatalysis (Zheng et al., 2018), but also because silver nanoparticles possess unique electrical, optical, and biological properties.

Several chemical and physical approaches have been developed for the production of core–shell-type ZnO-Ag nanocomposite (Zheng et al., 2007). However, most of these methods require hazardous materials, expensive types of equipment, and rigorous experimental conditions. Therefore, it is necessary to find a simple and eco-friendly method to synthesize ZnO-Ag nanocomposites with high performance and low cost. The biological approaches, using plants, microorganisms, and enzymes, have been proposed as possible eco-friendly alternatives to synthesizing nanoparticles. The plant bio-mediated nanoparticle synthesis is preferred because it is eco-friendly, simple, cost-effective for the biosynthesis process, and safe for biomedical application. However, a review of the literature revealed that the synthesis of metal oxides (Suresh and Sivasamy, 2020; Adhikari et al., 2015) and noble metallic nanoparticles (Ghosh et al., 2012; Khatami et al., 2018) in different sizes and shapes using plants has been explored in a number of studies, but the complete synthesis of metal oxides–metal nanocomposites by biomass has been reported in few works of literature (Elemike et al., 2017).

3.4 NANOBIOTECHNOLOGY

Biotechnology deals with the manipulation of DNA/RNA/proteins and metabolic products for the welfare of human beings and the environment (Hikal et al., 2021; Singer et al., 2018). The combination of biotechnology and nanotechnology has led to the rise of the concept of nanobiotechnology. The so-called nanobiotechnology led to immense development in contemporary biological issues as an alternative paradigm to mitigate the problems in life sciences.

In the present scenario, the emergence of microbial resistance and drug resistance in infectious and non-infectious diseases is a big challenge posed to the medical fraternity. At this juncture, nanobiotechnology can trigger a variety of scopes to curtail the resistance problem and target the cells precisely and minimize the toxic side effects (Algar et al., 2009). Nanobiotechnology offers some clinical applications such as immunomodulatory, drug delivery, imaging, and diagnosis (Fakruddin et al., 2012). Some more applications and devices are under progression and in clinical trials also. The aforementioned applications make an advantage in treating infectious and non-infectious diseases in a new dimension with greater efficacy and devoid of toxic side effects.

The potential application of nanobiotechnology is diagnosis; conventional methods involve specific antibodies and dyes to detect the probe (Jha et al., 2014). But now nanotechnology encompasses quantum dots and gold nanospheres to detect the target probes with greater efficacy and can be used for repeated cycles also (Zu and Gao, 2021). Further, sparse cells can be specifically sorted out by using nanoelectrodes.

Some standard drugs cannot easily reach the target sites; nanobiotechnology offers the delivery of drugs to the target site by nanoparticles, organic/inorganic polymers, capsules, and nanoshells. These therapeutic payloads target the drug at a specific site and minimize the drug toxicity. Likewise, viral vectors cause immunogenicity, while nanoparticles-based non-viral vectors transfer the gene into the cells effectively and are also less immunogenic (Reiss and Hutten, 2005). Another important advantage of nanobiotechnology is the development of biopharmaceuticals; nanoscale techniques enable modification of the target and develop the drugs. Despite the advantage, there are a lot of challenges imposed on the success of nanobiotechnology. Since the inception of nanotechnology, there are a lot of contrary perspectives regarding toxicity and its behavior in the environment (Eastman et al., 1996; Wong et al., 2009). There are no proper documented studies on release of nanoparticles in the aquatic ecosystem. Nevertheless, the impact of nanoparticles is gaining momentum and is applied in all sectors owing to their unique properties.

3.5 APPLICATION OF NPS/NCS

Numerous synthesis methods are either being developed or improved to enhance the properties and reduce production costs. Some methods are modified to achieve process-specific NPs and NCs to

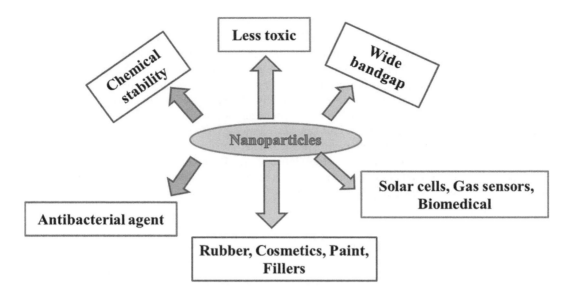

FIGURE 3.2 The nature of nanoparticles and their applications.

increase their optical, mechanical, physical, and chemical properties (Elemike et al., 2017). The NPs are synthesized by various methods that are categorized into bottom-up or top-down approach. In the bottom-up approach, NPs and NCs can be synthesized using chemical (chemical reduction) and biological methods (use of plants, microorganisms, etc.) by self-assembly of atoms to new nuclei which grow into a particle of nanoscale; while in the top-down approach, suitable bulk material breaks down into fine particles by size reduction with various lithographic techniques: grinding, milling, sputtering, thermal/laser ablation, etc. The various methods (physical and chemical) used for the synthesis of NPs and NCs are very hazardous due to the use of toxic chemicals that are responsible for various biological risks and are quite expensive (Figure 3.2) (Chandran et al., 2006).

Environment-friendly methods are becoming more popular in the field of chemistry and chemical technology when people are becoming more conscious of various environmental issues. The development of biologically inspired experimental processes for the syntheses of NPs and NCs is evolving as an important branch of nanotechnology. The biosynthesis of metal NPs and NCs is gaining more importance due to its simplicity, its rapid rate of formation of metal NPs and NCs, and eco-friendly approach. Various physical and chemical methods are more popular for NP synthesis; however, these methods are cost-intensive and the employment of toxic compounds limits their applications and are not environment-friendly (Lateef et al., 2015). Conventional synthesis of NPs can involve expensive chemical and physical processes that often use toxic materials with potential hazards such as environmental toxicity, cytotoxicity, and carcinogenicity (Atel et al., 2015). To overcome these issues, a safe eco-friendly green method is available (Thunugunta et al., 2015). Synthesizing NPs and NCs via biological entities acting as biological factories offer a clean, non-toxic, and environment-friendly method of synthesizing NPs with a wide range of sizes, shapes, compositions, and physicochemical properties. Another interesting feature of many biological entities is their ability to act as templates in the synthesis, assembly, and organization of nanometer-scale materials to fabricate well-defined micro- and macroscale structures. The advancements in green synthesis over chemical and physical methods are environment-friendly, cost-effective, and easily scaled up for large-scale syntheses of NPs; furthermore, there is no need to use high temperature, pressure, energy, and toxic chemicals (Figure 3.3) (Dauthal and Mukhopadhyay, 2015).

Interesting features of NPs have led to the entrance of several NPs-based therapeutics into the clinical trial stage during the last two decades (Kumar et al., 2007). The opportunity to modulate

Green Synthesis of Nanoparticles from Agro-waste

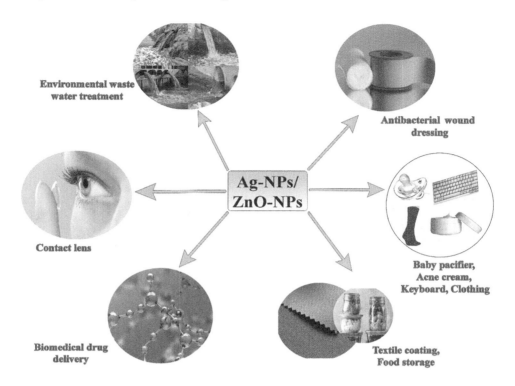

FIGURE 3.3 The major application of nanotechnology in the environment aspect.

numerous features of NPs has made them potent therapeutic vectors for cancer therapy. NPs can cause fewer side effects by improving the accumulation of drugs in diseased tissue, thereby reducing the dose required to achieve therapeutic efficacy. As an illustration, less than 0.01% of the injected dose of agents in the angstrom size range (e.g., antibodies) typically accumulates in the target region (Nair et al., 2013; Bankar et al., 2010), which is in part related to the small size of NPs and deregulated vascular structure and enhanced permeability and retention (EPR) effects (Yang et al., 2014). In recent years, the upcoming attempt toward biogenic substances coupled with nanomaterials shows higher efficiencies in the degradation and antibacterial efficacy. Agro-waste employed in the green synthesis of nanoparticles and nanocomposites reported in the literature is presented in Table 3.1.

3.6 FUTURE PERSPECTIVES

The green synthesis of nanoparticles by using agro-waste materials is facing challenges toward commercial success. Hence, the mechanism behind the synthesis of nanoparticles needs to be elucidated. Literature reports the plant constituents' alkaloids, phenols, proteins, carbohydrates, and terpenoids act as reducing as well as stabilizing agents for synthesis. This needs to be properly understood before bringing the ZnO-NPs and Ag-NPs or both based nanocomposites into biomedical applications. ZnO-NPs and Ag-NPs are essential elements for living organisms. The US Food and Drug Administration (FDA) already recommended the ZnO-NPs and Ag-NPs (Antonio et al., 2019). The acute and chronic toxicity of the pristine nanoparticles should be assessed and evaluated to predict their long-term toxicity caused by low doses. Furthermore, the internalization and accumulation of NPs and NCs inside the body cells should be determined. As a chemical-mediated synthesis of ZnO-NPs and Ag-NPs, it is reported that the toxicity of ZnO-NPs and Ag-NPs in animals

TABLE 3.1
Synthesis of Nanoparticles Using Agro-waste Materials

S. No.	Type of Agro-waste Materials	Nanoparticles/ Nanocomposites Produced	Applications	Reference
1	*Citrus aurantifolia* peel extract	Ag-NPs	Antibacterial activity	(Preeti and Mausumi, 2016)
2	Moringa oleifera	Ag-NPs	Antibacterial activity	(Prasad et al., 2011)
3	*Punica granatum*	Ag-NPs and Au-NPs	Catalytic activity against 4-nitrophenol	(Dash and Bag, 2014)
4	Sugarcane bagasse extract	Si-NPs	Photocatalytic dye degradation	(Ntalane et al., 2022)
5	Banana peel extract	Ag-NPs	Antifungal and antibacterial activity	(Ashok et al., 2015)
6	Grape waste	Au-NPs	Antibacterial activity	(Kiruba et al., 2018)
7	Sheep and goat fecal matter	ZnO-NPs	Antibacterial activity	(Mahesh and Shivayogeeswar, 2018)
8	Mango peel	Au-NPs	Antibacterial activity	(Ning and Wei-Hong, 2013)
9	Annona squamosa peel	Ag-NPs	Photocatalytic dye degradation	(Rajendran et al., 2012)
10	Timber industry waste	Ag-NPs	Antibacterial and antifungal activity	(Aishwarya et al., 2015)
11	*Panax ginseng* root extract	Ag-NPs and Au-NPs	Antibacterial and antifungal activity	(Priyanka et al., 2016)
12	*Fusarium oxysporum*	Ag-NPs	Photocatalytic dye degradation	(Birla et al., 2013)
13	*Nyctanthes arbor-tristis*	ZnO-NPs	Antifungal activity	(Pragati et al., 2018)
14	*Coriandrum sativum* leaf extract	ZnO-NPs	Anthracene photocatalytic degradation	(Hassan et al., 2015)
15	*Bilberry waste* (BW) and red currant waste (RCW) extracts	Ag-NPs	Antifungal activity	(Das et al., 2011)

is high (Cai et al., 2018). Therefore, the synthesis of NPs/NCs should be encouraged from various biogenic sources. At present, only very few limited studies are reported on the biomedical applications of NPs/NCs compared with other oxide nanoparticles. Furthermore, more intensive studies should be explored for other biomedical topics such as wound healing, antivirals, arthritis, etc.

3.7 CONCLUSION

Nanotechnology has moved into modern biological and medicinal implications for the advancement of the biomedical field. Recent developments in nanoparticle synthesis have focused on safe and environmentally sustainable approaches. Agro-waste-mediated nanoparticles are not only eco-friendly but also safe mode that utilizes biological resources for synthesis. The green approach to nanoparticle fabrication has opened up a new era of safe nanotechnology. The ZnO-NPs and Ag-NPs are well-known for their unique outstanding behavior and properties. In this chapter, we conceptualized the green synthesis of ZnO-NPs and Ag-NPs from bioresources. Furthermore, we have documented the effect of reactants and other kinetic conditions regarding the size, shape, and charge of the NPS/NCs. Additionally, we broadly present the antimicrobial and anticancer activities of green synthesized ZnO-NPs and Ag-NPs. Henceforth, this chapter will benefit the researchers and applied scientists who attempt to produce the green synthesis of ZnO-NPs and Ag-NPs and their subsequent applications in biomedicine.

ACKNOWLEDGMENTS

Krishnan Muthukumar is thankful to University Grants Commission, India, for providing Dr. D.S. Kothari Post-Doctoral Fellow (No. F.4–2/2006 (BSR)/ES/20–21/0014) to carry out his research work.

REFERENCES

Adhikari, S., Banerjee, A., Eswar, N. K., Sarkar, D., and Madras, G. 2015. "Photocatalytic inactivation of *E. coli* by ZnO–Ag nanoparticles under solar radiation". *RSC Advances*, 5, 51067–77. doi: 10.1039/C5RA06406F

Ahmad, Y., Wei, F., Syed, K., Tahir, A. U., Rehman, A., Khan, S., Ullah, S., and Yuan, Q. 2017. "The effects of bacteria-nanoparticles interface on the antibacterial activity of green synthesized silver nanoparticles". *Microbial Pathogenesis*, 102, 133–42. doi: 10.1016/j.micpath.2016.11.030

Ahmed, S., Ahmad, M., Swami, B. L., and Ikram, S. 2016. "A review on plants extract mediated synthesis of silver nanoparticles for antimicrobial applications: A green expertise". *Journal of Advanced Research*, 7(1), 17–28. doi: 10.1016/j.jare.2015.02.007

Aishwarya, D., Vidya Shetty, K., and Saidutta, M. B. 2015. "Timber industry waste-teak (Tectona grandis Linn.) leaf extract mediated synthesis of antibacterial silver nanoparticles". *International Nano Letters*, 5, 205–14. doi: 10.1007/s40089-015-0157-4

Akhayere, E., and Kavaz, D. 2022. "Synthesis of silica nanoparticles from agricultural waste". *Nanomaterials in Plant Protection*, 121–38. doi: 10.1016/B978-0-12-823575-1.00028-7

Algar, W. R., Massey, M., and Krull, U. J. 2009. "The application of quantum dots, gold nanoparticles and molecular switches to optical nucleic-acid diagnostics". *TrAC Trends in Analytical Chemistry*, 28(3), 292–306. doi: 10.1016/j.trac.2008.11.012

Antonio, Z., Annalaura, I., Stefano, N., and Roberto, L. 2019. "Green synthesis of silver nanoparticles using bilberry and red currant waste extracts". *Processes*, 7, 193–201. doi: 10.3390/pr7040193

Aragao, A. P., Oliveira, T. M., Quelemes, P. V., Perfeito, M. L. G., Araujo, M. C., Santiago, J. A. S., Cardoso, V. S., Quaresma, P., Leite, J. R., and Silva, D. A. 2019. "Green synthesis of silver nanoparticles using the seaweed *Gracilaria birdiae* and their antibacterial activity". *Arabian Journal of Chemistry*, 12, 4182–8. doi: 10.1016/j.arabjc.2016.04.014

Ashok, B., Bhagyashree, J., Ameeta, R. K., and Smita, Z. 2015. "Banana peel extract mediated novel route for the synthesis of silver nanoparticles". *Colloids and Surfaces A: Physicochemical and Engineering Aspects*, 368, 58–63. doi: 10.1016/j.colsurfa.2010.07.024

Aslam, B., Wang, W., Arshad, M. I., Khurshid, M., Muzammil, S., Rasool, M. H., and Baloch, Z. 2018. "Antibiotic resistance: A rundown of a global crisis". *Infection and Drug Resistance*, 11, 1645. doi: 10.2147/IDR.S173867

Atel, V., Berthold, D., Puranik, P., and Gantar, M. 2015. "Screening of cyanobacteria and microalgae for their ability to synthesize silver nanoparticles with antibacterial activity". *Biotechnology Reports*, 5, 112–19. doi: 10.1016/j.btre.2014.12.001

Azizi, S., Ahmad, M. B., Namvar, F., and Mohamad, R. 2014. "Green biosynthesis and characterization of zinc oxide nanoparticles using brown marine macroalga *Sargassum muticum* aqueous extract". *Materials Letters*, 116, 275–7. doi: 10.1016/j.matlet.2013.11.038

Azizi, S., Namvar, F., Mahdavi, M., Ahmad, M. B., and Mohamad, R. 2013. "Biosynthesis of silver nanoparticles using brown marine macroalga, *Sargassum muticum* aqueous extract". *Materials*, 6, 5942–50.

Bankar, A., Joshi, B., Kumar, A. R., and Ziniarde, S. 2010. "Banana peel extract mediated synthesis of gold nanoparticles". *Colloids and Surfaces B*, 80, 45–50. doi: 10.1016/j.colsurfb.2010.05.029

Bayda, S., Adeel, M., Tuccinardi, T., Cordani, M., and Rizzolio, F. 2019. "The history of nanoscience and nanotechnology: From chemical–physical applications to nanomedicine". *Molecules*, 25(1), 112. doi: 10.3390/molecules25010112

Bharathi, D., Vasantharaj, S., and Bhuvaneshwari, V. 2018. "Green synthesis of silver nanoparticles using *Cordia dichotoma* fruit extract and its enhanced antibacterial, anti-biofilm and photo catalytic activity". *Materials Research Express*, 5(5), 1–8. doi: 10.1088/2053-1591/aac2ef

Birla, S. S., Gaikwad, S. C., Gade, A. K., and Rai, M. K. 2013. "Rapid synthesis of silver nanoparticles from *Fusarium oxysporum* by optimizing physicocultural conditions". *Scientific World Journal*, 12. doi: 10.1155/2013/796018

Brayner, R., Ferrari-Iliou, R., Brivois, N., Djediat, S., Benedetti, M. F., and Fievet, F. 2006. "Toxicological impact studies based on *Escherichia coli* bacteria in ultrafine ZnO nanoparticles colloidal medium". *Nano Letters*, 6, 866–70. doi: 10.1021/nl052326h

Cai, L., Chen, J., Liu, Z., Wang, H., Yang, H., and Ding, W. 2018. "Magnesium oxide nanoparticles: Effective agricultural antibacterial agent against Ralstonia solanacearum". *Frontiers in Microbiology*, 9, 790. doi: 10.3389/fmicb.2018.00790

Chandran, S. P., Chaudhary, M., Pasricha, R., Ahmad, A., and Sastry, M. 2006. "Synthesis of gold nanotriangles and silver nanoparticles using aloe vera plant extract". *Biotechnology Progress*, 22, 577–83. doi: 10.1021/bp0501423

Chang, X., Li, Z., Zhai, X., Sun, S., Gu, D., Dong, L., Yin, Y., and Zhu, Y. 2016. "Efficient synthesis of sunlight-driven ZnO-based heterogeneous photocatalysts". *Materials & Design*, 98, 324–32. doi: 10.1016/j.matdes.2016.03.027

Das, R. K., Gogoi, N., and Bora, U. 2011. "Green synthesis of gold nanoparticles using *Nyctanthes arbortristis* flower extract". *Bioprocess and Biosystems Engineering*, 34, 61. doi: 10.1007/s00449-010-0510-y

Dash, S. S., and Bag, B. G. 2014. "Synthesis of gold nanoparticles using renewable *Punica granatum* juice and study of its catalytic activity". *Applied Nanoscience*, 4, 55–9. doi: 10.1007/s13204-012-0179-4

Dauthal, P., and Mukhopadhyay, M. 2015. "Agro-industrial waste-mediated synthesis and characterization of gold and silver nanoparticles and their catalytic activity for 4-nitroaniline hydrogenation". *Korean Journal of Chemical Engineering*, 32, 837–44. doi: 10.1007/s11814-014-0277-y

Dubey, S. P., Lahtinen, M., and Sillanpa, M. 2010. "Tansy fruit mediated greener synthesis of silver and gold nanoparticles". *Process Biochemistry*, 45(7), 1065–71. doi: 10.1016/j.procbio.2010.03.024

Eastman, J. A., Choi, U. S., Li, S., Thompson, L. J., and Lee, S. 1996. "Enhanced thermal conductivity through the development of nanofluids". *MRS Online Proceedings Library (OPL)*, 457. doi: 10.1557/PROC-457-3

Elemike, E. E., Onwudiwe, D. C., Ekennia, A. C., Ehiri, R. C., and Nnaji, N. J. 2017. "Phytosynthesis of silver nanoparticles using aqueous leaf extracts of *Lippia citriodora*: Antimicrobial, larvicidal and photocatalytic evaluations". *Materials Science and Engineering C*, 75, 980–9. doi: 10.1016/j.msec.2017.02.161

Fakruddin, M., Hossain, Z., and Afroz, H. 2012. "Prospects and applications of nanobiotechnology: A medical perspective". *Journal of Nanobiotechnology*, 10(1), 1–8. doi: 10.1186/1477-3155-10-40.

Gao, S., Jia, X., Yang, S., Li, Z., and Jiang, K. 2011. "Hierarchical Ag/ZnO micro/nanostructure: Green synthesis and enhanced photocatalytic performance". *Journal of Solid State Chemistry*, 184, 764–9. doi: 10.1016/j.jssc.2011.01.025

Ghosh, S., Goudar, V. S., Padmalekha, K. G., Bhat, S. V., Indi, S. S., and Vasan, H. N. 2012. "ZnO/Ag nanohybrid: Synthesis, characterization, synergistic antibacterial activity and its mechanism". *RSC Advances*, 2, 930–40. doi: 10.1039/C1RA00815C

Harinee, S., Muthukumar, K., Abirami, A., Amrutha, K., Dhivyaprasath, K., and Ashok, M. 2019. "UV-light photocatalytic activity of biocompatible nanoparticles provide multiple effects". *Advanced Materials Proceeding*, 4(3), 115–18. doi: 10.5185/amp.2019.0006

Hasnidawani, J. N., Azlina, H. N., Norita, N., Bonnia, N. N., Ratim, S., and Ali, E. S. 2016. "Synthesis of ZnO nanostructures using solgel method". *Procedia Chemistry*, 19, 211–16. doi: 10.1016/j.proche.2016.03.095

Hassan, S. S. M., El-Azab, W. I. M., Ali, H. R., and Mansour, M. S. M. 2015. "Green synthesis and characterization of ZnO nanoparticles for photocatalytic degradation of anthracene". *Advances in Natural Sciences: Nanoscience and Nanotechnology*, 6, 045012. doi: 10.1088/2043-6262/6/4/045012

Heiligtag, F. J., and Niederberger, M. 2013. "The fascinating world of nanoparticle research". *Materials Today*, 16, 262–71. doi: 10.1016/j.apsusc.2007.04.088

Hikal, W. M., Bratovcic, A., Baeshen, R. S., Tkachenko, K. G., and Said-Al, H. A. 2021. "Nanobiotechnology for the detection and control of waterborne parasites". *Open Journal of Ecology*, 11(3), 203–23. doi: 10.4236/oje.2021.113016

Hinton, M., Hedges, A. J., and Linton, A. H. 1985. "The ecology of *Escherichia coli* in market claves fed a milk-substitute diet". *Journal of Applied Microbiology*, 58, 27–35. doi: 10.1111/j.1365-2672.1985.tb01426.x

Hosseini, S. M., Sarsari, I. A., Kameli, P., and Salamati, H. 2015. "Effect of Ag doping on structural, optical and photocatalytic properties of ZnO nanoparticles". *Journal of Alloys and Compounds*, 640, 408–15. doi: 10.1016/j.jallcom.2015.03.136

Huang, H. X., and Wang, X. 2019. "Biomimetic fabrication of micro-/nanostructure on polypropylene surfaces with high dynamic superhydrophobic stability". *Materials Today Communications*, 19, 487–94. doi: 10.1016/j.mtcomm.2019.04.005

Inbakandan, D., Kumar, C., Bavanilatha, M., Ravindra, D. N., Kirubagaran, R., and Khan, S. A. 2016. "Ultrasonic-assisted green synthesis of flower like silver nanocolloids using marine sponge extract and its effect on oral biofilm bacteria and oral cancer cell lines". *Microbial Pathogenesis*, 99, 135–41. doi: 10.1016/j.micpath.2016.08.018

Ishwarya, R., Vaseeharan, B., Kalyani, S., Banumathi, B., Govindarajan, M., Alharbi, N. S., Kadaikunnan, S., Al-anbr, M. N., and Khaled, G. 2018. "Facile green synthesis of zinc oxide nanoparticles using *Ulva lactuca* seaweed extract and evaluation of their photocatalytic, antibiofilm and insecticidal activity". *Journal of Photochemistry and Photobiology B: Biology*, 178, 249–58. doi: 10.1016/j.jphotobiol.2017.11.006

Jeevanandam, J., Barhoum, A., Chan, Y. S., Dufresne, A., and Danquah, M. K. 2018. "Review on nanoparticles and nanostructured materials: History, sources, toxicity and regulations". *Beilstein Journal of Nanotechnology*, 9(1), 1050–74. doi: 10.3762/bjnano.9.98

Jeyabharathi, S., Kalishwaralal, K., Sundar, K., and Muthukumaran, A. 2017. "Synthesis of zinc oxide nanoparticles (ZnONPs) by aqueous extract of *Amaranthus caudatus* and evaluation of their toxicity and antimicrobial activity". *Materials Letters*, 209, 295–8. doi: 10.1016/j.matlet.2017.08.030

Jha, R. K., Jha, P. K., Chaudhury, K., Rana, S. V., and Guha, S. K. 2014. "An emerging interface between life science and nanotechnology: Present status and prospects of reproductive healthcare aided by nanobiotechnology". *Nano Reviews*, 5(1), 22762. doi: 10.3402/nano.v5.22762

Khandel, P., and Shahi, S. K. 2016. "Microbes mediated synthesis of metal nanoparticles: Current status and future prospects". *International Journal of Nanomaterials and Biostructures*, 6(1), 1.

Khatami, M., Varma, R. S., Zafarnia, N., Yaghoobi, H., Sarani, M., and Kumar, V. G. 2018. "Applications of green synthesized Ag, ZnO and Ag/ZnO nanoparticles for making clinical antimicrobial wound-healing bandages". *Sustainable Chemistry and Pharmacy*, 10, 9–15. doi: 10.1016/j.scp.2018.08.001

Kiruba, K., Hojatollah, V., and Valerie, O. 2018. "Value-adding to grape waste: Green synthesis of gold nanoparticles". *Journal of Food Engineering*, 142, 210–20. doi: 10.1016/j.jfoodeng.2014.06.014

Krishnaraj, C., Jagan, E. G., Rajasekar, S., Selvakumar, P., Kalaichelvan, P. T., and Mohan, N. 2010. "Synthesis of silver nanoparticles using *Acalypha indica* leaf extracts and its antibacterial activity against water borne pathogens". *Colloids and Surfaces B: Biointerfaces*, 76(1), 50–6. doi: 10.1016/j.colsurfb.2009.10.008

Kumar, S. A., Abyaneh, M. K., Gosavi, S. W., Kulkarni, S. K., Pasricha, R., Ahmad, A., and Khan, M. I. 2007. "Nitrate reductase-mediated synthesis of silver nanoparticles from AgNO$_3$". *Biotechnology Letters*, 29, 439–45. doi: 10.1007/s10529-006-9256-7

Kumar, V., Yadav, S. C., and Yadav, S. K. 2010. "*Syzygium cumini* leaf and seed extract mediated biosynthesis of silver nanoparticles and their characterization". *Journal of Chemical Technology & Biotechnology*, 85(10), 1301–9. doi: 10.1002/jctb.2427

Lateef, A., Azeez, M. A., Asafa, T. B., Yekeen, T. A., Akinboro, A., Oladipo, I. C., Ajetomobi, F. E., Gueguim-Kana, E. B., and Beukes, L. S. 2015. "*Cola nitida*-mediated biogenic synthesis of silver nanoparticles using seed and seed shell extracts and evaluation of antibacterial activities". *Bio Nano Science*, 5, 196–205. doi: 10.1007/s12668-015-0181-x

Li, G., He, D., Qian, Y., Guan, B., Gao, S., Cui, Y., Yokoyama, K., and Wang, L. 2012. "Fungus-mediated green synthesis of silver nanoparticles using *Aspergillus terreus*". *International Journal of Molecular Sciences*, 13, 466–76. doi: 10.3390/ijms13010466

Mahesh, M. C., and Shivayogeeswar, E. N. 2018. "Effect of sheep and goat fecal mediated synthesis and characterization of silver nanoparticles (AgNPs) and their antibacterial effects". *Journal of Nanofluids*, 7, 309–15. doi: 10.1166/jon.2018.1458

Martins, D. A. B., do Prado, H. F. A., Leite, R. S. R., Ferreira, H., de Souza, M. M., Moretti, R. D. S., and Gomes, E. 2011. "Agroindustrial wastes as substrates for microbial enzymes production and source of sugar for bioethanol production". In *Integrated waste management* (vol. II). IntechOpen. doi: 10.5772/23377

MuhammadIdris, N., Li, Z., Ye, L., Sim, E. K. W., Mahendran, R., Chi-Lui Ho, P., and Zhang, Y. 2009. "Tracking transplanted cells in live animal using up conversion fluorescent nanoparticles". *Biomaterials*, 30(28), 5104–13. doi: 10.1016/j.biomaterials.2009.05.062

Mustapha, T., Misni, N., Ithnin, N. R., Daskum, A. M., and Unyah, N. Z. 2022. "A review on plants and microorganisms mediated synthesis of silver nanoparticles, role of plants metabolites and applications". *International Journal of Environmental Research and Public Health*, 19(2), 674. doi: 10.3390/ijerph19020674

Muthukumar, K., Vignesh, S., Dahms, H. U., Santhosh, G. M., Palanichamy, S., Subramanian, G., and James, R. A. 2015. "Antifouling assessments on biogenic nanoparticles: A field study from polluted offshore platform". *Marine Pollution Bulletin*, 101, 816–25. doi: 10.1016/j.marpolbul.2015.08.033

Muthusamy, G., Thangasamy, S., Raja, M., Chinnappan, C., and Kandasamy, S. 2017. "Biosynthesis of silver nanoparticles from *Spirulina* microalgae and its antibacterial activity". *Environmental Science and Pollution Research*, 24, 19459–64. doi: 10.1007/s11356-017-9772-0

Nagarajan, S., and Kuppusamy, K. A. 2013. "Extracellular synthesis of zinc oxide nanoparticle using seaweeds of Gulf of Mannar". *International Journal of Nanobiotechnology*, *11*, 1–11. doi: 10.1186/1477-3155-11-39

Nair, V., Sambre, D., Joshi, S., Bankar, A., Kumar, R. A., and Zinjarde, S. 2013. "Yeast-derived melanin mediated synthesis of gold nanoparticles". *Journal of Bionanoscience*, *7*, 159–68. doi: 10.1166/jbns.2013.1108

Namra, M., Muhammad, F. A., Ammara, S., and Amjad, R. 2022. "Harmful Consequences of Proton Pump Inhibitors on Male Fertility: An Evidence from Subchronic Toxicity Study of Esomeprazole and Lansoprazole in Wistar Rats". *International Journal of Endocrinology*, *13*, 4479261, 13. doi: 10.1155/2022/4479261.

Naseer, B., Srivastava, G., Qadri, O. S., Faridi, S. A., Islam, R. U., and Younis, K. 2018. "Importance and health hazards of nanoparticles used in the food industry". *Nanotechnology Reviews*, *7*(6), 623–41. doi: 10.1016/j.jafr.2022.100270

Ning, Y., and Wei-Hong, L. 2013. "Mango peel extract mediated novel route for synthesis of silver nanoparticles and antibacterial application of silver nanoparticles loaded onto non-woven fabrics". *Industrial Crops and Products*, *48*, 81–8. doi: 10.1016/j.indcrop.2013.04.001

Ntalane, S. S., Raymond, T. T., and Lindiwe, K. 2022. "Extraction and synthesis of silicon nanoparticles (SiNPs) from *Sugarcane Bagasse Ash*: A mini-review". *Applied Science*, *12*(5), 2310. doi: 10.3390/app12052310

Obi, F. O., Ugwuishiwu, B. O., and Nwakaire, J. N. 2016. "Agricultural waste concept, generation, utilization and management". *Nigerian Journal of Technology*, *35*(4), 957–64.

Patil, S., and Chandrasekaran, R. 2020. "Biogenic nanoparticles: A comprehensive perspective in synthesis, characterization, application and its challenges". *Journal of Genetic Engineering and Biotechnology*, *18*(1), 1–23. doi: 10.1186/s43141-020-00081-3

Ponarulselvam, S., Panneerselvam, C., Murugan, K., Aarthi, N., Kalimuthu, K., and Thangamani, S. 2012. "Synthesis of silver nanoparticles using leaves of Catharanthus roseus Linn G. Don and their antiplasmodial activities". *Asian Pacific Journal of Tropical Biomedicine*, *2*(7), 574–80. doi: 10.1016/S2221-1691(12)60100-2

Pragati, J., Poonam, K., and Rana, J. S. 2018. "Green synthesis of zinc oxide nanoparticles using flower extract of *Nyctanthes arbor-tristis* and their antifungal activity". *Journal of King Saud University-Science*, *30*, 168–75. doi: 10.1016/j.jksus.2016.10.002

Prasad, T. N. V. K. V., and Elumalai, E. K. 2011. "Biofabrication of Ag nanoparticles using Moringa oleifera leaf extract and their antimicrobial activity". *Asian Pacific Journal of Tropical Biomedicine*, *1*(6): 439–442. doi: 10.1016/S2221-1691(11)60096-8.

Preeti, D., and Mausumi, M. 2016. "Phyto-synthesis and structural characterization of catalytically active gold nanoparticles biosynthesized using *Delonix regia* leaf extract". *3 Biotech*, *6*, 118–27.

Priyanka, S., Yeon, J. M., Chao, W., Ramya, M., and Deok, C. Y. 2016. "The development of a green approach for the biosynthesis of silver and gold nanoparticles by using *Panax ginseng* root extract, and their biological applications". *Artificial Cells, Nanomedicine, and Biotechnology*, *44*(4), 1150–7. doi: 10.3109/21691401.2015.1011809

Rajendran, K., Selvaraj, M. R., Arunachalam, P., Venkatesan, G. K., and Subhendu, C. 2012. "Agricultural waste Annona squamosa peel extract: Biosynthesis of silver nanoparticles". *Spectrochimica Acta Part A: Molecular and Biomolecular Spectroscopy*, *90*, 173–6. doi: 10.1016/j.saa.2012.01.029

Reiss, G., and Hutten, A. 2005. "Magnetic nanoparticles: Applications beyond data storage". *Nature Materials*, *4*(10), 725–6. doi: 10.1038/nmat1494

Seetharaman, P., Chandrasekaran, R., Gnanasekar, S., Mani, I., and Sivaperumal, S. 2017. "Biogenic gold nanoparticles synthesized using *Crescentia cujete* L. and evaluation of their different biological activities". *Biocatalysis and Agricultural Biotechnology*, *11*, 75–82. doi: 10.1016/j.bcab.2017.06.004

Senthilkumar, P., Santhosh Kumar, D. S. R., Sudhagar, B., Vanthana, M., Parveen, M. H., Sarathkumar, S., Thomas, J. C., Sandhiya Mary, S., and Kannan, C. 2016. "Seagrass-mediated silver nanoparticles synthesis by *Enhalus acoroides* and its α-glucosidase inhibitory activity from the Gulf of Mannar." *Journal of Nanostructure in Chemistry*, *6*, 275–80. doi: 10.1007/s40097-016-0200-7

Shanmugam, N., Rajkamal, P., Cholan, S., Kannadasan, N., Sathishkumar, K., Viruthagiri, G. and Sundaramanickam, A. 2013. "Biosynthesis of silver nanoparticles from the marine seaweed *Sargassum wightii* and their antibacterial activity against some human pathogens". *Applied Nanoscience*, *4*, 881–8. doi: 10.1007/s13204-013-0271-4

Singer, A., Markoutsa, E., Limayem, A., Mohapatra, S., and Mohapatra, S. S. 2018. "Nanobiotechnology medical applications: Overcoming challenges through innovation". *EuroBiotech Journal*, *2*, 146–60. doi: 10.2478/ebtj-2018-0019

Singh, R., and Nalwa, H. S. 2011. "Medical applications of nanoparticles in biological imaging, cell labeling, antimicrobial agents, and anticancer nanodrugs". *Journal of Biomedical Nanotechnology*, *7*(4), 489–503. doi: 10.1166/jbn.2011.1324

Subramanian, H., Krishnan, M., Hans-Uwe, D., Marimuthu, K., Sivanandham, V., Rajesh, J. B., Mahalingam, A., and Rathinam, A. J. 2019. "Biocompatible nanoparticles with enhanced photocatalytic and antimicrofouling potential". *International Biodeterioration and Biodegradation*, 145, 104790. doi: 10.1016/j.ibiod.2019.104790.

Subramanian, H., Krishnan, M., and Mahalingam, A. 2022. "Photocatalytic dye degradation and photoexcited anti-microbial activities of green zinc oxide nanoparticles synthesized via *Sargassum muticum* extracts". *RSC Advances*, 12, 985–97. doi: 10.1039/D1RA08196A

Suresh, M., and Sivasamy, A. 2020. "Fabrication of graphene nanosheets decorated by nitrogen-doped ZnO nanoparticles with enhanced visible photocatalytic activity for the degradation of methylene blue dye". *Journal of Molecular Liquids*, 317, 114112. doi: 10.1016/j.molliq.2020.114112

Thareja, R. K., and Shukla, S. 2017. "Synthesis and characterization of zinc oxide nanoparticles by laser ablation of zinc in liquid". *Applied Surface Science*, 253, 8889–95.

Thunugunta, T., Reddy, A. C., and Reddy, D. C. 2015. "Green synthesis of nanoparticles: Current prospectus". *Nanotechnology Reviews*, 4, 303–32. doi: 10.1515/ntrev-2015-0023

Vasyliev, G., and Vorobyova, G. V. 2020. "Valorization of food waste to produce eco-friendly means of corrosion protection and 'green' synthesis of nanoparticles". *Advances in Materials Science and Engineering*, 1–14. doi: 10.1155/2020/6615118

Wong, T. S., Brough, B., and Ho, C. M. 2009. "Creation of functional micro/nano systems through top-down and bottom-up approaches". *Molecular & Cellular Biomechanics: MCB*, 6(1), 1.

Xiang, Y. H., Ju, P., Wang, Y., Sun, Y., Zhang, D., and Yu, J. Q. 2016. "Chemical etching preparation of the Bi_2WO_6/BiOI p-n heterojunction with enhanced photocatalytic antifouling activity under visible light irradiation". *Chemical Engineering Journal*, 288, 264–75. doi: 10.1016/j.cej.2015.11.103

Yadav, L. S. R., Pratibha, S., Manjunath, K., Shivanna, M., Ramakrishnappa, T., Dhananjaya, N., and Nagaraju, G. 2019. "Green synthesis of Ag-ZnO nanoparticles: Structural analysis, hydrogen generation, formulation, and biodiesel applications". *Journal of Science: Advanced Materials and Devices*, 4, 425–31. doi: 10.1016/j.jsamd.2019.03.001

Yang, N., WeiHong, L., and Hao, L. 2014. "Biosynthesis of Au nanoparticles using agricultural waste mango peel extract and its in vitro cytotoxic effect on two normal cells". *Materials Letters*, 134, 67–70. doi: 10.1016/j.matlet.2014.07.025

Zare, M., Namratha, K., Alghamdi, S., Mohammad, Y. H. E., Hezam, A., Zare, M., Drmosh, Q. A., Byrappa, K., Chandrashekar, B. N., Ramakrishna, S., and Zhang, X. 2019. "Novel green biomimetic approach for synthesis of ZnO-Ag nanocomposite; antimicrobial activity against food-borne pathogen, biocompatibility and solar photocatalysis". *Science Reports*, 9, 1–15. doi: 10.1038/s41598-019-44309-w

Zhang, Z., and Feng, S. S. 2006. "The drug encapsulation efficiency, *in vitro* drug release, cellular uptake and cytotoxicity of paclitaxel-loaded poly(lactide)–tocopheryl polyethylene glycol succinate nanoparticles". *Biomaterials*, 27, 4025–33. doi: 10.1016/j.biomaterials.2006.03.006

Zheng, Y., Liu, W., Qin, Z., Chen, Y., Jiang, H., and Wang, X. 2018. "Mercaptopyrimidine-conjugated gold nanoclusters as nanoantibiotics for combating multidrug-resistant superbugs". *Bioconjugate Chemistry*, 29, 3094–103. doi: 10.1021/acs.bioconjchem.8b00452

Zheng, Y., Zheng, L., Zhan, Y., Lin, X., Zheng, Q., and Wei, K. 2007. "Ag/ZnO heterostructure nanocrystals: Synthesis, characterization, and photocatalysis". *Inorganic Chemistry*, 46, 6980–6. doi: 10.1016/j.molliq.2020.114112

Zu, H., and Gao, D. 2021. "Non-viral vectors in gene therapy: Recent development, challenges, and prospects". *The AAPS Journal*, 23(4), 1–12. doi: 10.1208/s12248-021-00608-7

4 Conversion of Waste Plastics into Sustainable Fuel

Mohanraj C., Senthilkumar T., Chandrasekar M., and Arulmozhi M.

CONTENTS

4.1 Introduction 41
    4.1.1 Resource Depletion 41
    4.1.2 Waste Plastic Production 43
4.2 Sources of Plastic Wastes 44
    4.2.1 Recycling, Reuse, and Energy Recovery of Waste Plastics 45
    4.2.2 Landfilling 46
    4.2.3 Mechanical Recycling 46
4.3 Thermochemical Recycling 46
    4.3.1 Incineration 47
    4.3.2 Pyrolysis 47
        4.3.2.1 Catalyst 47
    4.3.3 Residence Time 47
    4.3.4 Reaction Temperature 48
    4.3.5 Reactor Types 48
    4.3.6 Operating Pressure 49
4.4 Properties Analysis of Waste Plastic Oil and Its Blends 49
4.5 Use of Waste Plastic Oil on Diesel Engine 50
4.6 Conclusion 51
References 51

4.1 INTRODUCTION

4.1.1 Resource Depletion

Energy is highly needed for most of the activities happening in various sectors such as agriculture, construction, medical, power generation, transportation, etc. Generally, energy production is an important factor in the nation's development. At present, energy consumption is rapidly increasing in all the fields because of the population explosion with improved lifestyle. Due to this heavy energy demand, depletion of fossil fuel is taking place at a faster rate. This depletion of natural sources of crude oil, gases, and coal with the increasing energy demand causes a great concern for the future energy security (Panda et al, 2014). Petroleum crude oil is an unavoidable need to everyone in the world due to technological development. The world consumed 100–105 million barrels in 2017. Due to population growth, the demand and price of crude oil have increased consistently. In India, the energy needs have increased about 17 times in the last 5 years with increasing population growth rate. At this alarming rate of demand, sources of crude oil will not last for a long time as there is a large gap between the supply and demand.

    Energy consumption around the world approximates about 9741.1 million tons of oil equivalent. This total value includes 2331.9 million tons of natural gas, 3636.6 million tons of oil, 598.8 million tons of nuclear energy, 2578.4 million tons of coal, and 595.4 million of hydel energy. The mean energy consumption per person is 2.2 tons of coal equivalent (Ahmad et al, 2019). Around the world, most of the countries are underdeveloped which constitute about 80% of the total population. Developing countries also consume energy, contributing about 40% of the total energy consumption. People in some of the developing countries lack in energy access at reasonable prices. The regions like Asia, China, and India will be facing heavy demand for energy in future. The growth of expected energy need around the world is identified about 2.7% per year during 2001–2025. It is estimated that coal utilization will increase up to 2.3 billion tons in 2025 (Al-Salem et al, 2010). Also, oil can play a major role, contributing about 39% of the total energy consumption during this period. The growth rate of oil utilization is increased by 1.9% per year. Similarly, the natural gas usage growth rate is expected up to 2.2% per year until 2025. With this forecast, the renewable energy and alternative fuel participation is expected to contribute 23.6% of the total energy consumption.

Based on the hierarchy level, most of the wastes are either reused or recycled. Based on their quality, the applications are identified where they can be utilized. Because of this, unnecessary waste disposal in environments is reduced as well (Audisio et al, 1992). The wastes which cannot be reused or recycled are suggested for recovering the energy from it. The process like combustion, gasification, pyrolization, anaerobic digestion, and landfill gas (LFG) recovery is suggested to recover the energy from wastes. Owing to this, environmental damages are mostly reduced and depletion of natural resource is reduced significantly, along with sustainable developments.

Developments and environmental sustainability are two key elements in any planning and development to meet enjoyable future. Increasing resource scarcity leads to more concern about the circular economy. Many countries have started to implement stringent solid waste management regulation, from controlling emissions to energy recovery from wastes. As per environmental regulations, the wastes are classified into different hierarchy levels. Here, avoiding wastages is suggested strongly. The hierarchy options are reuse, recycling, and energy recovery. Finally, disposal of wastes through landfill is suggested. The waste management authorities have focused more on the waste hierarchy to meet the specific targets on waste reduction and utilization. The decision from waste management is based on not only environmental aspects but also on the economical concern. The larger quantity of waste generated and dumped or piled on landfill causes harmful health and ecological problems around the world. As such, the waste plastics are more challenging to waste management because of their non-degradable nature.

Rapid technological growth, increasing usage, ease of production, and design flexibility are major reasons for the plastic wastes to increase around the world. Plastics are produced from petroleum derivatives which consist of hydrocarbon with little additives such as colorants, antioxidants, and stabilizers. Next to food waste and paper waste, the plastic waste has important role in industrial and municipal waste. Due to its easy production and low cost, the country with low financial growth has more plastic wastes from various industries such as packing sector, construction, electrical and electronics, and other industries. Also, increasing plastic wastes in municipal solid waste assigns more responsibility to solid waste management. In countries which do not have better integrated solid waste management, a lot of plastic wastes are not collected and disposed properly. A recent survey says that 150 million tons of plastic wastes are produced per day around world, in which 40% of plastic wastes are uncollected and littered directly (Ayodhya et al, 2018). Continuous and significant increase of plastic usage among people causes great concern about disposal of the used plastics, and it is a big challenge to the researchers to find ways to dispose the plastics in an environment-friendly manner. Many plastics are non-degradable and non-recyclable, which creates rigorous air, water, and soil pollution. The waste plastics are disposed in several ways such as landfilling, incinerations, mechanical recycling, etc. Landfilling affects the soil quality and other environmental threats; meanwhile incineration induces air pollution due to emission of unwanted gases and soot particles. Sometimes mechanical recycling does not have good quality products and it has lower market values. Some other recycling techniques of waste plastics are to be developed so that all plastic wastes can be disposed properly without affecting the environment (Paul and Ganesan, 2010). Since major plastics are made from refined petroleum crude oil, it is a more possible and potentially suitable technique to convert liquid fuel oil which has similar properties like fossil fuel (Anand Kumar et al, 2005).

The process depolymerization is the degradation of bonds to break down into monomers. This process is utilized for the degradation of plastic to lower hydrocarbons. Chemical depolymerization has successfully been employed to recover monomers from PET, polyamides such as nylons, and polyurethanes. It has the ability to return a recovered resin to virgin resin-like quality, and the potential to recover a valuable feedstock from products that are economically challenging to recycle. Depolymerization is carried out in a specially designed pyrolysis reactor, in the absence of oxygen and in the presence of certain catalytic additives (zeolite, M-sand, TiO_2, etc.). The maximum reaction temperature is 350°C. The entire feed material is converted into either of the products, liquid RDF, gases, and solids (Singh and Sharma 2008). The solids can be reused as fuel in cement industries, while the gas is reused in the system as a fuel. The unused hot air from the reactor is released through the chimney.

4.1.2 Waste Plastic Production

Rapid technological growth, increasing usage, ease of production, and design flexibility are major reasons for increasing the plastic waste around the world. Plastics are produced from petroleum derivatives which consist of hydrocarbon with little additives such as colorants, antioxidants, and stabilizers. Due to non-degradable nature, plastic wastes become more challenging in solid waste management. Next to food waste and paper waste, the plastic waste has an important role in industrial and municipal waste. Due to easy production with low cost, the country with low financial growth has more plastic waste from various industries such as packing sector, construction, electrical and electronics, and other industries. Also, increasing plastic waste in municipal solid waste assigns more responsibility on solid waste management. In countries which do not have better integrated solid waste management, a lot of plastic wastes are not collected and disposed properly. According to a recent survey, 15,000 tons of plastic wastes are produced per day from India, of which 6000 tons of plastic wastes are uncollected and littered directly. Also, in India, Delhi, Chennai, Mumbai, Bangalore, and Hyderabad are the main cities producing major amount of plastic wastes in the country.

In this regard, some of the conversion processes are thermal pyrolysis, catalytic pyrolysis, and co-processing of waste plastics into fuel oil. Waste plastics are the most potential resources for fuel conversion process because they have high heating value and ease of availability. Moreover, the moisture content in plastics is excessively lower than in other alternative sources like crops, husks, and kitchen wastes. The raw waste plastic oil contains complex hydrocarbons mixture which can be extracted by pyrolysis process and purified through distillation. Generally, this kind of pyrolysis oil is used as a fuel for heating furnace in various industries. Recently, a few researches were done to utilize the pyrolysis oil for IC engines operations. Except for fuel extraction from waste plastics, few chemicals were separated like monomers. Commercially, many kinds of plastics are available for usage:

- Polyethylene teraphlate (PET)
- Polyvinyl chloride (PVC)
- Polyethylene (PE)
- Polypropylene (PP)
- Polystyrene (PS)

Prior to conversion process of waste plastic into fuel oil, it is important to analyze the properties of these materials, to check similarities between them to obtain optimized conversion process (Bensler et al, 2000). The most important similarities found are those of carbon and hydrogen which are formed in chains of carbon atoms. Normally, petroleum crude oil has a complex mixture of hydrocarbon which is processed through oil refineries for separation of by-products and distillation. Most of the crude oil products are used as fuels for transportation, furnace burner for heating, and power generation. All the separated products from the crude oil are not a single component, but are combination of different components to meet all the necessity fuel specifications in an economic manner. All these components have different ranges of chain lengths: gasoline has length between 3 and 10 carbon atoms, and diesel has between 5 and 18 carbon atmos. Both gasoline and diesel contain only hydrogen and carbon atoms. Plastic is the most common term among a wide range of polymers made from extremely refined fractions of petroleum crude oil, which are termed as monomers. The plastics are formed through the reaction of these various monomers between the length of 10 and 100 carbon atoms. Only less than 5% of crude oil is used around the world to produce the monomers such as ethane and propene (Kalargaris et al., 2017). These ethane and propene are used to make polymers such as polyethylene and polypropylene. Oxygen is found with some polymers oil (polyethylene terephthalate), whereas chloride is found with polyvinyl chloride (PVC).

4.2 SOURCES OF PLASTIC WASTES

The plastic applications are being extended to all areas of the products used in everyday life. Figure 4.1 shows the percentage of plastic wastes generated from various fields. One-third plastic wastes are generated from the packing industries, because it is lightweight with good strength and can consume lesser fuel while transporting the packed goods. Lesser fuel used while transporting produces lower emissions and, additionally, it reduces the total cost of each product.

Plastic packaging in food industries plays an important role, because of its durability and scalability. Plastics protect food from deterioration, which increases the life of any food item.

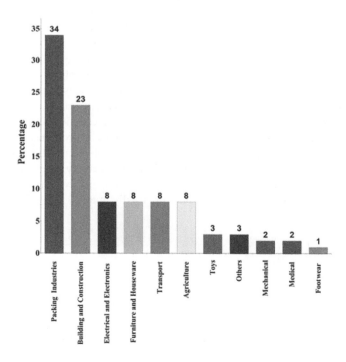

FIGURE 4.1 Plastic waste generation from various fields.

Another largest user of plastics is the building and construction sector. This sector consumes about 23% of the total usage. Plastic usage in the construction sector is continuously increasing owing to better strength-to-weight ratio, flexibility, low cost, durability, and lesser maintenance with larger corrosion resistance.

Most of the plastic products in construction fields are doors, windows, pipes, floor covers, and cables. Plastics are highly flexible for any environment, as it can be transported easily to any site. The requirement for electricity is increasing rapidly in all our lives, at home and in our jobs, at work and play. Due to good insulating properties, plastic is one of the main insulating materials in all the electrical and electronic equipment (EEE). Plastics in electrical equipment can make them lighter, safer, and more attractive. Plastics are divided into two different categories: thermosetting plastics which cannot be recycled or remelted; thermoplastics which can be recycled and remolded again. All kinds of transport need fuel, which is a part of the total cost of every product. Vehicle weight is an important parameter involved in total fuel consumption. Keeping this in mind, many metals in automotive vehicles are being replaced by plastics, for reducing the overall vehicle weight (Hamada et al, 2013). Many components in cars, buses, trucks, and train are replaced with plastics. A few plastic fittings are dashboards, seats, and flooring. Aeronautical and aerospace industries demand high manufacturing and design flexibility with minimal weight. Polymer reinforced composites are being used in rotor blades of helicopter, nacelles, wing skinks, and flaps for weight reduction. Due to better aesthetic properties, plastics are being used as interior products of commercial aircraft, such as stair units, seats, floorings, and bulkheads. Plastics can survive in marine and high-corrosive environment. The exterior and interior components of ships are manufactured with both thermoplastics and thermosetting plastics. In other cases, shredded plastics are mixed with aggregates in road constructions. These kinds of plastic-coated aggregates are being mixed with hot bitumen and the resulting mixture used for road construction.

4.2.1 Recycling, Reuse, and Energy Recovery of Waste Plastics

Plastic plays an important role in our life as a basic need. Due to its vast range of applications and non-degradable nature, it has become highly challenging in the solid waste management system. Also, recycled products are more hazardous than the virgin plastics as they are more toxic. It is because of the addition of color agents, stabilizers, and flame retardants during the first recycling process. Recycling of thermoplastics is limited to two to three times. After the third time, the strength of plastic would drastically reduce due to thermal degradation. According to a recent survey, only 30 vol% of plastic wastes are being recycled and the remaining are being discarded as waste. Thus, roughly 5.6 million tons per annum of plastic wastes are being disposed of in the country, which is about 15,342 tons per day.

The following environmental issues are being encountered when plastics are disposed:

(i) Fugitive emissions released during polymerization.
(ii) Different hazardous gases produced during manufacture of plastic products.
(iii) Soil pollution and erosion due to dumping of plastics on land.
(iv) Harmful products like chlorine, carbon monoxide, dioxin, nitrides, styrene, hydrochloric acid, and benzene which are released during external burning of plastics.
(v) Thermosetting plastics have disposal problems due to their non-recyclable nature.

Limited recycling and non-degradable nature of thermoplastics give a scope to research on conversion of waste plastics into useful fuel oil for the diesel engine. Waste plastics are subjected to a process called pyrolysis, which means heat addition in the absence of oxygen. The properties of extracted fuel oil have resemblance with standard petroleum products. Among various recycling techniques of waste plastics, this technique is very attractive from ecological perception.

4.2.2 LANDFILLING

During landfill, municipal solid waste (MSW) is broken down into methane, which could be converted into electricity through combustion. Landfilling is an effective method for MSW management. However, plastic waste generates endless problems. The larger amount of plastic waste is being used to landfill which is undesirable, due to generation of explosive gases, poor biodegradability, and legislative issues (must be reduced to about 28% within 2035). In concern with these hazards, many rules and regulations have been framed by different countries such as groundwater quality testing when waste leakage occurs. As the weight-to-volume ratio of plastic wastes is high, the expenditure involved in using it for landfill becomes high. As such, the landfill cannot be termed as an appropriate methodology for plastic waste management, and alternative methods which cause a lesser environmental impact should be identified.

4.2.3 MECHANICAL RECYCLING

Mechanical recycling is another method for reutilization of old plastics by melting of waste plastics and turning them into new products. But the recycled plastics have lower market value due to reduced quality and poor aesthetics. Sometimes this process is called downcycling of waste plastics. Due to the property of plastics, only thermoplastics can be recycled, and thermosetting plastics cannot undergo mechanical recycling. Moreover, mechanical recycling is not possible for all types of thermoplastics like plastic films and some types of polyethylene terephthalate (PET) bottles. Figure 4.2 shows mechanical recycling being carried out in two ways. Different types of plastics are isolated based on their types, after collection of mixed waste plastics. Then they are converted into new plastic products. In another method, mixed plastics are being directly converted into new products without segregation. From a quality perspective, products from segregated plastics are better compared to products made from mixed waste plastics.

4.3 THERMOCHEMICAL RECYCLING

Thermochemical conversion is a process of addition of heat and chemical catalysts, for breaking down the polymers of waste plastics. According to the type of plastics used, the heat input and catalysts are varied. Normally, thermochemical conversion is carried out through incineration and pyrolysis processes. Heat energy is recovered from incineration, whereas fuel oil is extracted by the pyrolysis process.

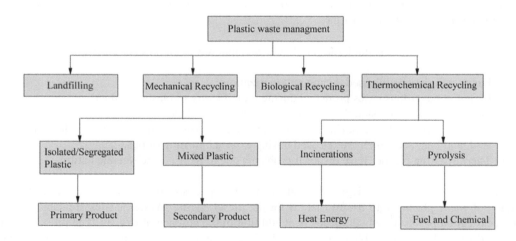

FIGURE 4.2 Plastic waste management.

4.3.1 Incineration

Incineration is done by combustion of inorganic–organic waste materials and converting it into heat, ash, and flue gas. The ash from inorganic waste substances is taken in the form of solid lumps or particulates. The flue gases which come out from the incineration process is cleaned properly and then dispersed into the atmosphere. Also, the heat recovered from this process is utilized for the generation of electrical energy. Incineration and gasification are similar technologies. Heat energy is derived from the incineration process, whereas combustible gas is the main output from the gasification process. Also, incineration and gasification are implemented for waste disposal, without energy and product recovery. Sometimes, incomplete incineration of waste plastics produces poisonous gases, which causes severe health issues on human beings.

4.3.2 Pyrolysis

Pyrolysis is a process which breaks down heavy polymer carbon chains into smaller molecular chains by using heat in the absence of oxygen. Similar to petroleum products (LPG, petrol, and diesel), all plastics such as PE (LDPE and HDPE), PP, and PS consist of hydrogen and carbon chains. The caloric value of various plastics and petroleum products are shown in Table 4.1. The important factors involved in this process are the type of catalysts used, residence time, reaction temperature, reactor type, operation pressure, and chemical composition of plastics materials.

4.3.2.1 Catalyst

To improve the quality of the pyrolysis process, various catalysts are used in research and industrial pyrolysis processes. The main purpose of using a catalyst is to reduce the carbon chain length and thus decrease the boiling point of waste plastics, which consume lesser energy compared to normal thermal pyrolysis process. Also, catalytic pyrolysis process reduces the activation energy for all reactions, whereas higher activation energy is needed for normal thermal pyrolysis. Catalysts are classified into homogeneous and heterogeneous catalysts. Homogeneous catalysts like Lewis acids ($AlCl_3$) are used for polyolefin pyrolysis. Heterogeneous catalysts such as conventional acid lumps, nanocrystalline zeolites, mesostructured catalysts, aluminum pillared clays, Gallo silicates, metals supported on carbon, and basic oxides are used for separation and energy recovery from the feedstock.

4.3.3 Residence Time

Residence time is the duration between the initial heating of waste plastic and the end of oil extraction. Fast, continuous, and slow pyrolysis are the three pyrolysis methods which are practiced

TABLE 4.1
Calorific Value Comparison

Materials	Calorific Value (MJ/kg)
Petrol	44.0
Diesel	43.0
Kerosene	43.4
Heavy oil	41.1
Gasoline	46.1
Coal	24.3
Polyethylene	46.3
Polypropylene	46.4
Polyvinyl chloride	18.0
Polystyrene	41.4

TABLE 4.2
Various Pyrolysis Based on Residence Time

Process	Residence Time	Heating Rate	Temperature	Yields
Fast pyrolysis	0.5–5 s	Up to 1,000 K/s	550–650°C	Gas and oil and char
Flash pyrolysis	Less than 0.5 s	Up to 10,000 K/s	450–900°C	Gas and oil and char
Slow pyrolysis	Up to 60 min	10–100 K/m	450–600°C	Gas and oil and char

commercially. Normally, fast pyrolysis consumes lower energy for extraction oil as compared to slow pyrolysis. But the quality of the oil was found to be better with slow pyrolysis than with fast pyrolysis. Moreover, longer residence time tends to produce light molecular weight hydrocarbon and non-condensable petroleum gases that are thermally stable. Rarely, long residence time produces more tar and char products due to engaging carbonization processes. Table 4.2 shows the distribution of the products through various pyrolysis based on residence time.

4.3.4 Reaction Temperature

Another important parameter is reaction temperature because it influences more than other parameters. It has not been confirmed that all polymers get cracked when fluctuations in temperature are witnessed. Van der Waals force is an intermolecular force which attracts the molecules collectively and prevents molecular structural damage. When heat is increased, the vibrations of molecules are more, to escape from the surface. The carbon chain is broken down due to the force induced by the polymer chain, which is greater than the enthalpy of C–C bond. This is one of the main reasons why larger molecular weight polymers are decomposed during the boiling process while heat is added. Also, the thermal cracking temperature of C–C bond in plastic was found to be constant all the time, but it varies based on different types of plastics. For example, the reaction temperature for PP is different from PE.

4.3.5 Reactor Types

Based on feeding and product removal, the reactors are classified into three types, namely, batch, semibatch, and continuous reactors. In a continuous reactor, feedstock would be continuously supplied to the reactor, and products from pyrolysis are removed continuously from another side. In a batch reactor, both feeding and product removal are done before the start of the process and after process completion. Semibatch reactor can adapt to the removal of the products continuously, where the feedstock could be filled before the initiation of the process. According to the mode of heat transfer and flow pattern, the reactors are classified as fluidized bed reactors, fixed bed reactors, and screw kiln reactors. In fixed bed reactors, the process is done in a fixed bed which has an easy design and operational procedure. Temperature distribution among feedstock is not uniform, which causes more energy consumption during the extraction process. It is difficult to feed in irregular shape feedstock during continuous pyrolysis. Gaseous products or inert gases are passed through bed into the feedstock in fluidized bed reactor. Due to gas flow through feedstock, the heat transfer rate is increased, which in turn induces a better chemical composition of yields. This type of reactor is used commercially for the extraction of oil from any waste material. Screw kiln reactor is being recently developed for better oil extraction process. The feedstock is passed through a screw-type reactor from feeder under the absence of oxygen. When the feedstock travels through the screw, separation of pyrolysis products and its collection are done. Highly viscous plastics or melted plastics can be used in this reactor as the screw is driven by the external prime mover.

4.3.6 Operating Pressure

The operating pressure influences the pyrolysis reaction as heat input at increased pressure causes the boiling point to shift to higher values. Therefore, heavy hydrocarbons are further pyrolyzed instead of vaporization under pressured pyrolysis process. Due to this, more power is required for further cracking. Also, it induces more non-condensable gaseous products which result in reduced liquid yields.

4.4 PROPERTIES ANALYSIS OF WASTE PLASTIC OIL AND ITS BLENDS

Different properties such as density, kinematic viscosity, calorific value, flash point, fire point, acid number, and cetane number for different waste plastic oil and their blends are discussed in the following section.

In Figure 4.3, two different peaks are noted between 2800 and 3000 cm^{-1} with all raw and distilled WPOs at stretching mode. This indicates the presence of alkanes such as (C–H$_3$, C–H$_2$, and

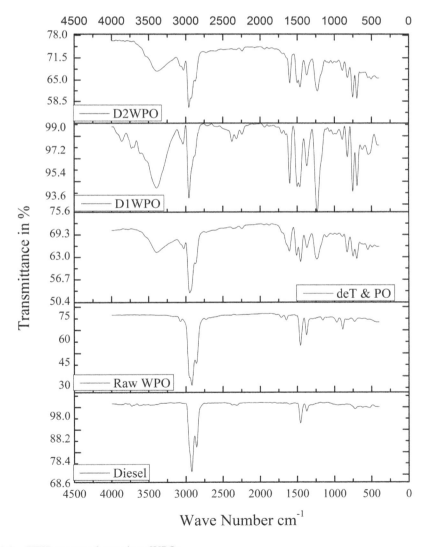

FIGURE 4.3 FTIR spectra for various WPOs.

C—H) in the samples. A sharp peak is found between 3020 and 3100 cm^{-1} except for diesel fuel and it also indicates the presence of alkenes (=CH$_2$ and =CH). At stretched mode, broad peaks are identified around 3200–3550 cm^{-1} in all the samples other than diesel. This peak indicates the presence of alcohol and phenols (O—H) in WPOs. During NH$_2$ scissoring, the 1° amines were produced, which is denoted by the peak between 1550 and 1650 cm^{-1} at the bending mode. Because of CH$_2$ and CH$_3$ relocation, alkanes were again produced, which are indicated by peak found around 1350–1470 cm^{-1} at the bending mode. A small peak at 729 cm^{-1} (C–Cl stretch) and 529 cm^{-1} (C–I stretch) confirm the presence of aliphatic iodo and chloro compounds in samples. In Figure 4.3, a peak between 2690 and 2840 cm^{-1} was identified and it indicates the medium intensity of C—H aldehyde at stretching mode in all the blends. Other than diesel, the sharp peak noted around 2240–2260 cm^{-1} confirms the presence of CN nitriles. The C—O stretch peak at 1231 cm^{-1}, especially with distilled blends, denotes the presence of ethers, esters, alcohols, and carboxylic acids. Around 690–900 cm^{-1}, an aromatic ring was found, particularly with distilled oil blends. The aromatic ring was viewed clearly with an increase in WPO and distillated WPO with diesel.

The density values of all the WPOs were recorded to be higher than that of diesel, especially distilled WPOs such as D1WPO and D2WPO. Higher density causes rapid pressure rise in fuel line, which affects the nozzle and leads to the earlier injection. The GCV distilled WPO and its blends were reported to be lower than that of diesel. The kinematic viscosity of all the samples was found to be less than diesel, particularly distilled WPOs. Sometimes fuel with lower kinematic viscosity led to leakages in pump and nozzles, which was found to reduce the quantity of fuel delivered during the injection. The flashpoint and fire point temperatures of distilled WPOs were found to be lower than diesel and raw WPOs. Safety measures should be considered during transport and storage of distilled WPOs. The acid values were found to be higher with raw WPO and deT&PO than that of diesel and distilled WPOs. Sulfur content was found to be more with raw WPO and deT&PO, and toxic emissions occurred after combustion. Distilled WPO blends were recommended with reduced amount of sulfur for engine operations. The cetane number of raw WPO and deT&PO were found to be higher than that of distilled WPOs. Fuel with lower cetane number increased the ignition delay. From FT-IR spectra, similar peaks were found in all the WPO and distilled WPO samples like diesel. Beyond this, the presence of alcohol and aromatic ring in samples was seen in distilled WPO samples.

4.5 USE OF WASTE PLASTIC OIL ON DIESEL ENGINE

From these experimental investigations, it was found that distilled waste plastic oil was better than raw waste plastic oil and mixed tire and waste plastic oil. BTE of raw waste plastic oil and mixed tire and waste plastic oil was about 12.9% greater at high load and 9.5% at low load than diesel, 16% at high load and 13% low load than WPO30. Lowered specific fuel consumption was observed with D1WPO30: about 9% at low load than diesel and 17% at high load and 13% at low load than WPO30. The cylinder pressure and heat release rate were found higher than diesel and WPO30. With regard to emission, NO$_x$, CO, and unburned hydrocarbon with distilled waste plastic oil were found to be lower than other categories of waste plastic oil blends. Rich blends of raw WPO affected the fuel filters severely due to the impurities present in it, whereas it has caused less impact with desulfurized T&PO blends. Also, rich blends of desulfurized fuel could not be used in diesel engine for a long term due to higher acid value, which might lead to corrosion and particles deposit.

From the above discussions, it was learnt that optimized D1WPO30 operated at retarded injection timing produced better performance when compared with standard injection timing. BTE was increased to about 4% at 20° bTDC and 11% at 17° bTDC during higher loads. The SFC was increased up to 9% at 20° bTDC and 13.5% at 17° bTDC during higher loads than standard injection timing. Considering two retarded injection timing of 20° bTDC and 17° bTDC, the 17° bTDC results were found better than 20° bTDC. The net heat release rate and cylinder pressure reduced with retarded injection timing when compared with normal operations. NO$_x$, CO, and

UBHC emissions were reduced significantly with retarded injection timing than standard injection timing. The smoke opacity increased to about 5% with retarded injection timings due to reduced cylinder temperature.

Direct injection engine was operated at retarded injection timing for 10% and 20% of EGR. The brake thermal efficiency was found to be higher than standard injection timing operation, but reduced up to 5% than the retarded injection timing operation. Specific fuel consumption was found to increase about 9% than other operations, due to reburning of unburned hydrocarbon. Both cylinder pressure and heat release rate were found to decrease with increase in EGR rates, which was mainly due to reduced cylinder temperature. Except for NO_x, other emission parameters like carbon monoxide and unburned hydrocarbon were found a little higher because of the replacement of oxygen concentration in the fresh mixture. The NO_x was reduced up to 5% with EGR operations due to lower cylinder temperature caused by increased heat capacity of the working mixture.

4.6 CONCLUSION

It is evident that the waste plastic had great potential for recycling, especially converting them into fuel for diesel engine. Many investigations were carried out to analyze the effect of various operational parameters such as injection timing and EGR adaptation for different blends of raw and distilled waste plastic oil blends on performance, emission, and combustion characteristics of direct injection engine at different loading conditions. Based on the experimental findings, the following conclusions were drawn. Before the experimental work, the obtained waste plastic oils were subjected to property study and FTIR analysis. It was observed that engine run using D1WPO30 had high BTE as 36.47%, low SFC as 0.25 kg/kWh, moderately low exhaust gas temperature as 315 °C, comparatively lower peak cylinder pressure of about 67.21 bar, moderately lower heat release rate, and low NO_x (1637 ppm), CO, and unburned hydrocarbon emission. It had moderately low smoke opacity. As far as injection timing is concerned, at 17° bTDC, high BTE as 40.81%, low SFC as 0.25 kg/kWh, and low emissions were observed. As far as exhaust gas recirculation is concerned, even though combustion of D1WPO30 with 10% EGR at 17° crank angle did not enhance the BTE or reduce SFC appreciably, a significant reduction in emissions was observed. The NO_x was noted to be about 1598 ppm.

REFERENCES

Ahmad, T, Danish, M, Kale, P, Geremew, B, Adeljou, SB, Nizami, M & Ayoub, M 2019, 'Optimization of process variables for biodiesel production by transesterification of flaxseed oil and produced biodiesel characterizations', *Renewable Energy*, vol. 139, pp. 1272–80.

Al-Salem, SM, Lettieri, P & Baeyens, J 2010, 'The valorization of plastic solid waste (PSW) by primary to quaternary routes: From re-use to energy and chemicals', *Progress in Energy and Combustion Science*, vol. 36, no. 1, pp. 103–29.

Anand Kumar, T, Mallikarjuna, JM & Ganesan, V 2005, 'Effect of intake port configuration and engine speed on flow field characteristics in a four-stroke gasoline engine – CFD approach using KIVA-3V code', Proceedings of National Conference on I.C Engines and Combustion, Annamalai University, Chidambaram, pp. 195–9.

Audisio, G, Bertini, F, Beltrame, PL & Carniti, P 1992, 'Catalytic degradation of polyolefins', *Makromolekulare Chemie: Macromolecular Symposia*, vol. 57, no. 1, pp. 191–209.

Ayodhya, AS, Lamani, VT, Bedar, P & Kumar, GN 2018, 'Effect of exhaust gas recirculation on a CRDI engine fueled with waste plastic oil blend', *Fuel*, vol. 227, no. 1, pp. 394–400.

Bensler, H, Buhren, F, Samson, E & Vervisch, L 2000, '3-D CFD analysis of the combustion process in a DI diesel engine using a flamelet model', SAE Technical Paper, pp. 1–10.

Hamada, K, Kaseem, M & Deri, F 2013, 'Recycling of waste from polymer materials: An overview of the recent works', *Journal of Polymer Degradation and Stability*, vol. 98, no. 12, pp. 2801–12.

Kalargaris, I, Tian, G & Gu, S 2017, 'Combustion, performance and emission analysis of a DI diesel engine using plastic pyrolysis oil', *Fuel Processing Technology*, vol. 157, no. 1, pp. 108–15.

Panda, AK, Singh, PK & Mishra, DK 2014, 'Thermolysis of waste plastics to liquid fuel: A suitable method for plastic waste management and production of value added products-a world prospective', *Renewable and Sustainable Energy Reviews*, vol. 14, no. 1, pp. 233–48.

Paul, B & Ganesan, V 2010, 'Flow field development in a direct injection diesel engine with different manifolds', *International Journal of Science and Technology*, vol. 2, no. 1, pp. 80–91.

Singh, B & Sharma, N 2008, 'Mechanistic implications of plastic degradation', *Polymer Degradation and Stability*, vol. 93, no. 3, pp. 561–84.

5 Innovations in Sludge-Conversion Techniques

Chithra K.

CONTENTS

5.1 Introduction 53
    5.1.1 Characteristics of Various Industrial Sludge 54
5.2 Biological Treatment 54
    5.2.1 Anaerobic Digestion 54
    5.2.2 Innovations in Technology 56
5.3 Thermochemical Treatment 57
    5.3.1 Incineration 60
        5.3.1.1 Multiple Hearth Furnaces 60
        5.3.1.2 Fluidized Bed Incinerators 60
        5.3.1.3 Bubbling Fluidized Bed Combustors 60
        5.3.1.4 Circulating Fluidized Bed 60
        5.3.1.5 Electric Infrared Incinerators 61
    5.3.2 Pyrolysis 61
        5.3.2.1 Slow Pyrolysis 61
        5.3.2.2 Fast Pyrolysis 61
        5.3.2.3 Flash Pyrolysis 61
    5.3.3 Pyrolysis Reactors 61
        5.3.3.1 Fixed Bed Reactor 61
        5.3.3.2 Fluidized Bed Reactor 62
        5.3.3.3 Bubbling Fluidized Bed Reactor 62
        5.3.3.4 Circulating Bed Reactor 62
        5.3.3.5 Ablative Reactor 62
        5.3.3.6 Auger Reactor 62
        5.3.3.7 Microwave-assisted Pyrolysis 62
    5.3.4 Gasification 62
        5.3.4.1 Gasifier Types 63
        5.3.4.2 Operating Conditions 63
    5.3.5 Innovations in Thermochemical Treatments 63
5.4 Conclusion 63
References 64

5.1 INTRODUCTION

Industries consume large volumes of water in their manufacturing process and produce wastewater as one of the by-products. This wastewater called effluent has to be treated before it is disposed of. Depending upon the effluent characteristics, different treatment systems are adopted. Waste sludge will be generated in most treatment techniques, and sludge disposal is again a primary concern.

DOI: 10.1201/9781003334415-5

As the wastewater sludge is classified as hazardous, waste treatment is required before disposal. On the other hand, biologically treated wastewater generates sludge rich in organic substances that can be converted to energy (Hakiki et al., 2018). Hence, sustainable sludge handling and treatment technologies are in great demand since conventional methods are expensive and limited. To reduce the impact of fossil fuel usage on the environment and its shortage in meeting the energy demand, technologies for waste-to-energy conversion have emerged in recent years indicating that waste sludge is an alternative energy source and no longer considered waste.

There are many options for the recovery of energy from sludge, like fuel gas, biogas, generation of electricity, biodiesel, bio-oil, nutrient, etc. The potentiality of sludge as an alternative energy source is based on the organic content present in them (Hakiki et al., 2018; Karayildirim et al., 2006). So researchers have reported various techniques that can reduce the negative impact of sludge on the environment and also reduce the volume of sludge by converting waste sludge to energy (Kurniawan et al., 2018).

The common features of waste sludge are the pollutant that demands stabilization and high moisture that makes it difficult to be removed. The characteristics of sludge are the key to the selection of treatment techniques for energy recovery. Table 5.1 gives the level of carbon content an important parameter that decides its ability to produce energy. Methods like pyrolysis, incineration, gasification, anaerobic digestion, and hydrothermal are the conventionally used methods for resource recovery from sludge (Tyagi & Lo, 2013). These techniques are depicted in Figure 5.1. However, these techniques are challenging as they are influenced by the chemical and physical properties of the sludge (Oladejo et al., 2019), particularly moisture, nitrogen, and heavy metal content.

This chapter will emphasize the various techniques for effective conversion of sludge to energy through thermochemical and biochemical treatment methodologies and innovations reported in the literature.

5.1.1 CHARACTERISTICS OF VARIOUS INDUSTRIAL SLUDGE

The elemental analysis of sludge is a major characteristic of sludge that provides the carbon, hydrogen, and oxygen content of the sludge that can act as a source of energy for the sludge. Various works of literature have predicted the elemental characteristics of various industrial sludges.

5.2 BIOLOGICAL TREATMENT

5.2.1 ANAEROBIC DIGESTION

Sludge is generated as a by-product during the treatment of wastewater from various industries. The sludge may be primary and secondary activated sludge, chemical-precipitated sludge, and digested sludge; so sludge treatment choice depends on its characteristics (Shi et al., 2018; Zhen et al., 2017). Sludge is considered a good source of nutrients, organic content, and energy; the choice of treatment technology to extract these depends on the characteristics of the sludge (Kacprzak et al., 2017). Due to the depletion of fossil fuels, climate change, and global warming, the need for a sustainable process for bioenergy recovery from sludge has increased.

Conventionally, anaerobic digestion (AD) is used to treat sludge. In an anaerobic digester, biochemical reactions (hydrolysis, acidogenesis, acetogenesis, and methanogenesis) and physicochemical reactions (ion association/dissociation, gas–liquid transfer, and precipitation) take place. The organic matter present in the sludge is stabilized through hydrolysis, acidification, and methanogenesis steps of AD. But the drawbacks of this process are longer hydraulic retention time and inefficiency of the microorganisms to degrade certain components present in the sludge.

TABLE 5.1
Chemical Composition of Various Sludge

S. No.	Industry/Sludge Type	C	H	N	S	O	Reference
1	Petroleum	51.4%	7.3%	3.3%	2.2%	35.08%	(Engel, 2014)
2	Petrochemical	64.23%	6.02%	0.8%	0.96%	27.99%	(Ayol & Yurdakoş, 2019)
3	Synthetic latex and rubber	46.7%	5.80%	6.81%	1.34%	–	(Lin et al., 2018)
4	Pharmaceutical	31.41%	4.44%	4.40%	1.95%	28.18%	(Liu et al., 2020)
5	Pulp and paper	44.9%	5.7%	7.9%	2.7%	24.4%	(Wang et al., 2021)
6	Municipal sewage sludge	43.8%	5.7%	4.75%	24.4%	1.18%	(Zaker et al., 2019)
7	Tannery sludge	23.85%	4.18%	1.17%	–	–	(Yang et al., 2020)
8	Tank bottom oil sludge	60%	25%	4.5%	4.39%	–	(Sivagami et al., 2021)
9	Dry sewage sludge with lime (Compiegne)	28.83 wt% db	5.04 wt% db	4.06 wt% db	0.68 wt% db	25.49 wt% db	(Ledakowicz et al., 2019)
10	Oily sludge	16.75%	2.21%	0.27%	0.25%	18.27%	(X. Zhang et al., 2022)
11	Anaerobically digested sludge	44.34%	7.54%	6.13%	1.46%	40.54%	(Kim et al., 2022)
12	Sewage sludge	40.77%	4.78%	3.71%	1.23%	49.51%	(Tong et al., 2021)
13	Wheat straw	42.04%	5.50%	0.79%	–	36.42%	(Li et al., 2022)
14	Sewage sludge	45.74%	5.62%	1.03%	1.23%	42.8%	(Zuo et al., 2021)
15	Industrial sludge	24.79%	2.90%	1.99%	0.53%	26.39%	(W. Zhang et al., 2022)
16	Dewatered sludge	34.65%	5.41%	7.22%	0.92%	25.58%	(A. Chen et al., 2021)
17	Oil scum	35.8%	6.5%	1.5%	1.1%	12.9%	(Lee et al., 2022)
18	Raw textile sludge	39.00%	6.09%	6.10%	1.17%	–	(Scheibe et al., 2022)
19	Pulp and paper industry's mixed sludge	42.4 ± 0.1%	5.8 ± 0.02%	1.2 ± 0.04%	0.4 ± 0.03%	50.2 ± 0.2%	(Hämäläinen et al., 2022)

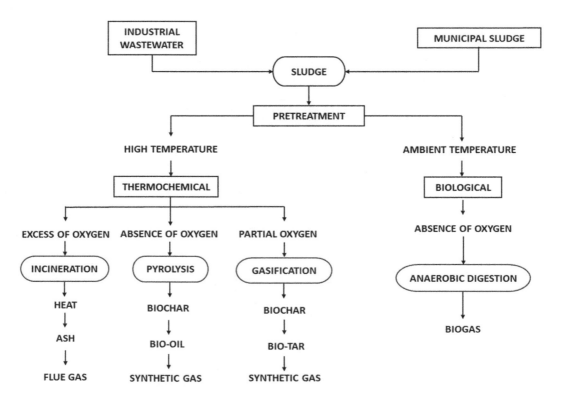

FIGURE 5.1 Techniques for energy recovery from sludge.

Source: Drawn based on the information gathered from Oladejio et al. (2019)

5.2.2 Innovations in Technology

To overcome the drawbacks of the conventional AD process, pretreatment of the sludge before the digestion of sludge is required to overcome these drawbacks. Pretreatment enhances biofuel production through structural modification, organic matter solubilization, and an increase in the surface area of the sludge. Anaerobic reactor modification and anaerobic co-digestion with organic waste are the alternative ways to enhance AD.

The pretreatment and combination of pretreatment methods like biological (enzymatic hydrolysis), mechanical (ultrasonic, high-pressure homogenizer), ozone, chemical, thermal, thermochemical, thermobiological, ultrasonic–biological, ultrasonic–ozone, and photocatalysis–biological are explored by various researchers recently (Jadhav et al., 2020).

It has been reported that various pretreatments like enzymatic hydrolysis (Tongco et al., 2020) of primary sludge, high-pressure homogenization (Nabi et al., 2019, 2021), hydrothermal carbonization (Kokko et al., 2022), anaerobic co-digestion of paper sludge (Gievers et al., 2022), thermal and ultrasonic combined pretreatment (Liu et al., 2021), ultrasound and microaerobic pretreatment (Rashvanlou et al., 2021), application of multistage process (Awasthi et al., 2019, 2020), mechanical cutting (X. Wang et al., 2021), microwave irradiation (Bozkurt & Apul, 2020), and pulse power (Kovačić et al., 2021) are effective in increasing the biodegradability of the sludge.

Energy recovery through anaerobic digestion with and without pretreatment of different sludge, namely, sewage sludge (Liu et al., 2021), tannery sludge (Priyadarshini et al., 2015); pulp and paper mill sludge (Veluchamy & Kalamdhad, 2017); slaughterhouse sludge (Oh & Yoon, 2017); and petrochemical (oil refinery) sludge (Roy et al., 2016) has been reported.

5.3 THERMOCHEMICAL TREATMENT

An efficient technique for producing energy from sludge generated from wastewater treatment is thermochemical technique (Manara & Zabaniotou, 2012). A brief description of the working principle of a few thermochemical technologies and the innovations done by researchers to improve the efficiency and energy are further outlined in Table 5.2.

TABLE 5.2
Recent Innovations in Thermochemical Treatment of Sludge

S. No.	Treatment Technique	Objective	Inference	Reference
1	Incineration	Emission and pollution control	The utilization of calcium and aluminum silicate–based additives to reduce the volatile heavy-metal chloride vapors. Competitive inhibition was observed when Ca and aluminum silicates were utilized during the process of combustion	(Zha et al., 2018)
2		Emission and pollution control	Utilization of CaO as conditioners for thermal incineration of textile dyeing sludge for emission control. The CaO addition inhibited HCN, NO, NO_2, COS, SO_2, CS_2, and SO_3 emissions. The CaO addition increased the Cu, Zn, As, and Pb retentions	(C. Xie et al., 2021)
3	Pyrolysis	To reduce and obtain paraffinic fuels from crude oil sludge	Co-pyrolysis of petroleum sludge with polyolefins (HDPE, LDPE, and PP) in the presence of different Y zeolites in a fixed bed reactor at 450 °C for 15 min yielded paraffinic products with a reduction of aromatic and cyclic compounds in the bio-oil	(Milato et al., 2020)
4		Bio-oil upgradation for NO_x pollution control	Step pyrolysis of industrial biowaste was used to regulate NO_x precursor emission. Comparatively, two-step pyrolysis could minimize NO_x precursor-N yield by 36–43%. A varying total yield of 20–45 wt% was observed for samples at high temperatures	(Zhan et al., 2018)
5		Development of industrial scale pyrolyzer	The process showed an oil yield of 17.1 wt% at 500 °C and complete conversion of feed was achieved	(Tang et al., 2019)
6		Product optimization	The use of distillery sludge and bio-compost mixed with coal samples in a 3:2 ratio, pyrolysis at 650 °C and gasification at 850 °C in a two-stage fixed bed system for the preparation of char, bio-oil, and syngas. The char produced contained high carbon content and calorific value in the range of 24.76–58.19% and 5849.83–to 7993.64 kcal/kg, respectively	(Dhote et al., 2022)
7		Energy optimization for microwave pyrolysis	By using a two-step method, the microwave power was significantly reduced (0–33%), and the energy recovery efficiency was slightly improved	(Lin et al., 2020)

(Continued)

TABLE 5.2 *(Continued)*
Recent Innovations in Thermochemical Treatment of Sludge

S. No.	Treatment Technique	Objective	Inference	Reference
8		Sewage sludge disposal by toxic metal ion encapsulation	Co-pyrolysis of sewage sludge and biomass waste is demonstrated to be a practical strategy for the disposal of sewage sludge in terms of both biochar sequestration and immobilization of the metals	(J. Zhang et al., 2020)
9		Bio-oil upgradation by feedstock optimization	Co-pyrolysis of microalgae biomass and sewage sludge resulted in an increase of C4 and C7 and a decrease in C9 carbon distribution. An excellent linear relationship between the H/C of feedstocks and the pyrolysis char was observed	(X. Wang et al., 2016)
10		Improved char with low heavy-metal toxicity	Industrial sludge co-pyrolysis with rice straw at different temperatures showed that the addition of rice straw can decrease the ash content and increase the thermal stability and pore structure of char. Simultaneously, decreasing heavy metal concentrations and enhancing the transformation of Cr, Zn, and Cd to a more stable fraction, with higher alkalinity, aromaticity, and specific surface area of char	(S. Xie et al., 2020)
11		Alternate fuel from paper pulp sludge after pretreatment	Hydrothermal carbonization (HTC) of paper sludge was investigated. The concentration of monocyclic aromatic hydrocarbons in the derived organic liquid fraction shows a positive correlation with the pyrolysis temperature. At 550 °C, the organic liquid fraction reached its highest yield at 13.7% with an oxygen level of 10.7 wt% and a higher heating value of 35.9 MJ/kg	(S. Wang et al., 2021)
12	Gasification	To improve the fuel properties for high-temperature gasification	Torrefaction-pretreated industrial sludge under CO_2 can be better applied to high-temperature gasification of coal. The results showed that the optimal torrefaction temperature of SL was 300 °C under CO_2 and its release intensity of gases was higher than that of Ar. –OH and C–H were decreased relative to Ar	(W. Zhang et al., 2022)
13		Enhanced hydrogen production by co-gasification	This research used sewage sludge and industrial wastewater sludge co-gasification in a pilot-scale fluidized bed gasifier with temperature controlled at (600–800 °C) using sludge addition ratio (0–60%), and steam-to-biomass ratio (0–1.0). Hydrogen production was increased from 9.1% to 11.94% with an increase in industrial wastewater sludge ratios from 0% to 60%	(Chen et al., 2021)
14		Process optimization	The co-gasification of bituminous coal and industrial sludge in the downdraft fixed bed gasifier was simulated by a validated steady-state model structured using the Aspen Plus software	(W. Zhang et al., 2022)

TABLE 5.2 *(Continued)*
Recent Innovations in Thermochemical Treatment of Sludge

S. No.	Treatment Technique	Objective	Inference	Reference
15		Gasification of wet sludge	The gas yield increased with the increasing temperature. Hydrogen and methane compositions in the gas was 42–57 vol%, while CO_2 was 30–40 vol%. The calorific value of the sludge was determined as around 16 MJ/kg, while after gasification the calorific value of the gas fuel produced was found to be 24.7 MJ/kg after 30 min reaction time in the presence of KOH	(Yildirir & Ballice, 2019)
15		A novel gasification process of sewage sludge	A novel drying and autothermal gasification scheme for sewage sludge was designed. Adding more torrefied biomass enhanced the quality of syngas. The addition of 0.1 kg torrefied biomass per kilogram sewage sludge was economical	(Huang et al., 2020)
17		Valorization of sewage sludge through catalytic subcritical and supercritical water gasification	The organic matter in sewage sludge was dissolved and hydrolyzed, leading to the production of H_2-rich syngas. The results also showed the order of catalyst effectiveness in favor of H_2 production. Compared to carbonate catalysts, hydroxide catalysts were more effective in improving the H_2 yield through water-gas shift reaction by intermediate formation of format salts	(Yan et al., 2020)
18		Co-hydrothermal gasification using novel solid catalyst derived from carbon-zinc battery waste	This study targets the research gap in co-hydrothermal gasification (HTG) process conditions and the composition of sewage sludge and microalgae biomass in hydrogen-enriched gas production. The uppermost hydrogen composition was 38.27 wt% in the gaseous product (40.7 wt%) at a catalyst dose of 4 wt% for a 2:1 ratio at a temperature of 440 °C. Optimization studies showed catalyst load of 2.3 wt%, a temperature of 426.36 °C, and a time of 70.22 min will result in 40 wt% of hydrogen gas yield	(Arun et al., 2020)
19		Steam co-gasification	In this study, horticultural waste and sewage sludge with different mass ratios were co-gasified with steam at different temperatures to investigate the product distribution, gas synergistic interaction, and optimal design for gas products from the co-gasification process. Results showed that with the increase of sludge ratio in blends, the H_2 content was increased and the syngas yield was decreased	(Hu et al., 2020)

5.3.1 INCINERATION

A widely accepted disposal method alternative to landfills is incineration. This technique is especially applied to hazardous waste and industrial sludge. Few merits of this technology are sludge volume reduction, less time consumption, effectiveness for sludge with high organic content, and capability of recovering energy. It converts organic carbon, sulfur, phosphorus, and nitrogen present in the sludge into gases and solid. Flue gas with high temperature is produced when solid waste is subjected to combustion under excess oxygen supply. Previously, the incineration process was done to reduce sludge volume and was not for energy recovery. But now the heat recovered from the flue gas can be directly and/or converted to electricity and reused. However, ancillary processes for the removal of SO_x and NO_x are required, which adds to the overall cost. In addition, there is a need for thermal drying of the sludge before incineration.

Different configurations of incinerators used are the fluidized bed, electric, rotary kiln, and multiple hearth furnaces (MHF), their function and advantages reported in the literature are discussed in the following sections.

5.3.1.1 Multiple Hearth Furnaces

Inside this furnace, there are three zones: the upper drying zone (425–760 °C) evaporates sludge moisture; a middle zone where the combustion of sludge takes place (925 °C), this zone is further subdivided into the upper-middle and lower-middle zones where solids, volatile gases, and fixed carbon are burnt, respectively; the third zone is the cooling zone where ash is cooled and the heat is reused. The use of catalyst and co-combustion to improve incineration, reuse options for ash formed, and minimization of emissions is the key area of research interest in recent years (Chen et al., 2018; Rong et al., 2017). Compared to multiple hearth furnaces, fluidized bed is good in performance and cost (Zaker et al., 2019).

5.3.1.2 Fluidized Bed Incinerators

In a fluidized bed incinerator, combustion is enhanced due to high turbulence created by the upward flow of airstream through a bed of sand or sand-like coarse material. This turbulence results in uniform mixing and heat transfer between sludge and gas (Cammarota et al., 2019). The rotating fluidized bed, circulating fluidized bed, and bubbling fluidized bed are the types of beds used to treat various types of sludge. The advantage of a fluidized bed incinerator (FBI) is that sludge with varying moisture (30–90%) content can be incinerated (Van Caneghem et al., 2012). The problem faced in the FBI is the agglomeration of clinkers resulting in increased bed height.

Advancements in fluidized bed incinerators are co-incinerating sludge with coarse waste having high calorific value and no pretreatment required. In these types, energy is used to remove water from sludge and to generate electricity.

5.3.1.3 Bubbling Fluidized Bed Combustors

In bubbling fluidized bed (BFB) combustors, primary air is passed through a distributor which keeps the solid particles in suspension with 0.5–3 m/s velocity. Feed size is important since either fine or larger the feed size will lead to carryover of particles and improper fluidization. This type is suited for low-volatile, high ash content, and high moisture content fuel.

5.3.1.4 Circulating Fluidized Bed

This type of reactor is mainly for combustion operation at a large-scale level. Here the entire vessel is utilized for the interaction of solid sludge and gas. The system effectiveness depends on riser geometry, operating conditions, heat transfer, and hydrodynamics. In circulating fluidized beds, unlike fluidized beds, there are two beds. In the first bed, pyrolysis takes place after which it is sent to the second unit to produce heat.

5.3.1.5 Electric Infrared Incinerators

Organic waste is combusted using silicon carbide rods which are driven by electricity in this type of incinerator. This thermal process is mobile and consists of four components: an infrared chamber, a gas-fired chamber, an emission control system, and a system for monitoring. For a smaller application, this type of incineration is more advantageous than MHF and fluidized beds.

5.3.2 Pyrolysis

The degradation of organic molecules at higher temperatures in an inert atmosphere to obtain value-added products of all three states (solid, liquid, and gas) is called pyrolysis. The degradation in sludge occurs because of thermal cracking of higher chain organics present at a temperature range of 300–600 °C. The pyrolysis process is completed in three major steps: initiation, propagation, and termination. The yield and the characteristics of the products obtained are dictated by the operating parameters of the process. Based on the gas residence time as well as on heating rates in the reactor, we have slow, fast, and flash pyrolysis.

5.3.2.1 Slow Pyrolysis

In slow pyrolysis, the maximum temperature attained by the feedstock is about 300 °C and the heating ramp range varies from 0.1 °C/s to 0.8 °C/s (Lu et al., 2020). The gaseous residence time varies from hours to days. The major product obtained from the process is biochar, as the feedstock is completely devolatized and all the light condensable vapors are converted into non-condensable gases.

5.3.2.2 Fast Pyrolysis

Fast pyrolysis aims at optimizing the yield of bio-oil, char, and gaseous products. The major yield is bio-oil as the gas residence time (0.5–2 s) in the reactor is maintained in such a way that it prevents the further thermal breakdown of the condensable vapors obtained (Sipra et al., 2018). The heating ramp of the process ranges from 10 °C/s to 100 °C/s. The process requires large heat-transfer rates, hence the feedstock requires to be grounded to a very fine size.

5.3.2.3 Flash Pyrolysis

Flash pyrolysis aims at converting the organics into tar, which can further be fractionated to obtain fuel at different grades. The heating rate of the process is above 1000 °C/s. The final temperature attained by the sample is around 700–900 °C and the residence of the gas is less than that of the fast pyrolysis.

5.3.3 Pyrolysis Reactors

The variations in the pyrolysis reactor play a major role in the degradation of sludge. Different pyrolysis reactors mentioned in the literature include fluidized bed reactors, fixed bed reactors, microwave-assisted reactors, and innovative solar or plasma reactors.

5.3.3.1 Fixed Bed Reactor

The pyrolysis process in the fixed bed reactor or packed bed reactor (PBR) is initiated by combustion or heat from an external source, which thermally breaks down the organics present in the sludge. The volatiles and light organics flow out of the reactor due to the thermal expansion of the gases and are collected as a gaseous product. In some modified PBR, a sweep of an inert gas like N_2 or Ar is used to remove the produced syngas from the reactor. This reactor produces biochar as a major product as it has a very slow heating rate. The major disadvantage of the reactor is that it has a long residence time with non-uniform heating.

5.3.3.2 Fluidized Bed Reactor

This reactor is used to assess the residence time and temperature impact on fast pyrolysis and its products formed. Complete mixing of feedstock and a high heating rate can be achieved in this type of reactor (D. Chen et al., 2014) The fluidized bed reactor is classified into bubbling fluidized bed reactor and circulating fluidized bed reactor.

5.3.3.3 Bubbling Fluidized Bed Reactor

The dried sludge of size 2–6 mm is fed with hot sand or other solid bubbling bed. Inert gas like N_2 is used to fluidize the bed. The sand bed is most widely used as an inert solid bed as it provides excellent control of temperature, aids in proper mixing, and provides the required heat transfer rate to the feedstock. The heat to the reactor is provided by combusting either char or syngas in a separate chamber. This reactor produces nearly 50–55% of bio-oil. Secondary cracking is avoided by removing the biochar.

5.3.3.4 Circulating Bed Reactor

A circulating fluidized bed reactor consists of a cyclone in an inner loop that circulates the solids. Riser ensures better temperature control and mixing. The inert gas velocity is much higher than that of the bubbling bed reactor. The advantage of this type of reactor is easy isolation of biochar, and reuse of heat from combustion through a loop seal back to the reactor.

5.3.3.5 Ablative Reactor

In this reactor, high pressure between feedstock particles and the reactor wall is formed. This pressure in transfer of heat from wall to particles results in the melting of feedstock liquid material. In this type of reactor, the rate of heat transfer is high and time is low; hence bio-oil yield can be expected up to 80% (Peacocke et al., 1994).

5.3.3.6 Auger Reactor

This reactor has a screw in its construction, hence called a screw reactor. It is tubular and continuous in operation. The feedstock is fed into the reactor due to the rotational motion of the screw. The screw ensures proper mixing and residence time of the feedstock in the reactor. The advantage of this reactor is that it can be used at the point of generation of the feedstock, thereby reducing the cost of transportation to the plant (Badger & Fransham, 2006).

5.3.3.7 Microwave-assisted Pyrolysis

Researchers have explored microwave-assisted pyrolysis in recent years for bio-oil production. This technology offers more advantages over other methods like less time, low energy, and uniform heating. Zaker et al. (2019) and Jia et al. (2020) have reported that the range of electromagnetic radiation in microwaves is 300 MHz to 300 GHz. The heating efficacy of microwave depends on the capacity of feedstock to absorb the microwave, hence good heating rate is possible if it contains high moisture, fixed carbon, and organic volatile matter.

5.3.4 GASIFICATION

The reactions involved in gasification are drying, pyrolysis, oxidation, and reduction. The drying phase starts at 100 °C and continues till 150 °C and is followed by the pyrolysis stage which occurs in the range of 200 °C to 500 °C. The combustion and cracking stages are in the range of 800 °C to 1200 °C while the reduction occurs between 650 °C to 900 °C. The end product is combustible gas rich in CO, H_2, and CH_4 (Ahmad et al., 2016) obtained through a controlled gasifying agent. This product gas called syngas is used in industries as a feedstock or as a fuel for electricity/heat production (McKendry, 2002; Ruiz et al., 2013). The composition of this gas depends on the design of the gasifier, operating conditions, gasifying agents, and feedstock. The effect of these on the gasification process will be discussed in the following sections.

5.3.4.1 Gasifier Types

Based on the operating mode, the gasifier is broadly classified into packed bed gasifiers (updraft gasifiers, downdraft gasifiers, cross-draft gasifiers), fluidized bed gasifiers (bubbling and circulating fluidized beds), and entrained-bed gasifiers. Due to its flexibility, robust nature, ability to handle a wide range of particles, and high heating rate, fluidized bed gasifiers are used. Few innovations are like solar steam fluidized bed gasifiers to maximize the carbon conversion by 79% (Lam et al., 2020; Li et al., 2020) steam/O_2 gasification with Ni-coated distributor that produced H_2-rich and less NH_3 content gases (Jeong et al., 2022), integrated air and external fired gas turbines gasification (Lee et al., 2022), coupled hydrothermal pretreatment and supercritical water gasification of sewage sludge (Gong et al., 2022), and distillery sludge and bio-compost mixed gasification (Dhote et al., 2022).

5.3.4.2 Operating Conditions

The operating condition influences the formation of the end product. Each reactor has a different composition of syngas formation. Producer gas composition and heating value depend on the operating temperature. The higher the bed temperatures, the lesser will be the char and tar formation. Ahmad et al. (2016) confirmed that the yield of CO and hydrogen increases with an increase in temperature while CO_2 decreases, thereby improving the quality of syngas and efficiency of gasification. Particle size is important in heat and mass transfer between the particles in the bed. If particles are small, then the surface area will be large, which in turn increases the rate of reaction and efficiency of gasification. The composition of syngas varies, depending on the gasifying agent used in the process. Studies have reported many gasifying agents like air, oxygen, steam CO_2, and a combination of these have been effective in improving the composition of syngas produced.

5.3.5 Innovations in Thermochemical Treatments

Major concerns for innovations in thermochemical treatments are based on pollution and emission control, increasing the quality and quantity of desired products, and scaling up the process into continuous industrial operations. The recent research inferences are listed in Table 5.2.

5.4 CONCLUSION

Energy recovery from sludge with high moisture content via thermochemical routes and biochemical is promising. This chapter has discussed AD, pyrolysis, gasification, and incineration in a comprehensive way. It can be concluded based on literature that various pretreatment of sludge prior to AD accelerates the hydrolysis, and subsequently AD performance. In thermochemical techniques, reactor configuration and operating parameters play a vital role in energy extraction. The process of incineration converts waste sludge to energy by complete combustion of organics and attains higher energy. The major disadvantage of the process is emissions, hence researchers' focus was to control emission and reduce air pollution impact. The process of pyrolysis thermally cracks the organics in sludge to produce three different fuels, char, bio-oil, and syngas, in an inert atmosphere. Recent innovations in this process aims at increasing the yield of alternative fuel oil, reduce the amount to thermal energy provided for pyrolysis, and to completely transform the process into a continuous industrial-scale production. The process of gasification converts the organic molecule to syngas by partial thermal oxidation. The recent research works in the field of gasification aimed at increasing in the volume of H_2 gas in the syngas mixture so that the calorific value can be increased. The recent works also suggested that combination of thermochemical treatments with another technique will increase the conversion efficiency and is also a better way in managing the solid wastes generated.

REFERENCES

Ahmad, Anis Atikah, Norfadhila Abdullah Zawawi, Farizul Hafiz Kasim, Abrar Inayat, and Azduwin Khasri. 2016. "Assessing the Gasification Performance of Biomass: A Review on Biomass Gasification Process Conditions, Optimization and Economic Evaluation." *Renewable and Sustainable Energy Reviews* 53: 1333–47. https://doi.org/10.1016/j.rser.2015.09.030.

Arun, Jayaseelan, Kannappan Panchamoorthy Gopinath, Dai Viet N. Vo, Panneer Selvam SundarRajan, and Mukundan Swathi. 2020. "Co-Hydrothermal Gasification of Scenedesmus Sp. with Sewage Sludge for Bio-Hydrogen Production Using Novel Solid Catalyst Derived from Carbon-Zinc Battery Waste." *Bioresource Technology Reports* 11 (May): 100459. https://doi.org/10.1016/j.biteb.2020.100459.

Awasthi, Mukesh Kumar, Surendra Sarsaiya, Anil Patel, Ankita Juneja, Rajendra Prasad Singh, Binghua Yan, Sanjeev Kumar Awasthi, et al. 2020. "Refining Biomass Residues for Sustainable Energy and Bio-Products: An Assessment of Technology, Its Importance, and Strategic Applications in Circular Bio-Economy." *Renewable and Sustainable Energy Reviews* 127 (December 2019): 109876. https://doi.org/10.1016/j.rser.2020.109876.

Awasthi, Mukesh Kumar, Surendra Sarsaiya, Steven Wainaina, Karthik Rajendran, Sumit Kumar, Wang Quan, Yumin Duan, et al. 2019. "A Critical Review of Organic Manure Biorefinery Models toward Sustainable Circular Bioeconomy: Technological Challenges, Advancements, Innovations, and Future Perspectives." *Renewable and Sustainable Energy Reviews* 111 (May): 115–31. https://doi.org/10.1016/j.rser.2019.05.017.

Ayol, Azize, and Özgün Tezer Yurdakoş. 2019. "Chemical and Thermal Characteristics of Petrochemical Industrial Sludge." *Desalination and Water Treatment* 172 (October 2018): 29–36. https://doi.org/10.5004/dwt.2019.24778.

Badger, Phillip C., and Peter Fransham. 2006. "Use of Mobile Fast Pyrolysis Plants to Densify Biomass and Reduce Biomass Handling Costs – A Preliminary Assessment." *Biomass and Bioenergy* 30 (4): 321–5. https://doi.org/10.1016/j.biombioe.2005.07.011.

Bozkurt, Yigit C., and Onur G. Apul. 2020. "Critical Review for Microwave Pretreatment of Waste-Activated Sludge Prior to Anaerobic Digestion." *Current Opinion in Environmental Science and Health* 14: 1–9. https://doi.org/10.1016/j.coesh.2019.10.003.

Cammarota, A., F. Cammarota, R. Chirone, G. Ruoppolo, R. Solimene, and M. Urciuolo. 2019. "Fluidized Bed Combustion of Pelletized Sewage Sludge in a Pilot Scale Reactor." *Combustion Science and Technology* 191 (9): 1661–76. https://doi.org/10.1080/00102202.2019.1605363.

Caneghem, J. Van, A. Brems, P. Lievens, C. Block, P. Billen, I. Vermeulen, R. Dewil, J. Baeyens, and C. Vandecasteele. 2012. "Fluidized Bed Waste Incinerators: Design, Operational and Environmental Issues." *Progress in Energy and Combustion Science* 38 (4): 551–82. https://doi.org/10.1016/j.pecs.2012.03.001.

Chen, Aixia, Ruirui Hu, Rong Han, Xiao Wei, Zheng Tian, and Li Chen. 2021. "The Production of Hydrogen-Rich Gas from Sludge Steam Gasification Catalyzed by Ni-Based Sludge Char Prepared with Mechanical Ball-Milling." *Journal of the Energy Institute* 99 (July): 21–30. https://doi.org/10.1016/j.joei.2021.07.015.

Chen, Guan Bang, Samuel Chatelier, Hsien Tsung Lin, Fang Hsien Wu, and Ta Hui Lin. 2018. "A Study of Sewage Sludge Co-Combustion with Australian Black Coal and Shiitake Substrate." *Energies* 11 (12). https://doi.org/10.3390/en11123436.

Chen, You Hsin, Thi Ngoc Lan Thao Ngo, and Kung Yuh Chiang. 2021. "Enhanced Hydrogen Production in Co-Gasification of Sewage Sludge and Industrial Wastewater Sludge by a Pilot-Scale Fluidized Bed Gasifier." *International Journal of Hydrogen Energy* 46 (27): 14083–95. https://doi.org/10.1016/j.ijhydene.2020.10.081.

Chen, D., L. Yin, H. Wang, and P. He. 2014. "Pyrolysis Technologies for Municipal Solid Waste: A Review." *Waste Management* 34 (12): 2466–86. https://doi.org/10.1016/j.wasman.2014.08.004. Elsevier Ltd.

Dhote, Lekha, Jerusha Ganduri, and Sunil Kumar. 2022. "Evaluation of Pyrolysis and Gasification of Distillery Sludge and Bio-Compost Mixed with Coal." *Fuel* 319 (July 2021): 123750. https://doi.org/10.1016/j.fuel.2022.123750.

Engel. 2014. "済無." *Paper Knowledge: Toward a Media History of Documents* (August): 1–7. https://doi.org/10.20944/preprints201708.0033.v1.

Gievers, Fabian, Meike Walz, Kirsten Loewe, Christian Bienert, and Achim Loewen. 2022. "Anaerobic Co-Digestion of Paper Sludge: Feasibility of Additional Methane Generation in Mechanical–Biological Treatment Plants." *Waste Management* 144 (April): 502–12. https://doi.org/10.1016/j.wasman.2022.04.016.

Gong, Miao, Aixin Feng, Linlu Wang, Mengqi Wang, Jinxiang Hu, and Yujie Fan. 2022. "ScienceDirect Coupling of Hydrothermal Pretreatment and Supercritical Water Gasification of Sewage Sludge for Hydrogen Production." *International Journal of Hydrogen Energy* 47 (41): 17914–25. https://doi.org/10.1016/j.ijhydene.2022.03.283.

Hakiki, R., T. Wikaningrum, and T. Kurniawan. 2018. "The Prospect of Hazardous Sludge Reduction Through Gasification Process." *IOP Conference Series: Earth and Environmental Science* 106 (1). https://doi.org/10.1088/1755-1315/106/1/012092.

Hämäläinen, A., Kokko, M., Kinnunen, V., Hilli, T., and Rintala, J. (2022). Hydrothermal carbonization of pulp and paper industry wastewater treatment sludges – characterization and potential use of hydrochars and filtrates. *Bioresource Technology, 355*. https://doi.org/10.1016/j.biortech.2022.127258.

Hu, Qiang, Yanjun Dai, and Chi Hwa Wang. 2020. "Steam Co-Gasification of Horticultural Waste and Sewage Sludge: Product Distribution, Synergistic Analysis and Optimization." *Bioresource Technology* 301 (December 2019): 122780. https://doi.org/10.1016/j.biortech.2020.122780.

Huang, Y. W., M. Q. Chen, and Q. H. Li. 2020. "Decentralized Drying-Gasification Scheme of Sewage Sludge with Torrefied Biomass as Auxiliary Feedstock." *International Journal of Hydrogen Energy* 45 (46): 24263–74. https://doi.org/10.1016/j.ijhydene.2020.06.036.

Jadhav, Amarsinh L., Rajendrakumar V. Saraf, and Aditya N. Dakhore. 2020. "Energy Recovery from Waste Water Treatment Plant Sludge." *Materials Today: Proceedings* 42: 1224–9. https://doi.org/10.1016/j.matpr.2020.12.871.

Jeong, Yong-seong, Tae-young Mun, and Joo-sik Kim. 2022. "Two-Stage Gasi Fi Cation of Dried Sewage Sludge : Effects of Gasifying Agent, Bed Material, Gas Cleaning System, and Ni-Coated Distributor on Product Gas Quality." *Renewable Energy* 185: 208–16. https://doi.org/10.1016/j.renene.2021.12.069.

Jia, Hongyu, Bingkun Liu, Xiuxia Zhang, Jie Chen, and Wenhai Ren. 2020. "Effects of Ultrasonic Treatment on the Pyrolysis Characteristics and Kinetics of Waste Activated Sludge." *Environmental Research* 183 (February): 109250. https://doi.org/10.1016/j.envres.2020.109250.

Kacprzak, Małgorzata, Ewa Neczaj, Krzysztof Fijałkowski, Anna Grobelak, Anna Grosser, Małgorzata Worwag, Agnieszka Rorat, Helge Brattebo, Åsgeir Almås, and Bal Ram Singh. 2017. "Sewage Sludge Disposal Strategies for Sustainable Development." *Environmental Research* 156 (June 1986): 39–46. https://doi.org/10.1016/j.envres.2017.03.010.

Karayildirim, Tamer, Jale Yanik, Mithat Yuksel, and Henning Bockhorn. 2006. "Characterisation of Products from Pyrolysis of Waste Sludges." *Fuel* 85 (10–11): 1498–508. https://doi.org/10.1016/j.fuel.2005.12.002.

Kim, Daegi, Gabin Kim, Doo Young, Kee-won Seong, and Ki Young. 2022. "Enhanced Hydrogen Production from Anaerobically Digested Sludge Using Microwave Assisted Pyrolysis." *Fuel* 314 (November 2021): 123091. https://doi.org/10.1016/j.fuel.2021.123091.

Kokko, Marika, Viljami Kinnunen, Tuomo Hilli, and Jukka Rintala. 2022. "Bioresource Technology Hydrothermal Carbonization of Pulp and Paper Industry Wastewater Treatment Sludges – Characterization and Potential Use of Hydrochars and Filtrates." *Bioresource Technology* 355 (April–May). https://doi.org/10.1016/j.biortech.2022.127258.

Kovačić, Đurđica, Slavko Rupčić, Davor Kralik, Daria Jovičić, Robert Spajić, and Marina Tišma. 2021. "Pulsed Electric Field: An Emerging Pretreatment Technology in a Biogas Production." *Waste Management* 120: 467–83. https://doi.org/10.1016/j.wasman.2020.10.009.

Kurniawan, Tetuko, Rijal Hakiki, and Filson Maratur Sidjabat. 2018. "Wastewater Sludge As an Alternative Energy Resource: A Review." *Journal of Environmental Engineering & Waste Management* 3 (1). https://doi.org/10.33021/jenv.v3i1.396.

Lam, C. M., Hsu, S. C., Alvarado, V., and Li, W. M. (2020). Integrated life-cycle data envelopment analysis for techno-environmental performance evaluation on sludge-to-energy systems. *Applied Energy, 266*. https://doi.org/10.1016/j.apenergy.2020.114867.

Ledakowicz, S., P. Stolarek, A. Malinowski, and O. Lepez. 2019. "Thermochemical Treatment of Sewage Sludge by Integration of Drying and Pyrolysis/Autogasification." *Renewable and Sustainable Energy Reviews* 104 (July 2018): 319–27. https://doi.org/10.1016/j.rser.2019.01.018.

Lee, Changmin, Seunghwan Kim, Man Ho, Young Su, Changweon Lee, Sungho Lee, Junmo Yang, and Jae Young. 2022. "Valorization of Petroleum Refinery Oil Sludges via Anaerobic Co-Digestion with Food Waste and Swine Manure." *Journal of Environmental Management* 307: 114562. https://doi.org/10.1016/j.jenvman.2022.114562.

Li, Aishu, Hengda Han, Song Hu, Meng Zhu, Qiangqiang Ren, Yi Wang, and Jun Xu. 2022. "A Novel Sludge Pyrolysis and Biomass Gasification Integrated Method to Enhance Hydrogen-Rich Gas Generation." *Energy Conversion and Management* 254 (October 2021): 115205. https://doi.org/10.1016/j.enconman.2022.115205.

Li, G., Z. Liu, F. Liu, Y. Weng, S. Ma, and Y. Zhang. 2020. "Thermodynamic Analysis and Techno-Economic Assessment of Synthetic Natural Gas Production via Ash Agglomerating Fluidized Bed Gasification Using Coal as Fuel." *International Journal of Hydrogen Energy* 45 (51): 27359–68. https://doi.org/10.1016/j.ijhydene.2020.07.025.

Lin, Kuo Hsiung, Nina Lai, Jun Yan Zeng, and Hung Lung Chiang. 2018. "Residue Characteristics of Sludge from a Chemical Industrial Plant by Microwave Heating Pyrolysis." *Environmental Science and Pollution Research* 25 (7): 6487–96. https://doi.org/10.1007/s11356-017-1003-1.

Lin, Kuo Hsiung, Nina Lai, Jun Yan Zeng, and Hung Lung Chiang. 2020. "Microwave-Pyrolysis Treatment of Biosludge from a Chemical Industrial Wastewater Treatment Plant for Exploring Product Characteristics and Potential Energy Recovery." *Energy* 199: 117446. https://doi.org/10.1016/j.energy.2020.117446.

Liu, Hongbo, Xingkang Wang, Song Qin, Wenjia Lai, Xin Yang, Suyun Xu, and Eric Lichtfouse. 2021. "Comprehensive Role of Thermal Combined Ultrasonic Pre-Treatment in Sewage Sludge Disposal." *Science of the Total Environment* 789: 147862. https://doi.org/10.1016/j.scitotenv.2021.147862.

Liu, Huidong, Guoren Xu, and Guibai Li. 2020. "The Characteristics of Pharmaceutical Sludge-Derived Biochar and Its Application for the Adsorption of Tetracycline." *Science of the Total Environment* 747: 141492. https://doi.org/10.1016/j.scitotenv.2020.141492.

Lu, Jia Shun, Yingju Chang, Chi Sun Poon, and Duu Jong Lee. 2020. "Slow Pyrolysis of Municipal Solid Waste (MSW): A Review." *Bioresource Technology* 312 (May): 123615. https://doi.org/10.1016/j.biortech.2020.123615.

Manara, P., and A. Zabaniotou. 2012. "Towards Sewage Sludge Based Biofuels via Thermochemical Conversion – A Review." *Renewable and Sustainable Energy Reviews* 16 (5): 2566–82. https://doi.org/10.1016/j.rser.2012.01.074.

McKendry, Peter. 2002. "Energy Production from Biomass (Part 3): Gasification Technologies." *Bioresource Technology* 83 (1): 55–63. https://doi.org/10.1016/S0960-8524(01)00120-1.

Milato, Jônatas V., Rodrigo José França, and Mônica R. C. Marques Calderari. 2020. "Co-Pyrolysis of Oil Sludge with Polyolefins: Evaluation of Different y Zeolites to Obtain Paraffinic Products." *Journal of Environmental Chemical Engineering* 8 (3). https://doi.org/10.1016/j.jece.2020.103805.

Nabi, Mohammad, Jinsong Liang, Panyue Zhang, Yan Wu, Chuan Fu, Siqi Wang, Junpei Ye, Dawen Gao, Fayyaz Ali Shah, and Jiaqi Dai. 2021. "Anaerobic Digestion of Sewage Sludge Pretreated by High Pressure Homogenization Using Expanded Granular Sludge Blanket Reactor: Feasibility, Operation Optimization and Microbial Community." *Journal of Environmental Chemical Engineering* 9 (1): 104720. https://doi.org/10.1016/j.jece.2020.104720.

Nabi, Mohammad, Guangming Zhang, Panyue Zhang, Xue Tao, Siqi Wang, Junpei Ye, Qian Zhang, Muhammad Zubair, Shuai Bao, and Yan Wu. 2019. "Contribution of Solid and Liquid Fractions of Sewage Sludge Pretreated by High Pressure Homogenization to Biogas Production." *Bioresource Technology* 286 (April): 121378. https://doi.org/10.1016/j.biortech.2019.121378.

Oh, Seung Yong, and Young Man Yoon. 2017. "Energy Recovery Efficiency of Poultry Slaughterhouse Sludge Cake by Hydrothermal Carbonization." *Energies* 10 (11). https://doi.org/10.3390/en10111876.

Oladejo, Jumoke, Kaiqi Shi, Xiang Luo, Gang Yang, and Tao Wu. 2019. "A Review of Sludge-to-Energy Recovery Methods." *Energies* 12 (1): 1–38. https://doi.org/10.3390/en12010060.

Peacocke, G. V. C., E. S. Madrali, C. Z. Li, A. J. Güell, F. Wu, R. Kandiyoti, and A. V. Bridgwater. 1994. "Effect of Reactor Configuration on the Yields and Structures of Pine-Wood Derived Pyrolysis Liquids: A Comparison between Ablative and Wire-Mesh Pyrolysis." *Biomass and Bioenergy* 7 (1–6): 155–67. https://doi.org/10.1016/0961-9534(94)00055-X.

Priyadarshini, R., L. Vaishnavi, D. Murugan, M. Sivarajan, A. Sivasamy, P. Saravanan, N. Balasubramanian, and C. Lajapathi Rai. 2015. "Kinetic Studies on Anaerobic Co-Digestion of Ultrasonic Disintegrated Feed and Biomass and Its Effect Substantiated by Microcalorimetry." *International Journal of Environmental Science and Technology* 12 (9): 3029–38. https://doi.org/10.1007/s13762-014-0688-7.

Rashvanlou, Reza Barati, Mahdi Farzadkia, Abbas Rezaee, Mitra Gholami, Majid Kermani, and Hasan Pasalari. 2021. "The Influence of Combined Low-Strength Ultrasonics and Micro-Aerobic Pretreatment Process on Methane Generation and Sludge Digestion: Lipase Enzyme, Microbial Activation, and Energy Yield." *Ultrasonics Sonochemistry* 73: 105531. https://doi.org/10.1016/j.ultsonch.2021.105531.

Rong, Hao, Teng Wang, Min Zhou, Hao Wang, Haobo Hou, and Yongjie Xue. 2017. "Combustion Characteristics and Slagging during Co-Combustion of Rice Husk and Sewage Sludge Blends." *Energies* 10 (4). https://doi.org/10.3390/en10040438.

Roy, Ratul, Laura Haak, Lin Li, and Krishna Pagilla. 2016. "Anaerobic Digestion for Solids Reduction and Detoxification of Refinery Waste Streams." *Process Biochemistry* 51 (10): 1552–60. https://doi.org/10.1016/j.procbio.2016.08.006.

Ruiz, J. A., M. C. Juárez, M. P. Morales, P. Muñoz, and M. A. Mendívil. 2013. "Biomass Gasification for Electricity Generation: Review of Current Technology Barriers." *Renewable and Sustainable Energy Reviews* 18: 174–83. https://doi.org/10.1016/j.rser.2012.10.021.

Shi, Shuai, Guoren Xu, Huarong Yu, and Zhao Zhang. 2018. "Strategies of Valorization of Sludge from Wastewater Treatment." *Journal of Chemical Technology and Biotechnology* 93 (4): 936–44. https://doi.org/10.1002/jctb.5548.

Sipra, Ayesha Tariq, Ningbo Gao, and Haris Sarwar. 2018. "Municipal Solid Waste (MSW) Pyrolysis for Bio-Fuel Production: A Review of Effects of MSW Components and Catalysts." *Fuel Processing Technology* 175 (February): 131–47. https://doi.org/10.1016/j.fuproc.2018.02.012.

Sivagami, Krishnasamy, Perumal Tamizhdurai, Shaikh Mujahed, and Indumathi Nambi. 2021. "Process Optimization for the Recovery of Oil from Tank Bottom Sludge Using Microwave Pyrolysis." *Process Safety and Environmental Protection* 148: 392–99. https://doi.org/10.1016/j.psep.2020.10.004.

Tang, Xinxin, Xuesong Wei, and Songying Chen. 2019. "Continuous Pyrolysis Technology for Oily Sludge Treatment in the Chain-Slap Conveyors." *Sustainability (Switzerland)* 11 (13). https://doi.org/10.3390/su11133614.

Tong, Yao, Tianhua Yang, Bingshuo Li, Xingping Kai, and Rundong Li. 2021. "The Journal of Supercritical Fluids Two-Stage Liquefaction of Sewage Sludge in Methanol-Water Mixed Solvents with Low-Medium Temperature." *The Journal of Supercritical Fluids* 168: 105094. https://doi.org/10.1016/j.supflu.2020.105094.

Tongco, Jovale Vincent, Sangmin Kim, Baek Rock Oh, Sun Yeon Heo, Joonyeob Lee, and Seokhwan Hwang. 2020. "Enhancement of Hydrolysis and Biogas Production of Primary Sludge by Use of Mixtures of Protease and Lipase." *Biotechnology and Bioprocess Engineering* 25 (1): 132–40. https://doi.org/10.1007/s12257-019-0302-4.

Tyagi, Vinay Kumar, and Shang Lien Lo. 2013. "Sludge: A Waste or Renewable Source for Energy and Resources Recovery?" *Renewable and Sustainable Energy Reviews* 25 (71): 708–28. https://doi.org/10.1016/j.rser.2013.05.029.

Veluchamy, C., and Ajay S. Kalamdhad. 2017. "Influence of Pretreatment Techniques on Anaerobic Digestion of Pulp and Paper Mill Sludge: A Review." *Bioresource Technology* 245 (June): 1206–19. https://doi.org/10.1016/j.biortech.2017.08.179.

Wang, Shule, Yuming Wen, Henry Hammarström, Pär Göran Jönsson, and Weihong Yang. 2021. "Pyrolysis Behaviour, Kinetics and Thermodynamic Data of Hydrothermal Carbonization–Treated Pulp and Paper Mill Sludge." *Renewable Energy* 177: 1282–92. https://doi.org/10.1016/j.renene.2021.06.027.

Wang, Xianbao, Chuyue Gao, Xuefei Qi, Yudi Zhang, Tiantian Chen, Yili Xie, Anlong Zhang, and Junling Gao. 2021. "Enhancing Sludge Fermentation and Anaerobic Digestion by Mechanical Cutting Pretreatment." *Journal of Water Process Engineering* 40 (November 2020): 101812. https://doi.org/10.1016/j.jwpe.2020.101812.

Wang, Xin, Bingwei Zhao, and Xiaoyi Yang. 2016. "Co-Pyrolysis of Microalgae and Sewage Sludge: Biocrude Assessment and Char Yield Prediction." *Energy Conversion and Management* 117: 326–34. https://doi.org/10.1016/j.enconman.2016.03.013.

Xie, Candie, Jingyong Liu, Jialin Liang, Wuming Xie, Fatih Evrendilek, and Weixin Li. 2021. "Science of the Total Environment Optimizing Environmental Pollution Controls in Response to Textile Dyeing Sludge, Incineration Temperature, CaO Conditioner, and Ash Minerals." *Science of the Total Environment* 785: 147219. https://doi.org/10.1016/j.scitotenv.2021.147219.

Xie, Shengyu, Guangwei Yu, Chunxing Li, Jie Li, Gang Wang, Shaoqing Dai, and Yin Wang. 2020. "Treatment of High-Ash Industrial Sludge for Producing Improved Char with Low Heavy Metal Toxicity." *Journal of Analytical and Applied Pyrolysis* 150 (April): 104866. https://doi.org/10.1016/j.jaap.2020.104866.

Yan, Mi, Dwi Hantoko, Ekkachai Kanchanatip, Rendong Zheng, Yingjie Zhong, and Ishrat Mubeen. 2020. "Valorization of Sewage Sludge through Catalytic Sub- and Supercritical Water Gasification." *Journal of the Energy Institute* 93 (4): 1419–27. https://doi.org/10.1016/j.joei.2020.01.004.

Yang, Yonglin, Hongrui Ma, Xiangping Chen, Chao Zhu, and Xiaojie Li. 2020. "Effect of Incineration Temperature on Chromium Speciation in Real Chromium-Rich Tannery Sludge under Air Atmosphere." *Environmental Research* 183 (January): 109159. https://doi.org/10.1016/j.envres.2020.109159.

Yildirir, Eyup, and Levent Ballice. 2019. "Supercritical Water Gasification of Wet Sludge from Biological Treatment of Textile and Leather Industrial Wastewater." *Journal of Supercritical Fluids* 146 (December 2018): 100–6. https://doi.org/10.1016/j.supflu.2019.01.012.

Zaker, Ali, Zhi Chen, Xiaolei Wang, and Qiang Zhang. 2019. "Microwave-Assisted Pyrolysis of Sewage Sludge: A Review." *Fuel Processing Technology* 187 (December 2018): 84–104. https://doi.org/10.1016/j.fuproc.2018.12.011.

Zha, Jianrui, Yaji Huang, Wenqing Xia, Zhipeng Xia, Changqi Liu, Lu Dong, and Lingqin Liu. 2018. "Effect of Mineral Reaction Between Calcium and Aluminosilicate on Heavy Metal Behavior During Sludge Incineration." *Fuel* 229 (April): 241–7. https://doi.org/10.1016/j.fuel.2018.05.015.

Zhan, Hao, Xiuzheng Zhuang, Yanpei Song, Xiuli Yin, Junji Cao, Zhenxing Shen, and Chuangzhi Wu. 2018. "Step Pyrolysis of N-Rich Industrial Biowastes: Regulatory Mechanism of nOx Precursor Formation via Exploring Decisive Reaction Pathways." *Chemical Engineering Journal* 344 (2): 320–31. https://doi.org/10.1016/j.cej.2018.03.099.

Zhang, Jin, Junwei Jin, Minyan Wang, Ravi Naidu, Yanju Liu, Yu Bon Man, Xinqiang Liang, et al. 2020. "Co-Pyrolysis of Sewage Sludge and Rice Husk/Bamboo Sawdust for Biochar with High Aromaticity and Low Metal Mobility." *Environmental Research* 191 (July): 110034. https://doi.org/10.1016/j.envres.2020.110034.

Zhang, Wenqi, Jianbiao Chen, Hua Fang, Guoxu Zhang, Zhibing Zhu, Wenhao Xu, Lin Mu, and Yuezhao Zhu. 2022. "Simulation on Co-Gasi Fi Cation of Bituminous Coal and Industrial Sludge in a Downdraft Fi Xed Bed Gasi Fi Er Coupling with Sensible Heat Recovery, and Potential Application in Sludge-to-Energy." *Energy* 243: 123052. https://doi.org/10.1016/j.energy.2021.123052.

Zhang, Xitong, Jiayu Xu, Shuai Ran, Ying Gao, Yinong Lyu, Yueshen Pan, Fei Cao, et al. 2022. "Experimental Study on Catalytic Pyrolysis of Oily Sludge for H_2 Production under New Nickel-Ore-Based Catalysts." *Energy* 249: 123675. https://doi.org/10.1016/j.energy.2022.123675.

Zhen, Guangyin, Xueqin Lu, Hiroyuki Kato, Youcai Zhao, and Yu You Li. 2017. "Overview of Pretreatment Strategies for Enhancing Sewage Sludge Disintegration and Subsequent Anaerobic Digestion: Current Advances, Full-Scale Application and Future Perspectives." *Renewable and Sustainable Energy Reviews* 69 (March): 559–77. https://doi.org/10.1016/j.rser.2016.11.187.

Zuo, Zongliang, Yan Feng, Xiaoteng Li, Siyi Luo, Jinshuang Ma, Huiping Sun, Xuejun Bi, et al. 2021. "Thermal-Chemical Conversion of Sewage Sludge Based on Waste Heat Cascade Recovery of Copper Slag: Mass and Energy Analysis." *Energy* 235: 121327. https://doi.org/10.1016/j.energy.2021.121327.

6 Scale-up of Microbial Fuel Cells
A Waste-to-Energy Option

Swathi S., Akanksha R., Karthick S., Sumisha A., Karnapa A., and Haribabu K.

CONTENTS

6.1 Introduction 69
6.2 Electrode Materials Used in MFCs 71
    6.2.1 Anode Materials for the MFC Design 72
    6.2.2 Cathode Materials for the MFC Design 73
6.3 Proton-exchange Membrane Selection 75
6.4 MFC Configuration Analysis 76
    6.4.1 Single- and Double-chamber Microbial Fuel Cells 76
    6.4.2 Microbial Fuel Cell Stacks 78
6.5 Novel Microbial Fuel Cell Configurations 80
6.6 Cost Analysis 81
6.7 Findings 83
6.8 Considerations for the Future 84
6.9 Conclusion 84
References 85

6.1 INTRODUCTION

The drawbacks of traditional wastewater treatment, particularly its high dependence on conventional energy sources, lead researchers to develop the first microbial fuel cell (MFC) prototype in 1911 (Potter, 1911). Since then, this bioelectrochemical hybrid has gained importance due to its benefits, such as nutrient recovery, decreased sludge generation, and energy conservation (Palanisamy et al., 2019; Sumisha et al., 2020). An MFC essentially converts chemical energy into electrical energy via the biocatalytic generation of electrons. Formerly, electron mediators or shuttles were used to carry out the conversion, but in more recent times, bacteria endowed with the property of extracellular electron transfer, collectively known as "exoelectrogens," are being utilized for the purpose (Logan and Regan, 2006). These exoelectrogens directly transfer electrons to the electrode via redox-active outer membrane proteins, like c-type cytochromes, or via type IV pili called nanowires, as illustrated in Figure 6.1. (M. Li et al., 2018; Logan et al., 2015).

    The generation of bioelectricity is achieved by an electron transfer between the anode and cathode, as illustrated in Figure 6.2. The transfer occurs when bacterial metabolic activity at the anode results in oxidation of an electron donor and subsequent release of electrons to the anode (electron acceptor). Protons generated from the oxidation process then travel to the cathode through usually a proton-exchange membrane (PEM), while the electrons travel to the cathode via an external circuit with applied resistance load (M. Li et al., 2018). The electron donor used in the process can be one of many substrates like simple carbohydrates such as glucose and acetate, or complex solutions such

DOI: 10.1201/9781003334415-6

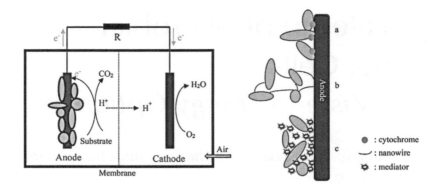

FIGURE 6.1 Schematic illustration of extracellular electron transfer (M. Li et al., 2018).

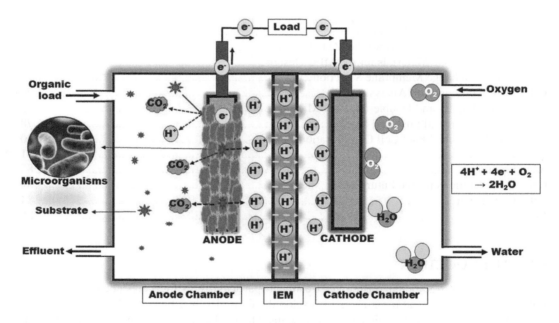

FIGURE 6.2 General functioning of an MFC (Palanisamy et al., 2019).

as synthetic wastewater, dye wastewater, and pulp and brewery wastewater, to name but a few (Pant et al., 2010). Based on the substrate used, MFCs can be used as biosensors, electrolyzers, and for multiple other low-power applications.

The following typical electrode reactions occur in an MFC when an organic substrate such as glucose is used:

Reaction at the anode:

$$C_6H_{12}O_6 + 6H_2O \rightarrow 6CO_2 + 24H^+ + 24e^- \tag{6.1}$$

Reaction at the cathode:

$$6O_2 + 24e^- + 24H^+ \rightarrow 12H_2O \tag{6.2}$$

Overall reaction:

$$C_6H_{12}O_6 + 6O_2 \rightarrow 6CO_2 + 6H_2O \tag{6.3}$$

The net output of this reaction in terms of power density depends on the MFC configuration, electrode characteristics, membrane properties, and substrate composition. The electrodes used in an MFC must be highly conductive, be resistant to corrosion, and possess considerable mechanical strength (Kaur et al., 2020). These electrode characteristics are most crucial in electron transfer and MFC performance. Properties that are desirable in the membrane used in an MFC include those that enable high proton conductivity as well as low oxygen transfer between the electrodes (Choudhury et al., 2017). The components and characteristics of the substrate used in the MFC directly influence the efficiency of the cell as the substrate is the primary energy and nutrient source of the cell. Therefore, the substrate should have characteristics suited for the efficient conversion of organic wastes to bioenergy (Pant et al., 2010).

Another factor that significantly influences the performance of an MFC is the mode of operation. MFCs are commonly operated either in a batch mode or in a continuous mode. The batch mode of operation is the most frequently used mode for MFC studies despite its various disadvantages, such as by-product toxicity, substrate depletion, subsequent nutrient reduction, and low power production. The continuous operation mode of MFCs, on the other hand, offers several advantages such as controlled secondary metabolite production, constant concentration of substrate and products, high productivity per unit volume, and power generation with simultaneous treatment of wastewater. Pasupuleti et al. (2016), during studies on making MFCs an energy-competitive system, found that a continuously operated MFC had a higher power density of 17.22 mW/m² than a batch operated MFC which yielded a power density of 0.75 mW/m². Similar studies performed on MFCs operated in both modes of operation have led to the conclusion that the continuous mode is more preferred for scale-up of MFCs (Pannell et al., 2016). Scaling up of MFC technology is primarily dependent on increasing the net reaction output of the cell while simultaneously decreasing the effective cost of the cell.

Based on the number of chambers, MFCs can be classified into two types: double-chamber microbial fuel cell (DCMFC) and single-chamber microbial fuel cell (SCMFC). DCMFCs have two chambers – the anode and cathode chambers – which are separated by a PEM that specifically permits the transfer of protons to the cathode while hindering the diffusion of oxygen from the cathode to the anode (Pannell et al., 2016). On the contrary, an SCMFC has only one chamber and can be further classified as SCMFCs without a membrane, which allows for direct movement and transfer of electrons and protons between the electrodes, and SCMFCs with a membrane, such as the air-cathode type wherein the cathode is directly exposed to the atmosphere. Other configurations of an MFC include benthic, microfluidic, and stack, which have been detailed in this chapter. Suitable optimization of all the aforementioned parameters is key to improving the performance, reducing the resistance to operating conditions, and enhancing the long-term stability of MFCs (Sumisha et al., 2022; Wei et al., 2011), especially when operated continuously.

This chapter details information regarding the prevalence and development of MFCs over the past several years. Potential shortcomings of an MFC are acknowledged, and different electrodes, membrane materials, and configurations are evaluated from the perspective of continuous long-term operation. In the end, the components and MFC configuration most suitable for scale-up are proposed based on the review performed. Additionally, considerations for future reviews and comparative studies between different types of MFCs have been put forth.

6.2 ELECTRODE MATERIALS USED IN MFCS

The electrode used is the main basis for the determination of the cost and performance of an MFC. Therefore, electrode design becomes a significant task in making MFCs a scalable and economical technology. The electrode materials in an MFC should possess general characteristics such as good conductivity and chemical stability, high mechanical strength, easy availability, and economic feasibility (Wei et al., 2011). For the improvement of the overall performance of an MFC, various electrode materials were studied. The electrode materials used in an MFC can be divided into two categories: anode and cathode materials.

6.2.1 Anode Materials for the MFC Design

The anode material is of great importance in enhancing MFC performance owing to its properties that support electron transfer, biofilm formation, and substrate oxidation (M. Li et al., 2018). To enhance the power generation in MFCs, the anode materials should have the following properties: (i) good biocompatibility, (ii) large surface area, (iii) good corrosion resistance, (iv) low electrical resistance, and (v) good chemical stability (Wei et al., 2011). Carbon is the most commonly used electrode material in MFCs. This includes graphite (Kumar et al., 2017), carbon nanotubes (Duan et al., 2016; Nguyen et al., 2013), carbon rods (Senthilkumar et al., 2020), plates and cloth (Ahmed et al., 2012), carbon fiber brush (Wang et al., 2011), carbon paper (Ghasemi et al., 2013; Kim et al., 2011), carbon sponge (Liu et al., 2015), carbon felt (Lv et al., 2019), reticulated vitreous carbon (RVC) (He et al., 2005), granular activated carbon (Jiang and Li, 2009), graphite rods (Gardel et al., 2012), graphite felts (Chi, 2013; Sumisha and Haribabu, 2018), graphite brush (Saikaly and Logan, 2017), and graphite granules (Rabaey et al., 2005). Figure 6.3 shows some of the carbon-based electrode materials. To further improve the MFC performance, the combinations of different materials, such as the addition of conductive polymers, metals, metal oxides, and hydrogels, have given rise to composite electrodes. Several transition metals, such as Mo, W, Fe, and Sn, and transition

FIGURE 6.3 Carbon-based electrode materials: (a) carbon paper; (b) graphite plate; (c) carbon cloth; (d) carbon mesh; (e) granular graphite; (f) granular-activated carbon; (g) carbon felt; (h) reticulated vitrified carbon; (i) carbon brush; (j) stainless steel mesh.

Source: Taken from Wei et al. (2011).

metal oxides are proven high-performance MFC materials for anode modification (Yamashita and Yokoyama, 2018). Titanium suboxide has been found to be an effective anode material with a power density reaching up to 1541 mW/m^2 (Ma et al., 2016). Graphene/poly(3,4-ethylenedioxythiophene) hybrid achieved a power density 15 times higher than unmodified carbon paper (Wang et al., 2013). NiWO$_4$/rGO-coated carbon cloth anode attained a power density 6.75 times higher than the unmodified one (Geetanjali et al., 2019). An inexpensive treatment of graphite brush with ammonia gas can also be employed for the anode (Valipour et al., 2016). Mink et al. (2012) reported the application of carbon nanotube (CNT) in MFCs. An anode of multiwalled carbon nanotubes (MWCNTs) with a nickel silicide (NiSi) was used in an MFC, and a high power density of 392 mW/m^3 was obtained. The MWCNTs provided the benefit of a high surface-to-volume ratio which in turn improved the charge transfer capacity of the anode. Another CNT-based nanocomposite was reported by Mehdinia et al. (2014) where glassy carbon electrode (GCE) modified by MWCNT/SnO$_2$ was used as an anode. This nanocomposite-coated GCE showed a power density of up to 1421 mW/m^2, which was 3.12 times higher than bare GCE. Various aromatic conducting polymers such as polythiophene (PTh) (Anappara et al., 2020), polypyrrole (Ppy) (Anappara et al., 2020; Sumisha and Haribabu, 2020; Zou et al., 2010), and polyaniline (PANI) have also been considered a good replacement for anode catalysts in MFC (Dutta and Kundu, 2014). A ternary PANI-TiO$_2$-GN (polyaniline, titanium dioxide, and graphene) nanocomposite was reported as a cost-effective alternative with a high power density (79.3 mW/m^2) (Han et al., 2018). Sumisha and Haribabu (2018) reported the high compatibility of polypyrrole- and polythiophene-modified graphene fiber (GF) anode. Both of the modified electrodes showed high power generation with a high power density of 1.22 W/m^2 for Ppy-NP-modified GF and 0.8 W/m^2 for PTh-NP-modified GF. Another conducting polymer, poly(3,4-ethylenedioxythiophene) (PEDOT), has been found to have higher electrochemical characteristics and electron transfer capacity (Kang et al., 2017, 2015). Along with these advantages, PEDOT modified anode was found to enhance MFC performance by improving the power densities and coulombic efficiency. Similarly, a self-supporting Ppy-CS-CNT conductive polypyrrole hydrogel anode material was investigated by Qi et al. (2020) and was found to have a power density of 364 mW/m^2, which is 1.36 times higher than the power density that was obtained using the Ppy anode. Several studies have supported the idea of incorporating hydrogel as an MFC electrode for better performance (Kumar et al., 2014; Liu et al., 2014; Tang et al., 2015; Wang et al., 2020). Various three-dimensional structured composites such as CNT-coated macroporous sponge (Xie et al., 2012), CNT-SnO$_2$ monoliths (Duan et al., 2016), graphene/MWCNTs/Fe$_3$O$_4$ foams (Song et al., 2016), hierarchical structured textile Ppy/NFs/PET (Tao et al., 2016), 3D graphene macroporous scaffold anodes (Ren et al., 2016), mesoporous polysulfone–carbon nanotube anode (Nguyen et al., 2013), etc. have shown outstanding results as 3D structures facilitate the growth of microbes, enhance the surface area, generate high conductivity, improve mass transport, and promote the interactions among microbes, organic substrates, and bioanodes.

Even after numerous modifications, in continuously operated MFCs, carbon- or graphite-based anodes like GAC (granular activated carbon) inside a stainless steel mesh (Linares et al., 2019), coal GAC (Liang et al., 2018), carbon felt (Mehravanfar et al., 2019) graphite fiber brushes (Koffi and Okabe, 2020), graphite rod (H. Li et al., 2018), carbon brush (Feng et al., 2014), and carbon cloth with carbon coating (Srikanth et al., 2016) are preferred due to their favorable electric properties and low cost.

6.2.2 Cathode Materials for the MFC Design

The cathode is another fundamental part in the design of an MFC as the performance of an MFC is directly affected by the oxygen reduction reaction (ORR) that occurs at the cathode. Moreover, cathode selection and modification are crucial not only for performance improvement but also for cost reduction (Kalathil et al., 2018). Although the cathode and anode materials used in an MFC are mostly similar, cathodes are often modified with a catalyst coating to better enable the redox reaction. The most commonly used cathode catalyst is platinum. It has very high ORR (oxygen reduction rate) kinetics, low over-potential, and high specific surface area. Despite its numerous advantages, the high cost involved in the use of platinum limits its application in continuous-flow MFCs (Abdallah et al., 2019). Therefore, several non-precious catalysts such as transition metal and their oxides, carbon-metal derivatives, conductive polymers, and hydrogels have been investigated to reduce the overall cost while enhancing the performance of MFCs (Karthick and Haribabu, 2020). Pyrolyzed carbon mixed iron-chelated ethylenediaminetetraacetic acid (PFeEDTA/C) catalyst–modified carbon cloth was proven a better substitute for the platinum catalyst as it produced a power density of up to 1122 mW/m^2, which is higher than Pt/C cathode (1166 mW/m^2) (Wang et al., 2011). The use of Co-naphthalocyanine (CoNPc) as a cathode catalyst was explored, and it was observed that this composite significantly improved ORR kinetics and catalytic activity, and in comparison to Pt/C (\$0.0447/cm^2), CoNPc (\$0.0114/cm^2) was low-cost (Kim et al., 2011). Inexpensive cobalt oxide nanoparticles–iron phthalocyanine (CoO$_x$–FePc) (Ahmed et al., 2012) and copper-phthalocyanine/C (Ghasemi et al., 2013) showed results comparable to Pt/C. MnFe$_2$O$_4$ NPs/PANI hybrid cathode catalyst not only possesses higher ORR activities but also improved the anodic half-cell potential with simultaneous cost reduction (Khilari et al., 2015). Conducting polymer polypyrrole and its composites have found good application in MFCs due to their high conductivity, chemical stability, ease of synthesis, and environment-friendly nature. Some of the reported Ppy composites are manganese cobaltite/polypyrrole (Khilari et al., 2014), manganese-polypyrrole-carbon nanotube (Lu et al., 2013), PPy/AQS (9,10-anthraquinone-2-sulfonic acid) and Ppy/ARS (Alizarin Red's) (Li et al., 2014), polypyrrole/carrageenan (Esmaeili et al., 2014), MnO$_2$/polypyrrole/MnO$_2$ nanotubes (NT-MPMs) (Yuan et al., 2015), CNT/polypyrrole (Ghasemi et al., 2016), polypyrrole/stainless steel (Pu et al., 2018), PPY-NPs/CC (Sumisha and Haribabu, 2020), polypyrrole-molybdenum oxide composite (Karthick and Haribabu, 2020), and tungsten oxide/polypyrrole composite (Karthick et al., 2020). Not only polypyrrole but also polyaniline and its composites such as vanadium oxide/PANI, Ni:Co/SPAni (Papiya et al., 2018) (MnFe$_2$O$_4$)/PANI (Khilari et al., 2015) are good alternatives of platinum. Iron oxide has also been reported as an inexpensive and suitable catalyst cathode for MFC with enhanced performance (Bhowmick et al., 2019; Ma et al., 2014). Nitrogen-doped graphene (Feng et al., 2011) and activated carbon-coated metal mesh have proven to be another green and cost-effective substitute as a cathode as the composite exhibits good catalytic activity and high ORR activity (Zhang et al., 2014). Yang et al. (2020) recorded that the Fe–N–C/AC catalyst derived from Fe(III)-chitosan hydrogel produced a power density of 2.4 W/m^2, which was 33% higher than plain AC with only a 6% increase in the material cost. Even after numerous effective anode/cathode modifications, these modifications cannot be applied for scaling up of MFCs due to the poisoning, high cost, high requirement of binder, loss in kinetics, and maintenance of catalyst on the electrode surface (stability). Therefore, unmodified carbonaceous materials are preferred for scale-up of MFC due to low cost, compatibility with bacterial growth, simple management (Goswami and Mishra, 2017), and large surface area, which allows cost-effective scale-up of MFCs (Linares et al., 2019). Various reports have proven the credibility of carbon-based electrodes in scale-up. Satyam et al. (2011) reported the cost-effective MFC scale-up with stainless steel as an anode and graphite plates as a cathode. Carbon fiber brush and carbon cloth were used as anode and cathode, respectively, for successful treatment of wastewater using continuous-flow MFC (Ahn and Logan, 2013). Carbon mesh and carbon brushes as cathode and anode, respectively, were used for stacking 250-L module MFC (Feng et al., 2014). Granular activated carbon has shown promising results (high power density and chemical oxygen demand [COD] removal) for continuous water treatment owing to its high surface area, good adsorption ability, and low cost when compared to platinum (Liang et al., 2018). Recently, Mehravanfar et al. (2019) used carbon felt for both anode and cathode to optimize the performance of an MFC operating under the continuous flow of real wastewater. Similarly, carbon cloth and granular activated carbon as the electrodes were used for the successful treatment of wastewater from a domestic household with five inhabitants (Linares et al., 2019).

6.3 PROTON-EXCHANGE MEMBRANE SELECTION

In an MFC, a proton-exchange membrane separates the anodic and cathodic compartments and selectively permits the transfer of protons from the anode to the cathode. This transfer of protons enables the production of electric current and water (ElMekawy et al., 2013). The materials most commonly used to make membranes in an MFC are Nafion, ceramic, clay, and other inorganic materials. Nafion is the most widely used membrane material in an MFC. Several of its shortcomings, such as high gas and substrate crossover, non-biocompatibility, low proton selectivity, high fabrication cost, and fouling (Koók et al., 2019), are overcome by doping. A few of the polymer additives used include nylon, polycarbonate, cellulose, and J-cloths (Palanisamy et al., 2019). Recently, clay has also been reported as a suitable PEM material for scale-up as it drastically reduces the cost of an MFC up to 79%, when compared to materials such as Nafion (Satyam et al., 2011). Another material that offers significant advantages is ceramic, as it is low in cost and offers great structural support (Winfield et al., 2016). Sulfonated membranes with inorganic additives and polymers such as sulfonated-oxy-polybenzimidazole, SPEEK/GO-X or SPEEK/DGO-X (Figure 6.4) (He et al., 2014), PVA-Nafion-borosilicate (MPN), sulfonated polystyrene-ethylene-butylene-polystyrene (SPSEBS), SSEBS-SeSiO$_2$, SPAEK/PW-mGO, 7.5% Fe$_3$O$_4$/SPEEK, SPEEK, sulfonated polyether ether ketone (SPEEK) TiO$_2$-SO$_3$H (Ayyaru and Dharmalingam, 2013), and SPEEK + 7.5% TiO$_2$ have also been found as suitable PEM materials owing to their low cost, high oxidative stability, and high water uptake (Palanisamy et al., 2019).

Some of the other cost-effective sulfonated polyvinyl alcohol (PVA)–based membranes are shown in Table 6.1 (Chakraborty et al., 2020). A zeolite (H-faujasite)–incorporated SPEEK membrane demonstrated high proton conductivity at low cost, restricted transfer of cations, and oxygen contributing to increase power density through the MFC (Narayanaswamy Venkatesan and Dharmalingam,

FIGURE 6.4 Image of sulfonated membranes (SPEEK, SPEEK/GO-5, and SPEEK/DGO-5) (He et al., 2014).

TABLE 6.1
Properties of Polyvinyl Alcohol–based Membrane

Membrane	Cost ($/m²)	Proton Conduction (S/cm)
0.5% GO-impregnated PVA–STA composite membrane	60	0.072
PVA sulfosuccinic acid composite membrane	2.15	0.0516
PVA with 4% glutaraldehyde composite membrane	NA	0.00012
SBC (sulfonated biochar)-600 with PVA	77	0.077

2015). Despite the availability of an assortment of such membrane materials, however, Nafion continues to be the most preferred membrane material for scale-up (Mitra and Hill, 2012).

6.4 MFC CONFIGURATION ANALYSIS

MFCs have deviated from their traditionally defined design and have adapted to a variety of new configurations. The primary goal of this adaptation is the fabrication of a low-cost system capable of sustained energy generation. Different configurations have been assessed for their different effects on power generation with respect to voltage, current, and power density (Kuchi et al., 2018). However, the ability of the configurations to withstand variation in factors such as substrate composition and temperature, as well as their maintenance requirements in the long haul, is yet to be studied in detail. The evaluation of these essential features, especially at a larger scale, necessitates pilot-scale studies. MFC configurations that have been tested on a pilot-scale and have shown to provide positive results when operated over long durations of time are detailed in the following sections.

6.4.1 Single- and Double-chamber Microbial Fuel Cells

Single- and double-chamber MFCs are the most conventionally used cell configurations for research, as detailed in Table 6.2. Among the two cell configuration types, the SCMFC is comparatively simpler and less costly as it does not have a separate cathode compartment. In air-cathode type SCMFCs, for example, the cathode is directly exposed to air, thereby eliminating the need for additional electron acceptors. Moreover, since there is a passive transfer of air to the cathode, expenditure on equipment such as air spargers is also reduced (Slate et al., 2019). Due to these advantages and features, SCMFCs are often preferred for scale-up applications. Ahn and Logan (2013) fabricated a scalable SCMFC with an air-cathode, several graphite fiber brush anodes, and additionally to minimize the spacing between the electrodes, a separator electrode assembly. It was observed that at a hydraulic retention time (HRT) of 8 h, a coulombic efficiency of 85% was attained and the cell voltage reached 0.21 ± 0.04 V. In a similar application using SCMFCs, Hiegemann et al. (2016) fabricated a 45-L MFC system, with four membrane-less SCMFCs, and integrated it into a full-scale wastewater treatment plant (WWTP). At an HRT of 22 h, the highest COD removal (24%), total suspended solids removal (40%), and nitrogen removal (28%) were obtained. Although SCMFCs such as the ones used in these applications are of simple design and are low in cost, some membrane-less SCMFCs have several disadvantages owing to the lack of a membrane separating the electrodes.

The absence of a membrane in an MFC can significantly decrease the coulombic efficiency of an MFC due to numerous reasons. The main reasons being the possibility of aerobic digestion of the substrate, the back-diffusion of oxygen to the anode, and unrestricted flow of substances in the substrate to the cathode (Christgen et al., 2015). These issues, however, can be mitigated in SCMFCs with membranes or in DCMFCs wherein the electrodes are separated by a membrane into two different chambers. A typical DCMFC comprises an anodic and a cathodic chamber; to separate these chambers, there may be an ion-exchange membrane, proton-exchange membranes, salt bridges, or even U-shaped glass tubes filled with salt-agar gels which act as proton-exchange membranes (Javed et al., 2018). Among these options, ion-exchange membranes and proton-exchange membranes are most commonly used in continuously operated MFCs owing to the unfavorable high resistance and low energy offered by the other options (Logan et al., 2006). Recently, studies on the application of nanofiltration membranes in MFCs have also gained pace. One such study conducted by Mengqian Lu et al. (Lu et al., 2017) used nanofiltration membranes in a 20-L MFC system containing two 10-L tubular DCMFCs that were integrated into a WWTP. The system was operated continuously for close to a year for the treatment of brewery wastewater and the highest COD removal rate obtained was $94.6 \pm 1.0\%$. The use of membranes in MFCs can, however, hinder their scale-up due to high costs and issues which include the movement of cationic species other than protons through

TABLE 6.2
Different DCMFC and SCMFC Setups

Cathode	Anode	Membrane	Configuration	Volume (mL)	Power Density (mW/m²)	Voltage (mV)	Reference
Nitrogen-doped AC	CC	–	SCMFC	28	1042	510	(Tian et al., 2018)
Mesoporous Co_3O_4	Graphite sheet	Nafion 117	DCMFC	240 × 3	347 ± 7	–	(Kumar et al., 2017)
Carbon-supported (NiPC)-MnO_x	Carbon felt	Nafion 117	DCMFC	38	8.02 W/m³	553	(Tiwari et al., 2017)
NiPC/C	Carbon felt	Nafion 117	DCMFC	38	6.97 W/m³	523	(Tiwari et al., 2017)
$FePO_4$ NPs/carbon black/CC	Saturated calomel electrode	–	SCMFC	15	46.4	657	(Zeng et al., 2017)
CC	Multiwalled MnO_2/PPy/MnO_2/CC	–	SCMFC	20	32.7 ± 3 W/m³	585	(Yuan et al., 2016)
CC	MnO_2 NTs	–	SCMFC	20	21.8 W/m³	545	(Yuan et al., 2016)
CC	MnO_2/PPy	–	SCMFC	20	29.7 W/m³	572	(Yuan et al., 2016)
N-G@CoNi/BCNT	Carbon brush	UltexCM I7000	DCMFC	140 + 130	2000	100	(Hou et al., 2016)
CNT/Ppy/CC	Carbon paper	Nafion 117	DCMFC	420	113.5	628	(Ghasemi et al., 2016)
Ppy/CC	Carbon paper	Nafion 117	DCMFC	420	69.12	447	(Ghasemi et al., 2016)
$RGO_{HI-AcOH}$/CC	Graphite brush treated with ammonia	–	SCMFC	28	1683	727	(Valipour et al., 2016)
RGO/Ni/CC	Graphite brush treated with ammonia	–	SCMFC	28	1015	683	(Valipour et al., 2016)
Petrocoke carbon	Carbon felts	–	SCMFC	28	1029.77	664.67	(Zhang et al., 2015)
Activated carbon	Carbon felts	–	SCMFC	28	833.53	620.33	(Zhang et al., 2015)
Anthraquinone disulfonate/polypyrrole (AQDS/PPY)	Graphite rods	Nafion 117	DCMFC	616 cm³ for each	0.35 W/m³	380–420	(Xu et al., 2015)
Fe_3O_4/PGC-CS	Graphite fiber brush	–	SCMFC	28 cm³	1443	610	(Ma et al., 2014)
Fe_3O_4/PGC-PS	Graphite fiber brush	–	SCMFC	28 cm³	1338	610	(Ma et al., 2014)
V_2O_5	Plain carbon paper	Nafion 117	H-shaped MFC	423 cm³	65.31	–	(Ghoreishi et al., 2014)
V_2O_5/PANI	Plain carbon paper	Nafion 117	H-shaped MFC	423 cm³	79.26	–	(Ghoreishi et al., 2014)
PANI	Plain carbon paper	Nafion 117	H-shaped MFC	423 cm³	42.4	–	(Ghoreishi et al., 2014)

the membrane as well as proton accumulation in the anode and subsequent reduction in microbial activity (Rozendal et al., 2006). Therefore, hybrid systems containing both SCMFCs and DCMFCs have been fabricated and operated. Wei Yang et al. (Yang et al., 2016) fabricated and compared different combinations of DCMFC and SCMFC stacks. The hybrid stacks containing both DCMFC and SCMFC were found to perform better with higher electrical output, self-sustained pH control, effective substrate conversion into electricity during a high power output, and long stable operation periods. Such hybrid systems, therefore, combine a majority of positive attributes of SCMFCs and DCMFCs; however, they do not address the issue of the comparatively higher cost of the DCMFCs used. Therefore, in most MFC systems, the SCMFC is often the more preferred configuration among the two for scale-up due to its simplicity and cost-effectiveness.

6.4.2 MICROBIAL FUEL CELL STACKS

Traditional DCMFC and SCMFC configurations have been found to show lower power outputs despite high voltages when operated continuously. This is primarily because, during scale-up, sufficient cathode area appropriate for the comparative increase in reactor volume is often lacking. This has led researchers to opt for the stacking of smaller cells with high scalability while maintaining low material costs. Rahimnejad et al. (2012) fabricated and operated an MFC stack with four unit cells containing glucose as the substrate (30 g/L), natural red (NR) as a mediator chemical in the anode (200 μmol/L), potassium permanganate as the oxidizing agent in the cathode (400 μmol/L), *Saccharomyces cerevisiae* as the actively working biocatalyst, and graphite. The MFC chambers had a total volume of 460 mL, of which 350 mL was the working volume. The maximum current and power generation in the stack MFC, when operated in a continuous mode, were 6447 mA/m^2 and 2003 mW/m^2, respectively. Estrada-Arriaga et al. (2018) found that 40 air-cathode MFC units capable of individually producing 0.08–1.1 V at open-circuit voltage could produce maximum current and power density of 500 mA/m^2 and 2500 mW/m^2, respectively, when connected in series. In a parallel connection, the maximum current and power density were found to be 24 mA/m^2 and 5.8 mW/m^2, respectively (Figure 6.5.). Aelterman et al. (2006) stacked six MFC units using series and parallel connections and produced a maximum average power output of 258 W/m^3 on an hourly basis, thereby corroborating the use of MFCs to generate useful energy. MFCs in a stacked arrangement can, therefore, acknowledge the shortcomings of a singular MFC by enhancing the overall power output and performance of the cell arrangement. Studies have shown the high efficacy of stacked MFC configurations in the removal of organic matter from wastewater. Liang et al. (2013) set up a 50-L oxic–anoxic two-stage biocathode MFC for nitrogen and organic matter removal from wastewater. Activated semi-coke-packed electrodes were used and the cell was operated in both batch mode and continuous mode. In the continuous mode, at HRTs of 6, 8, and 12 h, the average maximum power density was 38.2, 32.8, and 28.4 W/m^3, respectively. Subsequently, on the extension of HRT from 6 h to 18 h, the removal loads of COD, ammonia nitrogen, and total nitrogen in the effluent were 10, 0.37, and 0.4 kg, respectively. Dong et al. (2015) stacked five MFC modules to form a 90-L system for the treatment of brewery wastewater. The stacked configuration of MFCs was operated for 6 months using diluted wastewater and raw wastewater separately. A high net energy of 0.021 kWh/m^3 was obtained when the latter was used as the substrate. The integration of similar stack MFCs in wastewater treatment plants has been studied in recent years. Ge and He (2016) fabricated a modularized system of MFCs with a total volume of 200 L, comprising 96 MFC modules, and operated it in a local WWTP for the treatment of a primary effluent. The system was operated for a total of 300 days and produced approximately 200 mW of power which was used to drive a 60-W direct current (DC) pump for the recirculation of catholyte. Simultaneously, the MFC system had a total COD removal efficiency of 76.8% and a soluble COD removal efficiency of 55.5%. Liang et al. (2018) operated a similar MFC system of 1000 L with 50 individual modules for 1 year for the treatment of practical municipal wastewater. When the system was fed with artificial wastewater, a maximum power density of 125 W/m^3 was obtained. In a later stage, when the system was fed with

Scale-up of Microbial Fuel Cells

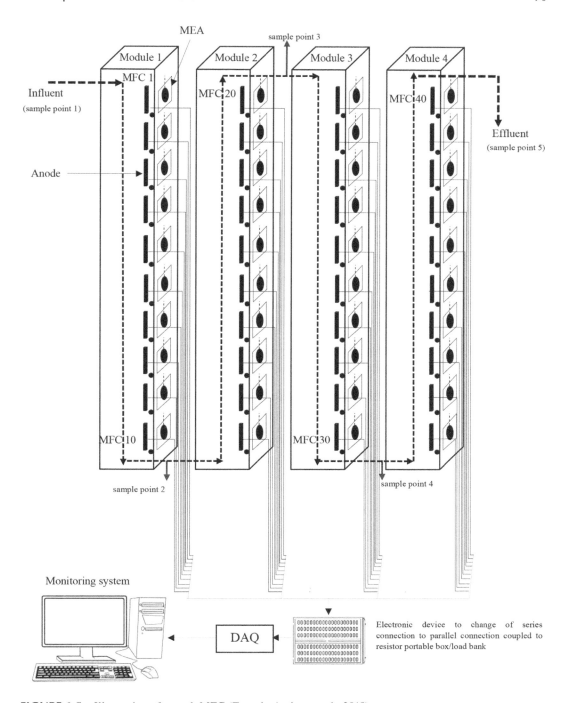

FIGURE 6.5 Illustration of a stack MFC (Estrada-Arriaga et al., 2018).

municipal wastewater, a maximum power density within the range of 7–60 W/m^3 was obtained. In another study, swine wastewater from an educational farm was treated using an MFC system that comprised 12 MFCs having a total volume of 110 L. The MFC system functioned for more than 210 days, at an HRT of 4 h. A maximum power density of 362 ± 52 mW/m^3 and a COD removal rate of approximately 5.0 kg/m³/day were obtained (Babanova et al., 2020). In a study performed by Gajda et al. (2020), urine was used as the substrate in a stacked ceramic MFC with a novel anode material containing powdered activated carbon (PAC) applied onto a carbon fiber scaffold. The cell was operated for 500 days and produced a maximum power of up to 21.1 W/m³. This stack cell is a prime example of a simple and low-cost MFC system that can be used for the simultaneous treatment of wastewater as well as energy generation over prolonged periods. Therefore, stack MFC systems are a viable means of MFC scale-up with the use of air-cathode MFCs in the stack yielding greater benefits.

6.5 NOVEL MICROBIAL FUEL CELL CONFIGURATIONS

New configurations of MFC systems have recently garnered attention due to their adaptable features and benefits over conventional configurations. For example, one novel configuration which is well-suited for continuous operation in marine ecosystems is the benthic microbial fuel cell (BMFCs). These are advanced cell systems capable of generating electricity from benthic or marine organic matter. The basic mechanism in a BMFC is similar to that of an MFC. The only difference being that in a BMFC, the anode is placed beneath the sediment layer and the cathode at the benthic zone, above the interface of the sediment and water, and therefore membranes are not required. The use of such BMFCs underwater is challenged by the low conversion efficiency of low-voltage energy to higher voltages required for practical use such as in modern-day electronics (Babauta et al., 2018). However, efforts are being made to make BMFCs a viable tool for underwater energy harvesting. For other substrates including marine water, configurations such as microfluidic microbial fuel cells (MMFCs) have been gaining widespread importance. MMFCs are miniature MFCs with a total cell volume in the range of 1–200 μL (Ren et al., 2012). An MMFC integrates all accessories of a standard MFC into a bio-chip and microfluidic chamber, as shown in Figure 6.6. This offers advantages such as a high surface-area-to-volume (SAV) ratio, rapid response to reactants, as well as precise manipulation (Luo et al., 2018). Luo et al. (2018) operated a serpentine microchannel MFC having a polymethylmethacrylate (PMMA) plate and two carbon paper electrodes (0.5 mm thickness). It achieved a peak power density of 360 mW/m² at a flow rate of 5 mL/h. Such MMFCs allow for quicker and accurate studies on how a large-scale MFC of a similar setup would work in a continuously operated mode.

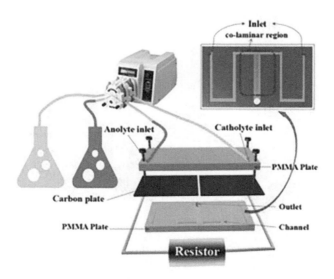

FIGURE 6.6 Schematic illustration of a serpentine MMFC (Luo et al., 2018).

Similarly, other configurations have been studied to address the limitations of conventional MFC configurations. Flat plate microbial fuel cells (FPMFC), for example, address the issue of transport losses due to the significant distance between the anode and cathode. It does so by placing the anode and cathode adjacently with a membrane vertically in between system (Kazemi et al., 2015). Min and Logan (2004) treated domestic wastewater for a combined period of 6 months, including 1 month for acclimation, using an FPMFC. An average power density of 56 ± 0 mW/m^2 was obtained for wastewater containing a COD level of 246 ± 3 mg COD/L, with an HRT of 2 h and an airflow rate of 2 mL/min with a 470 Ω resistance. These results show the viability of using an FPMFC for continuous scale-up.

The baffled microbial fuel cell (BAFMFC) configuration offers similar advantages by mimicking the design of an anaerobic baffled reactor. Feng et al. (2010) displayed this by fabricating a BAFMFC and operating it continuously to yield a maximum power density of 15.2 W/m^3 and a COD removal rate of 88.0% when the substrate used was glucose. A maximum power density of 10.7 W/m^3 and a COD removal rate of 89.1% was obtained when the substrate used was liquid from a corn stover steam explosion. Sonawane et al. (2013) implemented a similar configuration with multiple electrodes arranged in a fashion similar to that of a BAFMFC configuration and obtained a maximum power density of 427 mW/m^2, when operated in a stagnant mode and 597 mW/m^2 in a recirculation mode. Such an arrangement increased the solid retention time (SRT) of the organic matter in the wastewater, thereby decreasing the HRT of the setup. Similar to how the BAFMFC mimics the design of a baffled reactor, an upflow microbial fuel cell is an MFC (UMFC) configuration which couples the advantages of a traditional MFC with those of an upflow anaerobic sludge bed (UASB) system. A UASB works by treating wastewater, which is introduced vertically from below the reactor; this allows for the digestion of organic matter with reduced sludge generation (Daud et al., 2018). He et al. (2006) operated a UMFC using a U-shaped cathode inside an anode chamber, and sucrose as the substrate. Granular activated carbon was filled in both chambers. Maximum volumetric power of 29.2 W/m^3 was obtained when operating the cell with a continuous loading rate of 3.40 kg COD/m^3/day, at 35 °C. The low power output of the UMFC was linked to ohmic limitations of the cell system and kinetic resistances of the electrode chambers. In recent years, several new configurations such as the UMFC have been studied on a laboratory scale; a thorough analysis and comparison of all the configurations upon scale-up would be required to select the best possible MFC design.

6.6 COST ANALYSIS

Studies suggest that MFCs are highly cost-effective as compared to conventional WWTPs, such as those that use activated sludge technology. Conventional treatment plants are highly energy- and cost-intensive as their process often involves aeration and sludge treatment. Aeration alone can account for 45–75% of energy cost, while sludge treatment and its disposal can cost up to 60% of the operation cost of the plant. To put this into a better perspective, the annual expenditure of the United States alone, on domestic wastewater treatment, is around 25 billion dollars. MFCs are a type of self-sufficient technology that can provide a solution to reduce these costs while offering advantages such as minimal sludge production and low energy consumption (Huggins et al., 2013). Hence, research on MFCs is now focused on its scale-up and commercialization. In an experiment carried out by Patra (Patra, 2008), six MFC designs, as described in Table 6.3, were tested and their efficiency was evaluated on a power-to-cost ratio (PCR) basis. Among all the designs, Design 6 showed the best result with a final PCR of 0.42 mW/$. The cost incurred in the use of Design 6 was found to be lesser and 4.3 times more economical than the typical MFC design. The costs of the materials used for these designs are mentioned in Table 6.4. It was also estimated that it would lead to a savings of approximately $300,000/year if used for providing electricity for a city of 100,000 people (electricity charges = 9.8 c/KW). Another cost analysis based on the configuration of MFC was done by Palanisamy et al. (2019). This study drew the comparison between the contribution of

TABLE 6.3
Comparison of Different Designs and Their PCR (Patra, 2008)

Designs	Electrode Materials	Membrane	Setup	Power-to-cost Ratio (PCR) (mW/$)
Design 1	Cathode – platinum-coated CC with Kiwi mesh Anode – CC	Agar	Membrane and the metal cathode were connected through the epoxy glue	NA
Design 2	Cathode – platinum-coated CC with Kiwi mesh Anode – CC	Agar	Same as above but with extra agar layer to prevent leakage	NA
Design 3	Cathode – platinum-coated CC with Kiwi mesh Anode – CC	Agar	Similar to design 2 except plumbers putty was used instead of epoxy glue	NA
Design 4	Cathode – platinum-coated CC with Kiwi mesh Anode – CC	Agar	Similar to design 2, graphite from pencils were used to create carbon filament, which was then pulverized into powder and layered across the cathode and membrane	0.16
Design 5	Cathode – platinum-coated CC with Kiwi mesh Anode – CC	Gore-Tex	Membrane (Gore-Tex) was used in such a manner that it made a watertight seal across the cell	NA
Design 6	Cathode – platinum-coated CC with Kiwi mesh Anode – CC	Agar	Similar to design 4 but instead of cathode, anode was connected to the membrane	0.42

TABLE 6.4
Cost of the Materials Used in MFC (Patra, 2008)

S. No.	Materials	Cost
1	2″ PVC	$0.50
2	Platinum-coated CC	$2000/m^2
3	Kiwi mesh	$12.00/m^2
4	CC (carbon cloth)	$620/m^2
5	Nafion	$2500/m^2
6	Agar	$165/m^2
7	Gore-Tex	$82.5/m^2

electrode materials and separator (membrane) materials of tubular MFC ($) and SCMFC ($/m^3) to the total capital cost (Figure 6.7).

In a 20-L MFC system constructed for brewery wastewater treatment by Mengqian Lu et al. (described in the previous section), a detailed cost analysis was performed and the total cost for the construction of a system containing only two MFCs was found to be $3912.72. Additional capital costs associated with material and equipment for MFC system operation came up to $10,587.90 (Lu et al., 2017). The overall cost was found to be slightly higher than the conventionally used activated sludge system due to the use of additional machines and analyzing equipment. To avoid these additional costs while maintaining accuracy in analysis, optimization of the equipment and components used in an MFC must be done. For example, ceramic materials have been widely researched as a cost-effective alternative to costly PEMs, electrodes, and reactor designs and are currently

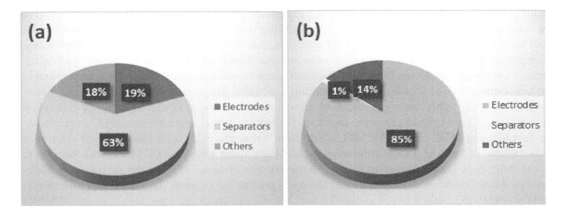

FIGURE 6.7 Cost analysis of (a) tubular MFC and (b) SCMFC on component basis (Palanisamy et al., 2019).

TABLE 6.5
Material Cost of the Stack System (40 Individual Cells) (Zhuang et al., 2012)

S. No.	Material	Amount	Price (US$)	Cost (US$)
1	Anode (Graphite felt)	0.57 m^2	26/m^2	14.8
2	Cathode (Gore-Tex cloth)	7.25 m^2	10/m^2	72.5
3	Conductive paint	232 g	0.004/g	0.9
4	Catalyst (MnO$_2$)	58 g	0.0015/g	0.09
5	Miscellaneous	–	–	16.0
	Total		104.3	

employed in MFCs (Gajda et al., 2015a, 2015b; Winfield et al., 2016). Satyam et al. (2011) presented another example of low-cost clay cylinder MFC for wastewater treatment. In this study, two MFCs with inner cylindrical anode chamber (made of earthenware) and outer cathode chamber were used. Here, the inner cylinder was used as PEM, thereby reducing the high cost associated with the separator. Stainless steel mesh and graphite plates were used as anode and cathode, respectively. This whole system was used to treat wastewater having a COD of about 500 mg/L. Both of the MFCs were found to have more than 90% COD removal efficiency and no significant difference in the power density was recorded when the system was scaled up from 0.6 L to 3.75 L. These results demonstrated the possibility of further scale-up along with 79% cost reduction with earthen material as membrane instead of conventionally used Nafion. Zhuang et al. (2012) reported a scale-up study where he extended 40 single tubular air-cathode fuel cell in a 3D alignment. The operational volume of the system was 10 L and this stack system was operated for more than 180 days. During this long-term performance, the stack series produced a maximum power density of 4.1 W/m^3 and a voltage of 23 V and was also capable of powering the LED panel. This study proved that the stack configurations have the advantages of ease in modularity, high scalability, and low costs. Table 6.5 represents the cost analysis of the stack system.

6.7 FINDINGS

Through this chapter, attempts to highlight the benefits of integrating MFC systems to WWTPs and address the limitations that hinder this integration were made. The most integral components of an MFC that were reviewed were the anode and cathode. Numerous modifications and newly synthesized materials were identified; however, unmodified carbonaceous materials were found to be most preferred due to their low cost, compatibility with bacterial growth, simple management, and large surface area, which allows cost-effective scale-up of MFCs. As for the proton-exchange membrane, Nafion was identified as the most suitable membrane to be used if MFCs with membranes are used for scale-up. Regarding the configuration, stacked air-cathode MFCs were found to be economically more feasible power-generating fuel cells. The reason being that most other configurations have higher internal resistances as well as other drawbacks that limit power generation at low costs. Ceramics and clay were also found to have potential application in MFC as electrode, membrane, and MFC casing. An MFC made from terracotta caves with carbon-based electrodes was capable of lighting a LED continuously for 7 days along with 92% COD removal. But due to some challenges such as loss of anolyte and long-term performance of ceramics, its full-scale applications are still restricted. In conclusion, the study performed suggests the stacking of air-cathode MFCs containing carbon-based electrodes and Nafion membrane for optimum scale-up of MFCs and further integration to fully functional WWTPs.

6.8 CONSIDERATIONS FOR THE FUTURE

Process and cost optimization of MFCs is crucial for its integration in existing WWTPs or its other applications. This is because the numerous advantages of MFCs over traditional WWTP technology are currently limited by the high cost of the MFC components and reactor design. The selection of the components of an MFC system must therefore be such that the capital cost is low while the cell performance is still high. Decisions regarding the selection of the configuration to be used in the MFC system would require a comparison of existing and novel MFC configurations operated in the same conditions. This comparative analysis must be coupled with a comprehensive study of the scalability and modularity of the configuration (Abdallah et al., 2019). Both experimental studies and simulations would be highly impactful in this analysis. As for the operational costs, since MFCs will have to be operated over prolonged periods, efficient conversion and transfer of energy from the MFC to the pumps or other equipment employed in the system must be devised to avoid additional costs. Further cost reduction must also be facilitated by the installation of adequate automated systems to detect and prevent large fluctuations in the performance of the MFC system. Since the MFC performance is largely dependent on the microbial culture used, avoiding fluctuations in operational parameters would be critical to its performance. In this regard, the selection of components and optimization of scaled-up MFC performance is the biggest challenge that needs to be addressed.

6.9 CONCLUSION

In recent years, extensive research work has been performed to utilize the dual benefits of energy generation and wastewater treatment that MFCs provide. However, researchers are yet to create a low-cost MFC system with high power production capacity for commercial use. This chapter presents principal details regarding MFC technology and the parameters that directly influence its performance, cost, and scalability. The parameters reviewed in this chapter include the mode of operation, the configuration, and the membrane and electrode materials used in an MFC. Through the review performed, it was found that in a majority of studies, the power output of continuously operated MFC systems is higher than batch-operated systems, thereby making it a more suitable option for scale-up and integration with WWTPs. Among the different configurations, the stack configuration of air-cathode MFCs was found to be most preferred for long-term operation. This has been attributed to the enhanced overall efficiency and cost-effectiveness of stack MFCs. Research on material cost reduction for scale-up of MFCs was found to be directed toward the use of modified low-cost carbon-based electrodes and membranes such as Nafion membranes. Recent research on new types of electrode and membrane materials necessitates further research on the scope of using these materials in an MFC. Experimental studies coupled with simulations must be performed to not only select the right material but also to optimize the MFC performance and reduce the overall cost by testing and comparing different setups. With dwindling non-renewable sources of energy, the need to commercialize MFC technology is on the rise. It is expected that with continuous research and sustained progress, MFC technology will be available for commercial use in the coming years.

REFERENCES

Abdallah, M., Feroz, S., Alani, S., Sayed, E.T., Shanableh, A., 2019. Continuous and scalable applications of microbial fuel cells: A critical review. *Reviews in Environmental Science and Biotechnology*. Springer. https://doi.org/10.1007/s11157-019-09508-x

Aelterman, P., Rabaey, K., The Pham, H., Boon, N., Verstraete, W., 2006. Continuous electricity generation at high voltages and currents using stacked microbial fuel cells. *Commun. Agric. Appl. Biol. Sci.* 71, 63–6.

Ahmed, J., Yuan, Y., Zhou, L., Kim, S., 2012. Carbon supported cobalt oxide nanoparticles-iron phthalocyanine as alternative cathode catalyst for oxygen reduction in microbial fuel cells. *J. Power Sources* 208, 170–5. https://doi.org/10.1016/j.jpowsour.2012.02.005

Ahn, Y., Logan, B.E., 2013. Domestic wastewater treatment using multi-electrode continuous flow MFCs with a separator electrode assembly design. *Appl. Microbiol. Biotechnol.* 97, 409–16. https://doi.org/10.1007/s00253-012-4455-8

Anappara, S., Kanirudhan, A., Prabakar, S., Krishnan, H., 2020. Energy generation in single chamber microbial fuel cell from pure and mixed culture bacteria by copper reduction. *Arab. J. Sci. Eng.* 45, 7719–24. https://doi.org/10.1007/s13369-020-04832-9

Ayyaru, S., Dharmalingam, S., 2013. Improved performance of microbial fuel cells using sulfonated polyether ether ketone (SPEEK) TiO 2–SO 3 H nanocomposite membrane. *RSC Adv*. 25243–51. https://doi.org/10.1039/c3ra44212h

Babanova, S., Jones, J., Phadke, S., Lu, M., Angulo, C., Garcia, J., Carpenter, K., Cortese, R., Chen, S., Phan, T., Bretschger, O., 2020. Continuous flow, large-scale, microbial fuel cell system for the sustained treatment of swine waste. *Water Environ. Res.* 92, 60–72. https://doi.org/10.1002/wer.1183

Babauta, J.T., Kerber, M., Hsu, L., Phipps, A., Chadwick, D.B., Arias-Thode, Y.M., 2018. Scaling up benthic microbial fuel cells using flyback converters. *J. Power Sources* 395, 98–105. https://doi.org/10.1016/j.jpowsour.2018.05.042

Bhowmick, G.D., Das, S., Verma, H.K., Neethu, B., Ghangrekar, M.M., 2019. Improved performance of microbial fuel cell by using conductive ink printed cathode containing Co3O4 or Fe3O4. *Electrochim. Acta*. 310, 173–83. https://doi.org/10.1016/j.electacta.2019.04.127

Chakraborty, I., Das, S., Dubey, B.K., Ghangrekar, M.M., 2020. Novel low cost proton exchange membrane made from sulphonated biochar for application in microbial fuel cells. *Mater. Chem. Phys.* 239, 122025. https://doi.org/10.1016/j.matchemphys.2019.122025

Chi, M., 2013. Graphite felt anode modified by electropolymerization of nano-polypyrrole to improve microbial fuel cell (MFC) production of bioelectricity. *J. Microb. Biochem. Technol.* 1, 10–13. https://doi.org/10.4172/1948-5948.s12-004

Choudhury, P., Uday, U.S.P., Mahata, N., Nath Tiwari, O., Narayan Ray, R., Kanti Bandyopadhyay, T., Bhunia, B., 2017. Performance improvement of microbial fuel cells for waste water treatment along with value addition: A review on past achievements and recent perspectives. *Renew. Sustain. Energy Rev*. 79, 372–89. https://doi.org/10.1016/j.rser.2017.05.098

Christgen, B., Scott, K., Dolfing, J., Head, I.M., Curtis, T.P., 2015. An evaluation of the performance and economics of membranes and separators in single chamber microbial fuel cells treating domestic wastewater. *PLoS One* 10, 1–13. https://doi.org/10.1371/journal.pone.0136108

Daud, M.K., Rizvi, H., Akram, M.F., Ali, S., Rizwan, M., Nafees, M., Jin, Z.S., 2018. Review of upflow anaerobic sludge blanket reactor technology: Effect of different parameters and developments for domestic wastewater treatment. *J. Chem*. 2018. https://doi.org/10.1155/2018/1596319

Dong, Y., Qu, Y., He, W., Du, Y., Liu, J., Han, X., Feng, Y., 2015. A 90-liter stackable baffled microbial fuel cell for brewery wastewater treatment based on energy self-sufficient mode. *Bioresour. Technol.* 195, 66–72. https://doi.org/10.1016/j.biortech.2015.06.026

Duan, T., Chen, Y., Wen, Q., Yin, J., Wang, Y., 2016. Three-dimensional macroporous CNT-SnO$_2$ composite monolith for electricity generation and energy storage in microbial fuel cells. *RSC Adv*. 6, 59610–18. https://doi.org/10.1039/c6ra11869k

Dutta, K., Kundu, P.P., 2014. A review on aromatic conducting polymers-based catalyst supporting matrices for application in microbial fuel cells. *Polym. Rev*. 54, 401–35. https://doi.org/10.1080/15583724.2014.881372

ElMekawy, A., Hegab, H.M., Dominguez-Benetton, X., Pant, D., 2013. Internal resistance of microfluidic microbial fuel cell: Challenges and potential opportunities. *Bioresour. Technol.* 142, 672–82. https://doi.org/10.1016/j.biortech.2013.05.061

Esmaeili, C., Ghasemi, M., Heng, L.Y., Hassan, S.H.A., Abdi, M.M., Daud, W.R.W., Ilbeygi, H., Ismail, A.F., 2014. Synthesis and application of polypyrrole/carrageenan nano-bio composite as a cathode catalyst in microbial fuel cells. *Carbohydr. Polym.* 114, 253–9. https://doi.org/10.1016/j.carbpol.2014.07.072

Estrada-Arriaga, E.B., Hernández-Romano, J., García-Sánchez, L., Guillén Garcés, R.A., Bahena-Bahena, E.O., Guadarrama-Pérez, O., Moeller Chavez, G.E., 2018. Domestic wastewater treatment and power generation in continuous flow air-cathode stacked microbial fuel cell: Effect of series and parallel configuration. *J. Environ. Manage.* 214, 232–41. https://doi.org/10.1016/j.jenvman.2018.03.007

Feng, L., Chen, Y., Chen, L., 2011. Easy-to-operate and low-temperature synthesis of gram-scale nitrogen-doped graphene and its application as cathode catalyst in microbial fuel cells. *ACS Nano* 5, 9611–18. https://doi.org/10.1021/nn202906f

Feng, Y., He, W., Liu, J., Wang, X., Qu, Y., Ren, N., 2014. A horizontal plug flow and stackable pilot microbial fuel cell for municipal wastewater treatment. *Bioresour. Technol.* 156, 132–8. https://doi.org/10.1016/j.biortech.2013.12.104

Feng, Y., Lee, H., Wang, X., Liu, Y., He, W., 2010. Continuous electricity generation by a graphite granule baffled air-cathode microbial fuel cell. *Bioresour. Technol.* 101, 632–8. https://doi.org/10.1016/j.biortech.2009.08.046

Gajda, I., Greenman, J., Ieropoulos, I., 2020. Microbial fuel cell stack performance enhancement through carbon veil anode modification with activated carbon powder. *Appl. Energy* 262, 114475. https://doi.org/10.1016/j.apenergy.2019.114475

Gajda, I., Greenman, J., Melhuish, C., Ieropoulos, I., 2015a. Simultaneous electricity generation and microbially-assisted electrosynthesis in ceramic MFCs. *Bioelectrochem.* 104, 58–64. https://doi.org/10.1016/j.bioelechem.2015.03.001

Gajda, I., Stinchcombe, A., Greenman, J., Melhuish, C., Ieropoulos, I., 2015b. Ceramic MFCs with internal cathode producing sufficient power for practical applications. *Int. J. Hydrogen Energy* 40, 14627–31. https://doi.org/10.1016/j.ijhydene.2015.06.039

Gardel, E.J., Nielsen, M.E., Grisdela, P.T., Girguis, P.R., 2012. Duty cycling influences current generation in multi-anode environmental microbial fuel cells. *Environ. Sci. Technol.* 46, 5222–9. https://doi.org/10.1021/es204622m

Ge, Z., He, Z., 2016. Long-term performance of a 200 liter modularized microbial fuel cell system treating municipal wastewater: Treatment, energy, and cost. *Environ. Sci. Water Res. Technol.* 2, 274–81. https://doi.org/10.1039/c6ew00020g

Geetanjali, Rani, R., Kumar, S., 2019. Enhanced performance of a single chamber microbial fuel cell using NiWO$_4$/reduced graphene oxide coated carbon cloth anode. *Fuel Cells* 19, 299–308. https://doi.org/10.1002/fuce.201800120

Ghasemi, M., Wan Daud, W.R., Hassan, S.H.A., Jafary, T., Rahimnejad, M., Ahmad, A., Yazdi, M.H., 2016. Carbon nanotube/polypyrrole nanocomposite as a novel cathode catalyst and proper alternative for Pt in microbial fuel cell. *Int. J. Hydrogen Energy* 41, 4872–8. https://doi.org/10.1016/j.ijhydene.2015.09.011

Ghasemi, M., Wan Daud, W.R., Rahimnejad, M., Rezayi, M., Fatemi, A., Jafari, Y., Somalu, M.R., Manzour, A., 2013. Copper-phthalocyanine and nickel nanoparticles as novel cathode catalysts in microbial fuel cells. *Int. J. Hydrogen Energy* 38, 9533–40. https://doi.org/10.1016/j.ijhydene.2013.01.177

Ghoreishi, K.B., Ghasemi, M., Rahimnejad, M., Yarmo, M.A., Daud, W.R.W., Asim, N., Ismail, M., 2014. Development and application of vanadium oxide/polyaniline composite as a novel cathode catalyst in microbial fuel cell. *Int. J. Energy Res.* 38, 70–7. https://doi.org/10.1002/er.3082

Goswami, R., Mishra, V.K., 2017. A review of design, operational conditions and applications of microbial fuel cells. *Biofuel* 7269. https://doi.org/10.1080/17597269.2017.1302682

Han, T.H., Parveen, N., Shim, J.H., Nguyen, A.T.N., Mahato, N., Cho, M.H., 2018. Ternary composite of polyaniline graphene and TiO2 as a bifunctional catalyst to enhance the performance of both the bioanode and cathode of a microbial fuel cell. *Ind. Eng. Chem. Res.* 57, 6705–13. https://doi.org/10.1021/acs.iecr.7b05314

He, Y., Wang, J., Zhang, H., Zhang, T., Zhang, B., Cao, S., Liu, J., 2014. Polydopamine-modified graphene oxide nanocomposite membrane for proton exchange membrane fuel cell under anhydrous conditions. *J. Mater. Chem. A* 2, 9548–8. https://doi.org/10.1039/c3ta15301k

He, Z., Minteer, S.D., Angenent, L.T., 2005. Electricity generation from artificial wastewater using an upflow microbial fuel cell. *Environ. Sci. Technol.* 39, 5262–7. https://doi.org/10.1021/es0502876

He, Z., Wagner, N., Minteer, S.D., Angenent, L.T., 2006. An upflow microbial fuel cell with an interior cathode: Assessment of the internal resistance by impedance spectroscopy. *Environ. Sci. Technol.* 40, 5212–17. https://doi.org/10.1021/es060394f

Hiegemann, H., Herzer, D., Nettmann, E., Lübken, M., Schulte, P., Schmelz, K.G., Gredigk-Hoffmann, S., Wichern, M., 2016. An integrated 45 L pilot microbial fuel cell system at a full-scale wastewater treatment plant. *Bioresour. Technol.* 218, 115–22. https://doi.org/10.1016/j.biortech.2016.06.052

Hou, Y., Yuan, H., Wen, Z., Cui, S., Guo, X., He, Z., Chen, J., 2016. Nitrogen-doped graphene/CoNi alloy encased within bamboo-like carbon nanotube hybrids as cathode catalysts in microbial fuel cells. *J. Power Sources* 307, 561–8. https://doi.org/10.1016/j.jpowsour.2016.01.018

Huggins, T., Fallgren, P.H., Jin, S., Ren, Z.J., 2013. Energy and performance comparison of microbial fuel cell and conventional microbial & biochemical technology energy and performance comparison of microbial fuel cell and conventional aeration treating of wastewater. *J. Microb. Biochem. Technol.* https://doi.org/10.4172/1948-5948.S6-002

Javed, M.M., Nisar, M.A., Ahmad, M.U., Yasmeen, N., Zahoor, S., 2018. Microbial fuel cells as an alternative energy source: Current status. *Biotechnol. Genet. Eng. Rev.* 34, 216–42. https://doi.org/10.1080/02648725.2018.1482108

Jiang, D., Li, B., 2009. Novel electrode materials to enhance the bacterial adhesion and increase the power generation in microbial fuel cells (MFCs). *Water Sci. Technol.* 59, 557–63. https://doi.org/10.2166/wst.2009.007

Kalathil, S., Abdullah, K., Arabia, S., 2018. *Microbial fuel cells: Electrode materials, encyclopedia of interfacial chemistry.* Elsevier. https://doi.org/10.1016/B978-0-12-409547-2.13459-6

Kang, Y.L., Ibrahim, S., Pichiah, S., 2015. Synergetic effect of conductive polymer poly(3,4-ethylenedioxythiophene) with different structural configuration of anode for microbial fuel cell application. *Bioresour. Technol.* 189, 364–9. https://doi.org/10.1016/j.biortech.2015.04.044

Kang, Y.L., Pichiah, S., Ibrahim, S., 2017. Facile reconstruction of microbial fuel cell (MFC) anode with enhanced exoelectrogens selection for intensified electricity generation. *Int. J. Hydrogen Energy* 42, 1661–71. https://doi.org/10.1016/j.ijhydene.2016.09.059

Karthick, S., Haribabu, K., 2020. Bioelectricity generation in a microbial fuel cell using polypyrrole-molybdenum oxide composite as an effective cathode catalyst. *Fuel* 275, 117994. https://doi.org/10.1016/j.fuel.2020.117994

Karthick, S., Sumisha, A., Haribabu, K. 2020. Performance of tungsten oxide/polypyrrole composite as cathode catalyst in single chamber microbial fuel cell. *J. Environ. Chem. Eng.* 8, 104520. https://doi.org/10.1016/j.jece.2020.104520

Kaur, R., Marwaha, A., Chhabra, V.A., Kim, K., Tripathi, S.K., 2020. Recent developments on functional nanomaterial-based electrodes for microbial fuel cells. *Renewable Sustainable Energy Rev.* 119, 109551. https://doi.org/10.1016/j.rser.2019.109551

Kazemi, S., Mohseni, M., Fatih, K., 2015. Passive air breathing flat-plate microbial fuel cell operation. *J. Chem. Technol. Biotechnol.* 90, 468–75. https://doi.org/10.1002/jctb.4325

Khilari, S., Pandit, S., Das, D., Pradhan, D., 2014. Manganese cobaltite/polypyrrole nanocomposite-based air-cathode for sustainable power generation in the single-chambered microbial fuel cells. *Biosens. Bioelectron.* 54, 534–40. https://doi.org/10.1016/j.bios.2013.11.044

Khilari, S., Pandit, S., Varanasi, J.L., Das, D., Pradhan, D., 2015. Bifunctional manganese ferrite/polyaniline hybrid as electrode material for enhanced energy recovery in microbial fuel cell. *ACS Appl. Mater. Interfaces* 7, 20657–66. https://doi.org/10.1021/acsami.5b05273

Kim, J.R., Kim, J.Y., Han, S.B., Park, K.W., Saratale, G.D., Oh, S.E., 2011. Application of co-naphthalocyanine (CoNPc) as alternative cathode catalyst and support structure for microbial fuel cells. *Bioresour. Technol.* 102, 342–7. https://doi.org/10.1016/j.biortech.2010.07.005

Koffi, N.J., Okabe, S., 2020. Domestic wastewater treatment and energy harvesting by serpentine up-flow MFCs equipped with PVDF-based activated carbon air-cathodes and a low voltage booster. *Chem. Eng. J.* 380. https://doi.org/10.1016/j.cej.2019.122443

Koók, L., Bakonyi, P., Harnisch, F., Kretzschmar, J., Chae, K., Zhen, G., Kumar, G., Rózsenberszki, T., Tóth, G., 2019. Biofouling of membranes in microbial electrochemical technologies: Causes, characterization methods and mitigation strategies. *Bioresour. technol.* 279, 327–38.

Kuchi, S., Sarkar, O., Butti, S.K., Velvizhi, G., Venkata Mohan, S., 2018. Stacking of microbial fuel cells with continuous mode operation for higher bioelectrogenic activity. *Bioresour. Technol.* 257, 210–16. https://doi.org/10.1016/j.biortech.2018.02.057

Kumar, G.G., Hashmi, S., Karthikeyan, C., GhavamiNejad, A., Vatankhah-Varnoosfaderani, M., Stadler, F.J., 2014. Graphene oxide/carbon nanotube composite hydrogels – versatile materials for microbial fuel cell applications. *Macromol. Rapid Commun.* 35, 1861–5. https://doi.org/10.1002/marc.201400332

Kumar, R., Singh, L., Zularisam, A.W., 2017. Mesoporous Co_3O_4 nanoflakes as an efficient and non-precious cathode catalyst for oxygen reduction reaction in air-cathode microbial fuel cells. *J. Taiwan Inst. Chem. Eng.* 78, 329–36. https://doi.org/10.1016/j.jtice.2017.06.026

Li, H., Song, H.L., Yang, X.L., Zhang, S., Yang, Y.L., Zhang, L.M., Xu, H., Wang, Y.W., 2018. A continuous flow MFC-CW coupled with a biofilm electrode reactor to simultaneously attenuate sulfamethoxazole and its corresponding resistance genes. *Sci. Total Environ.* 637–8, 295–305. https://doi.org/10.1016/j.scitotenv.2018.04.359

Li, Y., Liu, L., Liu, J., Yang, F., Ren, N., 2014. PPy/AQS (9, 10-anthraquinone-2-sulfonic acid) and PPy/ARS (Alizarin Red's) modified stainless steel mesh as cathode membrane in an integrated MBR/MFC system. *Desalination* 349, 94–101. https://doi.org/10.1016/j.desal.2014.06.027

Liang, P., Duan, R., Jiang, Y., Zhang, X., Qiu, Y., Huang, X., 2018. One-year operation of 1000-L modularized microbial fuel cell for municipal wastewater treatment. *Water Res.* 141, 1–8. https://doi.org/10.1016/j.watres.2018.04.066

Liang, P., Wei, J., Li, M., Huang, X., 2013. Scaling up a novel denitrifying microbial fuel cell with an oxic-anoxic two stage biocathode. *Front. Environ. Sci. Eng.* 7, 913–19. https://doi.org/10.1007/s11783-013-0583-3

Linares, R.V., Domínguez-Maldonado, J., Rodríguez-Leal, E., Patrón, G., Castillo-Hernández, A., Miranda, A., Romero, D.D., Moreno-Cervera, R., Camara-chale, G., Borroto, C.G., Alzate-Gaviria, L., 2019. Scale up of microbial fuel cell stack system for residential wastewater treatment in continuous mode operation. *Water (Switzerland)* 11, 1–16. https://doi.org/10.3390/w11020217

Liu, M., Zhou, M., Yang, H., Zhao, Y., Hu, Y., 2015. A cost-effective polyurethane based activated carbon sponge anode for high-performance microbial fuel cells. *RSC Adv.* 5, 84269–75. https://doi.org/10.1039/c5ra14644e

Liu, X.W., Huang, Y.X., Sun, X.F., Sheng, G.P., Zhao, F., Wang, S.G., Yu, H.Q., 2014. Conductive carbon nanotube hydrogel as a bioanode for enhanced microbial electrocatalysis. *ACS Appl. Mater. Interfaces* 6, 8158–64. https://doi.org/10.1021/am500624k

Logan, B.E., Hamelers, B., Rozendal, R., Schröder, U., Keller, J., Freguia, S., Aelterman, P., Verstraete, W., Rabaey, K., 2006. Microbial fuel cells: Methodology and technology. *Environ. Sci. Technol.* 40, 5181–92. https://doi.org/10.1021/es0605016

Logan, B.E., Regan, J.M., 2006. Electricity-producing bacterial communities in microbial fuel cells. *Trends Microbiol.* 14, 512–18. https://doi.org/10.1016/j.tim.2006.10.003

Logan, B.E., Wallack, M.J., Kim, K.Y., He, W., Feng, Y., Saikaly, P.E., 2015. Assessment of microbial fuel cell configurations and power densities. *Environ. Sci. Technol. Lett.* 2, 206–14. https://doi.org/10.1021/acs.estlett.5b00180

Lu, M., Chen, S., Babanova, S., Phadke, S., Salvacion, M., Mirhosseini, A., Chan, S., Carpenter, K., Cortese, R., Bretschger, O., 2017. Long-term performance of a 20-L continuous flow microbial fuel cell for treatment of brewery wastewater. *J. Power Sources* 356, 274–87. https://doi.org/10.1016/j.jpowsour.2017.03.132

Lu, M., Guo, L., Kharkwal, S., Wu, H., Ng, H.Y., Li, S.F.Y., 2013. Manganese-polypyrrole-carbon nanotube, a new oxygen reduction catalyst for air-cathode microbial fuel cells. *J. Power Sources* 221, 381–6. https://doi.org/10.1016/j.jpowsour.2012.08.034

Luo, X., Xie, W., Wang, R., Wu, X., Yu, L., Qiao, Y., 2018. Fast start-up microfluidic microbial fuel cells with serpentine microchannel. *Front. Microbiol.* 9, 1–7. https://doi.org/10.3389/fmicb.2018.02816

Lv, C., Liang, B., Zhong, M., Li, K., Qi, Y., 2019. Activated carbon-supported multi-doped graphene as high-efficient catalyst to modify air cathode in microbial fuel cells. *Electrochim. Acta* 304, 360–9. https://doi.org/10.1016/j.electacta.2019.02.094

Ma, M., Dai, Y., Zou, J.L., Wang, L., Pan, K., Fu, H.G., 2014. Synthesis of iron oxide/partly graphitized carbon composites as a high-efficiency and low-cost cathode catalyst for microbial fuel cells. *ACS Appl. Mater. Interfaces* 6, 13438–47. https://doi.org/10.1021/am501844p

Ma, M., You, S., Liu, G., Qu, J., Ren, N., 2016. Macroporous monolithic Magnéli-phase titanium suboxides as anode material for effective bioelectricity generation in microbial fuel cells. *J. Mater. Chem. A* 4, 18002–7. https://doi.org/10.1039/c6ta07521e

Mehdinia, A., Ziaei, E., Jabbari, A., 2014. Multi-walled carbon nanotube/SnO$_2$ nanocomposite: A novel anode material for microbial fuel cells. *Electrochim. Acta* 130, 512–18. https://doi.org/10.1016/j.electacta.2014.03.011

Mehravanfar, H., Mahdavi, M.A., Gheshlaghi, R., 2019. Economic optimization of stacked microbial fuel cells to maximize power generation and treatment of wastewater with minimal operating costs. *Int. J. Hydrogen Energy* 44, 20355–67. https://doi.org/10.1016/j.ijhydene.2019.06.010

Min, B., Logan, B.E., 2004. Continuous electricity generation from domestic wastewater and organic substrates in a flat plate microbial fuel cell. *Environ. Sci. Technol.* 38, 5809–14. https://doi.org/10.1021/es0491026

Mink, J.E., Rojas, J.P., Logan, B.E., Hussain, M.M., 2012. Vertically grown multiwalled carbon nanotube anode and nickel silicide integrated high performance microsized (1.25 μl) microbial fuel cell. *Nano Lett.* 12, 791–5. https://doi.org/10.1021/nl203801h

Mitra, P., Hill, G.A., 2012. Continuous microbial fuel cell using a photoautotrophic cathode and a fermentative anode. *Can. J. Chem. Eng.* 90, 1006–10. https://doi.org/10.1002/cjce.20605

Narayanaswamy Venkatesan, P., Dharmalingam, S., 2015. Effect of zeolite on SPEEK/zeolite hybrid membrane as electrolyte for microbial fuel cell applications. *RSC Adv.* 5, 84004–13. https://doi.org/10.1039/c5ra14701h

Nguyen, T.H., Yu, Y.Y., Wang, X., Wang, J.Y., Song, H., 2013. A 3D mesoporous polysulfone-carbon nanotube anode for enhanced bioelectricity output in microbial fuel cells. *Chem. Commun.* 49, 10754–6. https://doi.org/10.1039/c3cc45775c

Palanisamy, G., Jung, H.Y., Sadhasivam, T., Kurkuri, M.D., Kim, S.C., Roh, S.H., 2019. A comprehensive review on microbial fuel cell technologies: Processes, utilization, and advanced developments in electrodes and membranes. *J. Clean. Prod.* 221, 598–621. https://doi.org/10.1016/j.jclepro.2019.02.172

Pannell T.C., Goud R.K., Schell D.J., Borole A.P., 2016. Effect of fed-batch vs. continuous mode of operation on microbial fuel cell performance treating biorefinery wastewater. *Biochem. Eng. J.* 116, 85–95. https://doi.org/10.1016/j.bej.2016.04.029

Pant, D., Van Bogaert, G., Diels, L., Vanbroekhoven, K., 2010. A review of the substrates used in microbial fuel cells (MFCs) for sustainable energy production. *Bioresour. Technol.* 101, 1533–43. https://doi.org/10.1016/j.biortech.2009.10.017

Papiya, F., Pattanayak, P., Kumar, P., Kumar, V., Kundu, P.P., 2018. Development of highly efficient bimetallic nanocomposite cathode catalyst, composed of Ni:Co supported sulfonated polyaniline for application in microbial fuel cells. *Electrochim. Acta* 282, 931–45. https://doi.org/10.1016/j.electacta.2018.07.024

Pasupuleti, S.B., Srikanth, S., Dominguez-Benetton, X., Mohan, S.V., Pant, D., 2016. Dual gas diffusion cathode design for microbial fuel cell (MFC): Optimizing the suitable mode of operation in terms of bioelectrochemical and bioelectro-kinetic evaluation. *J. Chem. Technol. Biotechnol.* 91, 624–39. https://doi.org/10.1002/jctb.4613

Patra, A., 2008. Low-cost, single-chambered microbial fuel cells for harvesting energy and cleansing wastewater. *J. US, SJWP* 72–85.

Potter, M.C., 1911. Electrical effects accompanying the decomposition of organic compounds the decomtposition electrical effects accompanying of organic the fermentative activity of yeast and other organisms. *Cultures of. Proc. R. Soc. London* 84, 260–76.

Pu, K.B., Ma, Q., Cai, W.F., Chen, Q.Y., Wang, Y.H., Li, F.J., 2018. Polypyrrole modified stainless steel as high performance anode of microbial fuel cell. *Biochem. Eng. J.* 132, 255–61. https://doi.org/10.1016/j.bej.2018.01.018

Qi, L., Wu, J., Chen, Y., Wen, Q., Xu, H., Wang, Y., 2020. Shape-controllable binderless self-supporting hydrogel anode for microbial fuel cells. *Renew. Energy* 156, 1325–35. https://doi.org/10.1016/j.renene.2019.11.152

Rabaey, K., Clauwaert, P., Aelterman, P., Verstraete, W., 2005. Tubular microbial fuel cells for efficient electricity generation. *Environ. Sci. Technol.* 39, 8077–82. https://doi.org/10.1021/es050986i

Rahimnejad, M., Ghoreyshi, A.A., Najafpour, G.D., Younesi, H., Shakeri, M., 2012. A novel microbial fuel cell stack for continuous production of clean energy. *Int. J. Hydrogen Energy* 37, 5992–6000. https://doi.org/10.1016/j.ijhydene.2011.12.154

Ren, H., Lee, H.S., Chae, J., 2012. Miniaturizing microbial fuel cells for potential portable power sources: Promises and challenges. *Microfluid. Nanofluidics* 13, 353–81. https://doi.org/10.1007/s10404-012-0986-7

Ren, H., Tian, H., Gardner, C.L., Ren, T.L., Chae, J., 2016. A miniaturized microbial fuel cell with three-dimensional graphene macroporous scaffold anode demonstrating a record power density of over 10000 W m-3. *Nanoscale* 8, 3539–47. https://doi.org/10.1039/c5nr07267k

Rozendal, R.A., Hamelers, H.V.M., Buisman, C.J.N., 2006. Effects of membrane cation transport on pH and microbial fuel cell performance. *Environ. Sci. Technol.* 40, 5206–11. https://doi.org/10.1021/es060387r

Saikaly, P.E., Logan, B.E., 2017. The impact of new cathode materials relative to baseline performance of microbial fuel cells all with the same architecture and solution chemistry. *Envir. Sci.* 29–31. https://doi.org/10.1039/c7ee00910k

Satyam, B.S.R., Behera, M., Ghangrekar, M.M., 2011. *Performance and economics of low cost clay cylinder microbial fuel cell for wastewater treatment*. Proceedings of World Renewable Energy Congress, Linköping, Sweden, 57, 1189–96, 8–13 May. https://doi.org/10.3384/ecp110571189

Senthilkumar, K., Anappara, S., Krishnan, H., Ramasamy, P., 2020. Simultaneous power generation and Congo red dye degradation in double chamber microbial fuel cell using spent carbon electrodes. *Energy Sources, Part A Recover. Util. Environ. Eff.* https://doi.org/10.1080/15567036.2020.1781978

Slate, A.J., Whitehead, K.A., Brownson, D.A.C., Banks, C.E., 2019. Microbial fuel cells: An overview of current technology. *Renew. Sustain. Energy Rev.* 101, 60–81. https://doi.org/10.1016/j.rser.2018.09.044

Sonawane, J.M., Gupta, A., Ghosh, P.C., 2013. Multi-electrode microbial fuel cell (MEMFC): A close analysis towards large scale system architecture. *Int. J. Hydrogen Energy* 38, 5106–14. https://doi.org/10.1016/j.ijhydene.2013.02.030

Song, R. Bin, Zhao, C.E., Jiang, L.P., Abdel-Halim, E.S., Zhang, J.R., Zhu, J.J., 2016. Bacteria-Affinity 3D macroporous graphene/MWCNTs/Fe$_3$O$_4$ foams for high-performance microbial fuel cells. *ACS Appl. Mater. Interfaces* 8, 16170–7. https://doi.org/10.1021/acsami.6b03425

Srikanth, S., Kumar, M., Singh, D., Singh, M.P., Das, B.P., 2016. Electro-biocatalytic treatment of petroleum refinery wastewater using microbial fuel cell (MFC) in continuous mode operation. *Bioresour. Technol.* 221, 70–7. https://doi.org/10.1016/j.biortech.2016.09.034

Sumisha, A., Haribabu, K., 2018. Modification of graphite felt using nano polypyrrole and polythiophene for microbial fuel cell applications-a comparative study. *Int. J. Hydrogen Energy* 43, 3308–16. https://doi.org/10.1016/j.ijhydene.2017.12.175

Sumisha, A., Haribabu, K., 2020. Nanostructured polypyrrole as cathode catalyst for Fe (III) removal in single chamber microbial fuel cell. *Biotechnol. Bioprocess Eng.* 25, 78–85. https://doi.org/10.1007/s12257-019-0288-y

Sumisha, A., Harshini, V., Das, A., Haribabu, K., 2022. Single chamber membrane less microbial fuel cell for simultaneous energy generation and lead removal. *Russ. J. Electrochem.* 58, 143–50. https://doi.org/10.1134/S1023193522020094

Sumisha, A., Jiben, A., Aswathy, A., Karthick, S., Haribabu, K., 2020. Reduction of copper and generation of energy in double chamber microbial fuel cell using Shewanella putrefaciens. *Sep. Sci. Technol.* 55, 2391–9. https://doi.org/10.1080/01496395.2019.1625919

Tang, X., Li, H., Du, Z., Wang, W., Ng, H.Y., 2015. Conductive polypyrrole hydrogels and carbon nanotubes composite as an anode for microbial fuel cells. *RSC Adv.* 5, 50968–74. https://doi.org/10.1039/c5ra06064h

Tao, Y., Liu, Q., Chen, J., Wang, B., Wang, Y., Liu, K., Li, M., Jiang, H., Lu, Z., Wang, D., 2016. Hierarchically three-dimensional nanofiber based textile with high conductivity and biocompatibility as a microbial fuel cell anode. *Environ. Sci. Technol.* 50, 7889–95. https://doi.org/10.1021/acs.est.6b00648

Tian, X., Zhou, M., Li, M., Tan, C., Liang, L., Su, P., 2018. Nitrogen-doped activated carbon as metal-free oxygen reduction catalyst for cost-effective rolling-pressed air-cathode in microbial fuel cells. *Fuel* 223, 422–30. https://doi.org/10.1016/j.fuel.2017.11.143

Tiwari, B.R., Noori, M.T., Ghangrekar, M.M., 2017. Carbon supported nickel-phthalocyanine/MnO$_x$ as novel cathode catalyst for microbial fuel cell application. *Int. J. Hydrogen Energy* 42, 23085–94. https://doi.org/10.1016/j.ijhydene.2017.07.201

Valipour, A., Ayyaru, S., Ahn, Y., 2016. Application of graphene-based nanomaterials as novel cathode catalysts for improving power generation in single chamber microbial fuel cells. *J. Power Sources* 327, 548–56. https://doi.org/10.1016/j.jpowsour.2016.07.099

Wang, L., Liang, P., Zhang, J., Huang, X., 2011. Activity and stability of pyrolyzed iron ethylenediaminetetraacetic acid as cathode catalyst in microbial fuel cells. *Bioresour. Technol.* 102, 5093–7. https://doi.org/10.1016/j.biortech.2011.01.025

Wang, Y., Wen, Q., Chen, Y., Li, W., 2020. Conductive polypyrrole-carboxymethyl cellulose-titanium nitride/carbon brush hydrogels as bioanodes for enhanced energy output in microbial fuel cells. *Energy* 204, 117942. https://doi.org/10.1016/j.energy.2020.117942

Wang, Y., Zhao, C., Sun, D., Zhang, J., Zhu, J., 2013. A graphene/poly (3, 4-ethylenedioxythiophene) hybrid as an anode for high-performance. *Microbial Fuel Cell* 823–9. https://doi.org/10.1002/cplu.201300102

Wei, J., Liang, P., Huang, X., 2011. Recent progress in electrodes for microbial fuel cells. *Bioresour. Technol.* 102, 9335–44. https://doi.org/ 10.1016/j.biortech.2011.07.019

Winfield, J., Gajda, I., Greenman, J., Ieropoulos, I., 2016. A review into the use of ceramics in microbial fuel cells. *Bioresour. Technol.* 215, 296–303. https://doi.org/10.1016/j.biortech.2016.03.135

Xie, X., Ye, M., Hu, L., Liu, N., McDonough, J.R., Chen, W., Alshareef, H.N., Criddle, C.S., Cui, Y., 2012. Carbon nanotube-coated macroporous sponge for microbial fuel cell electrodes. *Energy Environ. Sci.* 5, 5265–70. https://doi.org/10.1039/c1ee02122b

Xu, L., Zhang, G.Q., Yuan, G.E., Liu, H.Y., Liu, J.D., Yang, F.L., 2015. Anti-fouling performance and mechanism of anthraquinone/polypyrrole composite modified membrane cathode in a novel MFC-aerobic MBR coupled system. *RSC Adv.* 5, 22533–43. https://doi.org/10.1039/c5ra00735f

Yamashita, T., Yokoyama, H., 2018. Molybdenum anode: A novel electrode for enhanced power generation in microbial fuel cells, identified via extensive screening of metal electrodes. *Biotechnol. Biofuels* 11, 1–13. https://doi.org/10.1186/s13068-018-1046-7

Yang, W., Li, J., Ye, D., Zhang, L., Zhu, X., Liao, Q., 2016. A hybrid microbial fuel cell stack based on single and double chamber microbial fuel cells for self-sustaining pH control. *J. Power Sources* 306, 685–91. https://doi.org/10.1016/j.jpowsour.2015.12.073

Yang, W., Wang, X., Rossi, R., Logan, B.E., 2020. Low-cost Fe–N–C catalyst derived from Fe (III)-chitosan hydrogel to enhance power production in microbial fuel cells. *Chem. Eng. J.* 380, 122522. https://doi.org/10.1016/j.cej.2019.122522

Yuan, H., Deng, L., Chen, Y., Yuan, Y., 2016. MnO_2/polypyrrole/MnO_2 multi-walled-nanotube-modified anode for high-performance microbial fuel cells. *Electrochim. Acta* 196, 280–5. https://doi.org/10.1016/j.electacta.2016.02.183

Yuan, H., Deng, L., Tang, J., Zhou, S., Chen, Y., Yuan, Y., 2015. Facile synthesis of MnO_2/Polypyrrole/MnO_2 multiwalled nanotubes as advanced electrocatalysts for the oxygen reduction reaction. *ChemElectroChem* 2, 1152–8. https://doi.org/10.1002/celc.201500109

Zeng, L., Li, X., Shi, Y., Qi, Y., Huang, D., Tadé, M., Wang, S., Liu, S., 2017. $FePO_4$ based single chamber air-cathode microbial fuel cell for online monitoring levofloxacin. *Biosens. Bioelectron.* 91, 367–73. https://doi.org/10.1016/j.bios.2016.12.021

Zhang, P., Liu, X.H., Li, K.X., Lu, Y.R., 2015. Heteroatom-doped highly porous carbon derived from petroleum coke as efficient cathode catalyst for microbial fuel cells. *Int. J. Hydrogen Energy* 40, 13530–7. https://doi.org/10.1016/j.ijhydene.2015.08.025

Zhang, X., Xia, X., Ivanov, I., Huang, X., Logan, B.E., 2014. Enhanced activated carbon cathode performance for microbial fuel cell by blending carbon black. *Environ. Sci. Technol.* 48, 2075–81. https://doi.org/10.1021/es405029y

Zhuang, L., Yuan, Y., Wang, Y., Zhou, S., 2012. Long-term evaluation of a 10-liter serpentine-type microbial fuel cell stack treating brewery wastewater. *Bioresour. Technol.* 123, 406–12. https://doi.org/10.1016/j.biortech.2012.07.038

Zou, Y., Pisciotta, J., Baskakov, I.V., 2010. Nanostructured polypyrrole-coated anode for sun-powered microbial fuel cells. *Bioelechem.* 79, 50–6. https://doi.org/10.1016/j.bioelechem.2009.11.001

7 Critical Role of Catalysts in Pyrolysis Reactions

Anjana P. Anantharaman

CONTENTS

7.1 Introduction 93
7.2 Catalytic Pyrolysis 94
7.3 Mechanism of Catalytic Reactions 95
7.4 Types of Catalysts for Pyrolysis Reaction 96
    7.4.1 Zeolite 96
    7.4.2 Metal Oxides..98
    7.4.3 Mesoporous Silica..99
    7.4.4 Carbonaceous Material...99
    7.4.5 Other Catalysts 99
7.5 Challenges in Catalyst 100
7.6 Conclusion 100
References.. 101

7.1 INTRODUCTION

The International Energy Agency (IEA) reports the rise in global energy demand by ~5% in 2021 and ~4% in 2022, with the majority of increase in the Asia-Pacific region. The production of energy from renewable sources is proportionally increased by ~8% in 2021 and is expected to grow in the future. However, the energy production from renewable energy alone is not sufficient enough to compensate the global electricity demand ("Electricity Market Report, July 2021" 2021). The projected data by the World Bank on municipal solid waste across globe reports 3.40 billion tons per day by 2050 in comparison to 2.01 billion tons produced in 2016. Among the waste generated, almost 33% is not properly handled in an environmentally safe manner (Kaza et al. 2018).

The evolution of waste-to-energy technology resolves simultaneously energy demand and waste management along with solution for environmental concerns. In this manner, converting linear economy into a circular economy and closed-loop recycling of waste assures sustainable development along with economic solutions. Waste-to-energy conversion is the technology to produce useful heat, electricity, and fuel by conversion of non-recyclable waste materials through thermal and biochemical methods. Thermal conversion of waste to energy includes gasification, incineration, and pyrolysis and biochemical conversion includes anaerobic digestion, ethanol fermentation, landfill, photobiological, dark fermentation, and microbial fuel cell. The choice of conversion method depends on type of the feedstock (Beyene, Werkneh, and Ambaye 2018). Municipal solid waste (MSW) consists of paper, plastic, food waste, textiles, metals, and glass that stores chemical energy in the chemical bond between carbon, hydrogen, and oxygen molecules. While breaking these chemical bonds, useful energy is released that results in biofuel production. MSW has significant potential to produce bioenergy resulting in heating value of ~20.57 MJ/kg (Sipra, Gao, and Sarwar 2018).

The pyrolysis technique is an efficient waste-to-energy conversion technique due to various operational and environmental benefits, flexible in handling both organic and inorganic raw materials, including municipal solid waste (MSW), biomass, plastic, and recycled waste. The key parameters

DOI: 10.1201/9781003334415-7

that influence the yield and quality of pyrolysis product include the method of pretreatment, temperature, heating rate, the type of carrier gas used for pyrolysis reaction, the type of the reactor used for pyrolysis reaction, and the type of the catalyst used for the pyrolysis reaction (Hu and Gholizadeh 2019). Thermochemical irreversible decomposition of organic matter at elevated temperature in the absence of air to produce useful products like non-condensable gases, condensable liquids generally called bio-oil and tars, and solid residue rich in carbon is defined as the pyrolysis process. Gaseous products include syngas, methane, short-chain hydrocarbon, and carbon dioxide; liquid products may contain aliphatic and aromatic compounds, phenols, aldehydes, levoglucosan, hydroxy-acetaldehyde, and water; and solid residue consists of impurities like aromatic compounds that are rich in carbon (Oyeleke, Ohunakin, and Adelekan 2021). The technical challenge that arises during pyrolysis are unsteady availability of feedstock, non-uniformity of the feedstock, pretreatment of the feedstock before pyrolysis process, the type of reactor, toxicity of the feedstock and products, and storage and stability of the products (Qureshi et al. 2020).

7.2 CATALYTIC PYROLYSIS

The bio-oil produced from pyrolysis reaction of lignocellulosic biomass contains a significant amount of reactive and oxygenated compounds which is a mixture of carbohydrates, aldehydes, ketones, organic acids, lignin fragments, aromatics, and alcohols. The presence of these compound lowers the heating value, increases viscosity, and destabilizes the fuel. Thus, the pyrolysis liquid is limited to the applications like combustion and gasification reactions. To use pyrolysis fuel directly in engines, it requires suitable upgradation technique. Different upgradation methods are steam reforming, hydro-deoxygenation, molecular distillation, esterification, and emulsification to supercritical fluids. Among the different upgradation techniques, catalytic hydro-deoxygenation is used to improve the bio-oil quality. The pyrolysis reaction taking place in the absence of any external agent is the conventional or thermal route; however, when an external agent (catalyst) is added to enhance the reaction, then it is a catalytic route. Catalyst usage in the pyrolysis reaction improves the yield, along with the quality of bio-oil. Higher cost of hydrogen limits the usage of catalytic hydrotreating over cost-effective catalytic cracking that can be conducted under atmospheric conditions for bio-oil upgradation. Catalyst presence modifies the reaction pathway by polymerization, aromatization, and alkyl condensation over normal reactions (Hu and Gholizadeh 2019; Tawalbeh et al. 2021).

The role of catalyst is to reduce the reaction temperature from 700 °C to around 400 °C, and to reduce time of feedstock retention in the reactor by improving the rate of cracking reaction that in turn increases the gas production and enhances the bio-oil quality. In comparison to the thermal pyrolysis which produces higher carbon compounds (C_5-C_{28}), catalytic pyrolysis produces quality liquid products with aromatic hydrocarbons (C_5-C_{12}) due to an enhanced decomposition reaction rate. Along with improved conversion rate even at lower temperatures, catalyst aids the removal of oxygenated compounds that leads to the higher heating value of oil products. Catalytic pyrolysis reduces the impurity concertation like nitrogen, sulfur, phosphorous, and halogens in liquid fuel (Miandad et al. 2016).

Based on the method by which the catalyst is mixed with biomass or pyrolysis vapors, catalytic pyrolysis can be classified into two methods: (i) catalyst and biomass are mixed in the reactor which is called in situ upgradation; (ii) catalyst is heated in a separate reactor known as ex situ upgradation. The vapor evolved from the reaction is allowed to diffuse through the pores for catalytic adsorption and further the catalytic cracking takes place in an in situ method. Zeolite, metal oxide, and carbon-based catalysts are widely used for the in situ method. The capital cost for the reactor setup is relatively less. However, when the catalyst and biomass are placed together, the possibility for coke formation is high and catalyst deactivates, and poor contact between catalyst and biomass reduces the heat transfer. On the other hand, the higher investment is required for the ex situ method since separate reactor for catalyst and pyrolysis and additional temperature controller assures better control of pyrolizer- and reactor-operating condition upgradation that results in higher selectivity of desirable aromatics (Norouzi et al. 2021).

Lignocellulosic biomass which is bulkier than petrochemical hydrocarbons are difficult to break down and thus the commercial catalyst having a narrow pore size won't be sufficient for the pyrolysis of biomass. Multidimensional structures like micro-, meso-, and macropores in one-, two- or three-dimension by coupling secondary levels of porosity in the commercial catalysts are most favored. The composite catalyst improves the diffusion inside the catalyst and increases the number of closely accessible active sites (Norouzi et al. 2021). The catalytic pyrolysis occurs in steps; initially, thermal cracking occurs on the catalyst surface and further the longer chain products are broken to small compounds at the interior porous structure with channels where product selectivity takes place. Catalysts with high surface area have better contact between the reactants and the catalyst assures enhanced cracking reaction rate. Smaller gaseous products are formed at the small internal pores and wax formation happens at the external catalyst sites. Catalyst influences the physical properties of the bio-oil products like lower viscosity, moisture content, pour point, and cetane index. Widely used catalysts for pyrolysis reactions are zeolite, metal oxides, silica, carbon, FCC catalysts, and clay (Miandad et al. 2016). Prior to the discussion on types of catalysts generally used for pyrolysis reaction, the mechanism of catalytic pyrolysis reaction is discussed.

7.3 MECHANISM OF CATALYTIC REACTIONS

The thermal degradation of long-chain hydrocarbons into small-chain hydrocarbons takes place at a temperature of 300–600 °C. Depending on the reaction conditions, the pyrolysis products also vary. Reactions with lower heating rate lead to char formation and that with higher heating rate volatile compounds are formed. When the reaction temperature is around 250–500 °C, depolymerization reaction rate is high, maximum bio-oil yield is achieved at a temperature of 450–550 °C, and better fragmentation with gaseous product is formed at temperature above 550 °C. Depending on the heating rate of pyrolysis and the residence time of feedstock in the reactor, the pyrolysis process is classified as slow pyrolysis, fast pyrolysis, and flash pyrolysis. Slow pyrolysis with longer residence time, and slow heating rate (0.1–2 °C/s) that occurs at 227–677 °C, may result majorly char and tar due to long residence time of gases and other processes that allows repolymerization and recombination reaction. Fast pyrolysis takes place at short residence time of 0.5–10 s, higher heating rate of 10–200 °C/s, and results primarily in liquid oil and gas at a temperature of 577–977 °C. Flash pyrolysis occurs at very high heating rate of ~2500 °C/s, short residence time of 0.1–0.5 s and moderate temperature range of 400–600 °C majorly produce liquid-phase products of higher yield. Another method of pyrolysis is microwave-assisted pyrolysis where microwave reaches the center of the biomass and transfers energy throughout the volume of the feedstock. Improved heating quality and production of value-added products by this technique do not require biomass shredding due to direct heating of feedstock, which is suitable for large-scale commercialization. Processes with low residence time and moderate temperature (350–500 °C) range ensure better biomass conversion into liquid products rather than charcoal, which is formed at low residence time and low temperature. The nature of pyrolysis product has direct influence on the reaction parameters and type of feedstock, and also depends on the type of the catalyst used (Sipra, Gao, and Sarwar 2018; Hu and Gholizadeh 2019).

Pyrolysis reaction consist of a series of reactions that include dehydration, depolymerization, isomerization, aromatization, decarboxylation, and charring. The reactions can be categorized as primary reactions consisting of char formation, depolymerization, and fragmentation. The products from the primary reaction being unstable may further undergo secondary reactions like cracking and recombination reaction (Hu and Gholizadeh 2019). In the presence of catalysts, the primary cracking reaction may take place on the external surface of catalyst where high weight molecules produce intermediates depending on the nature of the catalyst sites. Hydrogenation and dehydrogenation reactions are carried out on metallic sites, isomerization reaction is improved on acidic sites, and catalytic cracking on strong Brønsted acid site. Further, the selective cracking secondary reaction of long-chain hydrocarbons to shorter compounds takes place on the internal porous structure of the catalyst. Critical reaction steps like cracking, isomerization, and aromatization reactions are affected by the intercrystalline structure of catalyst (Oyeleke, Ohunakin, and Adelekan 2021). Desired reactions in pyrolysis like dehydration, hydrogenation, decarbonylation, decarboxylation, C—C coupling, and cracking will be enhanced while using the catalyst. The C—C bond cleavage of the polymer chain to form radicals through the initiation step, where it can take place as random scission or chain-end scission. Hydrogen abstraction and β-decomposition reaction take place through propagation by intermolecular abstraction or intramolecular abstraction (Sipra, Gao, and Sarwar 2018).

7.4 TYPES OF CATALYSTS FOR PYROLYSIS REACTION

7.4.1 Zeolite

Zeolite is a three-dimensional tetrahedral porous structure formed by aluminosilicate arrangement connected by oxygen bridges with complex pore structure consisting of channels and cavities. The active acidic sites along with porosity control the catalytic activity for pyrolysis of carbonaceous feedstocks like biomass and plastics leading to the production of light olefins over heavy fractions. Pore size, uniform pore diameter distribution, and open-pore structure with ion-exchange ability of zeolite act as a molecular sieve that has shape selectivity which influences the aromatic compound formation from pyrolysis vapors. Enhanced C—C cracking on the zeolite surface due to better pore surface area improves the thermal cracking of biomass at lower temperature. Active sites, solid acidity, shape selectivity, chemical and thermal stability of zeolite catalyst are widely used in pyrolysis reactions. Zeolite pore structure stabilizes the intermediates by adsorption and limits the repolymerization to form coke. Acidity of zeolite helps in catalytic transformation into aromatic hydrocarbons. Thus, the aromatic bio-oil selectivity is related to the interplay between the porosity and acidity (Liang, Shan, and Sun 2021).

Based on the pore size, zeolite is broadly categorized as small pore of <0.5 nm size, e.g., SAPO, A; medium pore of 0.5–0.6 nm size, e.g., ZSM-5, ZSM-11; and large pore of 0.6–0.8 nm size, e.g., Y, Beta, mordenite. Yield of the liquid and gaseous products from the pyrolysis reaction depends on the macroporous and microporous surface area of the catalyst. Large-pore-sized zeolite like Y and Beta zeolite has lower oxygenated compounds and higher coke yield. Medium-pore-sized zeolite leads to higher aromatics yield. Mesoporous zeolite catalyst improves the phenolic compounds in bio-oil and decreases the carboxylic acid and carbonyl yield. Y-zeolite with large pore size is reported to be beneficial for plastic pyrolysis resulting in aromatic hydrocarbon formation (Tawalbeh et al. 2021).

Apart from pore size, another important property of zeolite is acidity that is composed of both Brønsted and Lewis acid sites which interpret the density and strength aspects. The aromatic compound production in the bio-oil is enhanced with the acidic site concentration. The presence of Brønsted acidity originates from Si(OH)Al hydroxyl group in zeolite that results in aromatics production and Lewis acidity originates from tri-coordinated silicon defects which influence alkane production. Zeolite catalyst improves the volatile hydrocarbon production and deoxygenation of oxygenate compounds like benzenediol, phenol, cresols, etc. Zeolite shows better catalytic activity in the deoxygenation process and prevention of repolymerization of monocyclic aromatic compounds due to its shape selectivity and acidity. Acid strength, which interprets the basic probe molecule binding energy into the acidic site, is higher if Si/Al ratio in zeolite is lower, which improves the aromatization ability in zeolite catalyst. With the increase in Al content, more superficial hydrogen is required to satisfy the Al octet. Thus, a competition for superficial proton arises between the Brønsted acid sites and Al framework. With the excess proton content, acid sites become weaker, which affects the performance of zeolite (Bhoi et al. 2020). Zeolite produces higher ratio of branched carbon over straight-chain paraffin; thus, acidic zeolite catalyst is effective for plastic and biomass pyrolysis over mesoporous and amorphous catalysts. The lignin biomass containing large and bulky molecule is prone to catalyst deactivation by coke formation on the acidic zeolites due to mass transfer and diffusion limitation (Miandad et al. 2016).

Availability of high surface area results in higher contact between catalyst and feedstock that leads to higher cracking rate. The accessibility of active site depends on the porous structure of the catalyst and, thus, higher micropore structure results in higher yield of gaseous products and pore size influences the concentration of aromatics and cyclic compounds in the pyrolysis oil. The liquid products formed while using ZSM-5 has lower nitrogen, phosphorous, and sulfur compounds since the pollutant degradation may take place on the small pores of catalyst. Pyrolysis oil with better aromatic content and better specification is obtained by HZSM-5 and β-zeolite due to high acidity, pore size, and Si/Al (Li et al. 2020).

Zeolite catalyst deactivation is mainly caused by coke formation, by depositing on the catalyst surface (thermal coke) and filling the pores (catalytic coke). Cross-linking and polycondensation of biomass and secondary reaction of vapors by aldol condensation and Diels–Alder reaction lead to thermal coke formation. Catalytic coke is formed by carbonium ion chemistry and hydrogen transfer. The coke formation rate for linear plastics is slower on the zeolite catalyst surface due to Diels–Alder reaction. Coke deposition on the surface of zeolite catalyst limits its application. Excess active sites in ZSM-5 improve the oligomerization and polymerization of small molecules and generate coke. Coke formation is limited by reducing the acidity that covers the external active sites on the catalyst, metal-loaded zeolite catalyst that enhances the single-ring aromatics selectivity (Wang et al. 2020).

The addition of transition metals to zeolite improves the acidity and porosity since the presence of metal prefers decarboxylation and decarbonylation that result in more hydrogen availability for aromatic compound formation. Quality of bio-oil is improved with better yield of aromatic products, lower catalyst deactivation, and product selectivity by loading metals like Ni, Zn, Fe, Mo, and Ga on zeolite surface. Metal-loaded zeolite may function as bifunctional catalyst with activity affected by the pore size, acidity of zeolite, nature of metal species, loading content, and dispersion. The introduction of Ga species replaces the strong Brønsted acid sites with weak Lewis acid sites which reduces the intermediate cracking that results in higher aromatic selectivity by affecting the acidity and porosity. The presence of Ni species favors the oxygen elimination as CO/CO_2 that assures more hydrocarbon availability and promotes the transformation of phenols to aromatics and improves hydrothermal stability. However, excessive Ni content may block the acid site and decrease the aromatics yield. Modification of zeolite with Fe improves the acid site density that affects the yield of aromatics and light olefin. The chemical pathway of pyrolysis is modified by adding Fe species that results in aromatics like benzene and naphthalene and limits phenol production. Zn species affects the acidity of zeolite and regulates the pyrolysis product distribution by converting Brønsted acid sites to Lewis acid sites and stimulates the H-atom migration by C-H activation. Zeolite aromatization activity is affected by various Zn loading techniques (Liang, Shan, and Sun 2021).

Hydrogen-exchanged zeolite catalyst (H-ZSM) produced by calcination of zeolite catalyst allows the oxygenated products to react with Brønsted and Lewis acid sites that favor bio-oil and hydrocarbon yield. Bulk molecule cracking is difficult due to acid site distribution, microporous and homogeneous pore structure of H-ZSM that results in coke formation, short life span, and weak catalyst performance. On comparison of catalytic activity of ZSM-5 and HZSM-5, the latter favors pyrolysis of biomass due to better acidity, shape selectivity, and thermal and hydrothermal stability. Promotion of deoxygenation, decarbonylation, and decarboxylation reactions of oxygenated compounds by HZSM-5 improves the production of aromatics and reduces the formation of oxygenated lignin derivatives due to higher acidity and higher Si/Al ratio. These properties result in pyrolysis oil with better specifications like lower flash point, cetane index, and water content. Higher coke formation in β-zeolite can be related to the higher pore size. Similarly, fine-sized catalyst may lead to finer carbon formation on the catalyst surface. While using zeolite catalyst, the hydrocarbon content is slightly higher due to the cracking of char or cracking of long-chain hydrocarbon in the oil. Lower concentration of CO compared to CO_2 in the gaseous phase corresponds to better conversion of CO to CO_2 by decarboxylation reaction or oxygen radicals available from deoxygenation reaction. However, over time, the concentration of CO_2 and hydrocarbons in the gas phase is reported to reduce due to catalyst deactivation by coke deposition (Liang, Shan, and Sun 2021).

7.4.2 Metal Oxides

The metal-oxide-based catalysts are broadly classified as acidic metal oxides like Al_2O_3 and SiO_2, basic metal oxides like CaO and MgO, and transition metal oxides like ZnO, CuO, and Fe_2O_3 (Tawalbeh et al. 2021). The acidic sites in metal oxide catalysts favor the aromatization, dehydration, depolymerization, decarbonylation, and cracking reactions forming monocyclic compounds like benzene, toluene, and xylene that are stable in nature, along with CO_2 and H_2O. Bio-oil yield is higher; however, products formed are susceptible to hydrothermal deactivation and catalytic deactivation due to coke formation. Basic site catalyst favors the decarboxylation and dehydration reactions that result in CO and CO_2 formation. Basic site improves the oil stability and reduces olefin formation, and better gas production with higher H_2 yield over CO_2 (Tawalbeh et al. 2021). Metal oxide are highly active for deoxygenation reaction and affects the pyrolysis oil properties and composition. Catalyst presence improves the carbon content and oxygen-containing compounds in pyrolysis oil by higher cracking rate and deoxygenation reaction, resulting in oil with lower viscosity, density, and moisture content (Li et al. 2020).

Cheap and easily available Al_2O_3 is tested widely for pyrolysis reaction due to its high surface area. Lowering the activation energy for C—O bond compared to C—C bond in the feedstock improve the rate of dehydration, decarboxylation, aromatization, and alkylation reactions and thus produce higher rate of nitrogen- and aromatics-containing compounds in the pyrolysis oil. MoO_3 with higher tendency for reforming reaction can produce hydrogen in gas, higher aromatics, and better conversion of long-chain hydrocarbons to short-chain hydrocarbons. Presence of MoO_3 catalyst in pyrolysis reaction lowers oxygen-containing compounds in liquid and thus lowers the viscosity of the fuel due to enhanced cracking reaction (β-scission mechanism). MgO catalyst for pyrolysis reaction can convert the oxygen- and carbonyl-containing compounds by deoxygenation and cracking reaction like furans to olefins. The rate of aromatization reaction is enhanced that improves the mono-ring aromatic content in pyrolysis oil. MgO produces higher aliphatic compounds, lower pollutant emission, and reduced oxygen-containing compound. MgO produces significant ketones by ketonization reaction. CaO catalyst shows higher aromatization, deoxygenation reaction rate, and cracking rate since calcium cation neutralize the acid to form ketones and other hydrocarbons. Alkaline nature of calcium helps to react with acid, forms calcium carboxylates, and further reacts to form calcium carbonate and ketones by cracking reaction. CaO has higher tendency for deoxygenation reaction, and thus produces bio-oil with higher heating value. However, catalyst deactivation by coke formation and conversion of CaO to $CaCO_3$ may reduce the gas yield (Li et al. 2020). Recently, Tomishige et al. demonstrated Ru/CeO_2 as efficient heterogeneous catalyst at low temperature (473 K) and low H_2 pressure (2 MPa) catalytic pyrolysis of polyolefin to produce valuable chemicals, liquid fuels (C_5-C_{21}), and wax (C_{22}-C_{45}) with high yield. Selective dissociation of inner C—C bond in polyolefin without isomerization or aromatization by the specific catalyst is reported to enhance the pyrolysis performance (Li et al. 2020; Nakaji et al. 2021)

Lower cracking reaction rate and higher catalyst deactivation for metal oxide in comparison with zeolite are reported, even though aromatization reaction rate and cracking of large aromatics to light phenolic compounds are higher for metal oxides. However, catalyst activity and pyrolysis product selectivity can be significantly improved by adding secondary catalyst phase to zeolite by adding new Lewis sites on metal oxide phase. Transition metal oxides generally improves the alcohol, furan, ketone, acid, and phenol formation in bio-oil. Deoxygenation property is relatively less for metal oxides in comparison to zeolites (Norouzi et al. 2021).

7.4.3 MESOPOROUS SILICA

Mesoporous silica like MCM-41, MCM-48, MCM-50, SBA-X, HMS, and KIT-6 are widely used for different catalyst, adsorption, and sensing applications due to more open spaces, unique networks of channel, large molecule, and easy diffusion without blockage into the structure. The high surface area of 800–1400 m^2/g, meso-nanoscale pores of 2–50 nm, tunable morphology, and modification of pore size of mesoporous silica according to requirement assure wide application as catalyst. The high surface area corresponds to higher active site concentration and large pore size facilitates the mass transfer. Amorphous nature of the catalyst improves the thermal stability due to presence of silanol group on the pore wall. The mesoporous structure of silica can be effectively controlled by changing the reagent concentration, surfactant and co-polymer agent, temperature, and pH of the reaction. Mesoporous silica of 1D, 2D, or 3D configurations with ordered cubic, hexagonal, or lamellar arrangements can be prepared. Mesopores in the structure can be interconnected or isolated with similar or different size and shape. Catalyst stability can be further improved by surface modification or addition of other elements, increasing hydrophobicity or assuring the stronger silanol cross-linking groups. Si-OH groups on the surface provide limited weak acid sites and thus the cracking rate is lesser for pyrolysis reaction. Heteroatom addition to mesoporous silica results in sites with varying local environment due to the amorphous nature. Acidic properties can be modified by addition of Al into silica matrix which affects Si/Al ratio. Mg doping to silica reduces the coke production by facilitating hydrogen atom migration through C-H activation at the active site. La addition promotes the reaction efficiency due to enhanced dehydration and demethylation reaction of the oxygenated compound (Norouzi et al. 2021).

7.4.4 CARBONACEOUS MATERIAL

Carbonaceous materials derived from biomass has high porosity, high specific surface area, chemical inertness, and electron conductivity that act as good catalyst support. Biochar produced from the pyrolysis reaction can be reused as catalyst or catalyst support due to their high surface area. Carbon can be used as catalyst support by chemically activated or dispersed with metallic particles. Acid and base resistance, control over pore size for specific reaction, high structural stability, less expensive, control over hydrophobicity, recovery of active metals by burning carbon are the advantages of carbonaceous catalyst over other types of catalysts (Norouzi et al. 2021).

Carbon nanotubes (CNT) that are stronger with high thermal and electrical stability is a potential mechanical support. CNT can be single-walled, double-walled, or multiwalled, which provides strong intermolecular bond and van der Waals force between the carbon rings. Activated carbon derived from biomass has better adsorption and chemical reaction due to the pores. Both physical and chemical activation lead to carboxylic groups in pyrolysis products. Metals or other compounds are used to modify the acid sites on the activated carbon. Other forms of carbon like graphene and graphene oxide are widely used for pyrolysis reactions (Norouzi et al. 2021).

7.4.5 OTHER CATALYSTS

FCC catalyst is a form of silica-alumina catalyst. FCC catalyst has strong acidic sites, bonded by a non-zeolite with zeolite crystal structure with Lewis and Brønsted acid sites using ionizable hydrogen atoms that act as electron acceptors. Spent FCC catalysts that are used in cracking of heavy oil in petroleum refineries can be reused for pyrolysis reactions. On increasing the catalyst weight to polymer may affect the liquid oil production. High surface area, large pore size, good thermal stability, and low coke formation aid the formation of low carbon products from the reaction (Miandad et al. 2016).

The amorphous catalyst of silica-alumina consisting of Brønsted acid and Lewis acid sites affects the pyrolysis reaction. Better bio-oil yield corresponds to the lower acidic sites of the catalyst (Miandad et al. 2016).

Clay-based catalysts of montmorillonites and their analogies like beidellite, saponite, and hectorite outperform zeolite catalyst at high temperatures. Mild acidity which limits the over-crack and secondary reaction by hydrogen transfer results in high-molecular-weight hydrocarbons in bio-oil. Two-dimensional network of interconnected micropores is formed by layer structure in clay that helps in tailoring porous network by intercalating sheets with pillars. Brønsted acid is present in clay lattice hydroxyls and Lewis acid site is present in the pillars of clay material (Peng et al. 2022). Different types of catalysts generally used for pyrolysis reaction and their advantages and disadvantages are provided in Table 7.1.

7.5 CHALLENGES IN CATALYST

Deactivation of the active sites by coke formation reduces the aromatic yield by impeding the diffusion and capillary flow of the compounds. Depending on the catalyst topography and textural characteristics, coke formation takes place on the acidic sites available in the external surface of the catalyst. Polymerization of oxygenated intermediates results in thermal coke and accumulates on the external catalyst surface. Catalytic coke is thus obtained due to polyaromatic hydrocarbon formation and accumulates mostly on the catalyst channels. During pyrolysis, catalytic coke deposition results in catalyst deactivation that in turn reduces the aromatic hydrocarbon formation. When the catalyst used for in situ method has strong acidic sites with high density, coke formation reduces the catalyst life span and necessitates constant replacement, which in turn increases the operational cost. Diffusion of polymeric molecules on the catalyst pores limits the catalyst recovery. Coke deposition can be reduced by two methods: (i) access internal acidic sites by enhanced oxygenate diffusion since pore mouth poising initiates the deactivation and (ii) co-feed biomass with hydrogen-rich materials. Spent catalyst can be regenerated by heating at higher temperature in the presence of air to restore the catalytic activity (Liang, Shan, and Sun 2021).

Zeolite catalyst is a promising pyrolysis catalyst to produce fuel with low-oxygen content. However, coke formation or water attack and rapid deactivation, along with lower bio-oil yield and carbon deficiency, and ambiguous catalytic conversion mechanism limit its commercial application. The catalytic pyrolysis of biomass is associated with the water generation. When the zeolite catalyst comes in contact with hot water or other liquids, the crystalline framework collapses at a temperature of 150 °C. Coke deposition on zeolite channel leads to acid site blockage, which in turn reduces the aromatic hydrocarbon production. Co-pyrolysis of biomass with appropriate waste like plastic and tire waste at suitable conditions may possibly overcome the limitation of zeolite catalyst. To an extent, incorporation of metals in zeolite can improve the hydrothermal stability of zeolite (Liang, Shan, and Sun 2021; Hu and Gholizadeh 2019).

Precise design of efficient catalyst that can withstand acceptable working hours with suitable performance is the key challenge for catalytic pyrolysis reaction. Fundamental understanding of biomass degradation chemistry with degradation kinetics is essential to understand the catalyst activity. Reactivity of the catalyst and product selectivity are dependent on the pore size of the catalyst. The increase in pore size improves the hydrocarbon cracking; however, larger pore size tends to have higher activity initially and deactivates soon. Small pore size reduces coke and carbonaceous deposition. Optimum pore size is essential to achieve a specific activity.

7.6 CONCLUSION

Pyrolysis is an efficient and environment-friendly technique for waste-to-energy conversion. Pyrolysis is a thermochemical conversion of biomass/plastic into fuel-like liquid, gas, and solid residue under inert atmosphere at elevated temperature. The chapter elaborates the role of catalyst for pyrolysis of waste. Pyrolysis product with better quality and quantity is assured in the presence of catalyst due to enhanced cracking rate at lower temperature and lower residence time. Catalyst can be fed to the pyrolysis reactor by either in situ method or ex situ method depending on the product quality

TABLE 7.1
Types of Catalyst Used for Pyrolysis and Their Benefits

Classification	Catalyst Name	Advantages	Disadvantages
Zeolite	ZSM-5	High surface area Better modification by metals Liquid products with lower nitrogen, phosphorous, and sulfur compounds	High coke formation Lower aromatic selectivity
	HZSM-5	High deoxygenation rate and promotes aromatization	High coke formation
	Y-Zeolite	High acidity High cracking rate High hydrocarbon yield	High coke formation Lower bio-oil yield
	β-Zeolite	High rate of deoxygenation	Highest coke formation
Metal oxide	Al_2O_3	Easy availability High rate of cracking Pyrolysis oil contains nitrogen and aromatics compounds	Quick deactivation
	MoO_3^-	Higher reforming reaction Lower viscosity of oil	Not suitable for aromatic compounds
	MgO	Higher aromatization rate and better ketonization reaction	High coke formation
	CaO	High deoxygenation rate	High coke formation Lower gas yield
Mesoporous silica	MCM-41	High cracking rate High mono-ring aromatics yield	High coke formation
Carbonaceous material	CNT	High thermal and electrical stability	
FCC		High Si/Al ratio Low coke formation	

required and control over reaction. Zeolite, metal oxides, silica, carbon, FCC catalysts, and clay are widely used catalysts for pyrolysis reaction. Apart from nature of the catalyst, pyrolysis product yield is affected upon the reaction parameters like temperature, heating rate, and residence time. Zeolite is one of the widely used catalysts for pyrolysis reactions. Shape selectivity, open-pore structure, uniform pore size distribution, high surface area, and higher acid site concentration of zeolite catalyst correspond to higher aromatic hydrocarbon formation. Zeolite catalyst performance is further improved by adding metals of specific concentration or hydrogen-exchanged zeolite. Metal oxides are another most widely used catalyst due to the acidic sites and better deoxygenation reaction rate. Other types of catalysts are mesoporous silica, carbonaceous materials, FCC, silica-alumina, and clay. Major challenges in all type of catalysts include deactivation by coke formation, hydrothermal instability, and limited active site availability. In order to design the most suitable catalyst for the feedstock, detailed understanding about the reaction mechanism and kinetics is essential.

REFERENCES

Beyene, Hayelom Dargo, Adhena Ayaliew Werkneh, and Tekilt Gebregergs Ambaye. 2018. "Current Updates on Waste to Energy (WtE) Technologies: A Review." *Renewable Energy Focus* 24 (March): 1–11. https://doi.org/10.1016/j.ref.2017.11.001.

Bhoi, P. R., A. S. Ouedraogo, V. Soloiu, and R. Quirino. 2020. "Recent Advances on Catalysts for Improving Hydrocarbon Compounds in Bio-Oil of Biomass Catalytic Pyrolysis." *Renewable and Sustainable Energy Reviews* 121 (January): 109676. https://doi.org/10.1016/j.rser.2019.109676.

"Electricity Market Report, July 2021." 2021. https://doi.org/10.1787/f4044a30-en; https://www.oecd-ilibrary.org/energy/electricity-market-report-july-2021_f4044a30-en.

Hu, Xun, and Mortaza Gholizadeh. 2019. "Biomass Pyrolysis: A Review of the Process Development and Challenges from Initial Researches up to the Commercialisation Stage." *Journal of Energy Chemistry* 39: 109–43. https://doi.org/10.1016/j.jechem.2019.01.024.

Kaza, Silpa, Lisa Yao, Perinaz Bhada-Tata, and Frank Van Woerden. 2018. "What a Waste 2.0 A Global Snapshot of Solid Waste Management to 2050." https://www.wacaprogram.org/sites/waca/files/knowdoc/WhatAWaste2.0.pdf.

Li, Qingyin, Ali Faramarzi, Shu Zhang, Yi Wang, Xun Hu, and Mortaza Gholizadeh. 2020. "Progress in Catalytic Pyrolysis of Municipal Solid Waste." *Energy Conversion and Management* 226 (October): 113525. https://doi.org/10.1016/j.enconman.2020.113525.

Liang, Jie, Guangcun Shan, and Yifei Sun. 2021. "Catalytic Fast Pyrolysis of Lignocellulosic Biomass: Critical Role of Zeolite Catalysts." *Renewable and Sustainable Energy Reviews* 139 (March 2020): 110707. https://doi.org/10.1016/j.rser.2021.110707.

Miandad, R., M. A. Barakat, Asad S. Aburiazaiza, M. Rehan, and A. S. Nizami. 2016. "Catalytic Pyrolysis of Plastic Waste : A Review." *Process Safety and Environmental Protection* 102: 822–38. https://doi.org/10.1016/j.psep.2016.06.022.

Nakaji, Yosuke, Masazumi Tamura, Shuhei Miyaoka, Shogo Kumagai, Mifumi Tanji, Yoshinao Nakagawa, Toshiaki Yoshioka, and Keiichi Tomishige. 2021. "Low-Temperature Catalytic Upgrading of Waste Polyolefinic Plastics into Liquid Fuels and Waxes." *Applied Catalysis B: Environmental* 285 (December 2020): 119805. https://doi.org/10.1016/j.apcatb.2020.119805.

Norouzi, Omid, Somayeh Taghavi, Precious Arku, Sajedeh Jafarian, Michela Signoretto, and Animesh Dutta. 2021. "What Is the Best Catalyst for Biomass Pyrolysis?" *Journal of Analytical and Applied Pyrolysis* 158 (August): 105280. https://doi.org/10.1016/j.jaap.2021.105280.

Oyeleke, Oyetunji O., Olayinka S. Ohunakin, and Damola S. Adelekan. 2021. "Catalytic Pyrolysis in Waste to Energy Recovery Applications: A Review." *IOP Conference Series: Materials Science and Engineering* 1107 (1): 012226. https://doi.org/10.1088/1757-899x/1107/1/012226.

Peng, Yujie, Yunpu Wang, Linyao Ke, Leilei Dai, Qiuhao Wu, Kirk Cobb, Yuan Zeng, Rongge Zou, Yuhuan Liu, and Roger Ruan. 2022. "A Review on Catalytic Pyrolysis of Plastic Wastes to High-Value Products." *Energy Conversion and Management* 254 (December 2021): 115243. https://doi.org/10.1016/j.enconman.2022.115243.

Qureshi, Muhammad Saad, Anja Oasmaa, Hanna Pihkola, Ivan Deviatkin, Anna Tenhunen, Juha Mannila, Hannu Minkkinen, Maija Pohjakallio, and Jutta Laine-Ylijoki. 2020. "Pyrolysis of Plastic Waste: Opportunities and Challenges." *Journal of Analytical and Applied Pyrolysis* 152 (March). https://doi.org/10.1016/j.jaap.2020.104804.

Sipra, Ayesha Tariq, Ningbo Gao, and Haris Sarwar. 2018. "Municipal Solid Waste (MSW) Pyrolysis for Bio-Fuel Production: A Review of Effects of MSW Components and Catalysts." *Fuel Processing Technology* 175 (April): 131–47. https://doi.org/10.1016/j.fuproc.2018.02.012.

Tawalbeh, Muhammad, Amani Al-Othman, Tareq Salamah, Malek Alkasrawi, Remston Martis, and Ziad Abu El-Rub. 2021. "A Critical Review on Metal-Based Catalysts Used in the Pyrolysis of Lignocellulosic Biomass Materials." *Journal of Environmental Management* 299 (August): 113597. https://doi.org/10.1016/j.jenvman.2021.113597.

Wang, Guanyu, Yujie Dai, Haiping Yang, Qingang Xiong, Kaige Wang, Jinsong Zhou, Yunchao Li, and Shurong Wang. 2020. "A Review of Recent Advances in Biomass Pyrolysis." *Energy and Fuels* 34 (12): 15557–78. https://doi.org/10.1021/acs.energyfuels.0c03107.

8 Engineering Perspectives on the Application of Photosynthetic Algal Microbial Fuel Cells for Simultaneous Wastewater Remediation and Bioelectricity Generation

Baishali Dey, Nageshwari Krishnamoorthy, Rayanee Chaudhuri, Alisha Zaffer, Sivaraman Jayaraman, and Balasubramanian Paramasivan

CONTENTS

- 8.1 Introduction 103
- 8.2 Overview of Research Activity on Microbial and Photosynthetic Microbial Fuel Cells 105
- 8.3 Important Factors Influencing the Functioning of a PAMFC 107
    - 8.3.1 pH 108
    - 8.3.2 Temperature 108
    - 8.3.3 Light Intensity and Artificial Illumination 108
    - 8.3.4 Carbon Dioxide Concentration 109
    - 8.3.5 Oxygen Concentration 109
    - 8.3.6 Substrate Type 110
    - 8.3.7 Design Configuration 110
- 8.4 Most Commonly Used Algal and Bacterial Culture 112
- 8.5 Harvesting of Algae 113
- 8.6 Application of the Integrated Systems 114
- 8.7 Research Gap 115
- 8.8 Scope for Future Research 115
- References 116

8.1 INTRODUCTION

Renewable bioresources like solar, water, and wind hold immense potential, but their inefficient utilization along with high operational and installation costs is a significant drawback in harnessing them at full capacity. To overcome the above-mentioned drawbacks and improve the overall process yields, low-cost microbial fuel cells (MFCs) and photosynthetic microbial fuel cells (PMFC) were sought after. However, MFCs are often crippled with problems like CO_2 discharge to the environment and energy intensity due to mechanical aeration for oxygen supply, constituting 50% additional cost. Integrating this technology with algae can tackle the bottlenecks, as they consume

CO$_2$ during photosynthesis (reducing carbon footprint) and release oxygen (acting as an electron acceptor) which can help increase the power output. Figure 8.1 represents the working of a typical photosynthetic algal microbial fuel cell (PAMFC). Such algae-assisted MFCs were also reported to generate more power density compared to stand-alone MFCs (Zhang et al., 2019).

The fundamental mechanism of a PAMFC process is that the bacteria present in the anodic chamber utilize the organic matter of wastewater in an anoxic environment and release CO$_2$, electrons, and hydrogen ions (Kannan & Donnellan, 2021). The electrons pass through an external circuit to reach the cathodic chamber, where the algae take up sunlight and CO$_2$ (from anodic chamber) to photosynthesize and release oxygen, while generating biomass (Arun et al., 2020). This released oxygen acts as an electron acceptor for the generation of bioelectricity. Specifically, microalgae have great potential to be employed for PAMFC as they can utilize nitrogen and phosphate from wastewater, assisting in pollutant removal and thereby eutrophication (Arun et al., 2022). In addition, the algae can be harvested after each cycle to extract pigments and lipid for biofuel production. The biomass can also be used as substrate in the anode for subsequent processes to stimulate sustainability (Shukla & Kumar, 2018). Various factors like pH, temperature variations, substrate type, CO$_2$ and oxygen concentration, light intensity, and illumination devices affect the working of a PAMFC, thus affecting the power output. These factors need to be optimized for sustainable and efficient bioelectricity generation (Reddy et al., 2019).

Equations 8.1–8.3 are the biochemical reactions that take place in an electrochemical cell (He et al., 2014):

Anodic chamber:

$$C_6H_{12}O_6 + 6H_2O \rightarrow 6CO_2 + 24H^+ + 24e^- \tag{8.1}$$

Cathodic chamber:

$$nCO_2 + nH_2O \rightarrow \text{Algae} + \text{light}(CH_2O)(\text{Biomass}) + nO_2 \tag{8.2}$$

$$O_2 + 4H^+ + 4e^- \rightarrow 2H_2O \tag{8.3}$$

Researchers have conducted lab-scale studies using microalgae as a carbon source in the fuel cells, which solely focused on the metabolic role of algae in different compartments and the production of bioelectricity (Enamala et al., 2020). Addition of substrate to the photosynthetic and bacterial

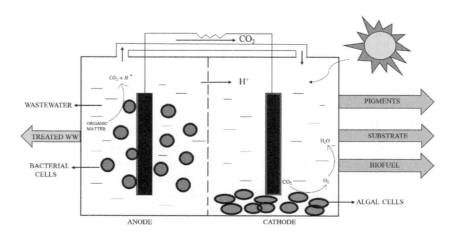

FIGURE 8.1 Schematic representation of a typical PAMFC setup.

components like glucose, acetate, and wastewater from different industries have been observed to enhance bioelectricity production (Arun et al., 2020; Eom et al., 2020). Especially, acetate and domestic wastewater have shown prominent results as substrate source in this aspect (Bhande et al., 2019; Firdous et al., 2018; Pant et al., 2010). However, wastewater treatment domain is not completely explored due to various limitations concerning efficiency, cost, etc., which has further prevented the scale-up of the technology. The knowledge gaps in simultaneous wastewater technology and production of value-added products have also limited the real-time use of this technology. The applications and various value-added products that can be obtained from an PAMFC system is showed in Figure 8.2.

The aim of this chapter is to analyze the research interest and trends of MFCs and its advancements since the very commencement of research in this area. Second, the chapter reviews the research activity conducted in the last 5 years to develop an understanding of the significant factors affecting the power density output in an PAMFC. Third, it throws light on the cost-affecting factors and the application of this integrated system in developing a zero-carbon output, sustainable wastewater treatment technology. In addition, the chapter deals with the efficiency of the PAMFC to produce electricity along with other benefits using algae in such bioeletrochemical systems. Lastly, the chapter addresses the setbacks in scaling up and commercialization of the technology and the future scope in developing into an efficient system.

8.2 OVERVIEW OF RESEARCH ACTIVITY ON MICROBIAL AND PHOTOSYNTHETIC MICROBIAL FUEL CELLS

The generation of electricity using bacteria was reported in the early twentieth century (Apollon et al., 2021). However, the number of publications on using algae as a biocathode material for simultaneous wastewater management and bioelectricity generation has increased only in the last 10 years. In order to get a comparative preview, the number of hits obtained from different online databases like Google Scholar, Web of Science (WOS), and Scopus, from 2001 to 2022, using various keyword combinations like "Microbial Fuel Cell," "Photosynthetic Microbial Fuel Cell," "Algal Microbial Fuel Cell," and "Bacterio-Algal Fuel Cell" are shown in Table 8.1. The result showed that "microbial fuel cell" got the highest number of hits from all the three databases, whereas photosynthetic/algal fuel cell was relatively a less explored area. To get a more detailed view of the research work conducted on the microbial fuel cells containing algae, all the bibliographical data collected using the previously mentioned keywords were fed to a scientometric analysis software, called Citespace 5.8.R3. In Figure 8.3, the number of published articles on microbial fuel cell was

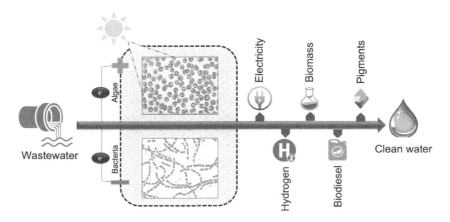

FIGURE 8.2 Various value-added products that can be obtained from an PAMFC system.

TABLE 8.1
Number of Publication Found with Different Keyword Combinations from Different Online Repositories from 2001 to 2022

Keywords Used	Number of Hits Obtained from Different Online Repositories		
Google Scholar		Scopus	Web of Science
Microbial fuel cell	21,000	12,268	12,216
Photosynthetic microbial fuel cell	5,380	242	323
Algal microbial fuel cell	4,800	177	229
Algal fuel cell	37	688	659
Photosynthetic algal microbial fuel cell	283	51	53
Bacterio-algal fuel cell	4	1	1

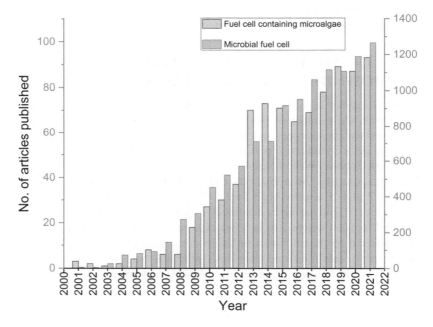

FIGURE 8.3 Number of articles published in the area of microbial fuel cell and microbial fuel cell containing microalgae in the last two decades.

compared with the number of articles on the microalgal fuel cell published over the last two decades. The result depicts that even though articles focusing on the microalgal fuel cell were less, it showed a steady growth over the years. Here it is worth mentioning that especially in the last decade, the rapid growth in the fuel cell containing photosynthetic microorganisms has gained more focus due to the realization of the advantages offered over the bacterial ones. Further, bibliographical analysis on the microbial fuel cell containing algae showed that 68% of the published documents are scientific research articles and 17% are review papers (Figure 8.4a). The subject distribution of the data set showed that 21% of the published documents fall under the area of "Environmental science," whereas other areas like "Energy," "Chemical Engineering," and "Biochemistry, Genetics

Application of Photosynthetic Algal Microbial Fuel Cells 107

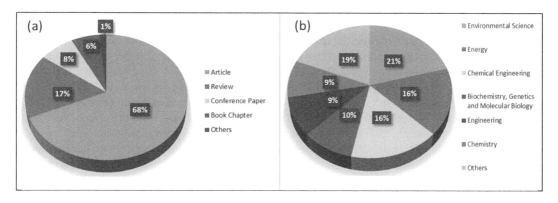

FIGURE 8.4 Distribution of the different types of published documents (a) in different areas and (b) in the field of algal microbial fuel cell.

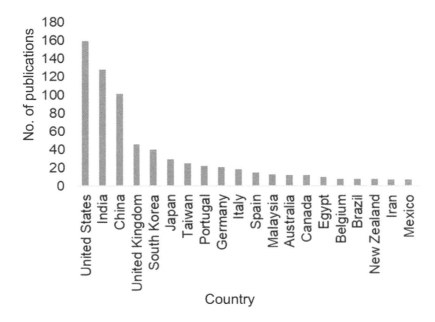

FIGURE 8.5 Top 20 contributing countries/territories in the field of algal fuel cell.

and Molecular Biology" contain 16%, 16%, and 10% of the total published documents, respectively (Figure 8.4b). This means that the latter aspects of the systems are yet to be explored. Among the different contributing countries in this field, the United States holds the first position followed by India and China (Figure 8.5). Apart from that, other Asian countries like South Korea, Japan, Taiwan, and Malaysia also showed significant contribution to this field.

8.3 IMPORTANT FACTORS INFLUENCING THE FUNCTIONING OF A PAMFC

For the efficient working of a PAMFC and maximum power output, various factors like pH, temperature, light intensity and artificial illumination devices, carbon dioxide and oxygen concentration, substrate type, and design configuration need to be studied closely.

8.3.1 pH

pH plays a crucial role in regulating the metabolic pathways taking place inside algae and bacteria, which aids in bioelectricity generation (Zhang et al., 2019). Microbes tend to perform optimally at low ion concentration and neutral pH. Thus, pH at anode should ideally vary between 6 and 7 and the pH at cathode can be neutral or slightly alkaline to be favorable for algal growth. Also, this condition is inevitable due to oxygen reduction (Arun et al., 2020). Even in single-chambered MFCs or with air-cathodes, both the rise and fall in pH affects the efficiency. Low pH affects the membrane efficiency (Saba et al., 2017), but a pH of 9.5 showed to generate a maximum power density of 0.66 W/m^3 (Kusmayadi et al., 2020). CO_2 generated by bacterial cells in an anode compartment can interact with water and form H_2CO_3, which can lower the pH. This issue can be prevented by inoculating algae at a high initial concentration for simultaneous CO_2 assimilation (Zhang et al., 2014). Another reason for the acidification of the anodic chamber may be due to the production of fatty acids by non-electrogenic organisms under an anaerobic environment (Zhang et al., 2019). Maintenance of pH can be brought about by buffers like MES (2[N-morpholino]ethane sulfonate), HEPES(4(2-hydroxyethyl)1-piperazine ethane sulfonic acid), PIPES (piperazine-N,N'-bis[2-ethane sulfonate]), caustic soda, carbonate, and zwitterionic buffers. For large-scale applications, carbonate buffers are most economical (Shukla & Kumar, 2018), but they can also accelerate the growth of methanogens and are non-biodegradable, resulting in eutrophication when released into waterbodies (Wang et al., 2018). However, zwitterionic buffers are stable, non-toxic, and do not adversely affect the metabolic pathways of the microorganisms (Kusmayadi et al., 2020). Ideally, algae in the cathode should utilize protons and electrons and generate water and oxygen. However, when non-noble carbon electrodes are used for being more economical, oxygen reduction in the cathode results in the production of hydroxyl ions. This observation was confirmed by Wang et al. (2018) that anode acidification can be balanced by the production of hydroxyl ions in the cathode area without the addition of buffers.

8.3.2 Temperature

Shukla and Kumar (2018) reported 35 °C as the optimum working temperature, which may vary with the organism used. Temperature and dissolved oxygen (DO) concentration are directly interdependent; as a result, an increase in temperature results in an increase in DO. Seasonal variations in temperature also affect the algal efficiency due to variation in photosynthetic activity (Saba et al., 2017). Higher operating temperature resulted in improved reaction kinetics and higher power density along with better nutrient removal due to faster rates of biochemical reaction. Higher reaction rates are because of rapid substrate absorption (Kusmayadi et al., 2020) and faster movement of electrons (Inglesby et al., 2013). Increased membrane conductivity was observed at higher temperatures ranging between 20 and 60 °C with proton-exchange membranes (Pérez-Page & Pérez-Herranz, 2011). However, extensively high temperature decreases the membrane conductivity due to water deficit in the membrane, as membrane hydration is an important factor in determining its durability and performance. Dehydrated membranes show higher ionic resistance and can even be damaged irreversibly (Shukla & Kumar, 2018). It was observed that with the temperature increase from 25 °C to 30 °C, the maximum power density increased somewhat from 1811.9 mW/m^3 to 2072.1 mW/m^3, but higher temperatures (38 °C) also lead to cell death (He et al., 2014).

8.3.3 Light Intensity and Artificial Illumination

In a photobioreactor, the most influencing factor affecting the growth of algal biomass is found to be light intensity and spectral composition (Greenman et al., 2019). Major emphasis has been laid on the light source and intensity as it impacts the chlorophyll content, stomatal opening, and photosynthesis. When light intensity varies between 3500 and 10,000 lux, algal biomass concentration increases gradually, but beyond this range, the PAMFC becomes saturated (Arun et al., 2020). The cultivation of biomass in the presence of substrate and light can be seen in Equation 8.4:

$$\text{Organics+ light} \rightarrow \text{Biomass+ Oxygen} \qquad (8.4)$$

Biomass and oxygen are produced during the light period and the algae consume the oxygen produced and reduce the organic matter during the dark period. Further, it was found that red light with high intensity (900 lux) gave better power output as compared to blue light with low intensity (100–600 lux) (Jaiswal et al., 2020; Mekuto et al., 2020). This occurs due to the high absorption of light energy by the photosynthetic system of algae at this wavelength. Jaiswal et al. (2020) further reported that continuous supply of light intensity is better than batch-mode operation. On the downside, the light intensity is a crucial parameter that needs to be monitored closely as low light can limit the growth of algae adversely affecting the power output. Also, high algal density can limit the penetration of light at deeper levels. High light intensity can slow down the photosynthesis rate, a phenomenon called photoinhibition. Thus reactor, designing needs to be done strategically to provide a maximum surface area (Lee et al., 2015). They have further cited that light intensity has an indirect effect on resistance in the fuel cell as well. DO increases with an increased rate of photosynthesis, which resulted in an increase of cathodic resistance.

In some cases, the anode department is covered to prevent the growth of algae along the anode (Mekuto et al., 2020). This system had a higher voltage and power density compared to uncovered anodic chamber (Jaiswal et al., 2020). The performance and efficiency of MFCs were enhanced when LED was used as a source of light as compared to conventional light sources such as compact fluorescent light (Arun et al., 2020). When an artificial illumination device was used, lower power (6–12 W compared to 12–18 W) generated higher power output. It is further reported that cathodic resistance could be decreased, and power density can be increased by optimizing the duration of the light and dark cycle (Lee et al., 2015).

8.3.4 Carbon Dioxide Concentration

As shown in Equation 8.1, the anaerobic breakdown of organic substances by anoxic bacteria leads to the formation of CO_2. In a typical microbial fuel cell, this CO_2 gets released into the environment, which leaves a significant carbon footprint. But in a photosynthetic microbial fuel cell, this CO_2 can be directed to the cathode chamber, using an external pipe (Liu et al., 2015), where the algae can utilize it during photosynthesis and help create a zero-carbon footprint sustainable system. As a consequence, this in situ carbon sequestration by algae acts as a carbon sink and aids in cleaning up the environment (Arun et al., 2020). Sparging CO_2 consumes extra energy and bears additional costs (Liu et al., 2015). It was also found by a group of researchers that in a double-chambered PAMFC, some concentration of the CO_2 passes to the cathodic side through the Nafion membrane present in-between. The result of growing heterotrophic microorganisms alongside autotrophic algae is the 100% assimilation of CO_2 from the anode (Saba et al., 2017). Thus, CO_2 is also an essential influencing parameter in the generation of bioelectricity and any changes in its rate or concentration can affect oxygen generation, which will ultimately affect the power output. In the initial stages of the wastewater treatment and bioelectricity generation cycle, low algal biomass concentration can lead to the dissolution of CO_2 into the water to produce H_2CO_3 which can decrease the pH. To prevent this from occurring, the initial algal biomass inoculum concentration should be high (Saba et al., 2017). At higher CO_2 concentrations, greater biomass density is achieved and algae produce 6% more lipid content, leading to a better quantity of biodiesel (Mehrabadi et al., 2016).

8.3.5 Oxygen Concentration

The anodic chamber needs to maintain an anaerobic environment for the anoxic breakdown of the substrate, whereas the cathodic chamber needs to be oxygenated as oxygen acts as the electron acceptor, necessary for bioelectricity generation (Shukla & Kumar, 2018). Oxygen can be supplied either through mechanical aeration by sparging air or pure oxygen or with the help of algae at the cathode. By applying a catalyst at the cathode, oxygen reduction can be enhanced. Other methods include adding potassium ferricyanide, rotating the electrode, or using an air-cathode (Saba et al., 2017). Thus, one of the major drawbacks of single-chambered algal MFCs is that during the process, the back diffusion of oxygen from the cathodic to the anodic side occurs. This situation can be fatal for obligate anaerobes or can lead to change in metabolic pathways in facultative anaerobes or can change the mixed microbial consortium (Saba et al., 2017). For example, Arun et al. (2020) cited an observation where a 53.4% decrease in power density was observed when the concentration of dissolved oxygen was increased from 7.8 mg/L to 9.5 mg/L in the cathodic chamber. During the beginning of each cycle, the power density falls as the initial concentration of oxygen released by the algal biomass is low, leading to a shortage of electron acceptor. But as the algal biomass density increases, power output improves due to an increase in oxygen concentration (Zhang et al., 2019). However, thick biofilm formation on the reactor surface can limit the movement of oxygen (Saba et al., 2017).

8.3.6 Substrate Type

Microorganisms breakdown the organic substrates in an anoxic environment in the anodic chamber to release electrons, protons, and CO_2 and reduce the overall chemical oxygen demand (COD) (Saba et al., 2017). In the cathode compartment, algae can be grown either in BG11 medium or in a wastewater source containing a high concentration of nitrogen and phosphorus or in a combination of both. PAMFCs can use a variety of substrates ranging from pure to complex ones, including wastewater. The easily digestible substrates, such as glucose, starch, molasses, volatile fatty acids, etc., have been extensively used in MFCs, but they are quite expensive (Saba et al., 2017). Moreover, easily degradable substances cannot sustain long-term bioelectricity generation and hence long-release substrates should be explored that will help in self-sustained bioelectricity production (Qi et al., 2018). Besides supplying oxygen in the cathode, algae can also be used as substrate at anode in the form of either live cells or powdered form or as dead biomass or pretreated biomass or after extraction of lipids (Shukla & Kumar, 2018), which will reduce the cost sufficiently and create a closed-loop system (Mekuto et al., 2020). As this setup is being tested for wastewater remediation with simultaneous bioelectricity generation, various wastewater sources have also been used as substrates like swine wastewater (Zhang et al., 2019), kitchen wastewater (Naina Mohamed et al., 2020), anaerobic sludge (Ma et al., 2017), paper recycling wastewater (Radha & Kanmani, 2017), food-based wastewater (Saba et al., 2017), dye wastewater (Enamala et al., 2020), brewery wastewater (Harewood et al., 2017), landfill leachate (Hernández-Flores et al., 2017), fermentation effluents (Dai et al., 2021), dairy wastewater (Choudhury et al., 2021), oil refinery wastewater (Ng et al., 2021), pharmaceutical wastewater (Nayak & Ghosh, 2019) etc. Before being used as substrates, sometimes these wastewater sources might need to be pretreated with methods like ultrasonication, heat treatment, alkali treatment, and microwave treatment in order to break down the more complex substances present for easier hydrolysis by bacteria (Shukla & Kumar, 2018). Some researchers are also trying to explore the potential of *Arthrospira maxima* as a substrate (Qi et al., 2018). Substrate feeding rate or the growth-limiting substance can act as the rate-limiting step (Greenman et al., 2019) as these factors determine the morphology of biofilm and bacterial growth (Kusmayadi et al., 2020). It was also reported by Nam et al. (2010) that an increase in substrate loading rate increased electricity production. Sometimes a mixture of substrates or wastewater sources with carbon sources like glucose (Enamala et al., 2020; Li et al., 2019) have also been tested for better power output. Also, the microbial population's assembly and activity are determined by the type of substrate used (Mekuto et al., 2020).

8.3.7 Design Configuration

PAMFCs resemble MFCs in many ways and can be single-chambered or double-chambered. In single-chambered PAMFCs (Figure 8.6), both electrodes are present in the same compartment, with

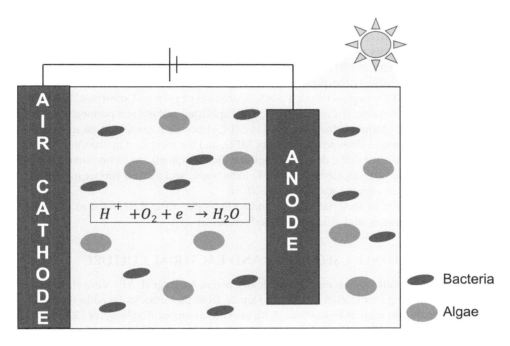

FIGURE 8.6 Diagrammatic representation of a typical single-chambered photosynthetic microbial fuel cell.

an air-cathode and no added oxygen supply unit. Algae and bacteria reside in the same chamber and grow synergistically (Saba et al., 2017). Thus, they can be easily scaled up due to simple design, low operational costs (no membrane required), and maintenance. But one major drawback is that pH changes occur simultaneously in both the cathode and anode regions and back diffusion of oxygen from cathodic to anodic region can affect power output. In a dual-chambered PAMFC, the electrodes are present in two separate chambers connected with an external circuit and separated with a membrane, which is most commonly a proton-exchange membrane (PEM) like Nafion. The light source is usually kept on the algal side and in some cases, the anode side maybe covered to prevent algal growth and avoid disturbance to the anoxic bacteria (Saba et al., 2017). Sometimes a photobioreactor might be connected to the cathode for easier maintenance and supply of algae. H-shaped reactors have also been designed, but narrow membrane area and low ion-exchange resulted in low power outputs. The major drawbacks of dual-chambered cells are membrane crossover, high internal resistance, and membrane fouling (Saba et al., 2017). The construction material of these fuel cells varies from plastic to acrylic to glass for easier penetration of light, promoting algal growth.

Electrode material plays an important role in the designing of a PMAFC. The material selection is based on cost, surface area, biocompatibility, anticorrosiveness, stability, and conductivity (Shukla & Kumar, 2018). It can be of two types: bio-electrodes, and chemical electrodes. Most common electrode materials are carbon cloth (Wang et al., 2018), stainless steel mesh (Bazdar et al., 2018), platinum, copper oxide (Liu et al., 2015), titanium (Kusmayadi et al., 2020), copper, aluminum, zinc, Ti-TiO$_2$ electrode (Taskan et al., 2014), aluminum–graphite (AL-C), iron–graphite (Fe-C), and copper–graphite (Cu-C) (Hou et al., 2016). But carbon electrodes find their usage in most large-scale applications due to their non-toxicity and low cost (Shukla & Kumar, 2018). They also help in forming stable bacterial biofilms (Saba et al., 2017). To prevent thick biofilm formation on the electrode, it is advised to make the electrode surface smooth rather than rough. In most studies, anode and cathode materials are the same, but in some studies, the electrodes are made of different materials to support biofilm formation in one and not the other. The distance between two electrodes needs to be spaced with careful consideration as it is crucial for power output. Placing the electrodes too close to each other results in decreased power density (Arun et al., 2020). Due to high tensile strength, ductility, and conductivity, copper wire is more frequently used than titanium.

The ion-exchange membrane can be of three types, namely, cation-exchange membrane (CEM), anion-exchange membrane (AEM), and bipolar membrane (BPM), based on the ionic groups attached. This, in turn, affects the passage of molecules from one chamber to another allowing protons, electrons, and CO_2 to pass through while restricting oxygen and substrate. The membrane should be featured to overcome internal resistance, pH splitting, and oxygen permeability to reduce the overall cost of MFC. Nafion is the most used in MFCs because it renders more specific conductivity for protons, increased columbic efficiency of MFC, and the increased microorganism lifetime. However, it has been reported that it does not function efficiently in alkaline conditions (Hernández-Flores et al., 2017). Clayware membrane, ceramic membrane, and natural rubber membrane are a few examples of low-cost membranes (Xu et al., 2015).

With the advancement in material science and technology, we look forward to new non-toxic buffers, cheaper membranes, and efficient electrode materials.

8.4 MOST COMMONLY USED ALGAL AND BACTERIAL CULTURE

In a typical microbial cell, anoxic electrochemically active bacteria (EAB) were initially used to generate electricity using an external circuit. To step up from the setbacks faced in the functioning of these cells, algae were used as biocathode, with or without immobilization, for CO_2 assimilation and in situ oxygen generation. Compared to other autotrophs, microalgae are fast-growing with tremendous photosynthetic efficiency and require lesser water than terrestrial plants (Angioni et al., 2018). Singular or mixed algal species can be utilized for power generation. The most frequently used algal species reported in the literature have been listed in Table 8.2. As discussed earlier, algae can also be used at the anode as a nutrient source for bacteria. It can be used in various forms like live algal biomass, dry powder form, pretreated or lipid-extracted, and in different concentrations. A pretreated powdered form of *Scenedesmus* was utilized by Liu et al. (2015) at the anode to obtain a maximum power density of 514.2 ± 19.4 mW/m². *Bacillus cereus*, *Pseudomonas aeruginosa*,

TABLE 8.2
Algal Species Used as Biocathode in a PAMFC and Their Corresponding Power Outputs

Algal Species	Power Density (W/m³)	References
Chlorella vulgaris	126 mW/m³	(Bazdar et al., 2018)
Scenedismus obliquus	153 mW/m²	(Kakarla & Min, 2014)
S. platensis	10 mW/m²	(Lin et al., 2013)
Fresh pond water	128 µW	(Gajda et al., 2015)
Ulva lactuca	0.98 W/m²	(Arun et al., 2020)
Chlamydomonas reinhardtii and *Pseudokirchneriella subcapitata*	2.2	(Xiao et al., 2012)
Chlorella and *Phormidium*	26 W	(Juang et al., 2012)
Microcystis aeruginosa and *Chlorella vulgaris*	4.14	(Wang et al., 2012)
Microcystis aeruginosa	0.058	(Cai et al., 2013)
Desmodesmus sp. *AS*	0.82	(Wu et al., 2014)

Cyanobacteria, *Geobacter sulfurreducens*, *Shewanella oneidensis* MR-1, *Clostridia*, *Geobacter metallireducens*, *Shewanella putrefaciens*, *Aeromonas hydrophila*, *Rhodoferax ferrireducens*, *Enterococcus faecium*, *Pseudomonas aeruginosa*, and *Betaproteobacteria* are a few examples of commonly used bacterial consortium. Using genetic engineering, new strains can be engineered with better electrogenic properties for optimum output.

Protons travel through the membranes to reduce oxygen and the electrons are taken up by the anode which then travel through an external circuit to cathode without the help of any mediators. Electron transfer can be either a direct or an indirect mode of transfer, depending on the type of bacterial species. The most common form of extracellular electron transfer occurs with the help of bacterial cytochromes (Shukla & Kumar, 2018) or redox molecules. In some cases, bacteria have appendages like nanopili or nanowire that aid in electron transfer (Lesnik & Liu, 2014). This direct form of electron transfer can be seen in *Shewanella oneidensis* and *Geobacter sulfurreducens*. Another system of indirect electron transfer is cyclic diffusion, where bacterial cells (like *Escherichia coli*, *Bacillus* sp., and *Clostridia* sp.) donate electrons to anode and in turn take up soluble compounds from solution via polymeric redox mediators or primary metabolites like hydrogen (Patil et al., 2012). Addition of external mediators like potassium ferrocyanide, platinum catalysts, methyl blue, neutral red, thionine, methyl viologen, and humic acid have also been reported. Algal cells from suspension form a biofilm with time and directly take up the electrons from the circuit, which penetrate the algal body.

8.5 HARVESTING OF ALGAE

Efficient harvesting of algae from cathode or anode for further extraction of value-added products like biofuels, pigments, etc. can help in offsetting the additional costs endured during the functioning of the fuel cells. Recently, production of biochar from microalgae has been gaining interest for application as adsorbent, fertilizer, etc. (Pathy et al., 2022). Harvesting of algae comprises 30% of downstream costs (Shukla & Kumar, 2018). One of the methods to reduce harvesting costs is the immobilization of algae into beads. Immobilization of algae not only reduces costs but also adds benefits like high algal density, stable operation, and faster reaction speed (He et al., 2014). Sand filtration and coagulation have also been proposed for harvesting algae, but they have not been proved to be efficient in removing algae due to their small size and low density. Direct electrolysis could be a potential option for effective harvesting if not for the enormous energy requirement (Monasterio et al., 2017). Modification of this electrochemical system by incorporating flocculation, combined with either flotation or sedimentation, can make it energy-efficient and cost-effective (Shin et al., 2017; Behera et al., 2020). These methods can be broadly classified as electrolytic coagulation, electrolytic flotation, and electrolytic flocculation (Richardson et al., 2014; Krishnamoorthy et al., 2021).

In the electrolytic coagulation method, sacrificial electrodes are used to release positively charged ions which form aggregates with negatively charged algal cells due to charge neutralization. The coagulates settle down and can be separated with the age-old method of sedimentation (Richardson et al., 2014; Nageshwari et al., 2021). During this process, oxygen and hydrogen bubbles maybe released from the cathode, which can help the already formed flocs to rise to the surface. This process is called electro-floatation (Bardone et al., 2018). The addition of flocculants like chitosan or modified starch can lead to the formation of algal flocs, called flocculation (Behera et al., 2019). It can also be induced with the help of electrolysis, ultrasound waves, or magnetic waves (Pirwitz et al., 2015). To maximize efficiency, these methods are used in combination with one another as electrocoagulation–flocculation (Fayad et al., 2017) and electrocoagulation–flotation (Parmentier et al., 2020; Shi et al., 2017).

An innovative method was suggested by Monasterio et al. (2017), in which microbial fuel cell was integrated with electrochemical removal of algae. MFCs were operated on wastewater for the generation of electricity and simultaneously, an electrolysis unit connected in series with four to five MFCs was operated for electrochemical removal of algae.

8.6 APPLICATION OF THE INTEGRATED SYSTEMS

Microbial fuel cells were the first electrochemical devices to utilize live organisms for wastewater treatment and to produce bioelectricity. But it faced a few bottlenecks such as non-eco-friendliness due to release of CO_2 produced by the bacteria into the atmosphere, high operational cost due to supplying of CO_2, pH alteration due to its dissolution into the water, and aeration costs up to 50% because of oxygen. Thus, adding algae into the cathode not only solved the bottlenecks but also gave added products that could offset the additional costs. A comparative study between MFC and PAMFC has been presented in Figure 8.7. The following are the advantages of using algae at cathode as reported in the literature:

(i) Value-added pigments production (chlorophyll, ß-carotene, canthaxanthin, lutein, zeaxanthin) (Arun et al., 2020)
(ii) Production of algal biomass (Shukla & Kumar, 2018)
(iii) Use as industrial filters (Mekuto et al., 2020)
(iv) Has a lesser water footprint compared to other plant-based fuel industries (Ma et al., 2017)
(v) Pollution control as discharge of biomass into waterbodies can cause eutrophication (Liu et al., 2015)
(vi) Wastewater treatment (COD, nitrogen, and phosphorus removal) (Ma et al., 2017)
(vii) Carbon fixation in turn reducing global warming (Li et al., 2019)
(viii) Biodiesel production (lipid extraction), biogas production (anaerobic digestion of algae), hydrocarbon fuel (hydrothermal liquefaction) (Shukla & Kumar, 2018)

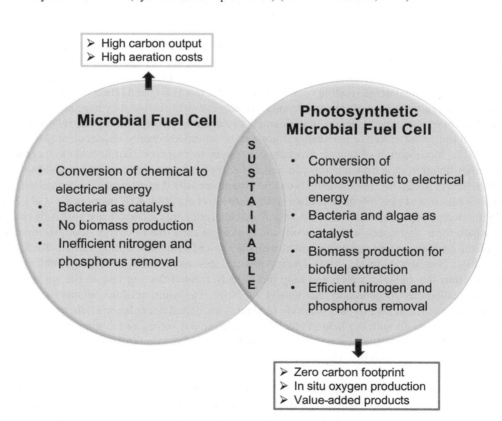

FIGURE 8.7 Comparison of characteristics between MFC and PAMFC.

(ix) Algal biomass can be also used as substrate in various forms:
  (a) Fresh algal biomass (Xu et al., 2015)
  (b) Algal powder (Shukla & Kumar, 2018)
  (c) Lipid extracted (Khandelwal et al., 2018)
  (d) Pretreated (sonication, autoclave, microwave, thermal, acid/alkali) (Arun et al., 2020)
(x) Algae grow in adverse ecological conditions as well as remove nutrients and toxic metals (Naina Mohamed et al., 2020)

8.7 RESEARCH GAP

1. Though articles have mentioned the direct use of algal biomass, some articles have also pointed out the need for pretreating the algal biomass as the cellulose present in the cell wall is not easy for the bacteria to hydrolyze (Xu et al., 2015). Requirement of pretreatments like thermal, sonication, alkaline, and acid treatment can increase the production costs. In addition, lipid-extracted algae were reported to contain harmful solvents and toxic chemicals, which can affect the growth of bacteria at the anode.
2. When the algal biomass density increases exponentially, they tend to form biofilms on reactor walls and the electrode surface. The thickness of cathodic biofilm influences oxygen diffusion, CO_2 transfer rate, and light penetration, which will ultimately affect electricity production. Studying biofilm formation in detail will help us develop efficient algal harvesting systems.
3. Buffer-less PAMFC systems need to be studied more in detail so that the pH can be regulated without the use of a buffer to limit the usage of chemicals and to prevent additional costs.
4. Algae being photosynthetic organisms require sunlight for biomass production and oxygen generation. Thus, the reactor setup should be such that it can receive the required amount of sunlight. Also, light penetration should not be affected even when the biomass becomes very dense at deeper levels of the reactor. In a year, the sunlight intensity varies according to season and to prevent it from affecting power generation, artificial illumination devices need to be set up to provide uninterrupted and consistent light throughout the year.
5. Membranes are expensive in nature. Thus, finding a cheaper alternative that will also prevent the unwanted diffusion of oxygen and substrate is required.
6. Stability, long-term performance, efficiency, and scaling up process from lab- to full-scale are the future challenges to be faced by many researchers.

8.8 SCOPE FOR FUTURE RESEARCH

The need for the development of robust technologies is critical because conventional systems consume a lot of energy. Reduction of CO_2 and greenhouse gases emission by these technologies is necessary. In addition, accomplishment of the permissible limits of nitrogen and phosphorus in the recycled wastewater needs to be achieved to prevent eutrophication and algal blooms. They also need to be low-cost to be installed in big to small cities, without the need of external supervision and excessive human resources. The development of an integrated system for the treatment of wastewater and bioelectricity generation using live organisms is still in its early phases and a lot of drawbacks need to be overcome before it can be utilized at its full potential at a commercial scale. Algal strains which can be fit for use at both electrodes need to be determined. The synergistic relationship between the bacteria and algae and the electron transfer mechanisms in addition to CO_2 uptake needs to be studied in detail for better control of growth and maximizing power output. In terms of bioelectricity generation, despite all the combinations of wastewater sources with bacterio-algal strains and electrodes being tried, the best combination is yet to be found for maximum power output. Various reactor configurations have also been tested over the years, but the optimum design

TABLE 8.3
Research Gaps and the Proposed Solution to Improve the Overall Efficiency and Commercialization of PAMFC

Research Gap	Proposed Solution
Use of buffers for pH regulation increases overall cost	Search for novel buffer-less systems
Low availability of reactor configurations in literature	Selection of optimum reactor design for maximum power output
Use of single algal species or in consortia	Selection of algal strains that can be used at both cathode and anode
Expensive membranes	Development of an efficient and cheaper membrane alternatives
Biofilm formation adversely affects power generation	Biofilm formation needs to be studied in detail to prevent occurrence
Lack of commercial scale plants	Optimization of parameters and standardization of influential factors to develop a sustainable technology

is yet to be determined. Ways to prevent thick biofilm formation needs to be discovered, so that power output does not get affected. Table 8.3 summarizes the research gaps along with their proposed solutions to develop an understanding of the need of this research area. Due to variations in wastewater or nutrient sources, microorganism strains, reactor environment, and operational setup, there have been discrepancies in data. This issue needs to be resolved for getting a consistent output and to develop field-based criteria. Also, the influencing parameters need to be optimized for keeping these factors in play.

REFERENCES

Angioni, S., Millia, L., Mustarelli, P., Doria, E., Temporiti, M. E., Mannucci, B., Corana, F., & Quartarone, E. (2018). Photosynthetic microbial fuel cell with polybenzimidazole membrane: Synergy between bacteria and algae for wastewater removal and biorefinery. *Heliyon*, *4*(3). https://doi.org/10.1016/j.heliyon.2018.e00560

Apollon, W., Luna-Maldonado, A. I., Kamaraj, S. K., Vidales-Contreras, J. A., Rodríguez-Fuentes, H., Gómez-Leyva, J. F., & Aranda-Ruíz, J. (2021). Progress and recent trends in photosynthetic assisted microbial fuel cells: A review. In *Biomass and Bioenergy* (Vol. 148, p. 106028). Elsevier Ltd. https://doi.org/10.1016/j.biombioe.2021.106028

Arun, S., Ramasamy, S., Pakshirajan, K., & Pugazhenthi, G. (2022). Bioelectricity production and shortcut nitrogen removal by microalgal-bacterial consortia using membrane photosynthetic microbial fuel cell. *Journal of Environmental Management*, *301*, 113871. https://doi.org/10.1016/J.JENVMAN.2021.113871

Arun, S., Sinharoy, A., Pakshirajan, K., & Lens, P. N. L. (2020). Algae based microbial fuel cells for wastewater treatment and recovery of value-added products. *Renewable and Sustainable Energy Reviews*, *132*, 110041. https://doi.org/10.1016/J.RSER.2020.110041

Bardone, E., Marzocchella, A., Keshavarz, T., Castellaños-Estupiñan, M. A., Sánchez-Galvis, M., García-Martínez, J. B., Barajas-Ferreira, C., Zuorro, A., & Barajas-Solano, A. F. (2018). Design of an electroflotation system for the concentration and harvesting of freshwater microalgae. *Chemical Engineering Transactions*, *64*.

Bazdar, E., Roshandel, R., Yaghmaei, S., & Mardanpour, M. M. (2018). The effect of different light intensities and light/dark regimes on the performance of photosynthetic microalgae microbial fuel cell. *Bioresource Technology*, *261*, 350–60. https://doi.org/10.1016/j.biortech.2018.04.026

Behera, B., & Balasubramanian, P. (2019). Natural plant extracts as an economical and ecofriendly alternative for harvesting microalgae. *Bioresource Technology*, *283*, 45–52.

Behera, B., Nageshwari, K., Darshini, M., & Balasubramanian, P. (2020). Evaluating the harvesting efficiency of inorganic coagulants on native microalgal consortium enriched with human urine. *Water Science and Technology*, *82*(6), 1217–26.

Bhande, R., Noori, M. T., & Ghangrekar, M. M. (2019). Performance improvement of sediment microbial fuel cell by enriching the sediment with cellulose: Kinetics of cellulose degradation. *Environmental Technology and Innovation*, *13*, 189–96. https://doi.org/10.1016/j.eti.2018.11.003

Cai, P. J., Xiao, X., He, Y. R., Li, W. W., Zang, G. L., Sheng, G. P., Hon-Wah Lam, M., Yu, L., & Yu, H. Q. (2013). Reactive oxygen species (ROS) generated by cyanobacteria act as an electron acceptor in the biocathode of a bio-electrochemical system. *Biosensors and Bioelectronics*, *39*(1), 306–10. https://doi.org/10.1016/j.bios.2012.06.058

Choudhury, P., Ray, R. N., Bandyopadhyay, T. K., Basak, B., Muthuraj, M., & Bhunia, B. (2021). Process engineering for stable power recovery from dairy wastewater using microbial fuel cell. *International Journal of Hydrogen Energy*, *46*(4), 3171–82. https://doi.org/10.1016/j.ijhydene.2020.06.152

Dai, H. N., Duong Nguyen, T. A., Le, L. P. M., Van Tran, M., Lan, T. H., & Wang, C. T. (2021). Power generation of Shewanella oneidensis MR-1 microbial fuel cells in bamboo fermentation effluent. *International Journal of Hydrogen Energy*, *46*(31), 16612–21. https://doi.org/10.1016/j.ijhydene.2020.09.264

Enamala, M. K., Dixit, R., Tangellapally, A., Singh, M., Dinakarrao, S. M. P., Chavali, M., Pamanji, S. R., Ashokkumar, V., Kadier, A., & Chandrasekhar, K. (2020). Photosynthetic microorganisms (Algae) mediated bioelectricity generation in microbial fuel cell: Concise review. *Environmental Technology & Innovation*, *19*, 100959. https://doi.org/10.1016/j.eti.2020.100959

Eom, H., Joo, H. J., Kim, S. C., & Kim, S. S. (2020). Properties of carbon-based nanofiber with Pd and its application to microbial fuel cells electrode. *Environmental Technology & Innovation*, *19*, 100800. https://doi.org/10.1016/J.ETI.2020.100800

Fayad, N., Yehya, T., Audonnet, F., & Vial, C. (2017). Harvesting of microalgae *Chlorella vulgaris* using electro-coagulation-flocculation in the batch mode. *Algal Research*, *25*, 1–11. https://doi.org/10.1016/j.algal.2017.03.015

Firdous, S., Jin, W., Shahid, N., Bhatti, Z. A., Iqbal, A., Abbasi, U., Mahmood, Q., & Ali, A. (2018). The performance of microbial fuel cells treating vegetable oil industrial wastewater. *Environmental Technology & Innovation*, *10*, 143–51. https://doi.org/10.1016/J.eti.2018.02.006

Gajda, I., Greenman, J., Melhuish, C., & Ieropoulos, I. (2015). Self-sustainable electricity production from algae grown in a microbial fuel cell system. *Biomass and Bioenergy*, *82*, 87–93. https://doi.org/10.1016/j.biombioe.2015.05.017

Greenman, J., Gajda, I., & Ieropoulos, I. (2019). Microbial fuel cells (MFC) and microalgae; photo microbial fuel cell (PMFC) as complete recycling machines. In *Sustainable Energy and Fuels* (Vol. 3, Issue 10, pp. 2546–60). Royal Society of Chemistry. https://doi.org/10.1039/c9se00354a

Harewood, A. J. T., Popuri, S. R., Cadogan, E. I., Lee, C. H., & Wang, C. C. (2017). Bioelectricity generation from brewery wastewater in a microbial fuel cell using chitosan/biodegradable copolymer membrane. *International Journal of Environmental Science and Technology*, *14*(7), 1535–50. https://doi.org/10.1007/s13762-017-1258-6

He, H., Zhou, M., Yang, J., Hu, Y., & Zhao, Y. (2014). Simultaneous wastewater treatment, electricity generation and biomass production by an immobilized photosynthetic algal microbial fuel cell. *Bioprocess And Biosystems Engineering*, *37*(5), 873–80. https://doi.org/10.1007/s00449-013-1058-4

Hernández-Flores, G., Solorza-Feria, O., & Poggi-Varaldo, H. M. (2017). Bioelectricity generation from wastewater and actual landfill leachates: A multivariate analysis using principal component analysis. *International Journal of Hydrogen Energy*, *42*(32), 20772–82. https://doi.org/10.1016/j.ijhydene.2017.01.021

Hou, Q., Nie, C., Pei, H., Hu, W., Jiang, L., & Yang, Z. (2016). The effect of algae species on the bioelectricity and biodiesel generation through open-air cathode microbial fuel cell with kitchen waste anaerobically digested effluent as substrate. *Bioresource Technology*, *218*, 902–08. https://doi.org/10.1016/j.biortech.2016.07.035

Inglesby, A. E., Yunus, K., & Fisher, A. C. (2013). In situ fluorescence and electrochemical monitoring of a photosynthetic microbial fuel cell. *Physical Chemistry Chemical Physics*, *15*(18), 6903–11. https://doi.org/10.1039/c3cp51076j

Jaiswal, K. K., Kumar, V., Vlaskin, M. S., Sharma, N., Rautela, I., Nanda, M., Arora, N., Singh, A., & Chauhan, P. K. (2020). Microalgae fuel cell for wastewater treatment: Recent advances and challenges. *Journal of Water Process Engineering*, *38*, 101549. https://doi.org/10.1016/j.jwpe.2020.101549

Juang, D. F., Lee, C. H., & Hsueh, S. C. (2012). Comparison of electrogenic capabilities of microbial fuel cell with different light power on algae grown cathode. *Bioresource Technology*, *123*, 23–9. https://doi.org/10.1016/j.biortech.2012.07.041

Kakarla, R., & Min, B. (2014). Photoautotrophic microalgae Scenedesmus obliquus attached on a cathode as oxygen producers for microbial fuel cell (MFC) operation. *International Journal of Hydrogen Energy*, *39*(19), 10275–83. https://doi.org/10.1016/j.ijhydene.2014.04.158

Kannan, N., & Donnellan, P. (2021). Algae-assisted microbial fuel cells: A practical overview. *Bioresource Technology Reports*, *15*, 100747. https://doi.org/10.1016/J.biteb.2021.100747

Khandelwal, A., Vijay, A., Dixit, A., & Chhabra, M. (2018). Microbial fuel cell powered by lipid extracted algae: A promising system for algal lipids and power generation. *Bioresource Technology*, *247*, 520–7. https://doi.org/10.1016/j.biortech.2017.09.119

Krishnamoorthy, N., Unpaprom, Y., Ramaraj, R., Maniam, G. P., Govindan, N., Arunachalam, T., & Paramasivan, B. (2021). Recent advances and future prospects of electrochemical processes for microalgae harvesting. *Journal of Environmental Chemical Engineering*, *9*(5), 105875.

Nageshwari, K., Baishali, D., Unpaprom, Y., Ramaraj, R., Maniam, G. P., Govindan, N., Thirugnanam, A., & Balasubramanian, P. (2021). Exploring the dynamics of microalgal diversity in high-rate algal ponds. *The Future of Effluent Treatment Plants*, 615–60.

Kusmayadi, A., Leong, Y. K., Yen, H. W., Huang, C. Y., Dong, C. D., & Chang, J. S. (2020). Microalgae-microbial fuel cell (mMFC): An integrated process for electricity generation, wastewater treatment, CO2 sequestration and biomass production. *International Journal of Energy Research*, *44*(12), 9254–65.

Lee, D. J., Chang, J. S., & Lai, J. Y. (2015). Microalgae–microbial fuel cell: A mini review. *Bioresource Technology*, *198*, 891–5. https://doi.org/10.1016/j.biortech.2015.09.061

Lesnik, K. L., & Liu, H. (2014). Establishing a core microbiome in acetate-fed microbial fuel cells. *Applied Microbiology and Biotechnology*, *98*(9), 4187–96. https://doi.org/10.1007/s00253-013-5502-9

Li, M., Zhou, M., Luo, J., Tan, C., Tian, X., Su, P., & Gu, T. (2019). Carbon dioxide sequestration accompanied by bioenergy generation using a bubbling-type photosynthetic algae microbial fuel cell. *Bioresource Technology*, *280*, 95–103. https://doi.org/10.1016/j.biortech.2019.02.038

Lin, C. C., Wei, C. H., Chen, C. I., Shieh, C. J., & Liu, Y. C. (2013). Characteristics of the photosynthesis microbial fuel cell with a *Spirulina platensis* biofilm. *Bioresource Technology*, *135*, 640–3. https://doi.org/10.1016/j.biortech.2012.09.138

Liu, T., Rao, L., Yuan, Y., & Zhuang, L. (2015). Bioelectricity generation in a microbial fuel cell with a self-sustainable photocathode. *Scientific World Journal*. https://doi.org/10.1155/2015/864568

Ma, J., Wang, Z., Zhang, J., Waite, T. D., & Wu, Z. (2017). Cost-effective Chlorella biomass production from dilute wastewater using a novel photosynthetic microbial fuel cell (PMFC). *Water Research*, *108*, 356–64. https://doi.org/10.1016/j.watres.2016.11.016

Mehrabadi, A., Craggs, R., & Farid, M. M. (2016). Biodiesel production potential of wastewater treatment high rate algal pond biomass. *Bioresource Technology*, *221*, 222–33. https://doi.org/10.1016/j.biortech.2016.09.028

Mekuto, L., Olowolafe, A. V. A., Pandit, S., Dyantyi, N., Nomngongo, P., & Huberts, R. (2020). Microalgae as a biocathode and feedstock in anode chamber for a self-sustainable microbial fuel cell technology: A review. *South African Journal of Chemical Engineering*, *31*(October 2019), 7–16. https://doi.org/10.1016/j.sajce.2019.10.002

Monasterio, S., Mascia, M., & Di Lorenzo, M. (2017). Electrochemical removal of microalgae with an integrated electrolysis-microbial fuel cell closed-loop system. *Separation and Purification Technology*, *183*, 373–81. https://doi.org/10.1016/j.seppur.2017.03.057

Naina Mohamed, S., Ajit Hiraman, P., Muthukumar, K., & Jayabalan, T. (2020). Bioelectricity production from kitchen wastewater using microbial fuel cell with photosynthetic algal cathode. *Bioresource Technology*, *295*, 122226. https://doi.org/10.1016/j.biortech.2019.122226

Nam, J. Y., Kim, H. W., Lim, K. H., & Shin, H. S. (2010). Effects of organic loading rates on the continuous electricity generation from fermented wastewater using a single-chamber microbial fuel cell. *Bioresource Technology*, *101*, S33–S37. https://doi.org/10.1016/j.biortech.2009.03.062

Nayak, J. K., & Ghosh, U. K. (2019). Post treatment of microalgae treated pharmaceutical wastewater in photosynthetic microbial fuel cell (PMFC) and biodiesel production. *Biomass and Bioenergy*, *131*, 105415. https://doi.org/10.1016/j.biombioe.2019.105415

Ng, F. L., Phang, S. M., Thong, C. H., Periasamy, V., Pindah, J., Yunus, K., & Fisher, A. C. (2021). Integration of bioelectricity generation from algal biophotovoltaic (BPV) devices with remediation of palm oil mill effluent (POME) as substrate for algal growth. *Environmental Technology and Innovation*, *21*, 101280. https://doi.org/10.1016/j.eti.2020.101280

Pant, D., Van Bogaert, G., Diels, L., & Vanbroekhoven, K. (2010). A review of the substrates used in microbial fuel cells (MFCs) for sustainable energy production. *Bioresource Technology*, *101*(6), 1533–43. https://doi.org/10.1016/J.BIORTECH.2009.10.017

Parmentier, D., Manhaeghe, D., Baccini, L., Van Meirhaeghe, R., Rousseau, D. P. L., & Van Hulle, S. (2020). A new reactor design for harvesting algae through electrocoagulation-flotation in a continuous mode. *Algal Research*, *47*, 101828. https://doi.org/10.1016/j.algal.2020.101828

Pathy, A., Krishnamoorthy, N., Chang, S. X., & Paramasivan, B. (2022). Malachite green removal using algal biochar and its composites with kombucha SCOBY: An integrated biosorption and phycoremediation approach. *Surfaces and Interfaces*, *30*, 101880.

Patil, S. A., Hägerhäll, C., & Gorton, L. (2012). Electron transfer mechanisms between microorganisms and electrodes in bioelectrochemical systems. *Bioanalytical Reviews*, *4*(2), 159–92. https://doi.org/10.1007/s12566-012-0033-x

Pérez-Page, M., & Pérez-Herranz, V. (2011). Effect of the operation and humidification temperatures on the performance of a pem fuel cell stack on dead-end mode. *International Journal of Electrochemical Science*, *6*(2), 492–505.

Pirwitz, K., Rihko-Struckmann, L., & Sundmacher, K. (2015). Comparison of flocculation methods for harvesting Dunaliella. *Bioresource Technology*, *196*, 145–52. https://doi.org/10.1016/j.biortech.2015.07.032

Qi, X., Ren, Y., Liang, P., & Wang, X. (2018). New insights in photosynthetic microbial fuel cell using anoxygenic phototrophic bacteria. *Bioresource Technology*, *258*, 310–17. https://doi.org/10.1016/j.biortech.2018.03.058

Radha, M., & Kanmani, S. (2017). Evaluation of paper recycling wastewater treatment accompanied by power generation using microbial fuel cell. *Global NEST Journal*, *19*(4), 682–6.

Reddy, C. N., Nguyen, H. T. H., Noori, M. T., & Min, B. (2019). Potential applications of algae in the cathode of microbial fuel cells for enhanced electricity generation with simultaneous nutrient removal and algae biorefinery: Current status and future perspectives. *Bioresource Technology*, *292*, 122010. https://doi.org/10.1016/J.BIORTECH.2019.122010

Richardson, J. W., Johnson, M. D., Lacey, R., Oyler, J., & Capareda, S. (2014). Harvesting and extraction technology contributions to algae biofuels economic viability. *Algal Research*, *5*(1), 70–8. https://doi.org/10.1016/j.algal.2014.05.007

Saba, B., Christy, A. D., Yu, Z., & Co, A. C. (2017). Sustainable power generation from bacterio-algal microbial fuel cells (MFCs): An overview. *Renewable and Sustainable Energy Reviews*, *73*, 75–84. https://doi.org/10.1016/j.rser.2017.01.115

Shi, W., Zhu, L., Chen, Q., Lu, J., Pan, G., Hu, L., & Yi, Q. (2017). Synergy of flocculation and flotation for microalgae harvesting using aluminium electrolysis. *Bioresource Technology*, *233*, 127–33. https://doi.org/10.1016/j.biortech.2017.02.084

Shin, H., Kim, K., Jung, J. Y., Bai, S. C., Chang, Y. K., & Han, J. I. (2017). Harvesting of Scenedesmus obliquus cultivated in seawater using electro-flotation. *Korean Journal of Chemical Engineering*, *34*(1), 62–5. https://doi.org/10.1007/s11814-016-0251-y

Shukla, M., & Kumar, S. (2018). Algal growth in photosynthetic algal microbial fuel cell and its subsequent utilization for biofuels. In *Renewable and Sustainable Energy Reviews* (Vol. 82, pp. 402–14). Elsevier Ltd. https://doi.org/10.1016/j.rser.2017.09.067

Taskan, E., Özkaya, B., & Hasar, H. (2014). Effect of different mediator concentrations on power generation in MFC using Ti-TiO$_2$ electrode. *International Journal of Energy Science*, *4*(1), 9–11. https://doi.org/10.14355/ijes.2014.0401.02

Wang, C. T., Huang, Y. S., Sangeetha, T., Chen, Y. M., Chong, W. T., Ong, H. C., Zhao, F., & Yan, W. M. (2018). Novel bufferless photosynthetic microbial fuel cell (PMFCs) for enhanced electrochemical performance. *Bioresource Technology*, *255*, 83–7. https://doi.org/10.1016/j.biortech.2018.01.086

Wang, H., Lu, L., Cui, F., Liu, D., Zhao, Z., & Xu, Y. (2012). Simultaneous bioelectrochemical degradation of algae sludge and energy recovery in microbial fuel cells. *RSC Advances*, *2*(18), 7228–34. https://doi.org/10.1039/c2ra20631e

Wu, Y. cheng, Wang, Z. jie, Zheng, Y., Xiao, Y., Yang, Z. Hui, & Zhao, F. (2014). Light intensity affects the performance of photo microbial fuel cells with *Desmodesmus sp*. A8 as cathodic microorganism. *Applied Energy*, *116*, 86–90. https://doi.org/10.1016/j.apenergy.2013.11.066

Xiao, L., Young, E. B., Berges, J. A., & He, Z. (2012). Integrated photo-bioelectrochemical system for contaminants removal and bioenergy production. *Environmental Science and Technology*, *46*(20), 11459–66. https://doi.org/10.1021/es303144n

Xu, C., Poon, K., Choi, M. M. F., & Wang, R. (2015). Using live algae at the anode of a microbial fuel cell to generate electricity. *Environmental Science and Pollution Research*, *22*(20), 15621–35. https://doi.org/10.1007/s11356-015-4744-8

Zhang, X., Xia, X., Ivanov, I., Huang, X., & Logan, B. E. (2014). Enhanced activated carbon cathode performance for microbial fuel cell by blending carbon black. *Environmental Science and Technology*, *48*(3), 2075–81. https://doi.org/10.1021/es405029y

Zhang, Y., Zhao, Y., & Zhou, M. (2019). A photosynthetic algal microbial fuel cell for treating swine wastewater. *Environmental Science and Pollution Research*, *26*(6), 6182–90. https://doi.org/10.1007/s11356-018-3960-4

9 Pyrolysis and Steam Gasification of Biomass Waste for Hydrogen Production

Prakash Parthasarathy, Tareq Al-Ansari, Gordon McKay, and K. Sheeba Narayanan

CONTENTS

9.1 Introduction 121
    9.1.1 Pyrolysis.. 121
9.2 Steam Gasification 123
9.3 Combined Slow Pyrolysis and Steam Gasification.. 124
9.4 Factors Influencing Products Yield in Combined Process 125
    9.4.1 Important Factors Influencing Char Production in Slow Pyrolysis of Combined Process... 125
        9.4.1.1 Biomass Type... 125
        9.4.1.2 Biomass Composition 125
        9.4.1.3 Biomass Particle Size... 125
        9.4.1.4 Pyrolysis Temperature 126
        9.4.1.5 Heating Rate 126
9.5 Factors Influencing Hydrogen Generation in Steam Gasification of the Combined Process... 127
    9.5.1 Gasification Temperature... 127
    9.5.2 Steam-to-biomass Ratio (*S/B*) 127
    9.5.3 Catalyst 128
    9.5.4 Sorbent-to-biomass Ratio 128
9.6 Products of Pyrolysis and Gasification 128
    9.6.1 Char 128
    9.6.2 Bio-oil 129
    9.6.3 Tar 129
    9.6.4 Syngas 129
9.7 Conclusion 130
References... 130

9.1 INTRODUCTION

9.1.1 Pyrolysis

Generally, pyrolysis and gasification are the two most common thermochemical processes for hydrogen production. They are entirely different from combustion processes. These processes increase the energy density of biomass to offer useful products, while combustion provides heat energy to deliver products of little use. Since ancient times, humans have been pyrolyzing and gasifying to generate energy, fuel, value-added chemicals, etc. (Demirbaş 2005). Pyrolysis is the thermal degradation of any carbonaceous material in the presence or absence of oxygen into carbon-rich char, hydrocarbon-rich liquids, and carbon-rich gases (Parthasarathy, Al-Ansari, et al. 2021). The process is illustrated in Figure 9.1.

Equation 9.1 gives the pyrolytic decomposition of biomass:

$$\text{Pyrolysis} \rightarrow H_2 + CO + CO_2 + HC \text{ gases} + Tar + Char \tag{9.1}$$

Pyrolysis is the precursor to all thermochemical processes, including combustion, gasification, and liquefaction. It is the only process that can produce all three forms of valuable products: solid–char, liquid–bio-oil, gas–syngas (AlNouss et al. 2021). Several operating parameters affect the conformation of the products, such as the temperature of pyrolysis, the heating rate, the particle size, the solid residence time, and the vapor residence time (Al-Rumaihi et al. 2021). Based on the process conditions, pyrolysis is sorted into three types and Table 9.1 gives the classification of pyrolysis process (Demirbas 2004).

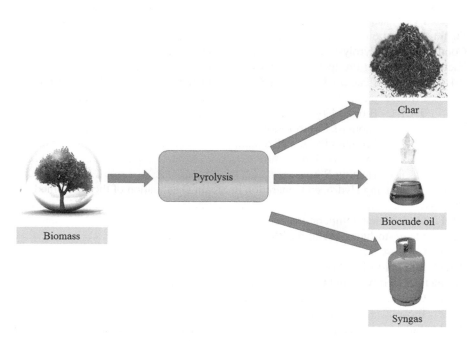

FIGURE 9.1 Schematics of pyrolysis phenomenon.

TABLE 9.1
Classification of Pyrolysis Process

Parameters	Flash	Fast	Slow
Primary product	Syngas	Pyrolytic oil	Char
Temperature (°C)	800–1000	600–1000	300–700
Heating rate (°C/s)	>1000	10–200	0.01–1
Feedstock size (mm)	<0.2	<1	5–50
Solid retention time (s)	<0.5	0.5–5	600–6000

While low temperatures, high heating rates, and a small gas retention time promote liquid yield, small heating rates, high temperatures, and a long gas retention time favor gas yield. Low operating temperature and low heating rates boost char yield.

9.2 STEAM GASIFICATION

Gasification is the transformation of carbonaceous materials into gaseous fuels and chemical feedstocks. It generally occurs at temperatures 800–1800 °C and in oxygen-depleted environments (Hernández et al. 2012). Nonetheless, it necessitates the use of a gasification medium. Air, O_2, steam, CO_2, or any combination of these can be used as a gasification medium (Parthasarathy, Fernandez, et al. 2021). Table 9.2 provides a quick comparison of different gasification techniques (Saxena et al. 2008).

The gasification process that uses steam as a gasifying agent is known as "steam gasification" (Table 9.3). This process is appropriate for biomass with a moisture content of less than 35% (De Lasa et al. 2011). In this process, carbonaceous material is converted into gases, including H_2, CO_2, CH_4, and CO, and light hydrocarbons, char, and tar (AlNouss et al. 2020). A simple schematic of the steam gasification phenomenon is presented in Figure 9.2.

TABLE 9.2
Difference between Different Gasification Processes

Parameters	Air Gasification	Steam Gasification	Oxygen Gasification
Typical gas constituents	CO, H_2, CO_2, N_2, water, hydrocarbon, tar	H_2, CO, CO_2, CH_4 hydrocarbon, tar	CO, H_2, CO_2, hydrocarbon
Typical syngas composition (%)	H_2-15, CO-20, N_2-48, CO_2-15, CH_4-2	H_2-40, CO-25, CO_2-25, CH_4-8, N_2-2	H_2-40, CO-40, CO_2-20
Syngas calorific (MJ/Nm³)	4–6	15–20	10–15
Costs involved	Economical		Expensive

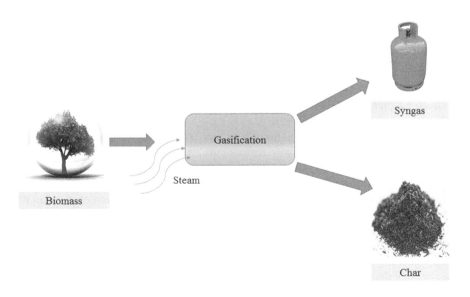

FIGURE 9.2 Illustration of steam gasification process.

TABLE 9.3
List of Reactions that Occur in Steam Gasification (Ahmed and Gupta 2009b; Yang et al. 2006)

Reaction Type	Reaction Equation	ΔH (kJ/mol)	Equation
Combustion	$H_2 + 0.5O_2 \rightarrow H_2O$	−242	(9.3)
Combustion	$CO + 0.5O_2 \rightarrow CO_2$	−283	(9.4)
Combustion	$CH_4 + 0.5O_2 \rightarrow CO + 2H_2$	−110	(9.5)
Dry reforming	$CH_4 + CO_2 \rightarrow 2CO + 2H_2$	+247	(9.6)
Steam reforming	$CH_4 + H_2O \rightarrow CO + 3H_2$	+206	(9.7)
Water-gas shift	$CO + H_2O \rightarrow CO_2 + H_2$	−40.9	(9.8)
Oxidation	$C + O_2 \rightarrow CO_2$	−393.5	(9.9)
Partial oxidation	$C + 0.5O_2 \rightarrow CO$	−123.1	(9.10)
Boudouard	$C + CO_2 \rightarrow 2CO$	+159.9	(9.11)
Water-gas reaction	$C + H_2O \rightarrow CO + H_2$	+118.5	(9.12)
Methane forming	$C + 2H_2 \rightarrow CH_4$	−87.5	(9.13)

The steam gasification process can be represented by Equation 9.2:

$$\text{Biomass} + \text{Steam} \rightarrow H_2 + CO + CO_2 + CH_4 + HC + \text{Tar} + \text{Char} \tag{9.2}$$

As steam gasification produces H_2-rich fuel gas with a higher heating value; steam gasification is regarded as a promising technology. According to reports, the maximum theoretical yield of H_2 from biomass via steam gasification is 17% (on a biomass weight basis). According to reports, the process can generate 50–55% H_2 (dry basis)-rich syngas (Kaushal and Tyagi 2012). Some of the advantages of steam gasification include providing an efficient method of producing renewable hydrogen, contributing to the highest yield of hydrogen from biomass, and producing a cleaner product with minimal environmental impact.

9.3 COMBINED SLOW PYROLYSIS AND STEAM GASIFICATION

Combined slow pyrolysis and steam gasification is the process of combining both slow pyrolysis and steam gasification techniques together. The process begins with slow pyrolysis of biomass, followed by steam gasification of char. Figure 9.3 depicts the process of combined slow pyrolysis and steam gasification.

The objective of slow pyrolysis of biomass is to increase the carbon content of char, which can be done by conducting pyrolysis at a low temperature, a long solid residence time, and a small heating rate. The pyrolyzed char is then gasified to produce hydrogen plenteous syngas. In steam gasification, the focus is on inducing hydrogen-generating reactions such as CO-shift, water-gas, steam–methane reforming, and tar reduction. The increased carbon content of char, as well as the stimulation of the aforementioned reactions, will undoubtedly better the conversion efficiency of biomass and the quality of gaseous products. The benefits of combining slow pyrolysis and steam gasification include improved syngas quality, increased H_2 yield, improved biomass conversion efficiency into syngas, low tar yield, and better product composition control (Parthasarathy and Sheeba 2015). The sections that follow go into greater detail about the combined slow pyrolysis and steam gasification of biomass.

FIGURE 9.3 Schematics of combined slow pyrolysis and steam gasification.

9.4 FACTORS INFLUENCING PRODUCTS YIELD IN COMBINED PROCESS

9.4.1 IMPORTANT FACTORS INFLUENCING CHAR PRODUCTION IN SLOW PYROLYSIS OF COMBINED PROCESS

9.4.1.1 Biomass Type
The type/nature of biomass used for slow pyrolysis process has a significant impact on product composition and yields. The quality of the soil in which the plant biomass is grown or cultivated, vegetation, farming methods, and composition of biomass components (extractives, moisture, cellulose, hemicellulose, lignin, and ash) in the biomass all can have an impact on the product composition. The same biomass does not have to have the same product composition. Many researchers have looked into different types of biomass.

9.4.1.2 Biomass Composition
The constituents of biomass, such as extractives, hemicellulose, cellulose, and lignin, behave in a different way during thermal disintegration and produce different products. Extractives decompose close to 200 °C, hemicellulose disintegrates around 150–350 °C range, cellulose degrades from 275 °C to 350 °C, and lignin decomposes at 250–500 °C (Yang et al. 2006; Apaydin-Varol, Putun, and Putun 2007). Because the composition of the biomass constituents and the temperature at which they degrade differ, they produce products of varying composition. On decomposition, cellulose and hemicellulose, for example, contribute predominantly to volatiles. Cellulose primarily produces condensable vapors and tar, whereas hemicellulose produces more non-condensable gases. In contrast, lignin primarily contributes gas and char. Lignin contributes 40% of the weight as char when heated slowly at 400 °C. It contributes only 10% of its weight to gaseous products. As a result, lignin decomposition is critical for obtaining the maximum yield of char in the product.

9.4.1.3 Biomass Particle Size
The biomass particle size is important in determining the yield and quality of char. Fine particles are generally chosen because they provide better heat and mass transfer. Coarser particles, on the other hand, are better suited for char generation. Finer particles, in general, produce more pores. The condensable produced during primary disintegration discharges through these pores, providing a small quantity of reactants for secondary cracking. It eventually leads to a higher liquid yield. Larger coarse particles, conversely, provide significant resistance to condensable vapor and cause it to stay inside the bed (biomass particle), providing a stage for secondary degradation favoring char generation.

Weerachanchai et al. investigated the influence of particle size on product composition/yields of palm kernel cake and cassava pulp residue at 700 °C and a heating rate of 20 °C/min (Weerachanchai, Tangsathitkulchai, and Tangsathitkulchai 2010). The particle size was varied between 0.71 and 3.56 mm. A reduction in char yield was observed up to 2 mm. However, there was an increase in yield of char thereafter. Guo and Lua investigated the pyrolysis behavior varying different feedstock particle sizes (Guo and Lua 2001). The particles smaller than 2 mm were reported to be influenced by the reaction temperature. For particles ranging in size from 2 to 50 mm, heat and mass transfer limitations played a major role determining pyrolysis products. It has been reported that heat transfer is less effective in larger particles than in smaller particles. Relatively larger particle sizes result in a slower pyrolysis rate and less conversion into products, primarily chars. At 675 °C, Demibras investigated the influence of particle size on char yield in corn cob (Demirbas 2004). The researcher discovered that increasing the particle size from 0.5 to 2.2 mm increased the char yield from 5.7% to 16.6%. There was a 65.7% increase in the yield of char. According to these findings, particle size must be optimal for char generation.

9.4.1.4 Pyrolysis Temperature

Temperature plays an important role in the pyrolysis process. Pyrolysis temperature is the temperature at which a biomass can be thermally decomposed (pyrolyzed) to yield useful products. To achieve effective thermal decomposition, biomass must be kept at pyrolysis temperatures for an extended period of time. Temperature and residence time have a considerable impact on char quality and quantity. In reality, higher temperatures produce more gaseous products, medium temperatures produce more liquid products, and lower temperatures produce more char.

Karaosmanoglu et al. investigated the slow pyrolysis of rapeseed seed straw and stalk at temperatures 350–650 °C and heating rates 10–30 °C/min (Karaosmanoglu, Tetik, and Gollu 1999). The effect of pyrolysis temperature on pyrolysis products was discovered to be more effective than heating rate. Temperature has been found to have a greater effect on pyrolysis products at higher heating rates than at lower heating rates. According to Asadullah et al. (2010), at low temperatures (<350 °C), the interlinked reactions of cellulose and lignin are superior, favoring the formation of char (Asadullah et al. 2010). Nonetheless, at high temperatures (>350 °C), depolymerization reactions dominate, resulting in the formation of volatiles. Şensöz and Can (2010) reported that low temperatures and small heating rates combined with a large amount of biomass sample promise significant char generation (Şensöz and Can 2010).

9.4.1.5 Heating Rate

The heating rate also has a significant impact on the yield and composition of the product. The two primary factors that largely influence product yields are temperature and heating rate. As pyrolytic decomposition entails several steps, the products of each step undergo transformation and decomposition over the entire temperature range. Regardless, heating rate alone does not define the product and its composition. The nature of the product is determined by the heating rate in conjunction with the product residence time. Instant heating to temperatures ranging from 400 °C to 600 °C yields more volatiles that can be condensed to generate liquids. While slow heating at moderate temperatures gradually releases volatiles, there is enough time for secondary reactions between char and volatiles to occur. This increased residence time allows for better char propagation. As a result, for maximum char yield, a low heating rate (0.01–2.0 °C/min) combined with a low temperature and a long residence time is required. High heating rate, moderate temperature (450–600 °C), and short solid residence time are required to achieve maximum liquid yield. Slow heating rate, high temperature (700–900 °C), and long residence time are required for maximum gas yield.

Slow pyrolysis experiments were carried out by Karaosmanoglu et al. at heating rates of 10 and 30 °C/min (Karaosmanoglu, Tetik, and Gollu 1999). It was discovered that a lower heating rate produced more char than a higher heating rate. According to the research, heating rates of 30 °C/min and above are suitable for higher liquid and gas yields, whereas heating rates of 10 °C/min or less are suitable for higher char yields. Gerçel pyrolyzed *Onopordum acanthium*. L. to investigate the influence of heating rate on pyrolysis products (Gerçel 2011). The effect was studied by varying two different heating rates of 7 and 40 °C/min while maintaining the same pyrolysis temperature. An 8% increase in char yield was noted when the heating rate was increased from 7 °C/min to 40 °C/min.

9.5 FACTORS INFLUENCING HYDROGEN GENERATION IN STEAM GASIFICATION OF THE COMBINED PROCESS

Some critical process parameters influence the yield of hydrogen in char steam gasification. Many researchers have studied the steam gasification of biomass to determine the effect of operating parameters such as temperature, residence time, steam-to-biomass ratio (*S/B*), particle size, biomass type, etc. on gasification performance.

9.5.1 Gasification Temperature

The most important factor in the steam gasification process is the gasification temperature. Several researchers have investigated the effect of temperature on product composition in steam gasification. In general, increasing the temperature accelerates the heating rate and favors more gaseous products. The increase in gasification temperature stimulates hydrogen-producing reactions such as water-gas, CO-shift, steam-methane reforming, and tar cracking. High temperatures also favor reactions such as Boudouard, which indirectly augments hydrogen generation. In general, high temperatures decompose biomass components more effectively and produce more gaseous products, which favors more hydrogen.

According to de Lasa, the optimal temperature for H_2 generation with minimal tar and CH_4 formation is 627–827 °C (De Lasa et al. 2011). According to Le Chatelier's principle, high temperatures induce reactants in exothermic reactions; however, they stimulate products in endothermic reactions. Since all H_2 generation reactions are endothermic, increasing the temperature increases H_2 yield considerably. According to Ahmed and Gupta (2009a), the increase in H_2 yield with temperature increase is caused by steam gasification of char and water-gas shift reaction (Ahmed and Gupta 2009a). Franco et al. discovered that H_2 yield increases with temperature (Franco et al. 2003). It was reported that the primary reasons for H_2 increase were the production of more gases in the initial pyrolysis stage (decomposition rate is faster at higher temperatures), favoring of endothermal char steam gasification at elevated temperatures, and an increase in gas yield due to steam reforming and cracking reactions.

9.5.2 Steam-to-biomass Ratio (*S/B*)

The *S/B* is another important factor that influences gas composition and yield. Many researchers have reported on the effect of *S/B* on product gas composition. In general, CH_4 is formed at low *S/B* values, while CO and H_2 are produced at moderate *S/B*. When the *S/B* ratio exceeds 1, C and CH_4 decompose further to produce H_2 and CO_2 at the expense of CO, promoting CO-shift reactions where CO is converted into CO_2. At the same time, the excess moisture in the steam lowers the temperature available in the reactor, thereby reducing the possibility of H_2 generation reactions like Boudouard, water-gas, CO-shift, methane formation, and tar cracking. It can be concluded that an excess amount of steam has a negative impact on H_2 generation. Hence, optimizing the supply of *S/B* is very critical for H_2 generation.

According to Gao et al., steam induces tar cracking, steam methane reforming, and char gasification reactions, which leads to increased gas yield (Gao, Li, and Quan 2009). However, it has been claimed that an excess of *S/B* will not favor H_2 generation. Acharya et al. discovered an increasing trend in the H_2 concentration and a downward trend in total gaseous yield when the steam pressure was increased (Acharya, Dutta, and Basu 2010). It was reported that the decrease in gaseous yield was caused due to the presence of surplus steam as it lowered the reaction temperature.

In their work, Li et al. discovered that an increase in steam initially resulted in an increase in total gas and H_2 yield. However, subsequent steam augmentation reduced total gas and H_2 yield (Li et al. 2009). When *S/B* was increased from 0.17 to 0.51, Abuadala and Dincer observed a 53% increase in H_2 yield (Abuadala and Dincer 2010). Sandeep and Dasappa carried out steam gasification experiments by varying *S/B* ratios. Higher *S/B* (1.5–2.7) produced a higher H_2 yield than lower *S/B* (0.75–1.5) (Sandeep and Dasappa 2014).

9.5.3 Catalyst

The role of catalysts in the study of pyrolysis and gasification reactions has recently gained traction. Researchers are working on natural and synthetic catalysts for thermochemical processes. Catalyst addition has been shown to promote gasification reactions such as water-gas, CO-shift, steam methane, and tar cracking, eventually leading to improved H_2 generation. The reaction time is greatly reduced due to the stimulation (by catalyst) of gasification reactions. Studies on the effect of catalyst on tar cracking are very encouraging. Tar cracking using catalyst is viewed as one of the effective methods of tar reduction. Catalyst presence also improves the conversion efficiency (biomass to gas products). All these effects put together results for an increased overall gas yield leading to higher H_2 yield.

Sutton et al. outlined some critical criteria for selecting catalysts (Sutton, Kelleher, and Ross 2001). A good catalyst should be able to effectively remove tar, inexpensive, readily available, highly reactive, and easily restored. Alkali metal catalysts, particularly potassium compounds, can aid in the production of H_2. However, these catalysts may have negative consequences such as ash deposition, ash slagging, ash fusion, and material surface corrosion. Moghtaderi investigated the role of Ni-based catalyst in the steam gasification of saw dust (Moghtaderi 2007). The catalyst's influence on H_2 yield was found to be insignificant at lower temperatures ranging from 200 °C to 450 °C. However, above 500 °C, the effect of the catalyst was significant. In another study, using a trimetallic catalyst, Li et al. observed a 74% increase in gas yield and nearly a 100% decrease in tar yield (Li et al. 2009).

9.5.4 Sorbent-to-biomass Ratio

If biomass used in steam gasification can be added to some sorbents to arrest CO_2, then it can be considered a CO_2 neutral fuel. CO_2 acceptors are the sorbents that are used to capture CO_2. Sorbent, in addition to removing CO_2, aids in the stimulation of H_2-generating reactions. CO_2 absorbed during gasification reactions promotes H_2-producing reactions (based on Le Chatlier's principle), such as water-gas, CO-shift, and steam-methane reforming reactions, ultimately improving H_2 yield.

Sorbent, like catalysts, must meet some basic criteria, which include easy availability, cost, and easy regenerative nature. In general, solid-based sorbents are more effective at capturing CO_2 than liquid sorbents. Some widely used sorbents are rhodium, aluminum oxide, nickel-based catalyst, dolomite, and CaO. There are both solid- and liquid-based sorbents. CaO sorbents are most commonly used because they are inexpensive and plentiful.

According to de Lasa et al., CaO sorbent can reduce gasification temperature by 150 °C (De Lasa et al. 2011). According to Florin and Harris, CaO sorbent can increase H_2 generation by up to 80% (Florin and Harris 2008). Kinoshota and Turn observed an increase in H_2 yield from 75% to 85% using an in situ CO_2 adsorption method (Kinoshita and Turn 2003). Acharya et al. discovered that increasing the CaO/biomass ratio from 0 to 2 increased H_2 yield by 2.7 times (Acharya, Dutta, and Basu 2010).

9.6 PRODUCTS OF PYROLYSIS AND GASIFICATION

9.6.1 Char

The solid by-product of pyrolysis is known as "pyrolytic char" or simply "char." Char has a variable carbon content, typically ranging from 60% to 90%. It also has some fixed carbon, volatiles (oxygen and hydrogen), and only a trace of inorganic ash. It is extremely porous and reactive. It does not cake, and because of this, it is easy to handle. The net calorific value (NCV) of biomass char is approximately 32 MJ/kg, which is significantly greater than the parent biomass (Diebold and Bridgwater 1997).

Char is a value-added product that can be used for a variety of purposes. Its primary application is as a fuel (charcoal). Because of its large microscopic surface area, char can also be used for filtration and pollutant adsorption. Char's adsorption capacity can be easily increased through physical or chemical activation, with the reactive product referred to as "activated carbon." More importantly, char has agricultural applications. "Biochar" refers to char that is intended for use as a soil amendment. Recently, the application of biochar to soil has gained popularity as a means of improving soil quality and sequestering carbon. Biochar improves the soil's ability to hold water and nutrients, reducing the need for fertilizers. It also helps to reduce the emissions of other greenhouse gases from the soil, such as N_2O and CH_4. Because of its high resistance to biological decomposition, biochar can directly store carbon for an extended period of time.

9.6.2 BIO-OIL

The two liquid products of biomass pyrolysis are bio-oil and tar. Aside from producing char, the pyrolysis process is also intended to produce bio-oil, whereas tar is an undesirable product. Aliphatic, aromatic, and naphthenic hydrocarbons, as well as oxygenated compounds such as furans, phenols, ethers, alcohols, aldehydes, ketones, and acids, make up the majority of bio-oil derived from biomass pyrolysis. Bio-oil constitutes the following composition: 20–25–30% water insoluble lignin, 25% water, 5–12% organic acids, 5–10% anhydrosugars, 5–10% non-polar hydrocarbons, and 10–25% of other oxygenated compounds (Shaw 2006). Bio-oil is typically darkish brown in color. The oxygenated bio-oils have a density of 1150–1300 kg/m3. Bio-oil has a pH of 2.5–3.0 and an absolute viscosity (500 °C) range of 40–100 cp. The HHV of bio-oil typically ranges from 16 MJ/kg to 19 MJ/kg (Mohan and Steele 2006).

Bio-oil can be used as a lubricant as well as to produce a variety of chemicals such as acetic acid, methanol, turpentine, phenols, levoglucosan, hydroxyacetaldehyde, and others. However, bio-oil cannot be used as a transportation fuel due to its low heating value, high viscosity, high water and oxygen content, corrosive nature, poor volatility, coking, and incongruity with conventional fuels. As a result, an upgrading process to convert bio-oil to palpable fuels is required (Vispute and Huber 2009).

9.6.3 TAR

Tar is an unwanted liquid product that occurs in a gasifier's low-temperature zones. It is a thick, black, and viscous liquid that is made up of a variety of hydrocarbons. Oxygenated chemicals, guaiacol, phenols, syringol, free fatty acids, and fatty acid esters are also present in tar (Razvigorova et al. 1994). The yield and composition of tar are influenced by several factors, the most important of which are the gasification temperature, reactor design and type, and feedstock type. Tar-laced syngas cannot be used in gas engines because it clogs the gas path and upsets the system. As a result, prior to entering the gas engine, the syngas must be tar reduced, which can be accomplished through proper gasifier design and the selection of operating variables such as reactor temperature and heating rate.

9.6.4 SYNGAS

The gaseous product of gasification is called "syngas," which typically comprises hydrogen, carbon monoxide, moisture, carbon dioxide, methane, aliphatic hydrocarbons, benzene, and toluene, and little traces of ammonia, hydrogen sulfide, and hydrochloric acid (Basu 2010). Due to its high energy content, syngas is used for numerous applications, including power production, ammonia synthesis in fertilizer industry, methanol synthesis in chemical industry, hydrogen and diesel gasoline production in refineries, etc. Syngas as such from a gasifier system contains impurities such as solid particulates, organic and inorganic substances, hence it needs to be cleaned and conditioned prior to its application.

9.7 CONCLUSION

This chapter discusses a novel method for producing hydrogen from biomass. The chapter delves into the combined slow pyrolysis and steam gasification process for hydrogen production, as well as the numerous factors that influence the combined slow pyrolysis and steam gasification process. It also discusses pyrolysis and gasification products.

Char generation in slow pyrolysis and hydrogen-inducing reactions in steam gasification must be encouraged to maximize hydrogen generation in the combined process. To be more specific, parameters favoring char quantity and quality should be encouraged in slow pyrolysis. These parameters include the following: biomass type, biomass composition, particle size, pyrolysis temperature, and heating rate. In order to promote hydrogen-liberating reactions in steam gasification, variables including gasification temperature, steam-to-biomass ratio, catalyst, and sorbent-to-biomass ratio must be better managed. This proposed combined slow pyrolysis of biomass followed by steam gasification of char would undoubtedly produce more hydrogen than the current thermochemical methods.

REFERENCES

Abuadala, A., and I. Dincer. 2010. "Efficiency Evaluation of Dry Hydrogen Production from Biomass Gasification." *Thermochimica Acta* 507–8: 127–34. https://doi.org/10.1016/j.tca.2010.05.013.

Acharya, B., A. Dutta, and P. Basu. 2010. "An Investigation into Steam Gasification of Biomass for Hydrogen Enriched Gas Production in Presence of CaO." *International Journal of Hydrogen Energy* 35 (4): 1582–9.

Ahmed, I., and A. K. Gupta. 2009a. "Syngas Yield During Pyrolysis and Steam Gasification of Paper." *Applied Energy* 86 (9): 1813–21. https://doi.org/10.1016/j.apenergy.2009.01.025.

Ahmed, I., and A. K. Gupta. 2009b. "Evolution of Syngas from Cardboard Gasification." *Applied Energy* 86 (9): 1732–40. https://doi.org/10.1016/j.apenergy.2008.11.018.

AlNouss, Ahmed, Prakash Parthasarathy, Hamish R. Mackey, Tareq Al-Ansari, and Gordon McKay. 2021. "Pyrolysis Study of Different Fruit Wastes Using an Aspen plus Model." *Frontiers in Sustainable Food Systems* 5 (February): 604001. https://doi.org/10.3389/fsufs.2021.604001.

AlNouss, Ahmed, Prakash Parthasarathy, Muhammad Shahbaz, Tareq Al-Ansari, Hamish Mackey, and Gordon McKay. 2020. "Techno-Economic and Sensitivity Analysis of Coconut Coir Pith-Biomass Gasification Using ASPEN Plus." *Applied Energy* 261 (March): 114350. https://doi.org/10.1016/j.apenergy.2019.114350.

Al-Rumaihi, Aisha, Prakash Parthasarathy, Anabel Fernandez, Tareq Al-Ansari, Hamish R. Mackey, Rosa Rodriguez, Germán Mazza, and Gordon McKay. 2021. "Thermal Degradation Characteristics and Kinetic Study of Camel Manure Pyrolysis." *Journal of Environmental Chemical Engineering* 9 (5): 106071. https://doi.org/10.1016/J.JECE.2021.106071.

Apaydin-Varol, Esin, Ersan Putun, and A. E. Putun. 2007. "Slow Pyrolysis of Pistachio Shell." *Fuel* 86: 1892–9. https://doi.org/10.1016/j.fuel.2006.11.041.

Asadullah, Mohammad, Shu Zhang, Zhenhua Min, Piyachat Yimsiri, and Chun Zhu Li. 2010. "Effects of Biomass Char Structure on Its Gasification Reactivity." *Bioresource Technology* 101 (20): 7935–43. https://doi.org/10.1016/J.BIORTECH.2010.05.048.

Basu, Prabir. 2010. *Biomass Gasification and Pyrolysis Practical Design and Theory*. First edition. Burlington: Academic Press.

Demirbas, Ayhan. 2004. "Effects of Temperature and Particle Size on Bio-Char Yield from Pyrolysis of Agricultural Residues." *Journal of Analytical and Applied Pyrolysis* 72 (2): 243–8. https://doi.org/10.1016/j.jaap.2004.07.003.

Demirbaş, Ayhan. 2005. "Hydrogen Production from Biomass via Supercritical Water Extraction." *Energy Sources* 27 (15): 1409–17. https://doi.org/10.1080/00908310490449379.

Diebold, J. P., and A. V. Bridgwater. 1997. "Overview of Fast Pyrolysis of Biomass for the Production of Liquid Fuels." In *Developments in Thermochemical Biomass Conversion*, edited by D. G. B. Boocock and A. V Bridgwater, 5–23. Dordrecht: Springer.

Florin, Nicholas H., and Andrew T. Harris. 2008. "Enhanced Hydrogen Production from Biomass with in Situ Carbon Dioxide Capture Using Calcium Oxide Sorbents." *Chemical Engineering Science* 63 (2): 287–316.

Franco, C., F. Pinto, I. Gulyurtlu, and I. Cabrita. 2003. "The Study of Reactions Influencing the Biomass Steam Gasification Process." *Fuel* 82 (7): 835–42. https://doi.org/10.1016/S0016-2361(02)00313-7.

Gao, Ningbo, Aimin Li, and Cui Quan. 2009. "A Novel Reforming Method for Hydrogen Production from Biomass Steam Gasification." *Bioresource Technology* 100 (18): 4271–7. https://doi.org/10.1016/j.biortech.2009.03.045.

Gerçel, Hasan Ferdi. 2011. "Bio-Oil Production from Onopordum Acanthium L. by Slow Pyrolysis." *Journal of Analytical and Applied Pyrolysis* 92 (1): 233–8. http://linkinghub.elsevier.com/retrieve/pii/S0165237011001112.

Guo, J., and A. C. Lua. 2001. "Kinetic Study on Pyrolytic Process of Oil-Palm Solid Waste Using Two-Step Consecutive Reaction Model." *Biomass and Bioenergy* 20 (3): 223–33. https://doi.org/10.1016/S0961-9534(00)00080-5.

Hernández, J. J., G. Aranda, J. Barba, and J. M. Mendoza. 2012. "Effect of Steam Content in the Air-Steam Flow on Biomass Entrained Flow Gasification." *Fuel Processing Technology* 99: 43–55.

Karaosmanoglu, F., E. Tetik, and E. Gollu. 1999. "Biofuel Production Using Slow Pyrolysis of the Straw and Stalk of the Rapeseed Plant." *Fuel Processing Technology* 59: 1–12.

Kaushal, Priyanka, and Rakesh Tyagi. 2012. "Steam Assisted Biomass Gasification-an Overview." *The Canadian Journal of Chemical Engineering* 90 (4): 1043–58. https://doi.org/10.1002/cjce.20594.

Kinoshita, C. M., and S. Q. Turn. 2003. "Production of Hydrogen from Bio-Oil Using CaO as a CO Sorbent." *International Journal of Hydrogen Energy* 28 (10): 1065–71.

Lasa, Hugo De, Enrique Salaices, Jahirul Mazumder, and Rahima Lucky. 2011. "Catalytic Steam Gasification of Biomass: Catalysts, Thermodynamics and Kinetics." *Chemical Reviews* 111 (9): 5404–33. https://doi.org/10.1021/cr200024w.

Li, Jianfen, Yanfang Yin, Xuanming Zhang, Jianjun Liu, and Rong Yan. 2009. "Hydrogen-Rich Gas Production by Steam Gasification of Palm Oil Wastes Over Supported Tri-Metallic Catalyst." *International Journal of Hydrogen Energy* 34: 9108–15.

Moghtaderi, B. 2007. "Effects of Controlling Parameters on Production of Hydrogen by Catalytic Steam Gasification of Biomass at Low Temperatures." *Fuel* 86 (15): 2422–30. http://linkinghub.elsevier.com/retrieve/pii/S0016236107000774.

Mohan, Dinesh, Charles U. Pittman, and Philip H. Steele. 2006. "Pyrolysis of Wood/Biomass for Bio-Oil: A Critical Review." *Energy and Fuels* 20 (3): 848–89. https://doi.org/10.1021/EF0502397.

Parthasarathy, P., T. Al-Ansari, H. M. Mackey, and G. McKay. 2021. "Effect of Heating Rate on the Pyrolysis of Camel Manure." *Biomass Conversion and Biorefinery 2021* (May): 1–13. https://doi.org/10.1007/S13399-021-01531-9.

Parthasarathy, P., A. Fernandez, T. Al-Ansari, H. R. Mackey, R. Rodriguez, and G. McKay. 2021. "Thermal Degradation Characteristics and Gasification Kinetics of Camel Manure Using Thermogravimetric Analysis." *Journal of Environmental Management* 287 (June): 112345. https://doi.org/10.1016/j.jenvman.2021.112345.

Parthasarathy, P., and K. N. Sheeba. 2015. "Combined Slow Pyrolysis and Steam Gasification of Biomass for Hydrogen Generation-a Review." *International Journal of Energy Research* 39 (2). https://doi.org/10.1002/er.3218.

Razvigorova, Maria, Maria Goranova, Venecia Minkova, and Jaroslav Cerny. 1994. "On the Composition of Volatiles Evolved During the Production of Carbon Adsorbents from Vegetable Wastes." *Fuel* 73 (11): 1718–22. https://doi.org/10.1016/0016-2361(94)90158-9.

Sandeep, K., and S. Dasappa. 2014. "Oxy–Steam Gasification of Biomass for Hydrogen Rich Syngas Production Using Downdraft Reactor Configuration." *International Journal of Energy Research* 38 (2): 174–88. https://doi.org/10.1002/ER.3019.

Saxena, R. C., Diptendu Seal, Satinder Kumar, and H. B. Goyal. 2008. "Thermo-Chemical Routes for Hydrogen Rich Gas from Biomass: A Review." *Renewable and Sustainable Energy Reviews* 12 (7): 1909–27. https://doi.org/10.1016/j.rser.2007.03.005.

Şensöz, Sevgi, and Mukaddes Can. 2010. "Pyrolysis of Pine (Pinus Brutia Ten.) Chips : 1. Effect of Pyrolysis Temperature and Heating Rate on the Product Yields." *Energy Sources* 24 (4): 347–55.

Shaw, Mark. 2006. "Pyrolysis of Lignocellulosic Biomass to Maximize Bio-Oil Yield: An Overview." *American Society of Agricultural and Biological Engineers* (November). https://doi.org/10.13031/2013.22063.

Sutton, David, Brian Kelleher, and Julian R. H. Ross. 2001. "Review of Literature on Catalysts for Biomass Gasification." *Fuel Processing Technology* 73 (3): 155–73. https://doi.org/10.1016/S0378-3820(01)00208-9.

Vispute, Tushar P., and George W. Huber. 2009. "Production of Hydrogen, Alkanes and Polyols by Aqueous Phase Processing of Wood-Derived Pyrolysis Oils." *Green Chemistry* 11 (9): 1433–45. https://doi.org/10.1039/B912522C.

Weerachanchai, Piyarat, Chaiyot Tangsathitkulchai, and Malee Tangsathitkulchai. 2010. "Comparison of Pyrolysis Kinetic Models for Thermogravimetric Analysis of Biomass." *Suranaree Journal of Science and Technology* 17 (4): 387–400.

Yang, Haiping, Rong Yan, Hanping Chen, Dong Ho Lee, David Tee Liang, and Chuguang Zheng. 2006. "Pyrolysis of Palm Oil Wastes for Enhanced Production of Hydrogen Rich Gases." *Fuel Processing Technology* 87 (10): 935–42. https://doi.org/10.1016/j.fuproc.2006.07.001.

10 Insight into Current Scenario of Electronic Waste to Nanomaterials Conversion

Menaka Jha, Sunaina, Sapna Devi, and Nausad Khan

CONTENTS

10.1 Overview of Electronic Waste Generated Worldwide 133
    10.1.1 Composition of e-Waste 134
    10.1.2 Diverse Materials in e-Waste 135
10.2 Conventional Methods for Treatment of e-Waste: Advantages and Disadvantages 136
    10.2.1 Landfilling 136
    10.2.2 Pyrolysis 136
    10.2.3 Repair 137
    10.2.4 Recycling 137
    10.2.5 Pyrometallurgical Processing 138
    10.2.6 Hydrometallurgical Processing 138
    10.2.7 Biometallurgical Processing 138
10.3 Alternative Approaches to Treat the Electronic Waste for Extracting Nanomaterials 139
    10.3.1 Batteries 140
    10.3.2 Discarded Machines and Instruments 140
10.4 Recent Advances in Research Related to Extraction of Nanomaterial from Electronic Waste 140
10.5 Challenges Related to Purity and Stability of the Extracted Nanomaterials Which Limit Their Commercialization 142
10.6 Conclusion and Future Prospects 143
References 144

10.1 OVERVIEW OF ELECTRONIC WASTE GENERATED WORLDWIDE

Advanced electrical and electronic devices become major part of the modern society. They symbolize the modern lifestyle, efficacy, prosperity, and comfort in many countries. However, after the end of their life span, they become electronic waste. Electronic waste (e-waste) is defined as "waste generated from the electrical and electronic devices that comprises of all components of electronic equipment and its sub-assemblies which are the part of the product at the time it is discarded" as per the European Union. Hence, e-waste includes all types of outdated, damaged, discarded, or unwanted electronic gadgets. Increased demand of portable electronic devices over the last decade has resulted in the huge increase of electronic waste. It makes e-waste a fastest growing sector in global economy. This inclination is affected by many factors such as urbanization, fast economic growth, industrial development, and increased demand for consumer goods (Babu, Parande, and Basha 2007). As per the Global E-waste Monitor 2020, 53.6 Mt e-waste was generated worldwide in 2019, which is almost six times since 2014. It is expected that almost 74.7 Mt e-waste will be generated by 2030. The interest of middle-class families has increased tremendously in developing countries toward the high-performing electronic appliances. Therefore, they are spending a lot on the purchase of electronic gadgets. In the electronic devices, there are lot of money spend on electronic instruments. They have many substances which are toxic in nature, such as heavy metals (lead, chromium, cadmium), polychlorinated biphenyls, brominated flame retardants, etc. Burning of these circuits/wires/gadgets releases chlorine, bromine, and other toxic products such as multiple chlorinated and brominated dioxin compounds, including mixed halogenated dibenzo-*p*-dioxins/dibenzofurans (Tue et al. 2013). Developed countries such as the United States, Japan, and European Union have possible infrastructure potential to recycle the e-waste, but there is legal or illegal transport of electronic waste to developing countries that have no regulations for the e-waste management (Ghimire and Ariya 2020). The report by Global E-waste Monitor 2020 stated that there is ~17.4% enhancement in e-waste since 2014 which formally collected or recycled. The unaccounted e-waste is generally illegally traded, incinerated, or landfilled. Figure 10.1 depicts the generation, recycling, and waste produced from e-waste worldwide in 2019. Recently, many developing countries have imposed many laws related to handling and managing electronic waste. Although 70% countries have some e-waste-related policies and laws, but their proper implementation is yet to achieved. The manhandling with no safety precautions put the environmental and public health at serious risks (Widmer et al. 2005). In recent years, e-waste is a hot topic in the field of sustainability as it involves energy, economy, waste management, technology, environment, human health, international affairs, and policy. Despite being a complex problem, there are still many opportunities for sustainable growth. In 2019, the raw materials extracted from the e-waste is assessed to be of nearly $57 billion. Therefore, there is great scope in conversion of e-waste to commercially important products. However, this opportunity has also many challenges such as complex structure of e-waste, economical issue, emission of toxic gases, disposal problem related to unwanted and hazardous residues, and lack of international collaboration. There is need of systematic approach to integrate the current knowledge, evaluate the challenges (economical, environmental, health, technological, governmental, policy and implementation, etc.), and explore sustainable solutions for better e-waste management (Weber and Kuch 2003).

10.1.1 Composition of e-Waste

Electronic waste is composed of a diverse range of materials, depending upon the type and category of electrical and electronic equipment (Needhidasan, Samuel, and Chidambaram 2014). Some of the

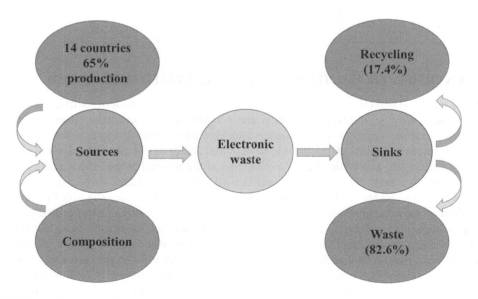

FIGURE 10.1 Schematic of e-waste generation and sinks.

materials found in e-waste are valuable metals (Au, Ag, Pt, Pd), transition and post-transition metals (Cu, Al, Ni, Sn, Zn, Fe, etc.) and metals of concern (Hg, Be, In, Pb, Cd, As, Sb, etc.), halogens and combustibles (plastics, organic fluids, etc.) (Hageluken 2006). Table 10.1 shows the contents of different materials in e-wastes obtained from different sources.

10.1.2 Diverse Materials in e-Waste

e-Waste contains more than thousands different materials, out of which many of elements are valuable and many are hazardous (Widmer et al. 2005). Burning the e-waste can segregate the plastics from the metals but simultaneously releases the toxic materials into the environment and cause pollution. Table 10.2 shows sources of typical toxic materials present in the e-waste and their adverse effects on humans.

TABLE 10.1
Composition in Different Types of Electronic Waste

e-Waste	Fe	Cu	Al	Pb	Ni	Plastic	Ag	Au	Pd	Reference
	Contents (w/w%)						Content (ppm)			
TV	28	10	10	1.0	0.3	28	280	17	10	(Hageluken 2006)
PCB	7	20	5	1.5	1	23	1000	250	110	(Hageluken 2006)
Mobiles	5	13	1	0.3	0.1	57	1340	350	210	(Hageluken 2006)
Calculators	3	3	5	0.1	0.5	0.2	260	50	5	(Hageluken 2006)
TV scrap	–	3.4	1.2	0.2	0.038	–	639	<10	<6	(Cui and Zhang 2008)
PCB scrap	–	10	7	1.2	0.85	–	280	110	–	(Hanafi et al. 2012)

TABLE 10.2
Typical e-Waste Sources and Their Health Implications

Components	Typical Sources	Effects on Humans	Reference
Mercury	Sensors, computers, printed circuit boards, cathode ray tubes, fluorescent lamps, etc.	Chronic damage to brain	(Grant et al. 2013)
Lead	Printed circuit boards, cathode ray tubes, light bulbs, and batteries	Affects the nervous system, blood system, and kidneys. Brain development of children	(Grant et al. 2013)
Cadmium	Switches, springs, printed circuit boards, xerox machines, cathode ray tubes, mobile phones, etc.	Respiratory irritation, chronic lung disease, toxicity to kidneys, etc.	(Grant et al. 2013)
Chromium	Anticorrosion coatings, data tapes, etc.	Strong allergic reactions such as asthmatic bronchitis, DNA damage	(Grant et al. 2013)
Beryllium	Power supply boxes, computers, X-ray machines, ceramic parts of electronics, etc.	May cause lung cancer and skin diseases	(Grant et al. 2013)
Polychlorinated biphenyls (PCBs)	Dielectric fluids, lubricants, coolants in generators, capacitors, and transformers, fluorescent lights, ceiling fan, dishwashers, and electric motors	Affects immune hormone, nervous, and enzyme systems. Carcinogenic for humans	(Grant et al. 2013)
Brominated flame retardants polybrominated diphenyl ethers	Fire retardants for electronic equipment	Affects growth hormones, sexual development, immune systems, and brain development in animals	(Grant et al. 2013)

10.2 CONVENTIONAL METHODS FOR TREATMENT OF E-WASTE: ADVANTAGES AND DISADVANTAGES

Nowadays, the most commonly used batteries are zinc-carbon dry cell and zinc-manganese dioxide, respectively. Both the batteries have been popularly used as power sources in household applications such as toys, radios, cameras, torches, etc. The main threatened use of these electronics is that it is openly discarded once it is discharged and ultimately leads to environmental hazards (Ghimire and Ariya 2020). In survey it is found that approximately 20,000 tons of primary battery are dumped annually, and 635,000 tons of batteries have been dumped in the environment in the last 43 years in Mexico (Bartolozzi 1990). Majorly these spent batteries consist of waste stream of metallic components such as Cd, Hg, and Pb, which is threat to the living organisms and is polluting natural reservoirs such as soils and groundwater. The following are the conventional methods for recycling e-waste with their merits and demerits.

The best way to treat electronic waste is repairing, reusing, or recycling the important material it contains. However, it is challenging too, as electronic waste contains toxic material also. Therefore, the ideal principle for e-waste treatment is recycling, transportation, and disposal of remaining waste with least side effects on health and the environment. There are various methods reported to manage e-waste and all these technologies vary between developed and developing nations. For instance, China still face challenges in processing PCBs and dealing with toxic substances in waste; on the other hand, EU has developed formal collection center for recycling many metals with high efficiency (Ghimire and Ariya 2020). However, the basic recycling method is not much efficient to recover all the metals, for example, copper and small traces of gold can be recovered in developing nations, but they are still not able to extract rare metals like indium, ruthenium, and palladium (Ghimire and Ariya 2020). The well-known conventional methods of e-waste handling are discussed in the following sections.

10.2.1 Landfilling

Disposal to landfilling is widely used methods for e-waste treatment as it is simple and easy in operation. It involves dumping of waste openly or in voids/pits, and this results in the formation of leachate in the landfill site which ultimately leads to change into wasteland and difficult to be exploited in the future (Ning et al. 2017). The drawback of this method is that e-waste releases toxic metals and polyhalogented organic contaminated material which are harmful to health as well as the environment (Ghimire and Ariya 2020). The harmful metals which are not disintegrated by physical processes remain as pollutants for a long time, disturbing the biogeochemical cycles, as it tends to accumulate within organism, including human, through food chains and causes negative effect on vital organs and sometimes even leads to death. Recent studies have shown that leachates and groundwater collected from the polluted landfills in Australia contained higher concentration of Pb, Al, As, Fe, and Ni. A comparison of landfills with e-waste and without e-waste reveals that landfills with e-waste contain higher toxic metals. Thus, landfilling is not the best approach to treat e-waste (Ghimire and Ariya 2020).

10.2.2 Pyrolysis

Pyrolysis or incineration process involves burning of combustible fraction of e-waste to obtain non-combustible metals. The benefits of incineration are faster rate of reaction and no tedious work of separation and recycling of metals, gases, and liquid obtained as a result of burning. These materials obtained after burning provide energy to self-sustain the process and thus serve as chemical or energy source (Kumar, Holuszko, and Espinosa 2017). Despite this, metal obtained from this process is chemically dependent on temperature; at low temperature, copper acts as catalyst and results in the release of polybrominated dibenzodioxins (PBDDs), polybrominated dibenzo-furans

(PBDFs), polychlorinated dibenzodioxins (PCDDs), polychlorinated dibenzofurans (PCDFs), fly ash, carbon oxides, hydrogen bromide, methane, ethylene, benzene, toluene, phenol, benzofuran, styrene, PAHs, bromophenols, etc. (Gullett et al. 2007). Other landfilling non-metallic products can result in secondary pollution produced by heavy metals and BFRs contaminated the groundwater (Ghimire and Ariya 2020). On the other hand, combustion at temperatures greater than 1200 °C helps to reduce CO formation as well as remove PBDD/Fs in the form of HBr or Br_2 and possesses lower toxicity (Ni et al. 2012); but at higher temperatures, formation of NO_x takes place above the standard emission level of 50 mg/nm³. Although HBr and Br_2, are less toxic, before releasing to environment they need additional treatment such as adsorption (Ning et al. 2017). Pyrolysis involves dehalogenation and pyrolyzing and showed prominent outcome for chlorinated plastics of e-waste, but also escalating the cost of recycling (Ghimire and Ariya 2020). The disadvantage is the remnants of pyrolysis, majorly termed as "bottom ash," contains high concentration of heavy metals such as Cu, Pb, and Cd, which is another concern for safer disposal (Long et al. 2013). Although these pollutants can be removed via iron oxide nanoparticle adsorbent, it is yet to be explored (Hu et al. 2014). It can be concluded that pyrolysis is tedious, expensive, and hazardous to environment.

10.2.3 Repair

For better sustainable waste management, reuse method is one of the best ways because it lowers the synthesis volume of EEE by reducing the amount of e-waste. For instance, a large volume of EEE such as mobile phones have created an opportunity to repair and reuse, but they are still not used in many countries. However, the limitation is change in product design, technology, and wireless services that cause difficulties to repair instruments (Wieser and Tröger 2016).

10.2.4 Recycling

This is one of the most appropriate methods for e-waste treatment because it reduces carbon emission and restricts many usable or hazardous substance. It was recorded in reports that a small fraction (17.4%) of total e-waste was generated in 2019 and collected from recycle methods. EEE consists of 38% of ferrous and 28% of non-ferrous materials, 19% plastics, 4% of glass, 1% of wood, and 10% of other materials. In case of printed circuit boards, which are part of electronic devices, are also rich in metal content, approximately 1 ton of circuit boards composed of 40–800 times excess gold and 3–40 times more copper than natural ores (Ghimire and Ariya 2020). These data revealed that e-waste is an excellent precursor of recycled metals and used as source material for electronic device production. The major drawback of e-waste is that it is non-homogeneous and very complex. On the other hand, PCB consists of heterogeneous mixture of metals, polymer, ceramics, and hazardous substances, and their composition is variable over time period. Thus, variation in composition causes difficulty in separation and recovery process (Kaya 2016).

For an effective recycling of e-waste, it requires complete information and characterization of the waste in terms of its composition, toxic substances, physical properties, and chemical properties (Ghimire and Ariya 2020). For analyzing valuable as well as hazardous materials, gamma activation analysis technique shows promising results with excellent sensitivity for elements such as Au, Ag, Cu, Sn, Br, and iodine (Sahajwalla et al. 2016). Recycling process involves three steps: dismantling, processing, and end processing. Dismantling involves separation of valuable and hazardous components manually or mechanically. In some countries labor cost is high, so they have mainly preferred mechanical method in which smasher is used to break e-waste, whereas cross-flow shedder is used to cut the scrap. Further, crushing and separation are carried out by hammer mills, magnetic separation, sieves, etc. These techniques deal with the production of metal concentrate that is further processed in metal mills and plastics. The third stage involves end processing that take place at different destinations depending on the desirable output; for example, ferrous materials are processed in steel plant to produce iron, while copper, circuit boards, etc. consist of precious metals that are processed in metal smelter and the remaining remnants are sent to disposal. The only disadvantage of recycling is that e-waste consists of 15 different polymer-based materials such as PBBs/PBDEs which make difficult to recycle plastics (Ghimire and Ariya 2020).

Recycling process has some pros and cons which are discussed in detail in the following sections.

10.2.5 PYROMETALLURGICAL PROCESSING

This method is mainly used to recover non-ferrous materials such as copper and precious metal with high efficiency by hydrometallurgical/electrometallurgical methods (Cui and Zhang 2008). This processing involves the following steps: smelting, combustion, and molten salt. First, e-waste is crushed and passed to molten bath at 1250 °C to remove plastics. The mixture is heated in air (up to 39% oxygen). At high temperatures, formation of ozone take place which converts impurities such as zinc, lead, etc. to their oxides and thus form silicate solution known as slag. It is then cooled and mined to extract precious metal before disposal. The molten copper is transferred to converters after extracting precious metals, and the liquid blister copper is refined in furnace and collected with 99.1% purity. Despite widely used, smelting methods have some drawbacks such as inability to recover Fe and Al, high energy consumption, and emission of hazardous by-products such as dioxane and halogen compound (Hubau et al. 2019).

10.2.6 HYDROMETALLURGICAL PROCESSING

This method is much superior than smelting processing owing to its high accuracy, predictability, and, most importantly, reduced environmental impacts (Ilyas et al. 2010). It involves conversion of e-waste into granular form, separation, purification, and recovery of metals by using alkaline or acidic media to leach metal from e-waste (Tuncuk et al. 2012). Most commonly utilized leaching agent is nitric acid for base metals, sulfuric acid, aqua regia for copper, cyanide for gold and silver, and sodium chlorate for palladium (Ilyas et al. 2010). The disadvantage of using hydrometallurgy processing is that it uses cyanide as a leaching agent for gold which causes toxicity and contamination of rivers and natural reservoirs (Birloaga et al. 2013). Although pretreatment of cyanide is possible before disposal, this is quite expensive (Ghimire and Ariya 2020). Possible alternatives for cyanide are thiosulfate, thiourea, and halides, which possess high gold recovery efficiency, are cheaper, and have less impact on the environment (Cui and Zhang 2008). Aqua regia, a 3:1 mixture of concentrated HCl and HNO_3, is another alternative to dissolve gold (Sheng and Etsell 2007), but use of chlorine causes toxicity of chlorine gas and strong corrosives of chlorine solution (Cui and Zhang 2008). Replacement of iodine reduces toxicity and is less dangerous to the environment, but the limitation is the high cost of solvent (Konyratbekova, Baikonurova, and Akcil 2015).

10.2.7 BIOMETALLURGICAL PROCESSING

This process involves microorganisms to use metal for their catalytic functions by binding metal ions to the cell surface or transferring them inside the cell for various intracellular functions (Cui and Zhang 2008). The different interaction includes sorption, reduction, and oxidation and sulfide precipitation (Ghimire and Ariya 2020). Most commonly explored microbes are cyanogenic bacteria, mesophilic chemolithotrophic bacteria, thermophilic bacteria, acidophilic bacteria, and fungi (Kaya 2016). Over the last decades, biometallurgy has gained interest from major industries and has emerged as a promising recycling process. Like hydrometallurgy, biometallurgy involves leaching process – the only difference is in the use of chemical produced by the microbe itself. Leaching helps metal in forming complexation or precipitation and then separated from culture broth for further refining. It is further divided into two categories: bioleaching and biosorption. In order to extract metal from sulfide ores enriched with base and precious metals, bioleaching process is carried out, and it is a bacteria-assisted method (Morin et al. 2006). This method involves oxidation of sulfide to water-soluble sulfate to recover desired metal by microbial oxidation. Microbial oxidation involves direct and indirect oxidation; direct oxidation involves oxidation of minerals and metals, and indirect oxidation involves conversion of ferrous to ferric, which further acts as an oxidizing agent for minerals. On industrial scale, to obtain copper, gold, and their ores, bioleaching process has been used (Ghimire and Ariya 2020). Furthermore, various studies have shown feasibility of bioleaching in the recovery of metals from e-waste (Natarajan and Ting 2015). The drawbacks of bioleaching process are that it is tedious, slow, and uncontrolled secondary reaction and precipitation of lead, tin, and other metals occur (Ilyas et al. 2010). Biosorption is another mode in biometallurgical processing that use biomass of microorganisms called biosorbent for sorption of metal (Chatterjee and Abraham 2017). Biosorbent are dead biomass of microorganisms such as algae, fungi, and bacteria (Cui and Zhang 2008). The advantages of biosorption are that it is eco-friendly, easy to operate, low cost, reduce chemical or biological sludge, and highly efficient in detoxifying eluents (Cui and Zhang 2008). The limitations are that it has tedious recovery process and is less effective to recycle metallic e-waste (Kumar, Holuszko, and Espinosa 2017). Although mineral processing has given hope for e-waste treatment in biometallurgy, the size of particles in e-waste treatment vary in the two processes (Cui and Zhang 2008). Despite this, reports revealed that biometallurgy can offer promising recycling technique of electronic waste. Research is still going on for effectiveness of this technology.

10.3 ALTERNATIVE APPROACHES TO TREAT THE ELECTRONIC WASTE FOR EXTRACTING NANOMATERIALS

Although recycling the e-waste is a conventional approach to manage e-waste, the extraction or recovery of nanomaterials was attempted much later (Figure 11.2). One of the pioneered groups in the field of waste management is that of Xi et al. (Xi, Li, and Liu 2004) who have established the synthesis method of magnetic Zn-Mn ferrite materials from waste batteries. Thereafter a lot of research has been conducted in the field of waste management for the recovery of various nanomaterials (Figure 10.2).

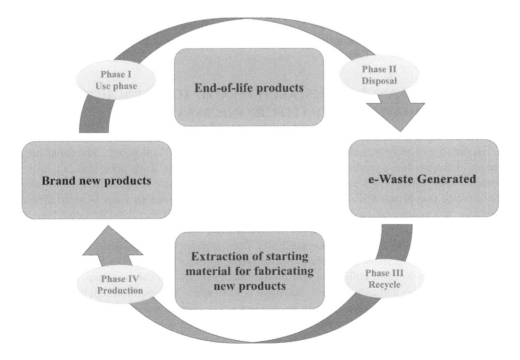

FIGURE 10.2 Closed-loop electronic material recycling.

The electronic waste generally used for recovery and extraction of valuable metals and their other compounds can be broadly classified into two categories: (i) battery waste and (ii) used machines and instruments.

10.3.1 Batteries

One of the most common recycled electronic waste is spent batteries. Batteries are portable power sources which have attracted the scientists working in the area of sustainable waste management due to their large usage. The batteries consumption was estimated to be 8 billion units per year in the United States and Europe, 6 billion units in Japan, and 1 billion unit in Brazil annually. Generally, these batteries are discarded causally as a domestic waste without taking proper measures. Batteries can be handled and recycled easily at laboratory level. Primary cells (such as Li-ion, alkaline Zn-MnO$_2$, Zn-carbon, etc. batteries) are widely used for metal extraction. Commonly, metallic Co, Li, Zn, Mn, Pb, Ni, Au, Pd, and rare earths are recycled from these spent batteries. In addition, various other commercially important nanomaterials such as ZnO, ZnMn$_2$O$_4$, Mn-Ni-Zn ferrites, MnO$_2$/graphene, and other perovskite have been synthesized successfully at laboratory scale (Duan et al. 2016; Xu et al. 2008).

10.3.2 Discarded Machines and Instruments

Other category of e-waste is discarded circuit boards used in computers, mobiles, laptops, automobiles, etc. Many metals such as Pb, Hg, Cu, Fe, Au, Ag, Pd, and Pt and other polymeric materials are also extracted from these e-wastes. A recent study has demonstrated a complete recovery of Ag by incinerating the organic solar cells. Other materials such as zero-valent copper particles, carbon nanotubes, Cu$_2$O, and Cu$_2$O/TiO$_2$ catalysts are also extracted from e-wastes. Very less research has been conducted on the e-waste other than the batteries for the recovery of metals or other nanomaterials. For an instance, printed circuit boards are electronic assemblies used in electric and electronic devices and are composed of Cu, Pb, Fe, Au and Hg etc for making electric connections. Overall, there is a remarkable opportunity for future investigation of waste printed circuit boards for the recovery of nanomaterials (Søndergaard et al. 2016; Cui and Anderson 2016).

A variety of commercially important metals and their oxides can be extracted from these e-wastes, which are listed in Table 10.3.

10.4 RECENT ADVANCES IN RESEARCH RELATED TO EXTRACTION OF NANOMATERIAL FROM ELECTRONIC WASTE

At present, the rapid escalation of electrical and electronic products has impacted everyone's life. These inevitable products include refrigerators, television, torches, laptop, etc. The rapid increase in volume of disposable e-waste is a threat to local community as it includes toxic metallic and non-metallic components. However, these e-wastes are the secondary resource of various precious metals; e-waste comprises 50% of Fe and steel and plastic material in form of polycarbonates. Furthermore, it contains various metals such as Cu, Zn, Ni, Au, and Pd (Natarajan and Ting 2014). Interestingly, it is found that the quantity of Cu in e-waste is 40 times higher than the natural ores (Nithya, Sivasankari, and Thirunavukkarasu 2021). Thus, e-waste is a pool of secondary resources of precious base metals. Figure 10.3 elucidates the recycling approach for metal recovery from e-waste; the process involves e-waste collection and pretreatment process. In the first step, it involves collection of e-waste from domestic as well industrial landfill sites. The pretreatment process involves dismantling of e-waste such as scrap, wire by size reduction, and de-soldering and fractionation for metallic and non-metallic components. Thus, in the final stage, the pretreated e-waste is subjected to metal recovery through pyrometallurgical, hydrometallurgical, and biohydrometallurgical ways. Several studies on practicing of these approach for efficient recycling and extraction of metals from

TABLE 10.3
List of Extracted Nanomaterials from Recyclable Waste

S. No.	Target Material	Recyclable Waste	Recycling Method	Recovery (%)	Reference
(i) Batteries					
1	ZnO	Zn-MnO$_2$ batteries	Hydrometallurgy-precipitation and liquid–liquid extraction	>98 (99% purity)	(Mantuano et al. 2006)
2	Zn$_x$Mn$_{1-x}$O	Zn-MnO$_2$ batteries	Hydrometallurgy	57 (yield)	(Qu et al. 2015)
3	Zn	Zn-Mn batteries	Thermal procedure-inert gas condensation and vacuum separation	>99 (99% purity)	(Deep et al. 2016)
4	Ni-Mn-Zn/Mn-Zn ferrites	Zn-Mn, Ni-MH, Zn-C batteries	Sol-gel, leaching, and precipitation	98% (Zn, Mn), 55% (Fe)	(Ma et al. 2014)
5	Polyaniline/graphite nanocomposites	Spent battery powder (SBP)	Oxidative polymerization	–	(Qu et al. 2015)
6	Lead iodide, perovskite NCs	Car batteries	Roasting at 500–600 °C and dissolution	–	(Wen et al. 2016)
7	Co ferrites	Li-ion batteries	Combined sol–gel and hydrothermal method, coprecipitation	99.9% (purity)	(Long et al. 2013)
8	MnO$_2$/graphene nanocomposites	Zn-MnO$_2$ acidic dry batteries	Precipitation method	–	(Xiu and Zhang 2012)
(ii) Used machines/instruments					
1	Nano-zero-valent Cu particles	Automobile shredder	Hydrometallurgy	–	(Singh and Lee 2016)
2	Cu$_2$O/TiO$_2$ catalyst, Cu$_2$O NPs	Printed circuit boards		90 (wt% Cu)	(Xiu and Zhang 2012)
3	Nano-Pb	Printed circuit boards	Vacuum separation, dynamic inert gas, condensation	98%	(Zhan et al. 2016)
4	Nano-Al$_2$O$_3$	Al electrolytic solution	Co-precipitation	99.9%	(Wu and Chang 2016)

e-waste are available (Barbieri et al. 2010). However, no effective operational reports are known for these approaches till date (Nithya, Sivasankari, and Thirunavukkarasu 2021).

Recycling of e-waste approaches with pretreatment is followed by the metal extraction by pyrometallurgy, hydrometallurgy, and biometallurgy. The pyrometallurgical process involves smelting of e-waste in high-temperature furnaces and then refractories are obtained as slag which is further processed as a precursor for extraction of metals (Cayumil et al. 2014). In hydrometallurgy process, selected lixiviates are utilized to leach out the metals by stable complexation (Garg et al. 2019). In biometallurgy, microorganisms are utilized for recovering metal either in the presence of cell or in the cell-free extract medium (Nithya, Sivasankari, and Thirunavukkarasu 2021). Of all the three processes, microbial-assisted process is relatively more efficient, low-cost, and required minimum industrial facilities utilized for metal recovery at low concentrations (Anjum, Shahid, and Akcil 2012).

In recent decades, the utilization of electrical equipment with advanced technology has increased rapidly; thus, every year the rate of e-waste generation is growing faster than the solid waste. The nations known for significant production of e-waste are China and America. However, different methods have been explored such as electrochemical technique, ion-exchange, etc. to treat the e-waste. At present, many metals are recovered through recycling approach, such as extract copper as copper nitrate through hydrometallurgy method, followed by conversion into copper oxide nanoparticle by green synthesis using leachates (Nithya, Sivasankari, and Thirunavukkarasu 2021). Further, copper

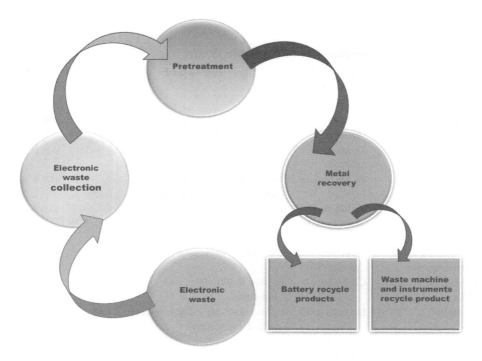

FIGURE 10.3 Different stages during the recycling of the electronic waste.

oxide nanoparticles are utilized as adsorbents for removal of methylene blue dye. A nanosized material can render useful solution to environmental issues owing to its small size, highly active surface area, and significant physicochemical properties (Thirunavukkarasu, Muthukumaran, and Nithya 2018). Many other nanomaterials can be recycled from electronic waste, which can be tailor-made and can be used for environmental remediation in the form of nanowires, nanotubes, and nano-quantum dots. Thus, the field of nanoscience and nanotechnology is becoming most explored area and has attracted many researchers.

To recycle nanomaterials from e-waste, many techniques were reported in literature which involved multiple step processing. But the most common used process is hydrometallurgical and pyrometallurgical processes, which work on the principles of precipitation and thermal treatments (Dutta et al. 2018).

10.5 CHALLENGES RELATED TO PURITY AND STABILITY OF THE EXTRACTED NANOMATERIALS WHICH LIMIT THEIR COMMERCIALIZATION

Several challenges are faced while regenerating e-wastes owing to their diversity. The formal recycling approach generally has a pretreatment step and few physical separation methods to reach the metal recovery stage. Pyrometallurgy falls under the formal recycling technique which facilitates extraction of precious metals (Au, Ag, Pd, Pt) by incinerating, smelting, drossing, smelting, and roasting. However, the major limitation of this process is the maintenance of the operational parameters (thermodynamics and usage of pure oxygen) for the total recovery of these complex e-wastes. Additionally, toxic gases, smoke, and dust are produced in this process due to the formation of slag, soot, and flue gases which have the remnant of hazardous chemicals and gases (dibenzo-*p*-dioxin, biphenyl, anthracene, polybrominated dibenzofurans, and polybrominated dibenzodioxins). These, in turn, can cause harm to human health and damage the ecological balance. Moreover, the pyrometallurgical processes pose a great disadvantage in the recovery of the plastic as it replaces the coke during combustion. Another recovery approach called hydrometallurgy facilitates the metal recovery process and has a comparatively high recovery rate and low investment than the previous technique. In this operation, leaching of the metallic elements is carried out in the pretreatment step. Various classes of lixiviants or extractants are used to dissolve the metals to leave them in the solution (leachate). The downstream of metals from the leachate depends on the types of e-waste, metals that can be recovered, and their purity and this can further be executed by solvent extraction, electrodeposition, ion-exchange, and adsorption procedures. This process is advantageous because of its high recovery rate of metals, less energy expenditure requirement, valid thermodynamical parameters, less toxic gas emission, and secondary waste generation. However, the main challenge of this process is to scale it up to the industrial level. Furthermore, another formal approach called biohydrometallurgy is reaching its peak because it facilitates metal recovery by the assistance of microbes which solubilize the elements from solid substances into the solutions which can then be recovered through several separation processes. Numerous metals like copper, silver, and aluminum had been recovered earlier through this process. Recently, this operation is executed more as it is a clean and green technology and has two distinct ways of operation – single-stage and double-stage operation. The former operation involves both microbes and metals together in nutritional media to benefit microbial growth. This further results in the dissolution of metals in the solution which limits the recovery efficiency due to induced cellular toxicity caused by the increase in the level of metal ions. The latter operation however has the association of microbes that are in the stationary phase of their growth cycle along with waste materials which are then added aseptically to enter the recovery process. Recent studies mostly include the usage of cell-free supernatant for metal extraction as it contains all the essential metabolites required for the separation process. Though the recovery rate is comparatively higher in the biohydrometallurgy method, the major challenges faced are the massive time of operation it demands and cytotoxicity effects on the microbial culture which constrains the scale-up and industrialization process. Additionally, the consumers of electronic products should understand the necessity for the proper disposal of these wastes. The end-users are equally responsible for the management of e-waste, hence proper awareness must be created as it can further impact human health and the environment. Besides, more recycling firms should be built to accommodate all the e-waste. In India, the Central Pollution Control Board has recently authorized 178 recycling firms which are not sufficient for the whole nation. Moreover, research is ongoing to find better and efficient ways to manage e-waste. Upcoming studies shall impart more knowledge on the process dynamics of base metals, their optimization, as well as environmental, economic, and social impacts of the e-waste. Furthermore, the high cost of recycling e-waste is also a huge concern. In order to carry out the whole process of collection, pretreatment, and disposal of e-waste, it requires a lot of sectors to administer at a time. Besides, with the increasing demand for electronic products globally, recycling e-waste has become necessary to save the environmental condition. Nanotechnology is also an emerging field in resolving some of these issues as these nanosized materials (1–100 nm) have the potential to increase the efficacy of downstream recovery of the metallic components. Similarly, a bottom-up approach can also be used and synthesis of nanomaterials from leachates has been observed.

10.6 CONCLUSION AND FUTURE PROSPECTS

There is no doubt regarding the importance of recycling the electronic waste for the extraction of valuable nanomaterials such as metallic nanoparticles, their oxides, metal ferrites. etc. However, there is still lack of research related to sustainable waste management and therefore, there is very less translation of laboratory-scale conversions to suitable technology. This is mainly due to lack of cost effectiveness and efficient recycling design. Furthermore, although the recycling of e-waste is important from the environment point of view, economically, it is difficult to set up large-scale processes without subsidies due to low or no profits. Conventional methods are causing pollution in the environment; therefore, alternative approaches are need of hour for sustainable development. Overall, most of the recycling techniques that are currently available require further refinement to create commercial-grade end products. Moreover, the problem related to the reproducibility of some recovery methods and the stability of the resulting products need to be addressed more extensively. The reusability of the extracted materials or their degradation following their application for various environmental remediation purposes is of critical importance, especially since the cost of production is generally high for high-purity nanomaterials. More emphasis should be put on the reproducibility of the results and the degradability or regeneration of recycled nanomaterials in order to place limits on the current research gaps. In addition, a systematic approach to assessing the risks involved in the use and disposal of key product–related nanomaterials is necessary for the nanoindustries to thrive. The development of methods to assess exposure and toxicity as well as models predicting the potential impacts on the environment and human health remains critical. Finally, applying the green principles to the production process may dramatically increase the merit of the pilot-scale projects and, therefore, should be taken into account.

REFERENCES

Anjum, Fozia, Muhammad Shahid, and Ata Akcil. 2012. "Biohydrometallurgy Techniques of Low Grade Ores: A Review on Black Shale." *Hydrometallurgy* 117–18: 1–12. https://doi.org/10.1016/j.hydromet.2012.01.007.

Babu, Balakrishnan Ramesh, Anand Kuber Parande, and Chiya Ahmed Basha. 2007. "Electrical and Electronic Waste: A Global Environmental Problem." *Waste Management and Research* 25 (4): 307–18. https://doi.org/10.1177/0734242X07076941.

Barbieri, Luisa, Roberto Giovanardi, Isabella Lancellotti, and Marco Michelazzi. 2010. "A New Environmentally Friendly Process for the Recovery of Gold from Electronic Waste." *Environmental Chemistry Letters* 8 (2): 171–8. https://doi.org/10.1007/s10311-009-0205-2.

Bartolozzi, M. 1990. "The Recovery of Metals from Spent Alkaline-Manganese Batteries: A Review of Patent Literature." *Resources, Conservation and Recycling* 4 (3): 233–40. https://doi.org/10.1016/0921-3449(90)90004-N.

Birloaga, Ionela, Ida De Michelis, Francesco Ferella, Mihai Buzatu, and Francesco Vegliò. 2013. "Study on the Influence of Various Factors in the Hydrometallurgical Processing of Waste Printed Circuit Boards for Copper and Gold Recovery." *Waste Management* 33 (4): 935–41. https://doi.org/10.1016/j.wasman.2013.01.003.

Cayumil, R., R. Khanna, M. Ikram-Ul-Haq, R. Rajarao, A. Hill, and V. Sahajwalla. 2014. "Generation of Copper Rich Metallic Phases from Waste Printed Circuit Boards." *Waste Management* 34 (10): 1783–92. https://doi.org/10.1016/j.wasman.2014.05.004.

Chatterjee, A., and J. Abraham. 2017. "Efficient Management of E-Wastes." *International Journal of Environmental Science and Technology* 14 (1): 211–22. https://doi.org/10.1007/s13762-016-1072-6.

Cui, Hao, and Corby G. Anderson. 2016. "Literature Review of Hydrometallurgical Recycling of Printed Circuit Boards (PCBs)." *Journal of Advanced Chemical Engineering* 6 (1): 1–11. https://doi.org/10.4172/2090-4568.1000142.

Cui, Jirang, and Lifeng Zhang. 2008. "Metallurgical Recovery of Metals from Electronic Waste: A Review." *Journal of Hazardous Materials* 158 (2–3): 228–56. https://doi.org/10.1016/j.jhazmat.2008.02.001.

Deep, Akash, Amit L. Sharma, Girish C. Mohanta, Parveen Kumar, and Ki Hyun Kim. 2016. "A Facile Chemical Route for Recovery of High Quality Zinc Oxide Nanoparticles from Spent Alkaline Batteries." *Waste Management* 51: 190–5. https://doi.org/10.1016/j.wasman.2016.01.033.

Duan, Xiaojuan, Jinxing Deng, Xue Wang, Jinshan Guo, and Peng Liu. 2016. "Manufacturing Conductive Polyaniline/Graphite Nanocomposites with Spent Battery Powder (SBP) for Energy Storage: A Potential Approach for Sustainable Waste Management." *Journal of Hazardous Materials* 312: 319–28. https://doi.org/10.1016/j.jhazmat.2016.03.009.

Dutta, Tanushree, Ki Hyun Kim, Akash Deep, Jan E. Szulejko, Kowsalya Vellingiri, Sandeep Kumar, Eilhann E. Kwon, and Seong Taek Yun. 2018. "Recovery of Nanomaterials from Battery and Electronic Wastes: A New Paradigm of Environmental Waste Management." *Renewable and Sustainable Energy Reviews* 82 (October): 3694–704. https://doi.org/10.1016/j.rser.2017.10.094.

Garg, Himanshi, Neha Nagar, Ganapathy Ellamparuthy, Shivakumar Irappa Angadi, and Chandra Sekhar Gahan. 2019. "Bench Scale Microbial Catalysed Leaching of Mobile Phone PCBs with an Increasing Pulp Density." *Heliyon* 5 (12): e02883. https://doi.org/10.1016/j.heliyon.2019.e02883.

Ghimire, Hem, and Parisa A. Ariya. 2020. "E-Wastes: Bridging the Knowledge Gaps in Global Production Budgets, Composition, Recycling and Sustainability Implications." *Sustainable Chemistry* 1 (2): 154–82. https://doi.org/10.3390/suschem1020012.

Grant, Kristen, Fiona C. Goldizen, Peter D. Sly, Marie Noel Brune, Maria Neira, Martin van den Berg, and Rosana E. Norman. 2013. "Health Consequences of Exposure to E-Waste: A Systematic Review." *The Lancet Global Health* 1 (6): e350–61. https://doi.org/10.1016/S2214-109X(13)70101-3.

Gullett, Brian K., William P. Linak, Abderrahmane Touati, Shirley J. Wasson, Staci Gatica, and Charles J. King. 2007. "Characterization of Air Emissions and Residual Ash from Open Burning of Electronic Wastes during Simulated Rudimentary Recycling Operations." *Journal of Material Cycles and Waste Management* 9 (1): 69–79. https://doi.org/10.1007/s10163-006-0161-x.

Hageluken, C. 2006. "Improving Metal Returns and Eco-Efficiency in Electronics Recycling." *Proceedings of the 2006 IEEE Conference* (May): 218–23. http://scholar.google.com/scholar?hl=en&btnG=Search&q=intitle:Improving+metal+returns+and+eco-efficiency+in+electronics+recycling+-#0.

Hanafi, Jessica, Eric Jobiliong, Agustina Christiani, Dhamma C. Soenarta, Juwan Kurniawan, and Januar Irawan. 2012. "Material Recovery and Characterization of PCB from Electronic Waste." *Procedia – Social and Behavioral Sciences* 57: 331–8. https://doi.org/10.1016/j.sbspro.2012.09.1194.

Hu, Zhenzhong, Maximilien Beuret, Hassan Khan, and Parisa A. Ariya. 2014. "Development of a Recyclable Remediation System for Gaseous BTEX: Combination of Iron Oxides Nanoparticles Adsorbents and Electrochemistry." *ACS Sustainable Chemistry and Engineering* 2 (12): 2739–47. https://doi.org/10.1021/sc500479b.

Hubau, Agathe, Alexandre Chagnes, Michel Minier, Solène Touzé, Simon Chapron, and Anne Gwénaëlle Guezennec. 2019. "Recycling-Oriented Methodology to Sample and Characterize the Metal Composition of Waste Printed Circuit Boards." *Waste Management* 91: 62–71. https://doi.org/10.1016/j.wasman.2019.04.041.

Ilyas, Sadia, Chi Ruan, H. N. Bhatti, M. A. Ghauri, and M. A. Anwar. 2010. "Column Bioleaching of Metals from Electronic Scrap." *Hydrometallurgy* 101 (3–4): 135–40. https://doi.org/10.1016/j.hydromet.2009.12.007.

Kaya, Muammer. 2016. "Recovery of Metals and Nonmetals from Electronic Waste by Physical and Chemical Recycling Processes." *Waste Management* 57: 64–90. https://doi.org/10.1016/j.wasman.2016.08.004.

Konyratbekova, Saltanat Sabitovna, Aliya Baikonurova, and Ata Akcil. 2015. "Non-Cyanide Leaching Processes in Gold Hydrometallurgy and Iodine-Iodide Applications: A Review." *Mineral Processing and Extractive Metallurgy Review* 36 (3): 198–212. https://doi.org/10.1080/08827508.2014.942813.

Kumar, Amit, Maria Holuszko, and Denise Crocce Romano Espinosa. 2017. "E-Waste: An Overview on Generation, Collection, Legislation and Recycling Practices." *Resources, Conservation and Recycling* 122: 32–42. https://doi.org/10.1016/j.resconrec.2017.01.018.

Long, Yu Yang, Yi Jian Feng, Si Shi Cai, Wei Xu Ding, and Dong Sheng Shen. 2013. "Flow Analysis of Heavy Metals in a Pilot-Scale Incinerator for Residues from Waste Electrical and Electronic Equipment Dismantling." *Journal of Hazardous Materials* 261: 427–34. https://doi.org/10.1016/j.jhazmat.2013.07.070.

Ma, Ya, Yan Cui, Xiaoxi Zuo, Shanna Huang, Keshui Hu, Xin Xiao, and Junmin Nan. 2014. "Reclaiming the Spent Alkaline Zinc Manganese Dioxide Batteries Collected from the Manufacturers to Prepare Valuable Electrolytic Zinc and LiNi0.5Mn1.5O4 Materials." *Waste Management* 34 (10): 1793–9. https://doi.org/10.1016/j.wasman.2014.05.009.

Mantuano, Danuza Pereira, Germano Dorella, Renata Cristina Alves Elias, and Marcelo Borges Mansur. 2006. "Analysis of a Hydrometallurgical Route to Recover Base Metals from Spent Rechargeable Batteries by Liquid-Liquid Extraction with Cyanex 272." *Journal of Power Sources* 159 (2): 1510–18. https://doi.org/10.1016/j.jpowsour.2005.12.056.

Morin, D., A. Lips, T. Pinches, J. Huisman, C. Frias, A. Norberg, and E. Forssberg. 2006. "BioMinE – Integrated Project for the Development of Biotechnology for Metal-Bearing Materials in Europe." *Hydrometallurgy* 83 (1–4): 69–76. https://doi.org/10.1016/j.hydromet.2006.03.047.

Natarajan, Gayathri, and Yen Peng Ting. 2014. "Pretreatment of E-Waste and Mutation of Alkali-Tolerant Cyanogenic Bacteria Promote Gold Biorecovery." *Bioresource Technology* 152: 80–5. https://doi.org/10.1016/j.biortech.2013.10.108.

Natarajan, Gayathri, and Yen Peng Ting. 2015. "Gold Biorecovery from E-Waste: An Improved Strategy Through Spent Medium Leaching with PH Modification." *Chemosphere* 136: 232–8. https://doi.org/10.1016/j.chemosphere.2015.05.046.

Needhidasan, Santhanam, Melvin Samuel, and Ramalingam Chidambaram. 2014. "Electronic Waste – An Emerging Threat to the Environment of Urban India." *Journal of Environmental Health Science and Engineering* 12 (1): 1–9. https://doi.org/10.1186/2052-336X-12-36.

Ni, Mingjiang, Hanxi Xiao, Yong Chi, Jianhua Yan, Alfons Buekens, Yuqi Jin, and Shengyong Lu. 2012. "Combustion and Inorganic Bromine Emission of Waste Printed Circuit Boards in a High Temperature Furnace." *Waste Management* 32 (3): 568–74. https://doi.org/10.1016/j.wasman.2011.10.016.

Ning, Chao, Carol Sze Ki Lin, David Chi Wai Hui, and Gordon McKay. 2017. "Waste Printed Circuit Board (PCB) Recycling Techniques." *Topics in Current Chemistry* 375 (2): 1–36. https://doi.org/10.1007/s41061-017-0118-7.

Nithya, Rajarathinam, Chandrasekaran Sivasankari, and Arunachalam Thirunavukkarasu. 2021. "Electronic Waste Generation, Regulation and Metal Recovery: A Review." *Environmental Chemistry Letters* 19 (2): 1347–68. https://doi.org/10.1007/s10311-020-01111-9.

Qu, Jiao, Yue Feng, Qian Zhang, Qiao Cong, Chunqiu Luo, and Xing Yuan. 2015. "A New Insight of Recycling of Spent Zn-Mn Alkaline Batteries: Synthesis of ZnxMn1-XO Nanoparticles and Solar Light Driven Photocatalytic Degradation of Bisphenol A Using Them." *Journal of Alloys and Compounds* 622: 703–7. https://doi.org/10.1016/j.jallcom.2014.10.166.

Sahajwalla, Veena, Farshid Pahlevani, Samane Maroufi, and Ravindra Rajarao. 2016. "Green Manufacturing: A Key to Innovation Economy." *Journal of Sustainable Metallurgy* 2 (4): 273–5. https://doi.org/10.1007/s40831-016-0087-z.

Sheng, Peter P., and Thomas H. Etsell. 2007. "Recovery of Gold from Computer Circuit Board Scrap Using Aqua Regia." *Waste Management and Research* 25 (4): 380–3. https://doi.org/10.1177/0734242X07076946.

Singh, Jiwan, and Byeong Kyu Lee. 2016. "Recovery of Precious Metals from Low-Grade Automobile Shredder Residue: A Novel Approach for the Recovery of Nanozero-Valent Copper Particles." *Waste Management* 48: 353–65. https://doi.org/10.1016/j.wasman.2015.10.019.

Søndergaard, Roar R., Yannick Serge Zimmermann, Nieves Espinosa, Markus Lenz, and Frederik Krebs. 2016. "Incineration of Organic Solar Cells: Efficient End of Life Management by Quantitative Silver Recovery." *Energy and Environmental Science* 9 (3): 857–61. https://doi.org/10.1039/c6ee00021e.

Thirunavukkarasu, Arunachalam, Karpagasundaram Muthukumaran, and Rajarathinam Nithya. 2018. "Adsorption of Acid Yellow 36 onto Green Nanoceria and Amine Functionalized Green Nanoceria: Comparative Studies on Kinetics, Isotherm, Thermodynamics, and Diffusion Analysis." *Journal of the Taiwan Institute of Chemical Engineers* 93: 211–25. https://doi.org/10.1016/j.jtice.2018.07.006.

Tue, Nguyen Minh, Shin Takahashi, Annamalai Subramanian, Shinichi Sakai, and Shinsuke Tanabe. 2013. "Environmental Contamination and Human Exposure to Dioxin-Related Compounds in e-Waste Recycling Sites of Developing Countries." *Environmental Sciences: Processes and Impacts* 15 (7): 1326–31. https://doi.org/10.1039/c3em00086a.

Tuncuk, A., V. Stazi, A. Akcil, E. Y. Yazici, and H. Deveci. 2012. "Aqueous Metal Recovery Techniques from E-Scrap: Hydrometallurgy in Recycling." *Minerals Engineering* 25 (1): 28–37. https://doi.org/10.1016/j.mineng.2011.09.019.

Weber, Roland, and Bertram Kuch. 2003. "Relevance of BFRs and Thermal Conditions on the Formation Pathways of Brominated and Brominated-Chlorinated Dibenzodioxins and Dibenzofurans." *Environment International* 29 (6): 699–710. https://doi.org/10.1016/S0160-4120(03)00118-1.

Wen, Xin, Xianliang Qiao, Xue Han, Libo Niu, Li Huo, and Guoyi Bai. 2016. "Multifunctional Magnetic Branched Polyethylenimine Nanogels with In-Situ Generated Fe3O4 and Their Applications as Dye Adsorbent and Catalyst Support." *Journal of Materials Science* 51 (6): 3170–81. https://doi.org/10.1007/s10853-015-9627-3.

Widmer, Rolf, Heidi Oswald-Krapf, Deepali Sinha-Khetriwal, Max Schnellmann, and Heinz Böni. 2005. "Global Perspectives on E-Waste." *Environmental Impact Assessment Review* 25 (5 Special Issue): 436–58. https://doi.org/10.1016/j.eiar.2005.04.001.

Wieser, Harald, and Nina Tröger. 2016. "Exploring the Inner Loops of the Circular Economy: Replacement, Repair, and Reuse of Mobile Phones in Austria." *Journal of Cleaner Production* 172: 3042–55. https://doi.org/10.1016/j.jclepro.2017.11.106.

Wu, Jun Yi, and Fang Chih Chang. 2016. "Recovery of Nano-Al2O3 from Waste Aluminum Electrolytic Solution Generated during the Manufacturing of Capacitors." *Desalination and Water Treatment* 57 (60): 29479–87. https://doi.org/10.1080/19443994.2016.1171171.

Xi, Guoxi, Yunqing Li, and Yu Min Liu. 2004. "Study on Preparation of Manganese-Zinc Ferrites Using Spent Zn-Mn Batteries." *Materials Letters* 58 (7–8): 1164–7. https://doi.org/10.1016/j.matlet.2003.08.029.

Xiu, Fu Rong, and Fu Shen Zhang. 2012. "Size-Controlled Preparation of Cu2O Nanoparticles from Waste Printed Circuit Boards by Supercritical Water Combined with Electrokinetic Process." *Journal of Hazardous Materials* 233–4: 200–6. https://doi.org/10.1016/j.jhazmat.2012.07.019.

Xu, Jinqiu, H. R. Thomas, Rob W. Francis, Ken R. Lum, Jingwei Wang, and Bo Liang. 2008. "A Review of Processes and Technologies for the Recycling of Lithium-Ion Secondary Batteries." *Journal of Power Sources* 177 (2): 512–27. https://doi.org/10.1016/j.jpowsour.2007.11.074.

Zhan, Lu, Xishu Xiang, Bing Xie, and Jie Sun. 2016. "A Novel Method of Preparing Highly Dispersed Spherical Lead Nanoparticles from Solders of Waste Printed Circuit Boards." *Chemical Engineering Journal* 303: 261–7. https://doi.org/10.1016/j.cej.2016.06.002.

11 Wastewater Treatment Using Nanoadsorbents Derived from Waste Materials

Menaka Jha, Sunaina, Arushi Arora, and Kritika Sood

CONTENTS

11.1 Introduction of Wastewater Treatment Need and Methodology 147
11.2 Advantages of Utilization of Waste Materials for Wastewater Treatment 150
11.3 Challenges in Extraction of Nanoadsorbents from Different Waste Materials 152
11.4 Utilization of Different Waste Materials for Extracting Efficient Nanoadsorbents 153
    11.4.1 From Agricultural Waste 153
    11.4.2 From Industrial Waste 154
    11.4.3 From Other Refuse Sources 155
11.5 Performance of Extracted Nanoadsorbents for Different Pollutants (Cations, Anions, Textile Waste, Pharmaceutical Waste, etc.) Present in the Wastewater Treatment 156
    11.5.1 Nanoadsorbents in Remediation of Cations 156
    11.5.2 Nanoadsorbents in Remediation of Anions 157
    11.5.3 Nanoadsorbents in Remediation of Dyes 157
    11.5.4 Nanoadsorbents in Remediation of Pharmaceuticals 158
11.6 An Overview of the Current Status of Low-cost Fabrication Methods of Nanoadsorbents from Waste Materials at an Industrial Scale 158
11.7 Conclusion and Future Prospects 160
References 160

11.1 INTRODUCTION OF WASTEWATER TREATMENT NEED AND METHODOLOGY

Water is a basic need of all living beings present on earth. The United Nations World Water Development Report addressed the crucial role of water as "Water for People Water for Life." Water is of prime importance in sustaining life on earth and has a chief role in maintaining the ecosystem (Sharma 1997). However, in recent years, access to clean water by a major part of population is still a challenge. Rapid growth in world population, huge industrialization, increase in agricultural activities to meet the food demand of increasing population, and other geological and environmental changes have led to the increase of pollutants in the waterbodies (El-Sayed Abdel-Raouf et al. 2019). As a result, the quality of water is deteriorating globally. Therefore, wastewater treatment is becoming a supreme challenge for world scientist, water regulatory authorities, and governmental agencies. Broadly, the water pollutants are classified as inorganic, organic, and biological pollutants (I. Ali, Asim, and Khan 2012). As per the International Water Association, approximately 80% of industrial and domestic effluent is directly discarded into freshwater bodies, which is the main cause of health-, environment-, and climate-related disasters. In the developed countries, the capacity of wastewater treatment is around 70%, whereas in developing countries it is ~8%, which is very less as compared to developed countries; therefore, they are facing more problems related to the water. In addition, the rapid growth in urbanization has also accelerated water pollution. This has not only caused the eutrophication and human health problems but also significantly contributes to greenhouse gas emission in the form of nitrous oxide and methane gas. Proper wastewater treatment of the sewer and other industrial plants plays a crucial role in sanitation and disease prevention. It is very important to channelize the community wastewater and sewer; otherwise, it can cause the contamination of local environment and drinking water supply, thereby enhancing the risk of disease transmission. Access to drinkable water, proper sanitation, and hygiene education can decrease health illness and death rate from diseases, thereby resulting in better global health and ultimately socioeconomic development. However, in many of the countries, due to lack of resources, funds, infrastructure, available technology, and space, proper management of wastewater is not practiced. Therefore, these countries are facing challenges to provide these basic needs to their people, leaving them at risk of contaminated water, poor sanitation and hygiene, and diseases related to them (M. E. Ali et al. 2020). The different sources of water contamination are shown in Figure 11.1.

Broadly, the wastewater can be divided into two main categories on the basis of its origin (El-Sayed Abdel-Raouf et al. 2019):

1. *Domestic wastewater:* The water which is discarded from the residential sources. It is generally generated by domestic activities such as cooking food, laundry, cleaning, and personal hygiene.

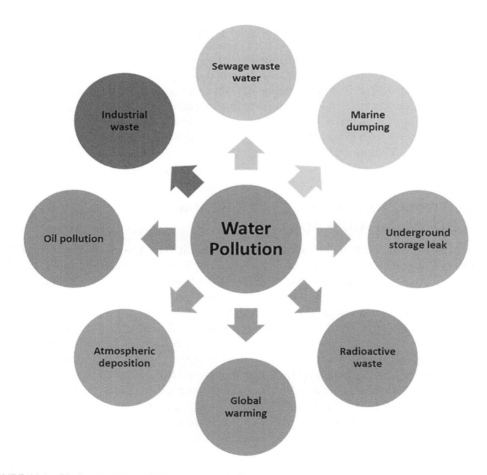

FIGURE 11.1 Various sources of water contamination.

2. *Industrial wastewater:* The water which is discharged from various industries generated during various manufacturing and processing steps. Some of the main industries causing water pollution are printing, food, pharmaceutical, beverage processing and production units, etc. The water discarded from these industries have a lot of chemical species in varying concentrations ranging from a few mg/L to g/mL. These impurities are usually referred to as micropollutants, which can be harmful for human health even in very small concentrations.

The main pollutants present in wastewater and the importance of their removal is listed in Table 11.1.

As discussed earlier, if wastewater is not properly treated, it will cause adverse effect on the environment and human health. These are also harmful for the aquatic life and wildlife populations. Some of the examples of the pollutants which are commonly found in the wastewater and their potential harmful impact on the environment is listed by Environment Canada (Roesler and Jenny 2021)

- The dissolved oxygen can be depleted by the decay of organic matter and debris present in wastewater in lakes and ponds which can cause deficiency of oxygen in fishes and other aquatic organisms, thereby affecting their growth and survival.
- The excess use of fertilizers can cause the leaching of chemicals such as phosphorus and nitrogen (including ammonia) into freshwater sources causing eutrophication which is toxic for the aquatic organisms, promote excessive plant growth, reduce available oxygen, harm spawning grounds, alter habitat, and lead to a decline of certain species.
- Chlorine-containing compounds and other inorganic chloramines can be harmful to aquatic invertebrates, algae, and fish.
- Bacteria, viruses, and disease-causing pathogens can contaminate beaches and infect the shellfish populations, causing limits on human recreation, drinking water consumption, and shellfish consumption.
- Metal contamination (such as mercury, lead, cadmium, chromium, arsenic) in water can cause chronic toxic effects on species.
- Some pharmaceuticals, personal care products, and other substances can also pollute the wastewater and cause threat to human health, aquatic life, and wildlife.

Various processes are utilized for achieving the desired quality of wastewater: separation, removal, and disposal of pollutants present in the wastewater (El-Sayed Abdel-Raouf et al. 2019). Generally, four basic technologies are used for the treatment of wastewater:

- *Physical methods:* These water treatment methods involve the use of tanks and other similar structure designs to contain and control the flow of wastewater to remove contaminants.

TABLE 11.1
Different Contaminants Present in Water and Their Hazardous Effects on the Environment

Contaminant	Reason for Their Removal
Pathogenic organisms	Found in wastewater can cause infectious diseases
Suspended solids(s)	Development of sludge deposits and anaerobic conditions when untreated wastewater is discharged to the aquatic environment
Refractory organics	Resist conventional wastewater treatment includes surfactants, phenols, and agricultural pesticides
Heavy metals	Highly toxic and environmentally hazardous species
Priority pollutants	Including organic and inorganic compounds, may be highly
Dissolved inorganic constituent	Such as calcium, sodium, and sulfate are often initially added to domestic water supplies and may have to be removed for wastewater reuse

- *Mechanical methods:* These methods involve the use of machines, both simple and complex, in design and operation.
- *Biological methods:* These methods use bacterial and other microorganism actions to treat wastewater. Some biological methods can remove stubborn impurities which are otherwise difficult to remove effectively by other means.
- *Chemical methods:* These methods enhance the activity of other methods by providing specialized treatment because of the addition of chemicals at various treatment stages.

11.2 ADVANTAGES OF UTILIZATION OF WASTE MATERIALS FOR WASTEWATER TREATMENT

We are well aware of the fact that modern lifestyle comes at a price, to live a luxurious and comfortable life. In present lifestyle conditions, frequent use of water for showering, cleaning, etc. is everyday routine. However, utilizing wastewater is of interest whether it is domestic wastewater or industrial wastewater. Our modern lifestyle provides us the luxury of using various products to make our life more comfortable and easier, but it comes at a price. A most common by-product of our modern lifestyle is wastewater, which can either be in the form of water running down the shower or runoff from wet roads. This wastewater is unfit for human consumption or use. Fortunately, we can make the wastewater potable and usable by employing wastewater treatment technologies that filter and treat the wastewater by removing contaminants such as sewage and chemicals (El-Sayed 2020). Four common ways to treat wastewater include physical water treatment, biological water treatment, chemical treatment, and sludge treatment.

Post–water treatment the adsorbents are likely to be dumped in the junkyard. The dumping action becomes environmentally challenging due to presence of inorganic adsorbents. In this regard, the advantageous waste adsorbents are in trend now, these can be regenerated for further use and can be used as value-added products such as for soil conditioning, cementitious materials, and different forms of biofuels. The following waste nanoadsorbents are used in wastewater treatment with a number of advantages:

(a) Activated charcoal is used for drinking water purification. Usage of commercial activated charcoal for purification of water is very expensive. The cost-effective activated charcoal can be synthesized from various wastes like agricultural residue, paper, sugar and mill industries, and petrochemical sludge. However, the activated charcoal loses its adsorption capability due to the presence of surface acid-oxygen. There are various methods that can be used to remove surface acid-oxygen to improve adsorption efficiency and regenerate the activation. The methods include heating treatment, pyrolysis, and electrochemical gas-phase oxidation treatment with N_2 or O_2 gases (Geethakarthi and Phanikumar 2011). Municipal water treatment was done with activated charcoal obtained from paper mill industry sludge to eliminate micropollutant by Jario and group, who have given the following equations for the calculation of adsorption efficacy:

$$E_{Ad} = \frac{QC_0 A}{1000 \times m} \quad (11.1)$$

$$A = \int_{t=t_0}^{t=t_f} \left(1 - \frac{C}{C_0}\right) dt \quad (11.2)$$

where E_{Ad} = adsorption capacity (mg/g), Q = flow rate (L/h), A = the total surface area (m²), m = dry weight of waste material adsorbents (g), C_0 = initial concentration of the wastewater (mg/L), C = wastewater concentration after adsorption (mg/L), t_0 = initial time (h), and t_f = total flow time (h) (Jaria et al. 2019). The activated charcoal was regenerated four times from date pits to remove Pb^{2+} ions from wastewater. Also, it has been observed that the granules of activated charcoal were regenerated through hydrothermal treatment and the obtained charcoal was capable to remove micropollutants, estrone, diclofenac, etc. even after being regenerated for five times (Krishnamoorthy et al. 2019).

(b) Soil fertility and soil nourishment can be improved by using waste adsorbents for wastewater treatment. According to nutrient discharge regulations, the set nutrient recovery and environmental goals can be controlled by improving soil nourishment. The amount of nitrogen and phosphorus are key components in soil fertility. The soil conditioning and fertility with micronutrients like N, P, Mg, Ca, and K have been used for many decades. These elements are present naturally in agricultural and industrial wastewater. The phosphorus is overused in agricultural sector which consequently leads to wastage. The nitrogen content is abundant in nature which has dwindled the reservoir of these elements. On the other hand, generating or inducing elements like Mg, Ca, K, and Zn are essential for soil nourishment, but it is costly to produce them as soil fertilizer. The problem is solved by recycling and reusing these elements from fertilizer industries and effluents from factories to treat wastewater. Many experimental studies are carried out in order to remove phosphate from wastewater and then releasing it to the soil; for example, Mor et al. produced activated rice husk from agricultural waste material (Mor, Chhoden, and Ravindra 2016). The activated rice husk ash was good enough to remove 89% of phosphate from wastewater. Wang et al. used bioadsorbent as a versatile slow-release fertilizer. A novel amphoteric wheat straw–based bioadsorbent has been applied to accumulate NH_4^+ and $H_2PO_4^-$ from wastewater for reuse as a multifunctional fertilizer compound. This fertilizer constitutes 95.1% nitrogen and 60% phosphorus release into soil within a month.

(c) Using chemical and agricultural solid wastes as wastewater adsorbent has potential for conversion to cementitious materials. Chemically cementitious materials are complex-bonded mixtures of calcium silicates and hydroxides and produce hydration products on reaction with water. Most of the commercial cements are manufactured with Portland cement combined with granulated blast furnace slag, limestones, alumina, etc. Nowadays a variety of biomass like saw, dust, husk, and others are utilized in cementitious materials. In the ash of agricultural residue, chemical compounds SiO_3, CaO, Al_2O_3, Fe_2O_3, TiO_2, SO_3, etc. are present; therefore, agricultural wastes are partially used as replacement for cement source. The chemical elements Ca, Mg, S, and Mn are present in the industrial and agricultural residue adsorbent during water treatment; they are present in the ashes obtained from biomass materials, which enhance the levels of cementitious material in the ashes (Wang et al. 2019).

(d) The wastewater adsorbent can be regenerated and used for energy generation applications either by hydrothermal carbonization or pyrolysis and gasification. After the waste materials are directly used as adsorbents, they can be dried to remove moisture and transformed into biochar through hydrothermal carbonization. The wastewater adsorbent can be used for electricity generation by transforming it to biochar via hydrothermal carbonization. Through minimal pretreatment like direct combustion, the biochar can be obtained for use in biofuel production (Tamirat et al. 2016). Earlier, it has been researched that sewage sludge metal-doped (such as Fe, S, and N) biochar has been used as an anode material for oxygen reduction and oxygen evolution activity in basic and acidic media. The successful conversion using biochar of chemical energy into electrical energy has been achieved via microbial catalysis (Hossain, Zaini, and Mahlia 2017). Additionally, bio-oil derived with same methodologies has been employed for binder tool for improvising asphalt placement and generation from this waste. Furthermore, the blending of bio-oil with asphalt results in long-term sustenance of roads and pavement paths. Traditionally, crude oils were used as asphalt mixture, but due to its high cost bio-oil obtained from waste biomass is preferred (Raman et al. 2015).

11.3 CHALLENGES IN EXTRACTION OF NANOADSORBENTS FROM DIFFERENT WASTE MATERIALS

For the wastewater treatment, adsorption technique is considered appropriate because of its simplicity and low-cost. The suitable adsorbent must be selected on the basis of nature of contaminants, its concentration in water, adsorbent efficacy, and selectivity for the pollutant.

Furthermore, these adsorbents must be non-toxic, stable, low cost, easily available, and can be easily regenerated. A large number of different kinds of waste such as agricultural wastes and residues, industrial by-products, and biomass materials have been utilized as nanoadsorbents for the wastewater treatment (Ahsan et al. 2001).

Industrial waste is composed of solid materials, liquids, gases, or mixtures, which are undesirable and emitted or discharged from any industrial process. Industrial waste is considered as one of the global problems which needs attention because of its environmental load and toxicity. The major challenge in this approach is that in most of the developing countries, there is no appropriate routes for the waste collection and waste disposal. There is hardly any training for the proper waste management and no qualified waste management professionals are available. The poor strategic planning, ignorance toward waste collection/segregation, and lack of government finance regulatory agenda are major factors which are responsible for the ineffective waste management. Apart from this, public attitude toward waste management is also responsible for this. Despite taking care, huge amount of industrial waste is generated. Recently, the recycling and reuse of industrial waste has been increasing significantly worldwide. This approach also has economic benefits in addition to environmental waste reduction (El-Ramady et al. 2020). Recovery of commercially important materials from the wastes is one of the key steps in waste management. Different types of waste materials such as sewage sludge, biomass, agricultural (Corncob, wood chips, rice husk, etc.) and industrial waste (scrap tire, mining discard, etc.) used for the synthesis of high-quality activated carbon and other products are proven to be excellent adsorbents for different impurities present in wastewater. A lot of efforts are being put into utilizing the waste generated from different sources for the fabrication of nanoadsorbents for the treatment of wastewater (Hossain et al. 2020). But still there are some significant challenges which limits their commercialization:

- Efficient nanoadsorbents can be made by mixing the different kinds of waste materials. The problem with mixing is that the adsorption capacity of different materials varies widely. Hence, the adsorption efficiency projection of blended materials will be challenging to determine for the overall process.
- The selection of waste material is also very crucial as their composition changes with different regions based on their geographical location, climate, and production rates of waste. Therefore, the protocols and processes to generate the nanoadsorbents from waste materials may vary greatly between the regions. Integration will need to be flexible to be fit-for-purpose according to the different scenarios.
- There is very less experimental studies available on selectivity of nanoadsorbents on the basis of different pollutants. Therefore, more experimental research is required to develop a better classification method.
- Despite some laboratory experiments performed for the usage of waste materials for water treatment, there is no integrated approach which can be employed for large-scale applications. Before implementing for the commercial application, a thorough experimental analysis and cost simulation on small scale is required to assess the possibility of large-scale production.

Although there are some limitations which restrict the use of waste materials for nanoadsorbents in wastewater management, researchers from all over the world are working toward this problem (Singh et al. 2018; Khajeh, Laurent, and Dastafkan 2013; Moharrami and Motamedi 2020).

Wastewater Treatment Using Nanoadsorbents

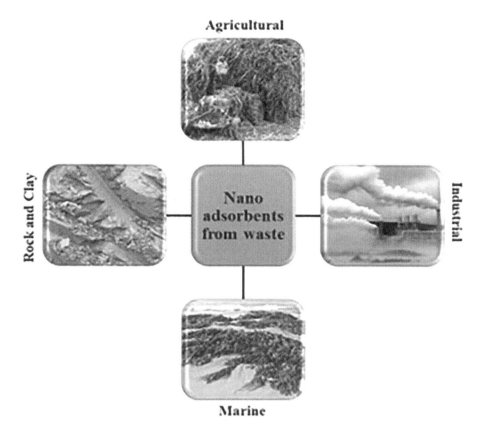

FIGURE 11.2 Different types of wastes utilized to extract nanoadsorbents.

11.4 UTILIZATION OF DIFFERENT WASTE MATERIALS FOR EXTRACTING EFFICIENT NANOADSORBENTS

The cost of adsorption process can be reduced if waste materials that are discarded as waste and are neglected are put to efficient use to remediate environmental pollutants. Most often, waste materials like agricultural, marine, and construction wastes have little or no economic value. The removal process of pollutants by these nanoadsorbents extracted from waste mainly depends upon the functionalization and surface area that results as a consequence of treatment that they undergo so as to form nanoadsorbents (Figure 11.2).

11.4.1 FROM AGRICULTURAL WASTE

Agricultural waste is an undesirable material, but being eco-friendly it is also a source of economic precursors which can be easily converted to products of value. The core components of agricultural waste materials are cellulose, lignin, sugars, starch, and hydrocarbons, which possess functional groups ideal for the adsorption processes of effluents. One of the ways by which agricultural waste can be put into use is by pretreating it by a suitable chemical process in order to convert it into a desired particle size or enhance functional groups on it. Heavy metal ions act as a target for removal by various waste materials.

Rice, which is the most popular food in Asia, harvesting produces large quantities of rice husk every year. About 18% of the total amount of husk remains as rice husk ash (RHA) whose important contents are silica (91.5%) and alumina (2%), both of which can be used in the synthesis of extremely efficient adsorbent, zeolites.

Santasnachok et al. (Santasnachok, Kurniawan, and Hinode 2015) were successful in synthesizing zeolites from rice husk ash and employed it in remediation of cadmium containing water. The removal capacities of Na-A and Na-X for cadmium are found to be 736.38 and 684.46 mg/g, respectively (dosage = 0.25 g/L). Kaur et al. (Kaur, Sharma, and Kumari 2019) synthesized nanocellulose from rice husk by mechanochemical method. The adsorbent dose was varied from 0.1 mL to 0.7g/50 mL. The maximum adsorption was found at 0.5 g/50 mL dose which was 97%. Daffalla et al. (Daffalla, Mukhtar, and Shaharun 2020a) synthesized developed rice husk–based adsorbents for phenol removal by thermal and chemical methods, both of which had different adsorbent active areas for phenol removal, thermal being higher. Thermally treated samples showed better phenol removal efficiency (36.4–64.9%) than the chemically treated sample (27.8%).

Lignocellulosic materials like sugarcane bagasse have a complex configuration with a variety of activated sites. They are often modified or tagged with other functionalities so as to increase the adsorption capabilities. Wannahari et al. (Wannahari et al. 2018) successfully prepared sugarcane bagasse–derived nanomagnetic adsorbent composite. The maximum adsorption capacity of synthesized nanoadsorbent was found to be 113.63 mg/g, with desorption efficiency up to 60% and reusability efficiency up to 80% for three consecutive cycles. Huang et al. (Huang et al. 2020) derived cationic dialdehyde cellulose (c-DAC) nanofibers from sugarcane bagasse. Nanofibrillated c-DAC adsorbent was made by modifying delignified bagasse with different amount of oxidizing reagent (sodium periodate) and cationizing reagent to adjust the surface charge. This excellent adsorbent material was used to adsorb Cr(VI) from water. Said et al. (Said et al. 2020) utilized tartaric acid–pretreated sugarcane bagasse in the adsorption of Congo red from industrial wastewater. The maximum sorption capacity (Q_m) was 8.670 mg/g according to Dubinin–Radushkevich adsorption model.

Other organic waste materials that are closely related to agricultural waste materials are also widely used. Mohammad (Mohammad 2013) reported the role of activated carbon generated from apricot stone as an ethoprophos pesticide adsorbent. The adsorption model fitted well with Langmuir equation, and maximum adsorption capability was found to be 20.04 mg/g and also fitted well with the second-order kinetics.

Oyewo et al. (Oyewo et al. 2021) were able to synthesize nanosized adsorbents from orange peels that were utilized in adsorption of Ce from simulated water. The formed nanoadsorbent showed excellent properties of high surface area, high porosity, narrow particle size distribution, as well as spherical morphology. The carboxylic and hydroxyl groups so introduced resulted in higher adsorption abilities toward Cerium, i.e., Langmuir model gave a sorption efficiency of 40 mg/g at 25 °C.

Weed is considered a much unwanted waste, but science has put that too into efficient use by converting it into a nanoadsorbent. Kamaraj et al. (Kamaraj et al. 2020) used parthenium weed–activated carbon loaded with zinc oxide nanoparticles to remove methylene blue, hexavalent chromium, and real industrial tannery wastewater.

11.4.2 From Industrial Waste

Industries are an important and indispensable part of a country's economy. They generate a huge amount of waste that poses disposal problems. Fly ash is an industrial effluent generated from combustion process. Fly ash includes significant amounts of silicon dioxide (SiO_2), aluminum oxide (Al_2O_3), and calcium oxide (CaO) (Dhmees, Khaleel, and Mahmoud 2018). He et al. (He et al. 2016) synthesized zeolites from coal fly ash by fusion method and was used for adsorption of heavy metals like Pb, Cu, Ni, and Mn in aqueous solution. Sivalingam and Sen (Sivalingam and Sen 2019) synthesized novel nanocrystalline zeolites from coal fly ash for adsorptive removal of crystal violet with 99% efficiency.

Wastewater Treatment Using Nanoadsorbents

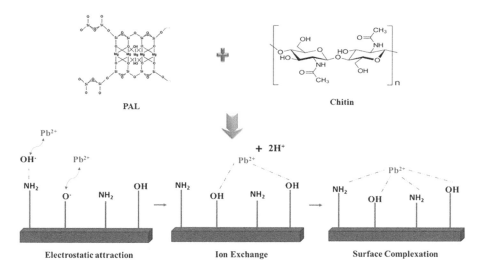

FIGURE 11.3 Credible adsorption mechanism of Pb(II) on nanoadsorbent.

Predescu et al. (Predescu et al. 2021) used ferrous mill scale refuse to form superparamagnetic iron oxide nanoparticles (Fe_3O_4 and $\gamma\text{-}Fe_2O_3$), which were superparamagnetic and had large surface area. These mixed-phase iron oxide nanoparticles were used for adsorption of metal ions nickel, copper, and cadmium with adsorption efficiency of 9.44, 11.12, and 19.15 mg/g, respectively. The efficiency of adsorption was found to be over 90% in just 10 min.

Afshar et al. (Afshar, Karimi, and Mozaffari 2016) converted steel industry waste to iron oxide nanoparticles (Fe_3O_4). Steel waste was leached to $FeCl_3$, which was further converted to magnetite that demonstrated an efficiency of 91% for adsorption of cobalt ions.

When molten iron slag is quenched, a by-product, blast furnace slag, is generated. This slag is a feasible source of silica as it primarily consists of SiO_2, Al_2O_3, and CaO. Dhmees et al. (Dhmees, Khaleel, and Mahmoud 2018) synthesized silica nanoparticles from commercial blast furnace slag which served as low-cost adsorbent for methylene blue from wastewater with adsorption efficiency of 109.8 mg/g.

Paper tends to be an important usage item for industries, schools, institutes, etc. Putro et al. (Putro et al. 2019) successfully converted paper waste to nanocrystalline cellulose by the process of deinking followed by acid hydrolysis. The so formed nanocrystalline cellulose demonstrated favorable adsorption of hydroxynaphthol blue and Congo red.

11.4.3 From Other Refuse Sources

Seaweed biomass from *Ulva fasciata* was used in solvent-free production of Fe-S-anchored graphene (FAG) by Mahto et al. (Mahto et al. 2018). As synthesized FAG had a pore size of 2.52 and 8.34 nm, it acted as a superadsorbent for Pb, Cr, and certain dyes. The material could be recycled well for over eight cycles. Mamah et al. (Mamah et al. 2020) synthesized palygorskite-chitin (PAL-chitin) hybrid where chitin was sourced from shrimps. Chitin possessed abundant amino and hydroxyl groups suitable for metal ion anchorage and PAL is inherently hydrophilic and environment-friendly. PAL-chitin (Figure 11.3) nanoadsorbent adsorbed Pb(II) ions with an efficiency of 92.9%. Waste rocks (Lv et al. 2021) from mining site were utilized for the adsorption of arsenic and fluoride, by ligand exchange reaction and adsorption mechanism, with an efficiency of 98.13% and 86.13%, respectively.

11.5 PERFORMANCE OF EXTRACTED NANOADSORBENTS FOR DIFFERENT POLLUTANTS (CATIONS, ANIONS, TEXTILE WASTE, PHARMACEUTICAL WASTE, ETC.) PRESENT IN THE WASTEWATER TREATMENT

Cations, anions, and organic pollutants like pharmaceuticals and dyes are harmful to health if they leach into waterbodies and ultimately enter the food chain. So, exploiting the nanoadsorbents with low cost and high removal efficiency is a boon for environment remediation. All the nanoadsorbents acquired from waste are activated or pretreated by different processes and hence according to the functionalities and degree of activation they perform diversely under different conditions.

11.5.1 Nanoadsorbents in Remediation of Cations

During the recent developmental times, heavy metals have found their ways into the waterbodies leached out of various electroplating, battery, vehicle manufacturing industries, etc. Metals such as cadmium, lead, mercury, nickel, and zinc as effluents from these industrial activities are non-biodegradable and tend to bioaccumulate, leading to various health problems and disorders. The World Health Organization has issued a minimum concentration limit for various metal ions. The maximum adsorption capacity (q_e) is the ultimate amount of a pollutant that a unit amount of adsorbent can take up. Table 11.2 summarizes nanoadsorbents derived from various waste materials that are used in cation remediation from wastewater.

TABLE 11.2
Different Waste-extracted Nanoadsorbents Utilized for Cation Removal from Wastewater

S. No.	Adsorbent	Target	Maximum Adsorption Capacity (mg/g)	Reference
1	Zeolites (X) from rice husk ash	Cd(II)	Na-X = 684.46 Na-A = 736.38	(Santasnachok, Kurniawan, and Hinode 2015)
2	Nanocellulose from sugarcane bagasse	Cr(VI)	80.5	(Huang et al. 2020)
3	Composite from sugarcane bagasse	Cu(II)	113.63	(Wannahari et al. 2018)
4	Orange peels	Ce(IV)	35.9	(Oyewo et al. 2021)
5	Nanocellulose from rice husk	La(III) Pb(II)	193.2 100.7	(Zhan et al. 2020)
6	Nanokaolin clay Nanobentonite	Zn(II) from battery effluent	167	(Maheswari, Sivakumar, and Thirumarimurugan 2019)
7	Ferrous mill waste	Cu(II) Cd(II) Ni(II)	11.12 19.15 9.44	(Predescu et al. 2021)
8	Oryza sativa husk	Pb(II)	6.101	(Kaur, Kumari, and Sharma 2020)
9	Cellulose nanocrystals	Pb(II)	6.4	(Vivian Abiaziem et al. 2019)
10	e-Waste acryl-butadiene styrene	Pb(II) Cu(II) Cd(II) Zn(II)	56.5 56.0 53.47 51.28	(Masoumi, Hemmati, and Ghaemy 2015)
11	Tea waste	Cr(VI)	94.34	(Cherdchoo, Nithettham, and Charoenpanich 2019)
12	Zeolites from coal fly ash	Ni(II)	47	(Y. Zhang et al. 2018)
13	Coal fly ash	Pb(II) Zn(II)	31.38 26.98	(Astuti et al. 2021)

11.5.2 Nanoadsorbents in Remediation of Anions

Anions are another group of hazardous materials in addition to metal cations which happen to be carcinogenic and toxic to the environment and human health. Their presence in waterbodies have caused serious health effects. For example, fluoride in drinking water has resulted in teeth distortions, DNA abnormalities, paralysis, etc. (Mondal, Bhaumik, and Datta 2015). The presence of phosphorus and nitrogen as phosphate and nitrate in waterbodies has become a grave concern of eutrophication, which is a serious concern to the ecosystem (Conley et al. 2009). Table 11.3 lists the major anionic species present in waterbodies and their removal by using various nanoadsorbents.

11.5.3 Nanoadsorbents in Remediation of Dyes

Dyes are used almost everywhere: clothes, paper, paints, etc. Synthetic dyes like methylene blue, reactive black, etc. are economic to create and give brighter hues. First, dyes have adverse effect on aquatic life as they restrict sunlight to pass through water. Many dyes as well as their by-products have proved to be mutagenic as well as carcinogenic. Textile dyes can cause diseases like dermatitis and lung illnesses along with itchy eyes, irritable skin, etc. Adsorption of dyes from wastewater has shown extremely encouraging results. Adsorbent materials can be obtained abundantly from waste with least effort and can be used in dye wastewater remediation (Table 11.4).

TABLE 11.3
Different Waste-extracted Nanoadsorbents Utilized for Anion Removal from Wastewater

S. No.	Adsorbent	Anion	Maximum Adsorption Capacity (mg/g)	Reference
1	Silica nanoadsorbent	F^-	12	(Pillai et al. 2020)
2	Rice husk biochar	F^-	21.7	(Goswami and Kumar 2018)
3	Silica from rice husk ash	NO_3^-	14.22	(To et al. 2020)
4	Biochar (embedding layered double hydroxide)	PO_4^{3-}	386	(M. Zhang et al. 2014)
5	MgO-biochar	PO_4^{3-}–NO_3^-	83595 mg/g	(M. Zhang et al. 2012)

TABLE 11.4
Different Waste-extracted Nanoadsorbents Utilized for Textile Dyes Removal from Wastewater

S. No.	Adsorbent	Dye	Maximum Adsorption Capacity (mg/g)	Reference
1	Rice husk ash	Organic dye	276.9	(Barbosa et al. 2018)
2	Sugarcane bagasse	Congo red	8.67	(Said et al. 2020)
3	Nano activated carbon	Methylene blue	28.09	(Shokry, Elkady, and Hamad 2019)
4	Mixed waste biomass	Alizarin red	42.58	(Gautam et al. 2020)
	Sugar beet pulp	Tartazine	68.78	
5	Cellulose nanocrystals	Crystal violet	2500	(Moharrami and Motamedi 2020)
		Methylene blue	1428	
6	Activated carbon from cherry tree	Methylene blue	41.49	(Ardekani et al. 2017)
7	Coal fly ash–derived zeolite	Methylene blue	345.36	(Sivalingam and Sen 2019)
8	Rice husk ash–derived zeolite	Crystal violet	125.43	(Sivalingam and Sen 2020)
9	Iraqi red kaolin clay	Methylene blue	240.4	(Jawad and Abdulhameed 2020)

TABLE 11.5
Different Waste-extracted Nanoadsorbents Utilized for Pharmaceutical Drugs Removal from Wastewater

S. No.	Adsorbent	Pharmaceutical	Maximum Adsorption Capacity (mg/g)	Reference
1	Grass nanocellulose	Ciprofloxacin Diclofenac	227.223 192.307	(Shahnaz et al. 2021)
2	Silica nanostructures from rice husk	Ciprofloxacin	190	(Nassar, Ahmed, and Raya 2019)
3	Activated carbon nanoparticles from vine wood	Amoxicillin Cephalexin Tetracycline Penicillin-G	–	(Pouretedal and Sadegh 2014)
4	Activated carbon from coconut shell (CS) and natural bentonite (BN)	Amoxicillin	233.775 44.662	(Budyanto et al. 2008)
5	Sugarcane bagasse Vegetable sponge	Paracetamol	120.5 37.5	(Ribeiro et al. 2011)

11.5.4 Nanoadsorbents in Remediation of Pharmaceuticals

Pharmaceutical drug disposal is a major problem having grave effects on the environment and human health. Globally, lakhs of tons of pharmaceutical products are consumed every year. They may enter into the environment and food chain either during the manufacturing process or during their use and disposal. Design and use of pharmaceutical compounds are done to stimulate a chemical response and most of them are too stable to be degraded on their own. Consequently, they leach into the food chain and can start accumulating up in an organism, who is effectively untargeted. Bioaccumulation of pharmaceuticals have resulted in increased resistance in an organism toward a particular strain of microbe, thereby increasing the potential risk of more dangerous diseases in organisms. Table 11.5 lists the major pharmaceutical drugs which have been treated using different absorbents derived from waste materials.

11.6 AN OVERVIEW OF THE CURRENT STATUS OF LOW-COST FABRICATION METHODS OF NANOADSORBENTS FROM WASTE MATERIALS AT AN INDUSTRIAL SCALE

The development of nanoadsorbents like activated carbon-mixed materials, zeolites, metals, polymeric nanoadsorbents, and metal oxides has provided a solution to environmental problems. Recently, nanoadsorbent materials have gained popularity in the adsorption chemistry due to their unique properties like high surface area, stability, and microbial effect. Adsorption can also be used for source reduction and reclamation for potable, industrial, and other purposes. As a result, much work has been carried out on water treatment by adsorption. Earlier, nanoadsorbents were derived from commercially available materials and these were expensive and infeasible. Many unsustainable approaches were there for removal of various dyes, heavy metal ions, and pharmaceuticals from ion exchange, precipitation, electroplating, membrane filtration, etc. Among all, the adsorption approach is the most positive one because of high efficiency and easy handling (Pouretedal and Sadegh 2014). At an industrial level, pollutants are remediated from water with the help of columns and contractors

Wastewater Treatment Using Nanoadsorbents

filled with suitable adsorbents, depending on available sources and regional variation. In recent years, low-cost fabricated nanoadsorbents are being used at large scale usage. However, fabricating an inexpensive and scalable nanoadsorbent at industrial level is still challenging. The waste-extracted nanoadsorbents are often used in water treatment plants and industries because they are easily accessible and cheap. To bring up the adsorption process at an industrial level, a protocol has been developed. In this protocol, an adsorption technology and applications for water treatment and recycling were introduced. It also includes reagents, equipment, preparation of adsorbents and their characterization, development of adsorption technology by batch and column processes on a laboratory scale, water treatment economically on a large scale by column operations, and applications of this technique (I. Ali and Gupta 2007). With high emphasis on sustainable approaches, research interest is being shifted toward eco-friendly renewable and recycled biobased adsorbent materials for water purification. As already discussed in Tables 11.2–11.4, various waste materials are utilized for useful and cheap production of adsorbents. These adsorbents proved effective in removing life threatening (i) heavy metal cations and anions, viz., Cd(II), Cu(II), Ce(V), La(III), Pb(II), Zn(II), F$^-$, NO^{3-}, and PO$_4^{3-}$; (ii) textile dyes, viz., Congo red, methylene blue, and crystal violet; and (iii) pharmaceutical drugs, viz., ciprofloxacin, diclofenac, amoxicillin, paracetamol etc. The ideology of waste-extracted nanoadsorbent is based on recovering value-added material from waste, which is an important stage of waste management. When it is not possible to reduce the volume or toxicity of a hazardous waste, it is recommended to recover or reuse it as a precursor for other processes and production. Recycling and conversion of various wastes, namely, rice husk ash, coal fly ash, mill scale, sugarcane bagasse, wine wood, etc. yields smart nanomaterials that are further characterized by X-ray, SEM, TEM, EDX, BET techniques (Gupta and Sharma 2003; Kazemipour et al. 2008). Previously, activated carbon, alumina, and silica have been used frequently for removal of heavy metal ions, but their high cost has reduced their use significantly. A variety of low-cost adsorbents have been developed and tested to remove heavy metal ions from aqueous solutions, including clay minerals. Mill scales wastes from steel industry have been utilized for successful synthesis of iron oxide and magnetite nanoparticles and for cobalt and lead ion absorbance and abatement. Rice husk–extracted zeolite adsorbent for the effective removal of phenol from wastewater was fabricated by Daffala et al. Rice husk comprises alumina and silica that are required in synthesizing negatively charged zeolites that have a tendency to attract cations (Figure 11.4) (Daffalla, Mukhtar, and Shaharun 2020b).

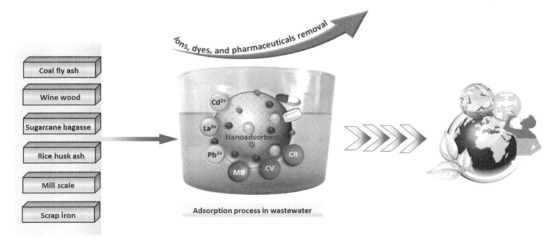

FIGURE 11.4 Schematic in vision of "Water for People Water for Life" with waste-extracted nanoadsorbents toward abatement of heavy metal ions (Cd^{2+}, La^{3+}, Pb^{2+}, Cr^{6+}, Zn^{2+}), textile dyes (MB = methylene blue, CV = crystal violet, CR = Congo red), and pharmaceuticals.

The aforementioned recycle and reuse strategies for fabrication of low-cost nanoadsorbents have contributed toward not only controlling environmental pollution problems but also saving us from toxic and contaminated water intake.

11.7 CONCLUSION AND FUTURE PROSPECTS

Water is the basic need for the survival of human being and other animals on earth. Apart from drinking and survival purposes, water is also crucial for domestic, agriculture, and industrial processes. This importance of water has made the scientific community to work continuously toward water recovery from the contaminated water to tackle the global problem of availability of safe water. There are many conventional and high-cost technologies which are employed for the wastewater treatment, but due to their high cost, they cannot be implemented at large scale. Therefore, to treat wastewater, some low-cost energy-efficient method is desired. Recently, utilization of waste materials for fabricating nanoadsorbents to treat the wastewater has gained a lot of attention from research fraternity. In addition to water treatment, this is also useful for effective waste management, circular economy, energy conservation, and sustainable environment impact. This concept of using the waste materials as a nanoadsorbent is important for reusing and recycling the waste. The main nanoadsorbents are derived from the agricultural residues, e.g., rice husk, fruit peels, egg shells, bagasse, etc. and the by-products are derived from the agriculture-based industries such as sugarcane, tealeaf, paper, etc., and straw and sewage sludge. There is effective removal of hazardous contamination from the municipal wastewater when treated with nanoadsorbent extracted from the construction material waste and marble polishing sludge–based industries.

There is scarcity of literature on the experiments performed on this concept; therefore, classification of these adsorbents for different contamination present in water is crucial. An integrated approach is essential for the fabrication of nanoadsorbents from waste materials, their regeneration, and end-processing. If this concept is applied for real, the water industries will play a key role in handling the waste globally and introduce a novel prospect for cost-effective wastewater treatment and circular economy.

REFERENCES

Afshar, Masume, Gholamreza Karimi, and Ezatollah Mozaffari. 2016. "Introducing a Recycling Method for Iron Oxide Nanoadsorbent from Mine Waste and Its Application in Wastewater Treatment." *Middle East Journal of Applied Science & Technology* (October): 627–32.

Ahsan, Shamim, Satoshi Kaneco, Kiyohisa Ohta, Takayuki Mizuno, and Keiko Kani. 2001. "Use of Some Natural and Waste Materials for Waste Water Treatment." *Water Research* 35 (15): 3738–42. https://doi.org/10.1016/S0043-1354(01)00047-1.

Ali, Imran, Mohd Asim, and Tabrez A. Khan. 2012. "Low Cost Adsorbents for the Removal of Organic Pollutants from Wastewater." *Journal of Environmental Management* 113: 170–83. https://doi.org/10.1016/j.jenvman.2012.08.028.

Ali, Imran, and V. K. Gupta. 2007. "Advances in Water Treatment by Adsorption Technology." *Nature Protocols* 1 (6): 2661–7. https://doi.org/10.1038/nprot.2006.370.

Ali, M. E., M. E. Hoque, S. K. Safdar Hossain, and M. C. Biswas. 2020. "Nanoadsorbents for Wastewater Treatment: Next Generation Biotechnological Solution." *International Journal of Environmental Science and Technology* 17. https://doi.org/10.1007/s13762-020-02755-4.

Ardekani, Payman Shirvani, Hajir Karimi, Mehrorang Ghaedi, Arash Asfaram, and Mihir Kumar Purkait. 2017. "Ultrasonic Assisted Removal of Methylene Blue on Ultrasonically Synthesized Zinc Hydroxide Nanoparticles on Activated Carbon Prepared from Wood of Cherry Tree: Experimental Design Methodology and Artificial Neural Network." *Journal of Molecular Liquids* 229: 114–24. https://doi.org/10.1016/j.molliq.2016.12.028.

Astuti, Widi, Achmad Chafidz, Ahmed S. Al-Fatesh, and Anis H. Fakeeha. 2021. "Removal of Lead (Pb(II)) and Zinc (Zn(II)) from Aqueous Solution Using Coal Fly Ash (CFA) as a Dual-Sites Adsorbent." *Chinese Journal of Chemical Engineering* 34 (II): 289–98. https://doi.org/10.1016/j.cjche.2020.08.046.

Barbosa, Tuany Ramos, Edson Luiz Foletto, Guilherme Luiz Dotto, and Sérgio Luiz Jahn. 2018. "Preparation of Mesoporous Geopolymer Using Metakaolin and Rice Husk Ash as Synthesis Precursors and Its Use as Potential Adsorbent to Remove Organic Dye from Aqueous Solutions." *Ceramics International* 44 (1): 416–23. https://doi.org/10.1016/j.ceramint.2017.09.193.

Budyanto, Sukamto, Suhariono Soedjono, Wenny Irawaty, and Nani Indraswati. 2008. "Studies of Adsorption Equilibria and Kinetics of Amoxicillin from Simulated Wastewater Using Activated Carbon and Natural Bentonite." *Journal of Environmental Protection Science* 2: 72–80.

Cherdchoo, Wachiraphorn, Srisuda Nithettham, and Jittima Charoenpanich. 2019. "Removal of Cr(VI) from Synthetic Wastewater by Adsorption onto Coffee Ground and Mixed Waste Tea." *Chemosphere* 221 (VI): 758–67. https://doi.org/10.1016/j.chemosphere.2019.01.100.

Conley, D. J., H. W. Paerl, R. W. Howarth, D. F. Boesch, S. P. Seitzinger, K. E. Havens, C. Lancelot, and G. E. Likens. 2009. "Controlling Eutrophication: Phosphorus and Nitrogen." *Science* 323: 1014–15.

Daffalla, Samah Babiker, Hilmi Mukhtar, and Maizatul Shima Shaharun. 2020a. "Preparation and Characterization of Rice Husk Adsorbents for Phenol Removal from Aqueous Systems." *PLoS One* 15 (12 December). https://doi.org/10.1371/journal.pone.0243540.

Daffalla, Samah Babiker, Hilmi Mukhtar, and Maizatul Shima Shaharun. 2020b. "Preparation and Characterization of Rice Husk Adsorbents for Phenol Removal from Aqueous Systems." *PLoS One* 15 (12 December). https://doi.org/10.1371/journal.pone.0243540.

Dhmees, Abdelghaffar S., Nagla M. Khaleel, and Sawsan A. Mahmoud. 2018. "Synthesis of Silica Nanoparticles from Blast Furnace Slag as Cost-Effective Adsorbent for Efficient Azo-Dye Removal." *Egyptian Journal of Petroleum* 27 (4): 1113–21. https://doi.org/10.1016/j.ejpe.2018.03.012.

El-Ramady, Hassan, Ahmed El-Henawy, Megahed Amer, Alaa El-Dein Omara, Tamer Elsakhawy, Heba Elbasiouny, Fathy Elbehiry, Doaa Abou Elyazid, and Mohammed El-Mahrouk. 2020. "Agricultural Waste and Its Nano-Management: Mini Review." *Egyptian Journal of Soil Science*. https://doi.org/10.21608/ejss.2020.46807.1397.

El-Sayed Abdel-Raouf, Manar, Manar Elsayed Abdel-Raouf, Nermine E. Maysour, Reem Kamal Farag, and Abdul-Raheim Mahmoud Abdul-Raheim. 2019. "Wastewater Treatment Methodologies, Review Article." *International Journal of Environment & Agricultural Science* (April): 18. www.researchgate.net/publication/332183222.

El-Sayed, Mohamed E. A. 2020. "Nanoadsorbents for Water and Wastewater Remediation." *Science of the Total Environment* 739: 139903. https://doi.org/10.1016/j.scitotenv.2020.139903.

Gautam, Pavan K., Pingali M. Shivapriya, Sushmita Banerjee, Amaresh K. Sahoo, and Sintu K. Samanta. 2020. "Biogenic Fabrication of Iron Nanoadsorbents from Mixed Waste Biomass for Aqueous Phase Removal of Alizarin Red S and Tartrazine: Kinetics, Isotherm, and Thermodynamic Investigation." *Environmental Progress and Sustainable Energy* 39 (2): 1–13. https://doi.org/10.1002/ep.13326.

Geethakarthi, A., and B. R. Phanikumar. 2011. "Industrial Sludge Based Adsorbents/Industrial Byproducts in the Removal of Reactive Dyes – A Review." *International Journal of Water Resources and Environmental Engineering* 3 (January): 1–9.

Goswami, Ritusmita, and Manish Kumar. 2018. "Removal of Fluoride from Aqueous Solution Using Nanoscale Rice Husk Biochar." *Groundwater for Sustainable Development* 7: 446–51. https://doi.org/10.1016/j.gsd.2017.12.010.

Gupta, Vinod K., and Saurabh Sharma. 2003. "Removal of Zinc from Aqueous Solutions Using Bagasse Fly Ash – A Low Cost Adsorbent." *Industrial and Engineering Chemistry Research* 42 (25): 6619–24. https://doi.org/10.1021/ie0303146.

He, Kuang, Yuancai Chen, Zhenghua Tang, and Yongyou Hu. 2016. "Removal of Heavy Metal Ions from Aqueous Solution by Zeolite Synthesized from Fly Ash." *Environmental Science and Pollution Research* 23 (3): 2778–88. https://doi.org/10.1007/s11356-015-5422-6.

Hossain, Nazia, Muhammed A. Bhuiyan, Biplob Kumar Pramanik, Sabzoi Nizamuddin, and Gregory Griffin. 2020. "Waste Materials for Wastewater Treatment and Waste Adsorbents for Biofuel and Cement Supplement Applications: A Critical Review." *Journal of Cleaner Production* 255. https://doi.org/10.1016/j.jclepro.2020.120261.

Hossain, Nazia, Juliana Haji Zaini, and T. M. I. Mahlia. 2017. "A Review of Bioethanol Production from Plant-Based Waste Biomass by Yeast Fermentation." *International Journal of Technology* 8 (1): 5–18. https://doi.org/10.14716/ijtech.v8i1.3948.

Huang, Xiangyu, Guilherme Dognani, Pejman Hadi, Mengying Yang, Aldo E. Job, and Benjamin S. Hsiao. 2020. "Cationic Dialdehyde Nanocellulose from Sugarcane Bagasse for Efficient Chromium(VI) Removal." *ACS Sustainable Chemistry & Engineering* 8 (12): 4734–44. https://doi.org/10.1021/acssuschemeng.9b06683.

Jaria, Guilaine, Vânia Calisto, Carla Patrícia Silva, María Victoria Gil, Marta Otero, and Valdemar I. Esteves. 2019. "Fixed-Bed Performance of a Waste-Derived Granular Activated Carbon for the Removal of Micropollutants from Municipal Wastewater." *Science of the Total Environment* 683: 699–708. https://doi.org/10.1016/j.scitotenv.2019.05.198.

Jawad, Ali H., and Ahmed Saud Abdulhameed. 2020. "Mesoporous Iraqi Red Kaolin Clay as an Efficient Adsorbent for Methylene Blue Dye: Adsorption Kinetic, Isotherm and Mechanism Study." *Surfaces and Interfaces* 18 (December 2019): 100422. https://doi.org/10.1016/j.surfin.2019.100422.

Kamaraj, M., N. R. Srinivasan, Gizachew Assefa, Amare T. Adugna, and Muluken Kebede. 2020. "Facile Development of Sunlit ZnO Nanoparticles-Activated Carbon Hybrid from Pernicious Weed as an Operative Nano-Adsorbent for Removal of Methylene Blue and Chromium from Aqueous Solution: Extended Application in Tannery Industrial Wastewater." *Environmental Technology and Innovation* 17: 100540. https://doi.org/10.1016/j.eti.2019.100540.

Kaur, Mandeep, Santosh Kumari, and Praveen Sharma. 2020. "Removal of Pb (II) from Aqueous Solution Using Nanoadsorbent of Oryza Sativa Husk: Isotherm, Kinetic and Thermodynamic Studies." *Biotechnology Reports* 25: e00410. https://doi.org/10.1016/j.btre.2019.e00410.

Kaur, Mandeep, Praveen Sharma, and Santosh Kumari. 2019. "Equilibrium Studies for Copper Removal from Aqueous Solution Using Nanoadsorbent Synthesized from Rice Husk." *SN Applied Sciences* 1 (9): 1–9. https://doi.org/10.1007/s42452-019-1024-0.

Kazemipour, Maryam, Mehdi Ansari, Shabnam Tajrobehkar, Majdeh Majdzadeh, and Hamed Reihani Kermani. 2008. "Removal of Lead, Cadmium, Zinc, and Copper from Industrial Wastewater by Carbon Developed from Walnut, Hazelnut, Almond, Pistachio Shell, and Apricot Stone." *Journal of Hazardous Materials* 150 (2): 322–7. https://doi.org/10.1016/j.jhazmat.2007.04.118.

Khajeh, Mostafa, Sophie Laurent, and Kamran Dastafkan. 2013. "Nanoadsorbents: Classification, Preparation, and Applications (with Emphasis on Aqueous Media)." *Chemical Reviews* 113 (10): 7728–68. https://doi.org/10.1021/cr400086v.

Krishnamoorthy, Rambabu, Bharath Govindan, Fawzi Banat, Velu Sagadevan, Monash Purushothaman, and Pau Loke Show. 2019. "Date Pits Activated Carbon for Divalent Lead Ions Removal." *Journal of Bioscience and Bioengineering* 128 (1): 88–97. https://doi.org/10.1016/j.jbiosc.2018.12.011.

Lv, Jin fang, Yong Xing Zheng, Xiong Tong, and Xiu Li. 2021. "Clean Utilization of Waste Rocks as a Novel Adsorbent to Treat the Beneficiation Wastewater Containing Arsenic and Fluorine." *Journal of Cleaner Production* 293: 126160. https://doi.org/10.1016/j.jclepro.2021.126160.

Maheswari, B. Uma, V. M. Sivakumar, and M. Thirumarimurugan. 2019. "Synthesis, Characterization of Modified Nanoadsorbents and Its Application in Removal of Zn2+ Ions from Battery Effluent." *Environmental Chemistry and Ecotoxicology* 1: 2–11. https://doi.org/10.1016/j.enceco.2019.05.001.

Mahto, Ashesh, Anshu Kumar, Jai Prakash Chaudhary, Madhuri Bhatt, Atul Kumar Sharma, Parimal Paul, Sanna Kotrappanavar Nataraj, and Ramavatar Meena. 2018. "Solvent-Free Production of Nano-FeS Anchored Graphene from Ulva Fasciata: A Scalable Synthesis of Super-Adsorbent for Lead, Chromium and Dyes." *Journal of Hazardous Materials* 353 (2010): 190–203. https://doi.org/10.1016/j.jhazmat.2018.03.054.

Mamah, Stanley Chinedu, Pei Sean Goh, Ahmad Fauzi Ismail, Mohamed Afizal Mohamed Amin, Nor Akalili Ahmad, Nur Diyana Suzaimi, and Yusuf Olobode Raji. 2020. "Facile Preparation of Palygorskite/Chitin Nanofibers Hybrids Nanomaterial with Remarkable Adsorption Capacity." *Materials Science and Engineering B: Solid-State Materials for Advanced Technology* 262 (February): 114725. https://doi.org/10.1016/j.mseb.2020.114725.

Masoumi, Arameh, Khadijeh Hemmati, and Mousa Ghaemy. 2015. "Structural Modification of Acrylonitrile-Butadiene-Styrene Waste as an Efficient Nanoadsorbent for Removal of Metal Ions from Water: Isotherm, Kinetic and Thermodynamic Study." *RSC Advances* 5 (3): 1735–44. https://doi.org/10.1039/c4ra10830b.

Mohammad, Somaia G. 2013. "Biosorption of Pesticide Onto a Low Cost Carbon Produced from Apricot Stone (Prunus Armeniaca).: Equilibrium, Kinetic and Thermodynamic Studies." *Journal of Applied Sciences Research* 9 (10): 6459–69.

Moharrami, Parisa, and Elaheh Motamedi. 2020. "Application of Cellulose Nanocrystals Prepared from Agricultural Wastes for Synthesis of Starch-Based Hydrogel Nanocomposites: Efficient and Selective Nanoadsorbent for Removal of Cationic Dyes from Water." *Bioresource Technology* 313 (April): 123661. https://doi.org/10.1016/j.biortech.2020.123661.

Mondal, Naba Kumar, Ria Bhaumik, and Jayanta Kumar Datta. 2015. "Removal of Fluoride by Aluminum Impregnated Coconut Fiber from Synthetic Fluoride Solution and Natural Water." *Alexandria Engineering Journal* 54 (4): 1273–84. https://doi.org/10.1016/j.aej.2015.08.006.

Mor, Suman, Kalzang Chhoden, and Khaiwal Ravindra. 2016. "Application of Agro-Waste Rice Husk Ash for the Removal of Phosphate from the Wastewater." *Journal of Cleaner Production* 129: 673–80. https://doi.org/10.1016/j.jclepro.2016.03.088.

Nassar, Mostafa Y., Ibrahim S. Ahmed, and Marwa A. Raya. 2019. "A Facile and Tunable Approach for Synthesis of Pure Silica Nanostructures from Rice Husk for the Removal of Ciprofloxacin Drug from Polluted Aqueous Solutions." *Journal of Molecular Liquids* 282: 251–63. https://doi.org/10.1016/j.molliq.2019.03.017.

Oyewo, Opeyemi A., Amanda Boshielo, Amos Adeniyi, and Maurice S. Onyango. 2021. "Evaluation of the Efficiency of Nanoadsorbent Derived from Orange Peels in the Removal of Cerium from Aqueous Solution." *Particulate Science and Technology* 39 (1): 43–51. https://doi.org/10.1080/02726351.2019.1658666.

Pillai, Parwathi, Swapnil Dharaskar, Manan Shah, and Rashi Sultania. 2020. "Determination of Fluoride Removal Using Silica Nano Adsorbent Modified by Rice Husk from Water." *Groundwater for Sustainable Development* 11 (March): 100423. https://doi.org/10.1016/j.gsd.2020.100423.

Pouretedal, H. R., and N. Sadegh. 2014. "Effective Removal of Amoxicillin, Cephalexin, Tetracycline and Penicillin G from Aqueous Solutions Using Activated Carbon Nanoparticles Prepared from Vine Wood." *Journal of Water Process Engineering* 1: 64–73. https://doi.org/10.1016/j.jwpe.2014.03.006.

Predescu, Andra Mihaela, Ecaterina Matei, Andrei Constantin Berbecaru, Maria Râpă, Mirela Gabriela Sohaciu, Cristian Predescu, and Ruxandra Vidu. 2021. "An Innovative Method of Converting Ferrous Mill Scale Wastes into Superparamagnetic Nanoadsorbents for Water Decontamination." *Materials* 14 (10). https://doi.org/10.3390/ma14102539.

Putro, Jindrayani Nyoo, Shella Permatasari Santoso, Felycia Edi Soetaredjo, Suryadi Ismadji, and Yi Hsu Ju. 2019. "Nanocrystalline Cellulose from Waste Paper: Adsorbent for Azo Dyes Removal." *Environmental Nanotechnology, Monitoring and Management* 12: 100260. https://doi.org/10.1016/j.enmm.2019.100260.

Raman, Noor Azah Abdul, Mohd Rosli Hainin, Norhidayah Abdul Hassan, and Farid Nasir Ani. 2015. "A Review on the Application of Bio-Oil as an Additive for Asphalt." *Jurnal Teknologi* 72 (5): 105–10. https://doi.org/10.11113/jt.v72.3948.

Ribeiro, Araceli Verónica Flores Nardy, Marciela Belisário, Rodrigo Moretto Galazzi, Daniele Cazoni Balthazar, Madson de Godoi Pereira, and Joselito Nardy Ribeiro. 2011. "Evaluation of Two Bioadsorbents for Removing Paracetamol from Aqueous Media." *Electronic Journal of Biotechnology* 14 (6). https://doi.org/10.2225/vol14-issue6-fulltext-8.

Roesler, Thomas A., and Carole Jenny. 2021. "Introduction to Treatment." *Medical Child Abuse*, 155–64. https://doi.org/10.1542/9781581105131-ch08.

Said, Abd El Aziz A., Aref A. M. Aly, Mohamed N. Goda, Mohamed Abd El-Aal, and Mohamed Abdelazim. 2020. "Adsorptive Remediation of Congo Red Dye in Aqueous Solutions Using Acid Pretreated Sugarcane Bagasse." *Journal of Polymers and the Environment* 28 (4): 1129–37. https://doi.org/10.1007/s10924-020-01665-3.

Santasnachok, Chawikarn, Winarto Kurniawan, and Hirofumi Hinode. 2015. "The Use of Synthesized Zeolites from Power Plant Rice Husk Ash Obtained from Thailand as Adsorbent for Cadmium Contamination Removal from Zinc Mining." *Journal of Environmental Chemical Engineering* 3 (3): 2115–26. https://doi.org/10.1016/j.jece.2015.07.016.

Shahnaz, Tasrin, V. Vishnu Priyan, Sivakumar Pandian, and Selvaraju Narayanasamy. 2021. "Use of Nanocellulose Extracted from Grass for Adsorption Abatement of Ciprofloxacin and Diclofenac Removal with Phyto, and Fish Toxicity Studies." *Environmental Pollution* 268: 115494. https://doi.org/10.1016/j.envpol.2020.115494.

Sharma, Y. 1997. "Water Pollution Control – A Guide to the Use of Water Quality Management Principles: Case Study I – The Ganga, India." https://www.academia.edu/9417981/Water_Pollution_Control_A_Guide_to_the_Use_of_Water_Quality_Management_Principles_Edited_by_Case_Study_I_The_Ganga_India.

Shokry, Hassan, Marwa Elkady, and Hesham Hamad. 2019. "Nano Activated Carbon from Industrial Mine Coal as Adsorbents for Removal of Dye from Simulated Textile Wastewater: Operational Parameters and Mechanism Study." *Journal of Materials Research and Technology* 8 (5): 4477–88. https://doi.org/10.1016/j.jmrt.2019.07.061.

Singh, N. B., Garima Nagpal, Sonal Agrawal, and Rachna. 2018. "Water Purification by Using Adsorbents: A Review." *Environmental Technology and Innovation* 11: 187–240. https://doi.org/10.1016/j.eti.2018.05.006.

Sivalingam, Sivamani, and Sujit Sen. 2019. "Efficient Removal of Textile Dye Using Nanosized Fly Ash Derived Zeolite-x: Kinetics and Process Optimization Study." *Journal of the Taiwan Institute of Chemical Engineers* 96: 305–14. https://doi.org/10.1016/j.jtice.2018.10.032.

Sivalingam, Sivamani, and Sujit Sen. 2020. "Rice Husk Ash Derived Nanocrystalline ZSM-5 for Highly Efficient Removal of a Toxic Textile Dye." *Journal of Materials Research and Technology* 9 (6): 14853–64. https://doi.org/10.1016/j.jmrt.2020.10.074.

Tamirat, Andebet Gedamu, John Rick, Amare Aregahegn Dubale, Wei Nien Su, and Bing Joe Hwang. 2016. "Using Hematite for Photoelectrochemical Water Splitting: A Review of Current Progress and Challenges." *Nanoscale Horizons* 1 (4): 243–67. https://doi.org/10.1039/c5nh00098j.

To, Phuong Kim, Hoa Thai Ma, Lam Nguyen Hoang, and Tan Tai Nguyen. 2020. "Nitrate Removal from Waste-Water Using Silica Nanoparticles." *Journal of Chemistry*. https://doi.org/10.1155/2020/8861423.

Vivian Abiaziem, Chioma, Akan Bassey Williams, Adedayo Ibijoke Inegbenebor, Chionyedua Theresa Onwordi, Cyril Osereme Ehi-Eromosele, and Leslie Felicia Petrik. 2019. "Adsorption of Lead Ion from Aqueous Solution unto Cellulose Nanocrystal from Cassava Peel." *Journal of Physics: Conference Series* 1299 (1). https://doi.org/10.1088/1742-6596/1299/1/012122.

Wang, Li, Yujiao Wang, Fang Ma, Vitus Tankpa, Shanshan Bai, Xiaomeng Guo, and Xin Wang. 2019. "Science of the Total Environment Mechanisms and Reutilization of Modi Fi Ed Biochar Used for Removal of Heavy Metals from Wastewater : A Review." *Science of the Total Environment* 668: 1298–309. https://doi.org/10.1016/j.scitotenv.2019.03.011.

Wannahari, R., P. Sannasi, M. F. M. Nordin, and H. Mukhtar. 2018. "Sugarcane Bagasse Derived Nano Magnetic Adsorbent Composite (SCB-NMAC) for Removal of Cu2+ from Aqueous Solution." *ARPN Journal of Engineering and Applied Sciences* 13 (1): 1–9.

Zhan, Chengbo, Priyanka R. Sharma, Hongrui He, Sunil K. Sharma, Alexis McCauley-Pearl, Ruifu Wang, and Benjamin S. Hsiao. 2020. "Rice Husk Based Nanocellulose Scaffolds for Highly Efficient Removal of Heavy Metal Ions from Contaminated Water." *Environmental Science: Water Research and Technology* 6 (11): 3080–90. https://doi.org/10.1039/d0ew00545b.

Zhang, Ming, Bin Gao, June Fang, Anne Elise Creamer, and Jeffery L. Ullman. 2014. "Self-Assembly of Needle-like Layered Double Hydroxide (LDH) Nanocrystals on Hydrochar: Characterization and Phosphate Removal Ability." *RSC Advances* 4 (53): 28171–5. https://doi.org/10.1039/c4ra02332c.

Zhang, Ming, Bin Gao, Ying Yao, Yingwen Xue, and Mandu Inyang. 2012. "Synthesis of Porous MgO-Biochar Nanocomposites for Removal of Phosphate and Nitrate from Aqueous Solutions." *Chemical Engineering Journal* 210: 26–32. https://doi.org/10.1016/j.cej.2012.08.052.

Zhang, Yixin, Jixiang Dong, Fanhui Guo, Zhongye Shao, and Jianjun Wu. 2018. "Zeolite Synthesized from Coal Fly Ash Produced by a Gasification Process for Ni2+ Removal Fromwater." *Minerals* 8 (3): 1–14. https://doi.org/10.3390/min8030116.

12 Environmental Sustainability
An Interdisciplinary Approach

M. Angkayarkan Vinayakaselvi

CONTENTS

12.1 Introduction 165
    12.1.1 State of the Environment.. 166
    12.1.2 The Global Scenario 166
    12.1.3 The Indian Scenario 167
12.2 Environmental Sustainability: Knowledge beyond Disciplines 167
12.3 Expanding the Boundaries by Combining Theory and Practice 168
12.4 Interdisciplinary Nature of Environmental Sustainability 168
12.5 Possibility of Including Humanities and Technology in Natural Resource Management 169
12.6 Role of SHG and NGO in Environmental Sustainability Self-Help Groups 170
    12.6.1 History of SHGs 170
    12.6.2 Non-governmental Organizations 171
    12.6.3 Media 172
    12.6.4 Awards 172
12.7 Role of Women in Environmental Protection 173
    12.7.1 Sanitation Practices 174
    12.7.2 Health Consciousness 174
12.8 Conclusion 174
References 175

12.1 INTRODUCTION

Sources have stated that the important factors such as the facilities from the ecosystem, materials required for survival, and the economic support that are required for all organic forms like humans and other living forms has always been provided by the environment (Naidoo and Olaniran 2014). Many causes such as changes in the climatic condition, pollution level in the environment, depletion of the resources available in nature, depletion of the ozone layer, and global warming frequently lead to problems like security issues of food and health. These issues are mainly caused due to the unselective works of humans leading to high stress on the nature. The degradation of the environment and the issues causing problems for better development are caused due to factors like increase in the population, economic activities, and consumption patterns of humans. Many solutions are being derived by the scientists for controlling the degradation of the environment leading to sustainable development. Currently, the condition of the environment is changing drastically than expected, which has led the scientists to gain more knowledge on the changes and find an opportunity to solve the issues (Kumar Das 2015). Factors that influence the correction in market failure and suppress expenses of guarding the environment such as rights of the property, market creation, deposits, and creation of bonds. The usage patterns and the service value of the nature is understood by various valuation techniques. Increase in the building capacity and education system has helped in the progress of gaining knowledge and also in the decision-making process, and these factors have led to exposure of environmental education worldwide (Panayotou 1994).

Indian undergraduate education has started environmental education as a main part of the education system since 1986. A conference held at Delhi in 2006 and attended by Y.K. Sabharwal, Chief Justice of India, on the topic "Environment-Awareness-Enforcement" mentioned national integrity on incorporating environmental education into National Curriculum Framework and this resulted in the discretion given by the Supreme Court of India in accordance to which the subject of environmental science has been made compulsory for the undergraduate education from class I to V in schools and measures are taken to overcome the environmental education problems. The main concern currently is whether environmental education should be combined with any other subject or should be taught separately. All these issues need to be sorted out at the earliest. The major concerns pointed out by various educationalist, policymakers, and planners are as follows: After 6 years did the situation improve? The introduction to environmental education as a topic from class I to XII and higher studies lead to hike in literacy rate of ecology. Is there any high efficiency in the mode of education of eco-literacy, its level, and aspects that affect the ecological education? What are the aspects and issues that are barriers for environmental education?

This chapter states that several measures have been taken to solve these issues. This has helped in decision-making by researchers and academicians, curriculum designers, and educators of environmental studies, authorities, and policymakers for enhancing the environmental studies and promoting the literacy rate in public, leading to awareness about challenges allied with the environment and development of the mandatory skills to address the problems and to motivate the commitments for decision-making and to take proper decision for a responsible action to live a life complied with nature (Knapp and Ferrante 2012).

12.1.1 STATE OF THE ENVIRONMENT

All the organic and inorganic things that frame and inspire the growth, advance, and existence of the organisms are described as an environment or a natural environment. It can also be stated as a mixture of components that gives a perfect structure to the daily living, like scientific, ethical, administrative, monetary, and social bond and organizations. A recent report mentions affirmation on the status of the environment state that the deprivation and defamation of the nature are mainly caused due to transformation, development, and urbanization along with heavy loss of natural resources causing issues like global warming, change in weather, depletion of the ozone, acid rain, and so on that are prompted and amplified by the increasing rate of population and illiteracy of the society (Evershed 2008).

12.1.2 THE GLOBAL SCENARIO

The intense effects on the nature are caused by the increase in population rate and monetary undertakings and intake forms that lead to heavy loss to nature, as mentioned in the Global Environment Outlook – 4 in 2007. It also specifies the huge increase in the energy consumption, mode of traveling, and various other usages, which have led to high rate of pollution in the atmosphere and exceptional emission of major anthropogenic GHG (Smil 2011). Land degradation is caused by the usage of non-usable lands that lead to depletion of nutrient components in the soil, depletion of water, massive changes in ecological cycles, and erosion of soil that reduces the yield, biodiversity, and many other services provided by the nature by causing huge changes in climatic variations. Development is highly affected due to shortage of water resources along with effects on the security of food, health of living beings, and other natural systems. This scarcity of water is caused mainly by the increasing rate of population and hike in the rate of consumption of wealth and reserves of nature. This highly affects the development of the humans in the upcoming times leading to reduction in future developments and will also cause many health issues (Lambin and Meyfroidt 2011).

12.1.3 THE INDIAN SCENARIO

Unsustainable practices like not adopting adequate soil conservation measures, improper crop rotation, careless use of agrochemicals like fertilizers and pesticides, improper planning and management of irrigation systems, and groundwater extraction that exceeds the recharge capacity all contribute to the issue. The good news is that India's forest cover is already progressively expanding and that more than two-thirds of the damaged land may recover. However, considering that India requires 33.3% more forest cover than it now has, it would be absurd to hope to get above the current 21%. The All-India Institute of Medical Sciences and the Central Pollution Control Board examined 50 cities throughout India and found that the amount of respirable suspended particle matter has increased in each of them (Beermann et al. 2016).

12.2 ENVIRONMENTAL SUSTAINABILITY: KNOWLEDGE BEYOND DISCIPLINES

Sustainable development presents significant challenges when assessed against a single indicator or when attempting to balance several indicators due to its interdisciplinary nature. Comprehensive frameworks are needed to evaluate the communal, monetary, ecological, and other elements of sustainability. The sustainable development goals (SDG) may be fluid and ever-changing among many stakeholders while being approved by the UN, they can be altered depending on the lens through which they are seen. In these circumstances, it is the researcher's duty to suggest a comprehensive justification of the specific explanation of the sustainable development goals they wish to examine. This definition may call for the clarification of interdependencies (Sarkis and Ibrahim 2022). Despite society's emphasis on sustainable development, the global environmental sustainability has become a conundrum, where climate change and biodiversity loss are of particular concern, and is far from being resolved (Ceballos et al. 2020). It demonstrates how important it is for us to alter our course and step up our efforts in order to attain sustainable development. We believe that operational methods to evaluate the environmental successes of current operations in relation to absolute environmental sustainability boundaries are important to drive a sustainable development trajectory. For achieving the sustainable development, it is important to include environmental sustainability boundaries in environmental evaluations (Muñoz and Gladek 2017).

All the 193 United Nations members have implemented the 17 SDGs in September 2015 as a follow-up to the Millennium Development Goals established in 2000 with the aim of eradicating the high rate of poverty, safeguarding the environment, and ensuring growth for everyone by 2030 (UN 2015). The SDGs were extended to add the agenda's focus on issues like change in climatic conditions, sustainable consumption, modernization, and the significance of peace and justice. They also made mandate action from all countries, even those with greater heights of growth. In order to improve people's lifestyle and improved development, high-quality education must be encouraged.

Complex issues like poverty, human rights, and climate change necessitate integrating knowledge and skills from several academic disciplines. The capacity to recognize complex situations and act appropriately is encouraged by interdisciplinarity, which is aligned with the intended outcomes of learning for sustainable development. The research states that there are numerous methods to add to interdisciplinary sustainable development learning, but it has shown to be challenging. The interdisciplinarity has its own known boons which will ultimately help the students to implement a viewpoint that incorporates communal, monetary, and ecological considerations. The creation of specialized fields of study unconnected to the core curriculum or only including sustainable development in environmental courses would not be sufficient to give students the information and abilities they require to deal with sustainability concerns in their everyday lives (Annan-Diab and Molinari 2017).

12.3 EXPANDING THE BOUNDARIES BY COMBINING THEORY AND PRACTICE

Sustainable development, also known as the Brundtland Report, was described as "development that satisfies the demands of the present without compromising the ability of future generations to satisfy their own needs" in the report of the United Nations World Commission on Environment and Development (WCED 1987). Since the report's release, sustainable development has gained widespread acceptance as humanity's chosen course (Robert, Parris, and Leiserowitz 2005). There are several tools and methods available now to describe, assess, and show the performance with regard to sustainable development. Over 800 initiatives are included in the Compendium of Sustainable Development Indicator Initiatives; however, they presumably make up a small part of the total. The literature on quantitative indicators of sustainable development, according to Parris and Kates (2003), is "slightly separate from the theoretical and essentially economic study of the theory and standards of defining sustainable development." This is due to the people reporting sustainable development in smaller scales like for small regions, industries, corporation and products, while the theatrical study usually concentrates on nation and global levels. This explains in part why practice and theory are not connected. Not many people acknowledge the applicability of current theory to those reporting at lesser sizes. This study demonstrates how a link based on assets may give a common framework for reporting on sustainable development, assist in the formulation of relevant indicators in various situations, and help identify when and how information can be merged. The asset-based strategy that is presented in this chapter is intended to organize continuous performance evaluations and reporting by existing bodies, such as Environmental Impact Assessments (EIA) and Social Impact Assessments (SIA). The methodology offers a foundation for determining and categorizing the effects of a planned activity in accordance with its contributions (both positive and negative) to sustainable development. It may be used, for instance, to complete the first three phases of Fischer's (1999) method for incorporating sustainability objectives and targets into evaluations of policies, plans, and programs.

Numerous strategies have been developed in an effort to show performance with regard to sustainable development. It sometimes seems as though the widespread use of quantitative indicators has nothing to do with any philosophy of sustainable development. One explanation for this gap is that although actual efforts to report on sustainable development typically function at smaller dimensions like regions, industries, and businesses, theoretical approaches have a tendency to concentrate on national and global levels. Not many people acknowledge the applicability of current theory to those reporting at lesser sizes. The theoretical exposition of sustainable development in terms of assets serves as the inspiration for the reporting strategy developed in this study. The strategy is demonstrated with a fictitious case. The asset-based approach offers a common framework for connecting what have previously been viewed as different ways to reporting on sustainable development and may be used in a range of settings (national, regional, industry, corporate, and product). Knowledge on how performance should be viewed and how information may be combined to report at different levels requires an understanding that different players are responsible for distinct asset mixes.

12.4 INTERDISCIPLINARY NATURE OF ENVIRONMENTAL SUSTAINABILITY

A sustainability trait may be applied to a variety of things, including goods, ecosystems, nations, and policy change. Sustainability is ill-defined and elusive (Brink, Hengeveld, and Tobi 2020), resulting in a large number of different tools and frameworks (Singh et al. 2009). Despite the fact that there is no "gold standard" for assessing sustainability, the Triple Bottom Line (TBL) model is popular. The TBL distinguishes between an economic, a social, and an environmental dimension. Sometimes, other dimensions are included. The TBL shows how the topic of environmental study had paved the way to sustainability science into an interdisciplinary discipline that includes the interplay between nature and civilization (Spangenberg 2011).

We concentrate on multidisciplinary measurement of sustainability in the manner of Kates et al. (2001) and Ostrom et al. Any study or collection of studies that are conducted by academics from two or more different scientific fields is referred to as interdisciplinary research. The research is grounded in a conceptual framework that connects or unifies theoretical frameworks from those disciplines, employs a study design and methodology that are not specific to one area of study, and necessitates the use of perspectives and expertise from the involved disciplines throughout various stages of the research process.

Although the development of interdisciplinarity is seen as positive conceptually, it may hinder practical research (Kajikawa 2008). The operationalization of the quality of sustainability into something measurable is the first obstacle (Tobi and Kampen 2018). Different measuring theories and reasoning are employed by various fields. The notions that different disciplines investigate and, consequently, quantify vary. Notions examined in the social sciences are typically seen as being less palpable than concepts studied in the scientific sciences. In the first, the researchers' preferred theory typically determines the measure of a notion, and various theories correspond to various measures and measurements. Generally, formal measurement theory, as it exists in the natural sciences, is absent in the social sciences (Kampen and Tobi 2011). Finkelstein (2005) asserts that social science ideas are often less "countable" and that measurement levels are typically lower. As a result, the social sciences' intellectual underpinnings and measuring standards are lower than those used by natural scientists. Therefore, a conceptual framework that takes these distinctions into account is necessary for methodological research on interdisciplinary measurement as well as studies that employ interdisciplinary measurement.

12.5 POSSIBILITY OF INCLUDING HUMANITIES AND TECHNOLOGY IN NATURAL RESOURCE MANAGEMENT

The environmental humanities are an emerging transdisciplinary endeavor that are rapidly growing in importance within the liberal arts and as a necessary element of the university of the twenty-first century. The Environmental Humanities have developed over the past 10 years into a significant collaborative scholarly endeavor, bringing together experts from a variety of environment-related topics in social sciences and humanities, such as ecocriticism (literature and environment studies), environmental history, environmental philosophy, environmental anthropology, and human geography. The majority of the focus of humanistic environmental studies is on cultural outputs, such as architecture, fiction and non-fiction writing, drama, music, visual arts, film, and other forms of media, as well as activism, politics, history, medical, and religious discourses. This focus on cultural items is partly due to their ability to organize or silence groups as well as to drastically alter environmental consciousness, for better or worse. Cultural artifacts frequently enable societies to imagine alternative outcomes and to brainstorm creatively about enacting changes that facilitate adaptation, boost resilience, reduce fear, adjust risk, and make the fight for resources more manageable, or at least less disastrous. By doing this, cultural products shed light on how societies, communities, and people perceive their surroundings and interact with environmental problems. Humanistic environmental studies should bring together academics from various fields of knowledge like social science, humanities which includes anthropology, history, art, ethics, economics, science, philosophy, literature, and medicine, psychology, religion, sociology, urban planning, and others. It should also draw on the expertise of individual humanists, social scientists, and others engaged in interdisciplinary work across world areas. Collaboration with academics in public humanities, medical humanities, and digital humanities is crucial as well, considering the detrimental impacts of environmental deterioration on human health. They show how people exploit, harm, and ruin their surroundings and how they cope with a terrible and uncertain future. According to Elizabeth DeLoughrey, Jill Didur, and Anthony Carrigan, humanistic environmental studies has the "radical potential to reshape our ecological futures" through systematically interacting with a variety of cultural artifacts (DeLoughrey, Didur, and Carrigan 2015).

12.6 ROLE OF SHG AND NGO IN ENVIRONMENTAL SUSTAINABILITY SELF-HELP GROUPS

Self-Help Group (SHG) is an unregistered association of microentrepreneurs from similar social and economic backgrounds who voluntarily join forces to save small amounts of money regularly and they agreed mutually to contribute a common fund, and rely on one another to meet their emergency needs. The group members rely on their pooled knowledge and peer pressure to guarantee ethical credit utilization and prompt payback. This process eradicates the need for warranty and is closely related to the solidarity lending which is widely used by microfinance institutions. Flat interest rates are utilized for the majority of loan calculations in order to keep the bookkeeping straightforward enough for the members to manage (Venkataramany and Bhasin 2011).

SHG is the mobilization of poor people organized and joined together for savings and income-generating activities for eradicating poverty. All of them agreed to have the common savings amount which accumulates as a common fund for later distribution among members as internal loan which solves the financial need of the household level. There are certain guidelines laid down to follow for the better outcomes of the SHGs ("Retailing of Self-Help Group (SHG) Products in India" 2019). SHGs are a wonderful technique to handle financial intermediation in India. The methodology consolidates admittance toward minimal effort budgetary administrations with a procedure of self-administration and empowerment of women. SHGs are shaped and bolstered at first by NGOs and now progressively by government offices. Connected not exclusively to banks, yet, additionally, to more extensive improvement programs, SHGs are believed to present numerous advantages, both monetary and social. SHGs empower women to develop their reserve funds and to get the credit and banks are progressively ready to lend the loan. SHGs are turning out to be network stages from which ladies are taking dynamic jobs in town undertakings, representing neighborhood political decision or making a move to address social or network issues (Prasad 2017). The SHGs that are shaped by NGOs, government organizations, or banks can be partitioned into three sorts. The NGO-advanced SHGs are presumably the most established. Money-related exercises may not be a piece of a task which could be topic-based like coordinated horticulture and associated exercises, watershed management, incorporated women and child improvement, and so forth. The accomplishment of this model has incited the legislature to embrace this as a methodology in all the improvement plans. The Bank-advanced groups were the ones solely intended for budgetary exercises (Rashtriya Mahila Kosh and ICICI for instance) and group business enterprise exercises. The current gatherings are those that have a blend of budgetary and non-money related exercises, where limit-building expertise advancement and individual and gathering exercises are empowered based on bunch quality. These are viably utilized by banks to broaden different kinds of help such as miniaturized scale credit, advances for taking up little undertakings, and so on. The special element of the SHG is its capacity to instill among its individuals sound propensities for frugality, reserve funds, and banking. Normal reserve funds, regular meetings, mandatory participation, and methodical preparation are the remarkable highlights of the SHG idea. Each group chooses one animator and two representatives among themselves. The animator has the responsibility to lead the group and maintain different registers. The representatives help the animator and keep up the financial balances of the group (Reddy 2007).

12.6.1 History of SHGs

India's SHG development has risen as the world's biggest and best system of women-based network-based microfinance establishment. Self-Help Group Bank Linkage program (SHG-BLP) is a milestone model started by the National Bank for a bank outreach program. SHG-BLP rose above itself into an all-encompassing project for Agriculture and Rural Development (NABARD) in 1992 to convey reasonable entryway step banking administrations and has to a great extent accomplished the expressed objectives of monetary incorporation; it is a home-developed self-improvement development with a target of making manageable occupation open doors for the rustic poor, begun as building money-related, social, monetary, and, recently, mechanical capital in provincial India (Niyonsenga et al. 2020). Until the development of Self-Help Group Movement in the latter part of the 1990s, monetary consideration was slippery as the normal man was out of the focal point of a formal financial framework despite nationalization of banks (in 1969 and 1980) and other arrangement activities like need segment loaning, lead bank plot, and so on. The huge range of store and credit results of Indian banks could not satisfy the money-related requirements of the rural poor. In this specific circumstance, NABARD led different examinations to convey a reasonable money-related model for the Indian rustic credit framework. It was uncovered in the investigations that the helpless needs better access to sufficient, convenient, and appropriate money-related items and administrations instead of modest credit. SHGs were considered as network-possessed, self-guided casual doorstep investment funds and credit conveyance instrument by a gathering of 10–20 individuals having a homogeneous financial foundation and originating from a little coterminous region, who work on the standard of self-improvement, solidarity, and shared premium. Accordingly, the SHGs were advanced by NABARD depending on its effective examination with Mysore Resettlement and Development Agency (MYRADA) in 1987. The gain from the Action Research drove NABARD to dispatch a pilot in 1992 with a pilot base of only 500 SHGs with strategy backing from the Reserve Bank of India encouraging banks to open bank accounts for the sake of Groups (Mukherjee, Mallik, and Thakur 2019).

> An unregistered organisation of micro-entrepreneurs with a uniform social and economic background who get together voluntarily to save regular little amounts of money is known as a Self-Help Group (SHG), jointly committing to contribute to a shared fund and relying on one another to cover their emergency requirements. In order to ensure responsible credit usage and prompt payback, the group members use their combined expertise and peer pressure. This technique, which does away with the requirement for collateral and is frequently employed by microfinance organisations, is closely connected to solidarity lending. For the majority of loan calculations, flat interest rates are utilised in order to keep the bookkeeping straightforward enough for the members to manage.
>
> (Navin et al. 2017)

12.6.2 Non-governmental Organizations

Volunteer groups, often known as NGOs or non-governmental organizations, have made a significant contribution to the dissemination of environmental education. Their primary goal is to raise environmental consciousness via academic, scientific, and real-world practical activities. They include organizations of professionals as well as the average layperson. The largest NGO, Bombay Natural History Society, was founded in Mumbai in September 1883 with the mission of preserving nature. Through lectures, field visits, expeditions, and publications, the organization spreads information about flora and wildlife (Kaushik et al. 2020). The Gandhi Peace Foundation's Environment Cell was established in Delhi primarily to support rural development organizations' environmental initiatives and to circulate eco-literacy knowledge by stating the issues in nature. They arrange seminars and workshops for professionals working on environmental concerns, policymakers, and individuals. Every October, the Kerala Sastra Sahitya or Parishad (KSSP), a significant NGO in the state, holds "Sastrakalayatha," a festival that features by showing public plays in the streets and spoofs on technical topics pertaining to health learning and the environs. The hydroelectric project in the state of Kerala at Silent Valley has led to dying out the evergreen rain forests and this project was opposed by this NGO. The Chandigarh-based Environment Society of India (ESI) launches creative initiatives and campaigns to raise environmental awareness. Since 1972, the Delhi-based Indian Environmental Society (IES) has supported efforts aimed at improving the environment in India (Sandhu and Liang 2021). It is involved in conservation of biodiversity, dissemination of information, solid waste management through eco-friendly technologies, and additionally provide environmental education. Through its Environment Education Program, the New Delhi-based World-Wide Fund (WWF) seeks to increase public knowledge of environmental issues and institutional and individual capacity for nature conservation and environmental preservation. It encourages nature groups and arranges nature camps to raise awareness. Society for Environment and Education (SEE) has created awareness among ladies and teenagers about the necessity of encouraging their fitness, sanitation, and a healthy surrounding. SEE also organizes training platforms on eco-literacy studies for kids and teenagers. By allowing the public to touch and handle snakes, Madras Snake Park and Sundervan in Ahmedabad employ wildlife animals to foster an appreciation of the environment. In Madhya Pradesh, Eklavya arranges environmental camps and creates environmental workbooks. Environmental awareness campaigns are organized by the Goan Research Institute for Development, the Madras Environmental Society, the Center for Environmental Concerns, Hyderabad, and SEARCH in Bangalore. Authorities at the national and international levels have recognized and praised the enormous contribution made by NGOs. The public really gives NGOs more trust than official entities (Blum 2009).

12.6.3 Media

The media is a significant component of society that actively promotes environmental education. Compared to street plays, exhibitions, lectures, seminars, competitions, etc., the press, publications, broadcasting, and TV have a considerably greater influence. The benefit of mass media is that they are far more successful at teaching the general public and can reach a huge audience in a short amount of time. Environmental education content is being produced by a large number of expert authors and creators. The BBC's "Earth file," the Discovery Channel, and Zee TV's "Hum-Zameen" have all demonstrated impressive success in raising environmental consciousness. The listeners of "Soolal kaappom," an everyday show broadcasted at 07:40 IST at Radio Station located at Tiruchirappalli, are kept informed on ecological problems faced (Mount 2022).

12.6.4 Awards

The Indira Gandhi Paryavaran Puraskar (IGPP) is given annually to citizens of India or any organization from India for noteworthy environmental accomplishments that are acknowledged as having a quantifiable effect on the preservation or general development of the ecosystem. Even after these promotions, it is seen that there is lack of eco-literacy rate, unfavorable pro-environment attitudes, and escalating ecological issues. This is due to the ineffectiveness of environmental education implementation. Consequently, on March 26, 2003, the Supreme Court issued an order mandating the implementation of ecological awareness studies at all grades of schools and universities beginning in the 2004–2005 academic year. Each board of education and the state governments were instructed to ensure that it was implemented gradually in the educational facilities (Sen 1999). From that point on, environmental education is a required component of every class' curriculum. Since there was no specific curriculum regarding environmental studies, the honorable Supreme Court has advised and directed the national framework to bring out and implement specific guidelines and curriculum from class I to XII. As a result, the way environmental education was taught in schools varied widely across the nation. It recommended using activities for classes I and II, environmental studies for classes III –V, infusion for courses VI–X, and project-based learning for classes XI and XII. In order to create a primary module pattern for eco-literacy for the courses in undergraduate university responding to the Supreme Court's instruction to the UGC (University Grants Commission), India's main policymaking and grant-giving organization for all higher education institutes located in India. The course consists of eight sections that are covered in 50 lectures, of which 45 are delivered in a classroom setting and 5 involve fieldwork. Exams are used for evaluation; they are given at the conclusion of the semester and are worth 100 marks, of which 25 are awarded for fieldwork. Since environmental education courses have been offered at the undergraduate level for 8 years, it is time to evaluate the program. In this regard, it is crucial to test the aspects that have effects on the rate of eco-literacy, way of adoption, and results of the education program in order to aid future scientists, environmental educationalists, syllabus designers, specialists, and policymakers in better defining the program (Gupta 2014).

12.7 ROLE OF WOMEN IN ENVIRONMENTAL PROTECTION

Environmentalism in recent years has become a dominant discourse. The various fields of this discipline collectively promote a number of issues regarding environment and development. Development requires an understanding and control over human relationship with nature, other wisely known as environmental management. Such management becomes meaningful only when women contribute and are allowed to contribute their full participation in the field. Development has fortunately given way to sustainable development, and growth has given way to "green growth." In India, as elsewhere, social science research has been closely linked to the rise and maturity of the environmental movement. After the Independence, there began an age of ecological innocence, when the urge to industrialize and "catch up" with the developed world relegated environmental concern to the background (Loomba* 2018). Nowadays, the concern for maintaining ecological balance emerged in the form of vocal and articulate social movement. Governments throughout the globe are increasingly assuming that a variety of environmental issues, including climate change, disposal of hazardous waste materials and ozone layer depletion, destruction of biological resources, and loss of forests, have a global scope for environmental management programs. The situation has become bad to worse. Now it is high time for taking appropriate measures to solve environmental problems. Numerous research on women and the environment have revealed that women play important roles in managing natural resources and significantly contribute to the preservation and restoration of the environment. Women have a strong understanding of the environment because of their close interaction with it. Women have a strong connection to their local environment through their work as farmers, water and firewood collectors, and other domestic tasks. Environmental issues often directly affect them the most (Bhat 2015). Thus, women have served as agriculturists, water resource managers, responsible domestic and household managers, health planners, forest managers, etc. Therefore, including women in environmental protection would aid in the socialization of responsibility. Additionally, it is necessary to keep a healthy balance between people, the environment, and natural resources. Women see the environment as one of humanity's life support systems. People are reliant on their surroundings since the environment is the source of all life. Healthy and peaceful living would not be possible without a nice atmosphere. The passion and commitment of women to their careers is a reflection of their unselfish love and care for the environment, for their children, and for the upcoming generations. In the era of globalization, industrialization, and e-communication and rapid changing scenario, India has emerged as the most intricate and one of the largest democratic countries in all over the world. This is particularly true in the current environment, in which the globe has become a global village, where territorial boundaries are constantly shifting, where new problems and linkages are waiting to be uncovered. Men and women both have done whatever good in this universe. Over the past few decades, women have contributed significantly to life and society by interrogating and exploring their own lives and that of other women (Khatri 2016). Today's Indian women deal with multiple issues concerning self and society. Women had for long been denied access to education, equal rights, right to work, and the freedom to choose. But in our societies the role of women outside the family has never been accepted in the same manner as it is in case of male. Before two decades, barring a handful of exceptions, the Indian women were not ready to penetrate and participate meaningfully in any sociopolitical decision-making mechanism. Moreover, the rigorous traditional values also confine women within the frontiers of family. Consequently, women are constrained to take the task of their distinctiveness in society. Findings of several studies on women indicate that a vast majority, if not most, of them are not properly aware of their role and right in the society and they prefer to remain at the periphery of the sociopolitical power (Vaish and Arrawatia 2021). The United Nations after declaring 1975–1985 as the decade for women's development noted that despite doing two-thirds of the world's work, they only receive 10% of its income and 1% of its production resources. Women are not allowed to occupy positions of authority within or outside of the home. They are not allowed to be a part in decision-making processes. Women's education and participation thus are essentially vital issues to rectify this imbalance and gender inequity. Encouraging engagement in the social, economic, and political spheres would improve the management of the environment. The Constitution of India has clearly stated the principle of gender equality in the Preamble, Fundamental Rights, Fundamental Duties and Directive Principles permitting the state to take necessary actions to provide positive judgment for the women. It is seen that the rural side of India has converse effects of absence of clean latrines on women's dignity and in order to overcome it, women must be given active chances to participate in the cleanliness campaigns. It is proven that women have better tendency to manage the household cleanliness than male. Women are usually not considered as those who give share to environmental destruction as they take responsibilities for the existence for the factors like water, diet, fodder, fuel, and other habitats that increase their problems which are different from that of men. These problems mainly relate to the societies, womankind, and natural resources and converse the growth of the harmful effect of current developmental prototypes (Vaish and Arrawatia 2021).

12.7.1 Sanitation Practices

In general, sanitation refers to the provision of facilities and services for the secured disposal of human urine and feces, according to the World Health Organization (2018): Guidelines on Sanitation and Health. Improving sanitation is proven to have a large positive influence on health both in families and across communities. Inadequate sanitation is a leading cause of disease worldwide. The term "sanitation" also refers to the upkeep of sanitary conditions through offerings like waste disposal and rubbish collection (World Health Organization), in particular, in rural regions and urban slums. Special consideration must be given to the requirements of women in the provision of sewage disposal, toilet facilities, and sanitation within reach of families. Women's participation in the conception, implementation, and upkeep of such services must be guaranteed (Kanungo et al. 2021).

12.7.2 Health Consciousness

Health consciousness describes an attitude in which one has an awareness of the healthiness of one's diet and lifestyle. The phrase emphasizes the increased danger of sickness and malnutrition that women suffer during all three crucial periods, including infancy and childhood, adolescence, and the reproductive stage. It is important to work to put an end to intra-household discrimination against girls and women in nutritional problems by using the proper tactics. And women are capable of doing this (McGuinness et al. 2020).

12.8 CONCLUSION

Earth and nature makes every life possible and how we treat them define who we are. It is therefore essential that we need to protect our earth in a very sustainable way. Environmental sustainability is the greatest mission that our world need to achieve right now in order to transform our world into a better place to live for our future generations. Within the implication of sustainable development goals, formulation of various treaties on natural and climatic changes, we are very well aware about the need for environmental sustainability. Humans should be more logical and rational in using our natural resource, even though we need development in each and every point in our life; development requires an understanding and control over human relationship with nature. For this it is important that we need to rely on renewable energy resources, and should also carry out technologies that would carry out development in an eco-friendly way. Interdisciplinary approach toward environmental sustainability connects various disciplinaries from natural science, social science, history, philosophy and so on, and thus provide an holistic solution for policymaking and decision-making.

The involvement of diversified fields with similar objective can facilitate the process of implementation in a better manner. It is not the primary duty of treaty-makers or policymakers to provide environmental protection and sustainable development; it is the responsibility of everyone, including media and governments, both at regional and global levels toward the creation of sustainable development in its most effective form.

REFERENCES

Annan-Diab, Fatima, and Carolina Molinari. 2017. "Interdisciplinarity: Practical Approach to Advancing Education for Sustainability and for the Sustainable Development Goals." *The International Journal of Management Education* 15 (2, Part B): 73–83. https://doi.org/10.1016/j.ijme.2017.03.006.

Beermann, Jan, Appukuttan Damodaran, Kirsten Jörgensen, and Miranda A. Schreurs. 2016. "Climate Action in Indian Cities: An Emerging New Research Area." *Journal of Integrative Environmental Sciences*. https://doi.org/10.1080/1943815X.2015.1130723.

Bhat, Rouf Ahmad. 2015. "Role of Education in the Empowerment of Women in India." *Journal of Education and Practice*. https://www.academia.edu/36826598.

Blum, Nicole. 2009. "Small NGO Schools in India: Implications for Access and Innovation." *Compare*. https://doi.org/10.1080/03057920902750491.

Brink, Matthijs, Geerten M. Hengeveld, and Hilde Tobi. 2020. "Interdisciplinary Measurement: A Systematic Review of the Case of Sustainability." *Ecological Indicators* 112: 106145.

Ceballos, Francisco, Samyuktha Kannan, Berber Kramer, Indranil Chakraborty, Prasenjit Maity, Christopher B. Barrett, April December, et al. 2020. "The Sustainable Development Goals Report." *World Development* 136 (2): 1–9.

DeLoughrey, Elizabeth, Jill Didur, and Anthony Carrigan. 2015. *Global Ecologies and the Environmental Humanities: Postcolonial Approaches*. Routledge.

Evershed, R. P. 2008. "Organic Residue Analysis in Archaeology: The Archaeological Biomarker Revolution." *Archaeometry*. https://doi.org/10.1111/j.1475-4754.2008.00446.x.

Finkelstein, L. 2005. "Problems of Measurement in Soft Systems." *Measurement* 38: 267–74. https://doi.org/10.1016/j.measurement.2005.09.002.

Fischer, T. B. 1999. "Benefits from SEA Application—A Comparative Review of North West England, Noord-Holland and EVR Brandenburg-Berlin." *Environmental Impact Assessment Review* 19: 143–73.

Gupta, Saurabh. 2014. "From Demanding to Delivering Development: Challenges of NGO-Led Development in Rural Rajasthan, India." *Journal of South Asian Development*. https://doi.org/10.1177/0973174114536099.

Kajikawa, Yuya. 2008. "Research Core and Framework of Sustainability Science." *Sustainability Science* 3 (2): 215–39.

Kampen, J. K., and H. Tobi. 2011. "Social Scientific Metrology as the Mediator between Sociology and Socionomy: A Cri de Coeur for the Systemizing of Social Indicators." In *Social Indicators: Statistics, Trends and Policy Development*, 1–26. Nova Science Publishers.

Kanungo, Suman, Pranab Chatterjee, Jayanta Saha, Tania Pan, Nandini Datta Chakrabarty, and Shanta Dutta. 2021. "Water, Sanitation, and Hygiene Practices in Urban Slums of Eastern India." *The Journal of Infectious Diseases*. https://doi.org/10.1093/infdis/jiab354.

Kates, R. W., et al. 2001. "Sustainability Science." *Science* 292 (27 April): 641–42.

Kaushik, Pritam, Jagrat Jaggi, Yash Jadhav, Badri Narayan Goswami, and Aditya Dhuri. 2020. "Parameters to Measure Performance of an NGO in India." *Asian Journal of Management*. https://doi.org/10.5958/2321-5763.2020.00055.4.

Khatri, Rita. 2016. "The Role of Education Towards Women Empowerment in India." *International Journal of Advanced Research*. https://doi.org/10.21474/ijar01/2117.

Knapp, Kenneth J., and Claudia J. Ferrante. 2012. "LIT_Ref1: Policy Awareness, Enforcement and Maintenance: Critical to Information Security Effectiveness in Organizations." *Journal of Management Policy and Practice* 13.

Kumar Das, Pradip. 2015. "An Introduction to the Concept of Environmental Management: Indian Context." *International Journal of Innovation and Economic Development*. https://doi.org/10.18775/ijied.1849-7551-7020.2015.24.2003.

Lambin, Eric F., and Patrick Meyfroidt. 2011. "Global Land Use Change, Economic Globalization, and the Looming Land Scarcity." *Proceedings of the National Academy of Sciences of the United States of America*. https://doi.org/10.1073/pnas.1100480108.

Loomba*, Dr. Shuchi. 2018. "Role of Microfinance in Women Empowerment in India." Unknown Publication.

McGuinness, Sarah L., Joanne O'Toole, S. Fiona Barker, Andrew B. Forbes, Thomas B. Boving, Asha Giriyan, Kavita Patil, et al. 2020. "Household Water Storage Management, Hygiene Practices, and Associated Drinking Water Quality in Rural India." *Environmental Science & Technology*. https://doi.org/10.1021/acs.est.9b04818.

Mount, Liz. 2022. "'Funding Does Something to People': NGOs Navigating Funding Challenges in India." *Development in Practice*. https://doi.org/10.1080/09614524.2021.1911938.

Mukherjee, Shrabani, Subhadri Sankar Mallik, and Debdulal Thakur. 2019. "Tracking Financial Inclusion in India: A Study of SHG Initiatives." *Indian Journal of Human Development*. https://doi.org/10.1177/0973703019839807.

Muñoz, Sabag and E. Gladek. 2017. One planet approaches: methodology mapping and pathways foreward, report commissioned by WWF and IUCN.

Navin, T., N. Srinivasa Rao, Shashank Singh, and Rajendra Singh Gautam. 2017. "SHG Federations as Livelihood Support Organizations." *IRA-International Journal of Management & Social Sciences (ISSN 2455-2267)*. https://doi.org/10.21013/jmss.v6.n2.p11.

Naidoo, S., and Olaniran, A. O. 2014. "Treated Wastewater Effluent as a Source of Microbial Pollution of Surface Water Resources." *International Journal of Environmental Research and Public Health* 11.

Niyonsenga, Theo, Danish Ahmad, Itismita Mohanty, Laili Irani, and Dileep Mavalankar. 2020. "Participation in Microfinance Based Self Help Groups in India: Who Becomes a Member and for How Long?" *PLoS One*. https://doi.org/10.1371/journal.pone.0237519.

Panayotou, Theodore. 1994. "Economic Instruments for Environmental Management and Sustainable Development." United Nations Environment Programme's Consultative Expert Group Meeting.

Parris, T. M., and R. W. Kates. 2003. "Characterizing and Measuring Sustainable Development." *Annual Review of Environment and Resources* 28: 559–86. doi: 10.1146/annurev.energy.28.050302.105551.

Prasad, K. V. S. 2017. "A Review on Performance of Shg- Bank Linkage Programme in India." *International Journal of Mechanical Engineering and Technology* 8.

Reddy, C. S. 2007. *SHG Federations in India*. APMAS.

"Retailing of Self-Help Group (SHG) Products in India." 2019. *Journal of Applied Business and Economics*. https://doi.org/10.33423/jabe.v21i5.2273.

Robert, Kates W., Thomas M. Parris, and Anthony A. Leiserowitz. 2005. "What Is Sustainable Development? Goals, Indicators, Values, and Practice." *Environment: Science and Policy for Sustainable Development* 47 (3): 8–21.

Sandhu, Monisha Vaid, and Zhanming Liang. 2021. "Competency Assessment of Project Managers of a National NGO in India." *Journal of Health Management*. https://doi.org/10.1177/09720634211035248.

Sarkis, Joseph, and Sherwat Ibrahim. 2022. "Building Knowledge beyond Our Experience: Integrating Sustainable Development Goals into IJPR's Research Future." *International Journal of Production Research*. https://doi.org/10.1080/00207543.2022.2028922.

Sen, Siddhartha. 1999. "Some Aspects of State-NGO Relationships in India in the Post-Independence Era." *Development and Change*. https://doi.org/10.1111/1467-7660.00120.

Singh, Rajesh Kumar, H. Ramalinga Murty, S. Kumar Gupta, and A. Kumar Dikshit. 2009. "An Overview of Sustainability Assessment Methodologies." *Ecological Indicators* 9 (2): 189–212.

Smil, Vaclav. 2011. "Harvesting the Biosphere: The Human Impact." *Population and Development Review*. https://doi.org/10.1111/j.1728-4457.2011.00450.x.

Spangenberg, Joachim H. 2011. "Sustainability Science: A Review, an Analysis and Some Empirical Lessons." *Environmental Conservation* 38 (3): 275–87.

Tobi, Hilde, and Jarl K. Kampen. 2018. "Research Design: The Methodology for Interdisciplinary Research Framework." *Quality & Quantity* 52 (3): 1209–25.

United Nations (UN). 2015. "Transforming our World: The 2030 Agenda for Sustainable Development 2015." https://sdgs.un.org/publications/transforming-our-world-2030-agenda-sustainable-development-17981

Vaish, Ritu, and Mini Amit Arrawatia. 2021. "Role of Women Entrepreneurship In Promoting Women Empowerment." *Elementary Education Online* 20.

Venkataramany, Sivakumar, and Balbir B. Bhasin. 2011. "Path to Financial Inclusion: The Success of Self-Help Groups-Bank Linkage Program in India." *International Business & Economics Research Journal (IBER)*. https://doi.org/10.19030/iber.v8i11.3181.

WCED, Special Working Session. 1987. "World Commission on Environment and Development." *Our Common Future* 17 (1): 1–91.

World Health Organization (WHO). 2018. "Guidelines on Sanitation and Health 2018." World Health Organization, Geneva. Licence: CC BY-NC-SA 3.0 IGO.

13 Environmental Movements and Law in India
A Brief Introduction

M. Angkayarkan Vinayakaselvi and Abinaya R.

CONTENTS

13.1 Introduction 177
13.2 The Need for Knowing about the Sociopolitical Environmental Movements 178
13.3 Movements during the 1970s in India 179
    13.3.1 Chipko Movement 179
    13.3.2 The Silent Valley Movement 180
13.4 Movements during the 1980s in India 181
    13.4.1 Appiko Movement 181
13.5 Movements during the 1990s in India 181
    13.5.1 Narmada Bachao Andolan 181
13.6 Movements of the New Millennium 182
    13.6.1 Sea Turtles and Their Conservation 182
    13.6.2 Chilika Bachao Andolan 183
13.7 Legal Perspectives in Ensuring Environmental Sustainability in India 183
    13.7.1 Water and Air Act 184
    13.7.2 Environment Protection Act 184
    13.7.3 Coastal Regulation Zone 184
    13.7.4 Forest Laws and Policies of India 185
    13.7.5 Waste Management 185
13.8 Conclusion 185
References 186

13.1 INTRODUCTION

In India, environmental movements, as any other movement, are collective actions of several members from different strata of the society (Omvedt 1984). It is an outcome of socialistic and communistic ideologies of working-class people who are united by a shared belief to resist the changes in existing social order (Guha and Martinez-Alier 1997; Bardhan 1984). The symbiotic relationship between human beings and nature is the trajectory point for environmental movements (Sachs 1993). Guha and Gadgil define the environmental movement as an organized social activity consciously directed toward promoting sustainable use of natural resources, preventing the environmental abuse (Guha 1988), and preserving the environment and restoring it to its original order (Gadgil and Guha 1993; Kothari 1989). Almost all the environmental movements are the expression of activist's concern to protect and improve the environment for the present and the future generation (Wignaraja 1993). Technologists and scientists of every nation need to have a better understanding of social–cultural history of environmental movements as both are concerned with preserving the environment (Khoshoo 1984; Roy 2003; Kumar 2014). All social movements contribute ideas to the public opinion. Hence, environmental movement also becomes a social movement where like-minded people join together and voice to protect environment (Pandey 1991) and strive to bring in relevant changes in the already existing environmental policies and practices. In India, environmental movements have been shaped by the broader sociocultural practices, political constraints, and opportunities unique to the national context (Gadgil and Guha 1994). Gadgil and Guha have identified four major groups of environmentalists in India. Most of environmental activists in India follow Gandhian ideology (Swain 1997; Omvedt 1984) and vision which emphasizes on the importance of moral necessity to prevent exploitation and to provide justice to all, including the poor and marginalized people, whereas the Marxist thinkers recommend equality in social order and usage of environment (Martinez-Alier 2002). Scientists advocate the usage of suitable technologies within the given geological and temporal context, whereas the local community people at village level always work for protecting local forests (Poffenberger and McGean 1997; Pathak 1994) and fights to pursue environment-friendly agricultural practices. In India, there are forest- and land-based environmental issues which are addressed by Chipko, Appiko, and Tribal Movements that articulated the right of access to non-commercial use of natural resources, forest resources, prevention of land degradation, and, above all, maintaining social justice and human rights in terms of forest usage (Guha 2000). Movements like National Fisherman's Forum in Kerala and Chilika Bachao Andolan in Orissa engage with marine resources and fisheries and aquaculture and advocate the ban on trawling, preventing the commercialization of shrimp and prawn culture, protection of marine resources, and implementation of coastal zone regulations. Movements like Zahiro Gas Morchan in Bhopal, Ganga Mukti Andolan in Bihar, and movements against pollution of Sone river in Madhya Pradesh and Chaliyar river in Kerala spoke against the industrial pollution and it worked for establishing stringent norms pollution control and to prevent encroachment of industries by challenging the livelihood issues of the local population. Silent Valley Movement, Narmada Bachao Andolan, and Tehri Movement resisted the destructive developments in the forms of dam construction and irrigation projects (Karan 1994) which advocated protection of tropical forests and ecological balance as their primary vision but never thought about the rehabilitation measures of displaced people. Jan Andolan in Dabhol, Koe-Karo Jan Sanghatana in Bihar, and Anti-Mine project in Doon Valley spoke against mining as they depleted natural resources. Movements like Ekjoot in Bhimashankar region of Maharashtra and Shramik Mukti Andolan in Sanjay Gandhi National Park, Bombay, voiced against creation of wildlife sanctuaries and national parks as they lay the foundation for displacement of the native people and loss of their livelihood. There are movements in the forms of local groups like Society for Clean Cities, Bombay Natural History Society, Centre for Science and Environment, Delhi, research and documentation organizations such as Bombay Environmental action group, Save Bombay Committee, championed policy input, stricter measures for clean environment, clear policy on national park and wildlife sanctuaries, community-based environmental management, and publication on environmental issues.

13.2 THE NEED FOR KNOWING ABOUT THE SOCIOPOLITICAL ENVIRONMENTAL MOVEMENTS

Most of the researchers and historians trace the origin of environmental movements from Stockholm Conference in the 1970s in the United Nations (Rai 2013; Nabhi 2006). The Brundtland Commission report has made the environmental movements global and its policymaking. During the last four decades, India has also witnessed the emergence of environmental movements to save water, natural resources like land, forest and wildlife, air, aquatic wealth, and fossil fuel from pollution and overexploitation. These movements focus on natural resource management and the distribution of resources within the framework of sustainability, equality and social justice. Vandana Shiva, the acclaimed environmental activist, stated that the nature and the forest have always been treated as a teacher and the forest always resonate the message of interconnectedness and diversity, renewability and sustainability, integrity and pluralism. (Shiva 1988) During the classical period, in the Dravidian culture the geographical regions have been divided on the basis of landscapes into five major divisions. Geographical regions are associated with specific naturally occurring landscapes. They are divided into five types: Kurunji, Mullai, Marutham, Neithal, and Palai. Kurunji is a mountainous and hilly region, whereas Mullai denotes the forests and its related landscapes. Marutham is the cropland, and Neithal is the coastal areas. Palai implies the wasteland or desert areas. They believe that there is an interconnectedness between the outer landscape and the inner universe. In Vedic scriptures and epics (Nabhi 2006), Nature is viewed as a power that was perceived with respect for its benevolent nature toward mankind. In India, environment is a part of the cultural memory as its philosophy and culture believe that the human beings are constituted by the five basic life-supporting systems – sky, air, fire, water, and earth – and such a constitution ensures the ecological balance and livelihood on our earth. Hence, there is interdependent relationship between the environment and human beings irrespective of ethnicity, community, race, caste, religion, and gender. In India, religious scriptures train the mind of individuals to treat nature respectfully, whereas earth is referred to as the mother goddess. According to the Indian philosophy, destruction of nature is a self-defeating act, whereas the protection of nature is viewed as an esteemed goal. There are festivals and rituals in India to thank water, air, and the land. Hence, it is not strange to state that unlike other movements, environmental protection has been initiated by the tribal people, forest-dwellers, peasants, fisher folks, and women.

Globally, the Unites Nations Conference on Human Environment (UNCHE) held in Stockholm in 1972 is the beginning point in environmental awareness (Rai 2013; Nabhi 2006), As far as India is concerned, this conference is a turning point as the former prime minister, Indira Gandhi, expressed her concern for environmental protection in developing countries (Kohli and Bruce 1990). Her viewpoints gain prominence because she distinguished between the views of the First World and Third World on the idea of development itself. According to her, unlike the developed countries, industrialization in the developing countries never imply environmental degradation. Industrialization does not imply destruction, but rather it is viewed as one of the primary means of improving food and water supply and guaranteeing sanitation and protective shelter (Bandyopadhyay 1985).

13.3 MOVEMENTS DURING THE 1970S IN INDIA

13.3.1 Chipko Movement

In the developing countries like India, greater focus was given to sustain and improve the standards of living of the common folks. The former prime minister, Indira Gandhi, proclaimed the importance of development and conservation (Nabhi 2006) and cautioned the importance of preservation of natural resources for the future generation. During the 1970s, the environmental movements were formed and supported by voluntary and non-governmental organizations. They, in consultation with government officials, enhanced people's awareness and attitudes toward environment and its protection (Jenkins and Klandermans 1995; Frank et al. 2000). They played a significant role in identifying the most polluted areas and involved themselves in giving environmental education and awareness about the natural resources conservation, pollution control, afforestation, proper water utilization, and ecological preservation and development (Agarwal and Narain 1985). This part of the chapter will contribute on the emergence, agenda, and outcome of Chipko Movement. Though India wanted to give adequate attention to the preservation of environment, encroachment of forest had been occurring periodically. Resistance of such encroachment happened in the form of Chipko Movement which had its roots during colonial period itself. There were organized protests against colonial forest policy during the early decades of the twentieth century, Chipko Movement started in 1973 at Mandal of Chamoli District of Garhwal division of Uttar Pradesh (Weber 1988). The organizers adopted the ideology of non-violence (Rai 2013) as propagated by Mahatma Gandhi and Vinoba Bhave, This movement primarily resisted the ecological destabilization in the hills (Mawdsley 1998). It was a reaction against the decision of State Forest Department to give a lease of the forest trees to Simon Company, the manufacturers of sports equipment in Allahabad.

Apart from cutting of the trees, laying of roads, building dams, generating hydroelectric power, and increasing tourism were the potential threats to the hills and forests. They identified entrepreneurs, customers of newly emerging companies, political patrons as the beneficiaries of the rapid growth where the working-class people were exploited in the name of development. As there were many noticeable floods in the place, people started reacting to it in the form of Chipko Movement. As the word "Chipko" means "to hug," the villagers protested by hugging the tree and preventing the company from cutting the trees. The major agenda of Chipko Movement is not only to protect timber, fodder, and wood but also to preserve soil and water, Sunderlal Bahuguna was called as the Chipko messenger as he visited the entire region spreading the Chipko message. The villagers never allowed to cut the movement became successful as it enforced a 15-year ban to cut the trees for commercial purposes. Women were active members in the movement. Through this movement, people became conscious of the values of forest, involvement of indigenous strategies (Poffenberger and McGean 1997) in the progressive measures of modernization, and importance of every individual in preserving the environment. Unlike the western models of development where women were marginalized to do only the products delivery, the Chipko Movement drew women folk into the center where they became producers of subsistence goods. They were the creators and preservers of traditional ecosystem who saw conservation of forests as an important means of living and survival. It involved women in the decision-making and in questioning the public power, authorities, and policymakers. India through Chipko Movement emerged as one of the pioneering countries that took initiatives to protect the environment. It is India's most powerful and successful ecological movements. It gave awareness about the interconnectivity among deforestation, soil erosion, floods, and the fragility of mountain ecosystem. Internationally, Chipko was recognized as India's cultural response toward love for their environment. It has emerged as a powerful model for all other future movements to come. Chipko Movement propagated two major ideas to the world: ecology as economy and the important benefits of forest as soil, water, and air and not timber. This movement inspired other protests against commercial logging throughout the Himalayan region. This movement initiated replantation of multipurpose broad-leafed tree species in the deforested lands. It also resisted plantation of tree species like eucalyptus and teak that benefited only rich merchants and industrialists. The Chipko activists recommended trees whose leaves, roots, fruits, and nuts may be useful as the timber itself. They demanded complete stoppage of cutting trees for commercial purposes, recognition of traditional rights on the basis of minimum needs of the people, involvement of common people in tree cultivation, formation of local village committee to manage mountains and forests, development of home-based industries that use forest yielded raw materials, and, above all, prioritization of afforestation considering the local needs and conditions.

13.3.2 The Silent Valley Movement

The Silent Valley that is located in the Palghat District, in the Malabar region of Kerala is known for its flora and fauna and for its rich biodiversity. Originally, during the colonial period, the Britishers conceived the idea of constructing a dam on the river Kunthipuja in 1929 and after surveying the technical feasibility in 1958, the Planning Commission of Government of India sanctioned the project in 1973. It basically aimed at generating hydroelectric power, irrigation of cropland, and creating jobs for people during the construction period. As it was situated away from the prominent urban centers and highways, the Valley was not exploited for timber-cutting. The natives of Kerala were initially ignorant of the ecological consequences of the project. Though the Government of Kerala in 1978 passed an ordinance to protect the ecological balance in the Silent Valley, there raised many opposition to the project. The Task Force report (1977) declared forest as a natural reserve and recognized its renewable and sustainable nature and generated large-scale disagreement against the project. With the initiatives taken by the International Unions for conservation of nature and natural resources and the Kerala Forest Research Institute, the Silent Valley was declared as a biosphere reserve in 1978. In due course, the Kerala Shastra Sahitya Parishad registered a society called Protection of

Silent Valley. The Kerala Shastra Sahitya Parishad is a network of rural school teachers and local citizens that promoted environmental scientific projects in the villages (Karan 1994). The movement challenged the immediate purpose of the project, the generation of hydroelectric power. Their study proved that the benefit would not reach the local people but rather disrupted by sharing the generated energy with the surrounding states and industrialized areas and the state capital city, Trivandrum. The struggle between the environment groups and the state government made the Indian government to appoint a high-level committee to examine he feasibility. The committee recommended the abandonment of the project and the project was turned down based on the technoeconomic feasibility and socioeconomic possibilities (Rai 2013; Nabhi 2006). The state government accepted to abandon it in 1983. It was a spontaneous and natural and unorganized movement that turned to be a properly organized movement later. His movement propagated the lesson that any development-oriented project should not displace people and destruct the base of natural resource (Bandyopadhyay and Shiva 1988). This movement became successful as they brought awareness among the downtrodden people about ecology and the environment. During the later period, the Movement forced the State of Kerala to generate environment-friendly small-scale hydropower project. Campaigning and petitioning were the main strategies adopted by the members of Silent Valley Movement. This project was one of the typical examples that showcases Indian dilemma between environment and development.

13.4 MOVEMENTS DURING THE 1980S IN INDIA

13.4.1 Appiko Movement

Appiko Movement emerged from the Uttar Karnataka region in September–November 1983 and it was influenced by Chipko Movement. Appiko Movement was initiated to protect the forest from contractors from felling the timber extraction. The forest-dwellers were prevented from collecting fuel wood raw materials like twigs, fodders, and non-timber forest products. The native people's customary rights were denied by the commercial contractors. In September 1983, women and youth expressed their resistance by walking for 5 miles in the surrounding areas of Saklani and protested against the fellers and contractors of the state government to stop cutting the tree by hugging, similarly to Chipko Movement. It was a 38-day protest and the intensity of the protest made the state government to yield to the demands of the people and withdraw the sanctioned order for cutting the trees in Saklani forest area. The movement was extended to the neighboring places and the second phase of movement began in October 1983 in Bengon forest where the hilly terrain filled with semi-evergreen type of trees. The lives of the native inhabitants of the tribal forest-dwellers primarily depended on the resources of the forest. They used forest resources like bamboo tress as their raw materials to make baskets and mats. The natural resources like the bamboo trees and honey were the main source of income for them. As they were uprooted from the forest for commercial purposes, Appiko Movement emerged as a spontaneous reaction against such exploitative decision. People also started similar movements in other areas of Karnataka. The main objectives of Appiko movement are protection of the existing forest cover, regeneration and replantation of trees in the denuded land, and utilizing forest wealth with proper consideration to conserve the natural resources by adopting Gandhian ideology of non-violence to ensure the balance between men and nature. It created awareness among the villagers throughout the Western Ghats and sensitized people about the dangers caused by the commercial and industrial markets (Chandhoke and Ghosh 1995).

13.5 MOVEMENTS DURING THE 1990S IN INDIA

13.5.1 Narmada Bachao Andolan

Save Narmada Movement was initiated to stop the two major projects – Sardar Sarovar Project and Narmada Sagar Project (Roy et al. 1992). Sardar Sarovar Project aimed at constructing a dam over the river Narmada, the fifth major river in India. The government had a proposal of making the world's largest man-made lake. The Narmada Valley is situated in Vindhya and Satpura ranges in Central India. The proposed projects created issues related to Narmada water dispute. It was projected as one of the world's largest multipurpose water projects (Paranjypye 1990). As Narmada has 51 tributaries, it was aimed to construct 30 large dams and many small dams over the river. It was also aimed to increase the food production and hydropower generation in the three major states of Gujarat, Madhya Pradesh, and Maharashtra. Though the project had many developmental visions, it challenged the livelihood of one million people and posed threat to submerge vast hectares of agricultural lands. Hence the tribal people opposed to it as it would not only submerge the cultivation land but also displace people from more than 250 villages in the neighborhood. The World Bank funded the project and approved to sanction loan in 1985 with the prescription of the following rehabilitation guidelines and providing resettlement measures to the community people by ensuring social and environmental protection. As the central and the state government could not meet the guidelines suggested by the World Bank (Roy et al. 1992), it ceased to fund the project in 1997, but the Indian government never stepped back from its plan. The Narmada Basin has a tropical climate with higher variations of rainfall, temperature, and humidity. In addition, it had two world-famous national parks like Kanha and Satpura and five sanctuaries. Moreover, the Narmada Basin received several migratory birds flying from North to South. The deforestation caused by the construction of dams would affect not only the livelihood of human beings but also the feeding and breeding of the wildlife (Roy et al. 1992; Rai 2013). The compensatory measures proposed by the government did not address its impacts on the ecosystem, ecological pressures, and micro-climatic change. Hence, in the 1980s, the Narmada Movement was initiated to struggle for the resettlement and rehabilitation of the displaced people and to struggle for preserving the environmental integrity and the natural ecosystem of the valley (Baviskar 2004). In 1988, the Narmada Bachao Movement demanded the stopping of work in the Narmada Valley (Roy et al. 1992). In September 1989, around 50,000 people all over India demonstrated protests by branding the project as destructive. The World Bank was formed to create an independent review committee called the Morse Commission and the Morse report was published in 1992 and in 1994 it sent Pamela Cox Committee for further review. Both the reports endorsed the issues raised by the movement. Originally the project was designed to benefit the people of Maharashtra, Gujarat, and Madhya Pradesh, but the field survey done in 1993 revealed the people of Madhya Pradesh opposed the project as most of them were the residents of the basin. The case remained the same in Gujarat as the government did not provide adequate compensation for the lands and homes taken away from them. There was also a discrepancy in the settlement of compensation where the rich received more and the poor received less compensation. The movement achieved its one of its major goals with the withdrawal of funding by the World Bank in 1993 and it halted the Sardar Sarovar Construction between 1994 and 1999 and it also prevented the foreign investors between 1999 and 2001. The Save Narmada Movement is distinctive movement in the history of environmentalism in India. It was successful in mobilizing hundreds of people all over India and it gained international reputation due to media coverage. It also followed Gandhian non-violence. Unlike other movements, thinkers and activists too participated in this movement like Baba Amte, Medha Patkar, and Arundhati Roy (Nabhi 2006) who strongly pronounced the negative impacts of the projects. This movement has focused on the issues of eco-development through intellectual debates, political mobilization, and grassroots activism. It gained attention of the activists at both the national and international levels and it is projected as the major political discourse of alternative development in India (Khondker 2001; Khagram 2004; Guha 2014).

13.6 MOVEMENTS OF THE NEW MILLENNIUM

13.6.1 Sea Turtles and Their Conservation

In Indian, sea turtles are considered to be a part of the cultural history and it is an endangered marine species. The Bhitar Kauika Marine Sanctuary in Kendrapara district of Orissa attracts only

Olive Ridley sea turtles every year, and it is the largest nesting ground of them in the world. Their lives are challenged due to the unrestricted traveling activities. The Orissa government launched a programming operation Kachhapa every year as mass turtle death occurs on the Orissa coast. Due to the mechanized trawling, there were recommendations to declare the entire Orissa coast as a "sea turtle conservation zone" (Nabhi 2006). Similarly, in Kerala, at the Kolavipalam beach in Calicut, the turtles come to nest every year. The villagers and government took proper initiatives to construct a turtle hatchery in the form of Marine National Park to protect the coast and for turtle conservation and mangrove afforestation. Sea turtles are killed in nets at every stage of its life. They are taken as eggs for making beverages, and killed as adults for crafting souvenirs and shell products. They are also killed for food industries. As they are endangered, they need to be protected. In March 2004, an international conference on marine turtle conservation was organized where officials, scientists, and experts from 25 countries of South-east Asia participated and decided to declare the year 2006 to organize awareness campaigns. This is a milestone in the protection of environment.

13.6.2 Chilika Bachao Andolan

The Chilika Lake is the largest brackish water lagoon connected to the Bay of Bengal and it is situated 100 km from Bhubaneshwar in Orissa,. It habitats diversified species, including some rare, vulnerable, and endangered species. As a coastal terrain, it is the largest inland lake in India with numerous islands.

One of the major concerns in the area is the Shrimp Farm Project – agreement of the government of Orissa and the Tata Group. It is also known as Chilika Integrated Shrimp Farming Project. It consists of a shrimp farm, a hatchery, a shrimp mill, and a processing plant. The project was claimed to promote economic development of the region. On the contrary, the project threatened the livelihood of the fishermen, and the lake was polluted through the discharge of the effluents from the project (Nabhi 2006).

The movement Chilika Bachao Andolan became an umbrella movement under which numerous protests from varied organizations were held: the student forum like Meet the Student's Group, Krantadarshi Yuva Sangam, Chilika Suraksha Parishad, Chilika Matsyajibi Mahasangha; civil society organizations like Ganatantrik Adhikar Suraksha Sanghathan, Orissa Khrushak Mahasangha, and so on. The movement began as a grassroots agitation in a few villages against the Tata Project and it spread to the other states with a similar vigor of Chipko Movement. Though it is a lesser known movement in Eastern India, it became a mass movement. The Shrimp project was stopped because of non-clearance from the Union Ministry of Environment and Forest. The movement addressed various issues like environmental degradation, pollution of the lake ecosystem, deforestation, soil erosion, and silting of the lake. Apart from these issues, the movement represents the importance of traditional and cultural rights of the people.

13.7 LEGAL PERSPECTIVES IN ENSURING ENVIRONMENTAL SUSTAINABILITY IN INDIA

During the British rule, Forest Department was established to legitimize the exploitation of natural resources (Gadgil and Guha 1993). The conservation of the forest was claimed under the grounds of climate and physical factors. The major focus of the policies and acts were commercial supply-based forest areas and resources with an additional agenda of controlling land use.

The policies of the pre-Independence period failed to recognize the rights of forest-dwellers (Agrawal 2005) without any provision to wildlife protection and private forests. It allowed unchecked exploitation by the government and conversion of forest land to agriculture and plantation. The Forest Act 1927 widened the power of government toward village and private forests and common properties with prohibitions to commercial products like teak and sandalwoods (Arora 1994). The main work of the forest department was to implement the provision of act and conservation of forest.

With the 42nd Amendment to the Indian Constitution, provisions on environmental protection were added in 1976 for the first time (EGyanKosh). The Indian Constitution is one of the few in the world to contain specific provisions on environmental protection due to its participation in Stockholm Conference 1972 (Ahmad 2001). Provisions to make laws by the parliament under Constitution were added in the State List under Articles 249, 250, 252, 253, and 356/357.

The first board for environmental protection was constituted in Maharashtra in 1970. Under the Water Act 1974, Maharashtra Pollution Control Board was formed. At present, all the states in India have State Pollution Controls Boards (SPCB) and the union territories have committees or boards.

13.7.1 Water and Air Act

The Water Act 1974 provides for the prevention and control of water pollution and the maintaining or restoring of wholesomeness of water includes central government, state government, Central Pollution Control Board, and State Pollution Control Board which are the regulators of this act. The functions of the Pollution Control Boards under the Water Act are concerned with prevention, control, or abatement of water pollution and regulatory – to set up and modify effluent standards, to inspect and grant permission toward the sewage or trade effluent, and to promote standards of cleanliness of waterbodies by laying standards. Other functions include data management, information dissemination, and capacity-building of the stakeholders – regulator, polluter, and receptor. The law also sets provisions for the stakeholders. The Air Act 1981 includes similar provisions and functions of the Water Act with an agenda toward the preservation of the quality of air and control of air pollution (Ahmad 2001).

13.7.2 Environment Protection Act

The Bhopal episode is widely considered to have constituted the Environment Protection (EP) Act 1986. This acts as an umbrella legislation to provide a framework for environmental regulations. This covers the industrial and infrastructure activities along with other provisions (Ahmad 2001). The EP Act works as a regulator providing authority to the central government. The important functions include laying source- and location-specific standards for emission or discharge of environment pollutants, and the quality of environment. It also includes planning and execution of programs toward the prevention, control, and abatement of rules. This act facilitates the central government with the creation of authorities that can be both location- and source-specific and with power to delegate and effect other laws. The central government has the power to make rules with ratification by parliament. All of the above acts facilitate and provide for research and development.

13.7.3 Coastal Regulation Zone

The objectives of CRZ are to conserve and protect coastal areas and to encourage scientific development considering dangers of natural hazards and provide provision for the well-being of the regional communities. The CRZ regulations under the Environment (Protection) Rules 1986 cover the coastal stretches of seas, bays, estuaries, creeks, rivers, and backwaters. Based on the notifications of CRZ in the legislation, it imposes relevant restrictions on the setting up and expansion of industries, operations, or processes in the CRZ. Within the provided classification of CRZ like National and Marine Parks, the specified notifications of the Government of India impose various prohibitive activities and set up regulations of permissible activities and provide guidelines for coastal zone management plans. The activities that are not listed in the Environment Impact Assessment (EIA) notification 2006 require clearance from the Ministry of Environment and Forests (MoEF).

13.7.4 FOREST LAWS AND POLICIES OF INDIA

India's forest contains enormous variety with 16 major types and 221 forest types providing habitat for diversified flora and fauna along with the regional population of the country. The management of Indian forests earlier was under the Indian Forest Act 1927. The creation and implementation of consequent acts and policies contributed to the systematic management of forests. Forest Policy 1894 focused on timber harvesting and prohibited the exploitation of commercial crops and also encouraged permanent cultivation. Forest Policy 1952 proposed a conservative approach that one-third of the land area of the country to be retained under forest and tree cover. The Forest Conservation Act 1980 provides provisions to preserve the forest ecosystem by limiting the rate of deforestation. The major objectives of National Forest Policy 1988 are to ensure environmental stability through management of ecological balance, and thus protecting the indigenous resources.

Wild Life (Protection) Act 1972 provides for the prevention and control of poaching, smuggling, and illegal trade in wildlife and its derivatives. This act was amended many times toward strict measures to preserve the wildlife. The act also facilitates the declaration of national parks and sanctuaries with essential ecological systems.

Scheduled Tribes and Other Traditional Forest Dwellers (Recognition of Forest Rights) Acts 2006 is also known as the Forest Rights Act (FRA). Ministry of Tribal Affairs is the nodal ministry for the implementation of the act. Ownership or title rights, use rights, relief and development rights, and forest management rights are the provisions provided under the act.

13.7.5 WASTE MANAGEMENT

Biomedical waste (Management & Handling) Rules 1988 and Bio-medical Waste Management Rules 2016 with further amendments were notified by the Ministry of Environment Forests & Climate Change under the EP Act 1986. The rule aims to promote the proper processing, treatment, and disposal of biomedical wastes as per the guidelines provided by the legislation toward a proper environmental management. The steps for the management and implementation of the rules include segregation of waste at source, pretreat laboratory and highly infectious waste, collection and storage of segregated waste in color-coded bags or containers, intramural transportation from generation site to central storage area, storage at central facility, and treatment and final disposal.

Hazardous and Other Wastes (Management) Rules 2016 was constituted under the EP Act 1986. The important stakeholders of the rules are the authorities, occupier, operator of facility, importer, exporter, and transporter. The objective of the rules in the sequence of priority is prevention, minimization, reuse, recycling, recovery, co-processing, and safe storage and disposal.

Solid Waste Management (SWW) Rules 2016 focuses on source segregation, recovery, reuse, and recycle, which is applicable beyond the municipal areas replacing the Municipal Solid Wastes (Management and Handling) Rules 2000. The rule considers all domestic, institutional, commercial, and any other non-residential solid waste generators with the exception of industrial waste, hazardous waste, hazardous chemicals, biomedical wastes, e-waste, and lead acid batteries and radioactive waste that are covered under separate rules constituted under the EP Act 1986.

E-Waste (Management) Rules 2016 – Under EP Act 1986 aims to promote environmentally safe recycling toward efficient e-waste collection mechanism. It also aims to minimize illegal recycling and to reduce the hazardous substances in the e-waste components. It also notifies regulations for the stakeholders, producers, dismantler, recycler, manufacturer and refurbisher, and dealer of the e-waste.

13.8 CONCLUSION

Environmental movements have been playing a vital role in the representation of environment issues and promoting sustainability. They are also milestones in the reassurance of the faith toward democracy. The participation and organization of the environmental movements is diversified, ranging across varied social strata like gender, caste, class, and social and educational organizations. The participation includes the representative narratives, research, and media outlets. These environmental movements are not political movements but are sociocultural movements. Scientists, technologists, and inventors are expected to be sensitive toward such sociocultural environmental movements and also to have adequate knowledge about the laws related to environmental protection and waste management. The environmental movements help to channelize them toward understanding the impacts of their invention on the environment and also alert them toward environmental law and sustainability.

REFERENCES

Agarwal, Anil, and Sunita Narain. 1985. *The State of India's Environment, 1984–85: The Second Citizens' Report*. Centre for Science and Environment.
Agrawal, Arun. 2005. *Environmentality: Technologies of Government and the Making of Subjects*. Duke University Press.
Ahmad, Furqan. 2001. "Origin and Growth of Environmental Law in India." *Journal of the Indian Law Institute*, vol. 43, no. 3, pp. 358–87. *JSTOR*, www.jstor.org/stable/43951782.
Arora, Dolly. 19 March 1994. "From State Regulation to People's Participation: Case of Forest Management in India." *Economic and Political Weekly*, vol. 29, no. 12, pp. 691–8.
Bandyopadhyay, J. 1985. *India's Environment: Crises and Responses*. Nataraj Books.
Bandyopadhyay, J., and Vandana Shiva. 1988. "Political Economy of Ecology Movements." *Economic and Political Weekly*, vol. 23, no. 24, pp. 1223–32. *JSTOR*, www.jstor.org/stable/4378609.
Bardhan, Pranab K. 1984. *The Political Economy of Development in India*. Blackwell.
Baviskar, Amita. 2004. *In the Belly of the River: Tribal Conflicts Over Development in the Narmada Valley*. Oxford University Press.
Chandhoke, Neera, and Ashish Ghosh. 1995. *Grassroots Movements and Social Change*. DCRC.
EGyanKosh, egyankosh.ac.in/bitstream/123456789/42059/1/Unit-3.
Frank, David J., et al. 2000. "The Nation-State and the Natural Environment over the Twentieth Century." *American Sociological Review*, vol. 65, no. 1, p. 96.
Gadgil, Madhav, and Ramachandra Guha. 1993. *This Fissured Land: An Ecological History of India*. University of California Press.
Gadgil, Madhav, and Ramachandra Guha. 1994. "Ecological Conflicts and the Environmental Movement in India." *Development and Change*, vol. 25, no. 1, pp. 101–36.
Guha, Ramachandra. 2000. *The Unquiet Woods: Ecological Change and Peasant Resistance in the Himalaya*. University of California Press.
Guha, Ramachandra. December 1988. "Ideological Trends in Inidan Environmentalism." *Economic and Political Weekly*, vol. 23, no. 49, pp. 2578–81.
Guha, Ramachandra. 2014. *Environmentalism: A Global History*. Penguin.
Guha, Ramachandra, and Joan Martinez-Alier. 1997. *Varieties of Environmentalism: Essays North and South*. Oxford University Press.
Jenkins, J. C., and Bert Klandermans. 1995. *The Politics of Social Protest: Comparative Perspectives on States and Social Movements*. Routledge.
Karan, P. P. 1994. "Environmental Movements in India." *Geographical Review*, vol. 84, no. 1, pp. 32–41. *JSTOR*, https://doi.org/10.2307/215779.
Khagram, Sanjeev. 2004. *Dams and Development: Transnational Struggle for Water and Power*. Oxford University Press.
Khondker, Habibul H. 2001. "Environmental Movements, Civil Society and Globalization: An Introduction." *Asian Journal of Social Science*, vol. 29, no. 1, pp. 1–8.
Khoshoo, T. N. 1984. *Environmental Concerns and Strategies*. Indian Environmental Society.
Kohli, Atul, and David K. E. Bruce. 1990. *Democracy and Discontent: India's Growing Crisis of Governability*. Cambridge University Press.
Kothari, Rajni. 1989. *State Against Democracy: In Search of Humane Governance*. Apex Press.
Kumar, K. 2014. "Environmental Movements in India: Re-Assessing." *International Journal of Scientific Engineering and Research (IJSER)*, vol. 2, no. 2.
Martinez-Alier, Juan. 2002. *The Environmentalism of the Poor: A Study of Ecological Conflicts and Valuation*. Edward Elgar.

Mawdsley, Emma. 1998. "After Chipko: From Environment to Region in Uttaranchal." *Journal of Peasant Studies*, vol. 25, no. 4, pp. 36–54.

Nabhi, Uma S. January 2006. "Environmental Movements in India: An Assessment of Their Impact on State and Non-State Actors." *India Quarterly*, vol. 62, no. 1, pp. 123–45.

Omvedt, Gail. 3 November 1984. "Ecology and Social Movement." *Economic and Political Weekly*, vol. 19, no. 44, pp. 1865–7.

Pandey, Shashi R. 1991. *Community Action for Social Justice: Grassroots Organizations in India*. Sage.

Paranjpye, Vijay. 1990. *High Dams on the Narmada: A Holistic Analysis of the River Valley Projects*. INTACH.

Pathak, Akhileshwar. 1994. *Contested Domains: The State, Peasants, and Forests in Contemporary India*. Sage.

Poffenberger, Mark, and Betsy McGean. 1997. *Village Voices, Forest Choices: Joint Forest Management in India*. Oxford University Press.

Rai, B. 2013. "New Social Movements in India: An Aspect of Environmental Movements." *International Journal of Science and Research (IJSR)*, vol. 4, no. 9.

Roy, Kartik C., et al. 1992. *Economic Development and Environment: A Case Study of India*. Oxford University Press.

Roy, Sovan. 2003. *Environmental Science: A Comprehensive Treatise on Ecology and Environment*. Syndicate.

Sachs, Wolfgang. 1993. *Global Ecology: A New Arena of Political Conflict*. Zed Books.

Shiva, Vandana. 1988. *Staying Alive: Women, Ecology and Survival in India*. Kali for Women.

Swain, Ashok. September 1997. "Democratic Consolidation? Environmental Movements in India." *Asian Survey*, vol. 37, no. 9.

Weber, Thomas. 1988. *Hugging the Trees: The Story of the Chipko Movement*. Viking.

Wignaraja, Ponna. 1993. *New Social Movements in the South: Empowering the People*. Zed Books.

14 Co-pyrolysis of Biomass with Polymer Waste for the Production of High-quality Biofuel

Dineshkumar Muniyappan and Anand Ramanathan

CONTENTS

14.1 Introduction 189
14.2 Significance of Co-pyrolysis Process 190
14.3 Biomass to Plastic Co-pyrolysis Mechanism 191
14.4 Biomass with Plastic Waste Co-pyrolysis Processes 191
14.5 Synergistic Effect in Co-pyrolysis 191
    14.5.1 Bio-oil Quality Improvement 191
    14.5.2 Increase in Bio-oil Yield 192
14.6 Influence of Operating Conditions on Co-pyrolysis 192
    14.6.1 Residence Time and Reaction Temperature 193
    14.6.2 Feedstock Particle Size 193
    14.6.3 Effect of Heating Rate 194
    14.6.4 Effect of Carrier Gas Flow Rate 194
14.7 Types of Reactors Used in Biomass–Plastic Co-pyrolysis 195
    14.7.1 Auger Reactor/Screw Reactor 195
    14.7.2 Fixed Bed Reactor 196
    14.7.3 Rotary Kiln Reactor 196
    14.7.4 Fluidized Bed Reactor 197
    14.7.5 Tubular Reactor 198
    14.7.6 Plasma Pyrolysis Reactor 199
    14.7.7 Microwave Reactor 199
14.8 By-products Obtained from Co-pyrolysis Process 200
    14.8.1 Gaseous Product 200
    14.8.2 Co-pyrolysis Char 200
14.9 Conclusion 200
References 201

14.1 INTRODUCTION

Due to the limitation of the fossil fuel, several countries have started to investigate alternative energy sources in order to fulfill the growing need for sustainable and clean energy. Plastic is a petrochemical substance that is extensively employed in most applications, and it will serve as a more promising and superior substitute to fossil fuel (Mwanza and Mbohwa 2017). Furthermore, as a result of greater research and development efforts into plastic manufacturing technology, yearly worldwide output has expanded from around 1.5 million tons from 1950 to 359 billion tons in 2018 ("Plastic Waste Worldwide – Statistics & Facts | Statista" 2022). Biowaste energy is created from the sun, which is then absorbed, converted, and stored through photosynthesis process. It is a large source of renewable energy that is accessible in a variety of forms, creates carbon-neutral effects, and has economic advantages (Nanduri, Kulkarni, and Mills 2021; Kumar et al. 2020). Pyrolysis may convert biomass to liquid fuels, and liquid fuels are believed a substitute to petroleum (J. Chen et al. 2021). However, despite being ecologically beneficial, biomass pyrolysis has a number of drawbacks, the most glaring of which is the biomass waste pyrolysis oil, commonly known as bio-oil. Due to its higher oxygen and water concentration, bio-oil is unstable, viscous, and it has a poor calorific value (Engamba Esso et al. 2022; Salvilla et al. 2020). Hence, it is required to increase the quality of pyrolysis oil and the desired compounds by the use of scientific and economical methods. Incorporating a co-reactant with a greater hydrogen-to-carbon ratio in biomass pyrolysis helps ensure the production of high-quality products (Singh et al. 2021). Most of the plastic objects are intended for single use and will be discarded after a short period of time. This new breakthrough has had serious environmental implications, causing widespread worry throughout the world (Jiang and Wei 2018; Ansari et al. 2021). Plastic recycling processes include pyrolysis, co-liquefaction, hydrothermal, biological conversion, and co-gasification. However, since it does not emit dangerous compounds into the environment, co-pyrolysis has lately received a lot of interest. The simultaneous thermal degradation of two forms of trash in an inert atmosphere is known as co-pyrolysis. According to a research, co-pyrolysis of plastic and waste biomass in the presence of a catalysts resulted in better product composition and quality than pyrolysis of biomass/plastic alone. The co-pyrolysis of polymer with biowaste enables the modification of the chemical composition of the char, pyrolysis oil, non-condensable gas, and their adaptation as a liquid fuel. Furthermore, co-pyrolysis with plastic waste and biomass waste may help to cut cost of production, improve waste disposal choices, and minimize environmental consequences (Zhang et al. 2018).

In this chapter, the biomass to plastic co-pyrolysis mechanism, synergistic effect in co-pyrolysis, and influence of operating conditions on co-pyrolysis product yields are discussed. Additionally, the types of reactors used in the biomass to plastic co-pyrolysis process are also discussed. In addition, details about by-products obtained from a co-pyrolysis process such as gaseous product and char are presented in this chapter.

14.2 SIGNIFICANCE OF CO-PYROLYSIS PROCESS

Plastics are used to lower the oxygenate content and increase the aromatic content in the pyrolysis oil. However, the idea behind blending biomass oil with waste plastic oil seems impossible and might increase the operational cost of the process. The oil produced from a biomass source cannot be mixed entirely with plastic oil due to bio-oils polarity. If oils from both the feedstocks are used, an unstable mixture is formed. If biomass and waste plastic pyrolysis occur separately, then higher energy and greater cost for production are required. The co-pyrolysis process is a feasible alternative to obtain homogeneous bio-oil similar to the "oil blending" method. Radical interactions during co-pyrolysis will promote the development of chemically stable pyrolysis oil, thereby avoiding liquid separation (phase separation) (Martínez et al. 2014).

During the co-pyrolysis process, several radical reactions can take place: initiation of depolymerization of plastic and the secondary radical formation, which include the formation of monomer, depolymerization, the favorable reaction of hydrogen transfer, intermolecular hydrogen transfer, isomerization through the vinyl group, and radical recombination (Önal, Uzun, and Pütün 2014). Also, the main advantage of using this technique is that biomass or plastic waste volume decreases significantly by utilizing more waste products as feedstocks for pyrolysis. Thus, the process reduces landfilling, thereby saving waste management costs and resolving numerous environmental issues and increasing future energy security (Garforth et al. 2004). Additionally, the co-pyrolysis process has proven to be quite economical compared to other thermochemical techniques.

14.3 BIOMASS TO PLASTIC CO-PYROLYSIS MECHANISM

The reaction mechanism related to co-pyrolysis of waste biomass with polymer is needed to investigate to grasp knowledge on occurrence of synergistic effect. By considering co-pyrolysis with plastic and biomass, the decomposition mechanism is comparatively different. The biomass will undergo a series of endothermic and exothermic reactions, while plastic wastes undergo a sequence of radical mechanisms such as initiation, followed by propagation and finally termination (Abnisa and Wan Daud 2014). It is noted that the thermal degradation stability for waste biomass is lesser than the plastic during co-pyrolysis process. The components of free radicals from biomass degraded will enhance the plastic decomposition and its macromolecules. To summarize, the secondary formation of radicals involves depolymerization, transfer of hydrogen in the reaction, and intermolecular transfer of hydrogen (Jakab, Várhegyi, and Faix 2000).

14.4 BIOMASS WITH PLASTIC WASTE CO-PYROLYSIS PROCESSES

Co-pyrolysis process contains mainly three fundamental steps such as preparation of the sample, co-pyrolysis process, and condensation and characterization. In the first step, the waste biomass and plastics are collected and cleaned in order to avoid contamination and other materials present in the sample. Then size reduction was performed using griding machine for both plastic and biomass samples and proper screening will be performed to select the uniform size for better heat transfer between the feedstock particles. In the second step, co-pyrolysis process will be performed by mixing both biomass and plastic feedstock in a different proportion and the product yield will be recorded. During co-pyrolysis process, the operating factors such as temperature, nitrogen flow rate, heating rate, and retention time will be varied to predict the optimum operating condition to produce maximum amount of desired product. Third step is condensation and characterization; in this step, the volatile gas released from pyrolysis reactor will be condensed to produce a liquid product such as pyrolysis oil and aqueous phase. The remaining non-condensable gas will pass through the condenser to gas chromatography to analyze potential gaseous product present in the sample. The pyrolysis oil produced from the process will be collected and it will be tested for all the physiochemical characterization such as viscosity, density, flash point, fire point and Conradson carbon residue test, and calorific value in order to access its potential for being utilized as the substitute of conventional fuel.

14.5 SYNERGISTIC EFFECT IN CO-PYROLYSIS

The primary cause of all quantity and quality improvement in a pyrolysis oil is synergistic effect. When the co-pyrolysis reaction occurs, radical interactions have synergistic effects. The negative or positive synergy is determined by the contact and type of components, the retention time of pyrolysis, the heating rate and temperature, and the addition of catalyst and solvents. Among these aspects, the types of mixing the feedstock will play a vital role in altering the synergistic effect (Fei et al. 2012).

14.5.1 Bio-oil Quality Improvement

Synergistic effects also have a substantial effect on the quality of pyrolysis oil produced. These advancements are visible, particularly in terms of fuel characteristics. Generally, with high oxygen concentration, oil derived only from the wood-based biomass pyrolysis has a reduced calorific value and this issue can be solved by performing pyrolysis with the biomass and plastics. The previous studies show that co-pyrolysis process exhibited a higher calorific value. This is due to higher concentration of hydrocarbon content in the oil. Önal et al. (Önal, Uzun, and Pütün 2014) studied the pyrolysis oil production through co-pyrolysis with high-density polyethylene and almond shell. Their result revealed that the liquid product produced from co-pyrolysis had a more energy density as compared to almond shell pyrolysis alone. It is due to more concentration of HDPE to the biomass, which improved the hydrogen-to-carbon ratio from 1.70 to 2.28. Another study was conducted using waste tire with pine chips and the results reported that H and C concentration in pyrolysis oil improved with the increase of waste tire proportion and at the same time oxygen content was decreased. They also observed that the lower heating value will change from 14.9 to 19.0 with the ratio blend of 90/10 and 80/20, respectively (Martínez et al. 2014).

14.5.2 Increase in Bio-oil Yield

The previous studies reported that the presence of volatile compounds favors the generation of more amount of pyrolysis oil. The volatile matter provides benefits of higher volatile reactivity, which favors pyrolysis oil production. The ash concentration also plays an important role in influencing the quantity of liquid product in the pyrolysis of biomass. However, higher ash concentration contribute in decreasing the liquid yield and it influences the non-condensable gas and char formation in higher level (Fahmi et al. 2008). The cellulose is highly volatile as compared to hemicellulose. Hence, the wood-based feedstock with more amount of cellulose favors the generation of more volatile matter and that leads to improvement in bio-oil yield. All types of polymers have higher volatile content, which means all types of polymers have good potential to generate higher amount of liquid yield by pyrolysis process. Based on the characterization study, the volatile content higher than 99 wt% was found for LDPE, HDPE, polyamide, and PS. A recent study revealed that during pyrolysis of PP, PE, and PS, the results observed was that PS yield more pyrolysis oil 89.5 wt% as compared to PP and PE (Demirbas 2004).

14.6 INFLUENCE OF OPERATING CONDITIONS ON CO-PYROLYSIS

Co-pyrolysis processes are influenced by operating factors of the reactor apart from the type of feedstock. Some operating parameters are temperature, time for completion of the reaction, and feedstock particle size. In addition to these factors, the process also depends on other conditions inside the reactor, like the nature of the gas inside the reactor before the start of the process. A detail about different operating parameters influencing the co-pyrolysis process is shown in Figure 14.1. Also, a description of these factors is given in the following sections.

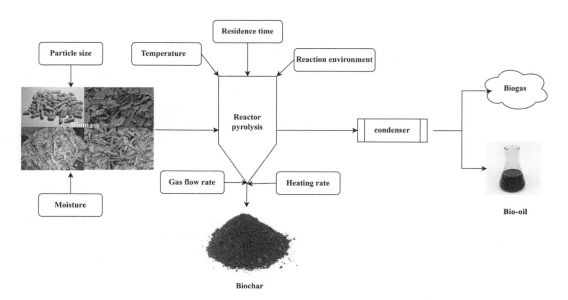

FIGURE 14.1 Different operating parameters influencing the co-pyrolysis process.

14.6.1 Residence Time and Reaction Temperature

Shorter residence times are advantageous to produce liquid products. In general, residence times ranging from a few seconds to a few minutes are optimum for the production of liquid products. However, if residence times are too low, biomass conversion will be incomplete (Olukcu et al. 2002). Residence time is often not the sole factor dictating product yields. Reaction temperature also plays a major role here. Medium range temperatures of 400–550 °C are suitable for the production of liquid products as higher temperatures promote secondary cracking reactions in tars. Higher temperatures result in higher yields of syngas and lead to secondary char formation as well (Olukcu et al. 2002). On the other hand, higher temperatures coupled with prolonged residence times reduce liquid yield, and also the bio-oil oxygen content. Thus, a simple reduction of residence time is not a solution to optimize the bio-oil quality. Optimization of residence times is done taking other external factors into account.

14.6.2 Feedstock Particle Size

Biomass is a poor conductor of heat; hence it introduces additional heat transfer challenges during pyrolysis. An important parameter influencing heat transfer is feedstock size. An understanding of the correlation between feedstock particle size and temperature is useful to optimize residence times for specific applications (Scott et al. 1999; Pütün et al. 2007). Smaller particles heat up uniformly. Particles with a larger diameter conduct heat to their core poorly. This lower heat conductivity results in overall poor average temperature, thereby decreasing the yield of liquid products. Therefore, feedstocks with smaller particle sizes are preferred for better liquid product yields, particularly in rapid pyrolysis systems. Larger particle sizes translate to higher activation energies. The DTG curves of smaller particles are on the low-temperature end and bigger particles on the high-temperature end (Haykiri-Acma 2006). One reason for the relationship between temperature and particle size might be due to surface area increase. Particle size is inversely proportional to the surface area. Owing to its large surface area, the decomposition of matrix begins at comparatively lower temperatures. The higher surface area paves the way for better interaction with the pyrolysis environment leading to the formation of volatile products inhibiting secondary reactions. Different pyrolysis processes have different particle size requirements to optimize liquid yields. For fluidized bed pyrolysis, less than 2 mm is the optimum particle size; whereas for rotating cone pyrolysis, the optimum size is 200 mm. The optimum size is less than 6 mm for circulating fluidized beds.

Particle sizes might also vary with specific processes and specific feedstocks. Song et al. (Song et al. 2002) performed pyrolysis in size ranges of 0.425–1.25 mm. The particle size of 0.425 mm diameter resulted in maximum bio-oil yield. Sensöz et al. (Şensöz, Angin, and Yorgun 2000) reported that particle diameters of 0.425–0.6 mm resulted in a maximum yield of bio-oil as opposed to particle size less than 0.425 mm and particle size greater than 0.85 mm. Aylón et al. (Aylón et al. 2008) took feedstocks of the same size (2 mm) for fixed bed and moving bed reactors. Their findings further confirmed the previous research, suggesting that the yield of products can change with various pyrolysis processes even if the feedstock particle size was maintained constant. It means that larger biomass particles can also produce high bio-oil yield with a suitable method of pyrolysis. There are a few biomass feedstocks wherein the maximum product yield is obtained for intermediate feedstock sizes (Şensöz, Angin, and Yorgun 2000; Encinar et al. 1996). Biomass with higher oxygen content, such as sunflower and rapeseed (Şensöz, Angin, and Yorgun 2000), seems to be independent of particle size with regard to bio-oil yield. The selection of biomass particle size must be made keeping in mind the cost considerations. A smaller particle size means that the cost of grinding increases. A larger particle size means that researchers would have to adopt different pyrolyzers.

14.6.3 Effect of Heating Rate

The heating rate is an additional important factor in co-pyrolysis processes. Fast pyrolysis has a greater heating rate as opposed to conventional pyrolysis. Higher heating rates facilitate the breaking down of biomass and the formation of volatile products (Gibbins-Matham and Kandiyoti 1988). Heating rates as high as 1000 °C/min were reported in the literature. Higher temperatures and heating rates are conducive to gaseous product formation (Bohn and Benham 1984; Strezov, Moghtaderi, and Lucas 2003). Higher biochar yield is possible with slow pyrolysis at lower heating rates of 6–48 °C/min (Pütün et al. 2007; Debdoubi et al. 2006). However, once heat transfer limitations are overcome, product yields do not significantly change with increased heating rates. In other words, liquid yields display an initial rise with an increase in heating rates. However, once the heating rate reaches a certain threshold, a further increase in liquid yield is not noticeable.

Sukiran et al. (Sukiran, Bakar, and Chin 2009) observed higher bio-oil yields with an increase in heating rates from 0 °C/min to 50 °C/min than the increase in yields when heating rates increased from 50 °C/min to 100 °C/min. Thus, heating rates shift optimum temperatures and yields. Higher heating rates result in higher optimum temperatures for bio-oil yield (Bohn and Benham 1984; Tsai, Lee, and Chang 2007). This was expected as higher heating rates, as mentioned earlier, increase biomass fragmentation, and thus volatile product yield rises with temperature rise, which in turn lead to temperature shifts. However, a significant temperature shift was not observed for an increase in heating rates for lower heating rates (Strezov, Moghtaderi, and Lucas 2003).

14.6.4 Effect of Carrier Gas Flow Rate

The surrounding environment in the reactor also impacts the final product's composition. Char formation can be accelerated with secondary reactions of pyrolysis vapors in the surrounding environment. To optimize the formation of volatile products, the environment inside the reactor has to be inert. Other significant factors influencing the composition of products include the rate of purging of pyrolysis vapors and the rate of quenching of vapors. An inert environment inside the reactor is created by purging nitrogen or argon vapors. The passing of these vapors is also known to reduce residence time (Acikgoz, Onay, and Kockar 2004; Abnisa and Wan Daud 2015).

On the contrary, inert gas use in the reactor only manages to keep the pyrolysis vapors out of the reaction zone. To prevent secondary reactions of these vapors, rapid quenching is often required. Researchers have observed an insignificant rise in the liquid product yield with a rise in inert gas flow rates. Studies reported in the literature have reported an increase of 3.3% in liquid yield by increasing inert gas flow rates from 50 cm^3/min to 150 cm^3/min (Bohn and Benham 1984; Strezov, Moghtaderi, and Lucas 2003). Current research indicates that fairly lower velocities and gas flow rates are sufficient to produce optimum liquid yields. Ekinci et al. (Ekinci et al. 1992) performed fixed bed pyrolysis with steam and nitrogen at low velocities. They found that steam was more effective than nitrogen at lower velocities. The results indicated that optimum liquid yields were possible with velocities as low as 0.3 m/s. A possible explanation for this result is that higher gas velocities promote the yield of gaseous products. At higher gas velocities, the gaseous products formed escape without condensing.

Steam is more suitable for the production of liquid yields as it sweeps hot pyrolysis vapors and facilitates reactions between solid char and gaseous products leading to condensation (Minkova et al. 2001). However, the addition of steam in the reactor atmosphere produces bio-oil with a high amount of oxygen, leading to a higher heating value. Therefore, the type of sweeping gas inside the reactor is dependent on the application. Minkova et al. (Minkova et al. 2001) compared pyrolysis products with the introduction of steam and nitrogen. They noted that the yield of liquid tars was higher in the presence of steam, while nitrogen promoted the char and non-condensable gas yields. The flow rates of sweeping gas not only affect the product of pyrolysis but also the composition of the products. At higher steam flow rates, aliphatic and aromatic compounds were promoted, while polar compounds were not developed. The influence of sweeping gas varied with the biomass type also. Gas flow rate had a substantial influence on product yield if the biomass was compact and the oxygen content in the biomass was low. An inert environment was not effective for reactive and loosely structured biomass feedstocks (El Harfi, Mokhlisse, and Chanâa 1999).

14.7 TYPES OF REACTORS USED IN BIOMASS–PLASTIC CO-PYROLYSIS

Various reactors are used in co-pyrolysis processes. All pyrolyzers comprise a reactor, a condenser, and a cyclone. Feedstock materials and catalysts (with the exception of ex situ catalytic co-pyrolysis) are fed into the reactor. Separation of solid and gaseous products takes place through the cyclone. The condenser is employed to quickly quench the gaseous products to produce liquids. The decision to use a particular reactor is contingent upon other process parameters and the requirement of the researchers. Slow pyrolysis processes use reactors such as drum, screw, rotary kilns, and auger reactors. On the other hand, faster pyrolysis processes employ reactors like a vacuum, ablative, rotating cone, and fluidized and fixed bed reactors. Slow pyrolysis and the corresponding reactors are preferred when maximum biochar yield is necessary. If a higher volatile yield is necessary, then faster pyrolysis is preferred. The heating rates of individual reactors are also different (Azargohar et al. 2013; Crombie and Mašek 2014). A description of a few major reactors used in co-pyrolysis is provided in the following sections.

14.7.1 Auger Reactor/Screw Reactor

Auger reactors are often used in both fast and slow pyrolysis processes. Auger reactors are also employed for efficient bio-oil production. These reactors are used on a small scale. Here, the feedstock is moved up by using an auger. Then, it is heated externally or by a carrier such as an iron sphere or sand (Cruz 2012). The detailed schematic representation of Auger co-pyrolysis reactor is shown in Figure 14.2. This simplicity in design is a reason for their increasing popularity. This simplicity in design also leads to easier separation of the solid components as they slide down into a collection vessel due to gravity. However, the reactor has higher residence time and is difficult to scale (Campuzano, Brown, and Martínez 2019). Screw feeder has a high torque input; thus, it is capable of handling a variety of biomass

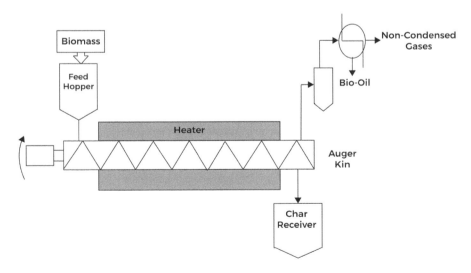

FIGURE 14.2 Auger co-pyrolysis reactor.

Source: Pandey et al. 2019.

feedstocks, including herbaceous biomass (Cruz 2012). Biomass feedstocks that are bound together can affect the capabilities of a regular reactor. Hence, a high torque configuration like the screw feeder is essential for pyrolyzing feedstocks (Azargohar et al. 2013; Crombie and Mašek 2014).

14.7.2 Fixed Bed Reactor

A fixed bed reactor is ideal for a lesser heating rate, and this can be explained by its low heat transfer coefficient. When a fixed bed reactor is used with a larger sample, the volumetric temperature in the core of the sample is not identical. Hence, different areas of the feedstock are decomposed at different temperatures at the same time. A fixed bed reactor is really inefficient and rarely used in scale-up facilities. The detailed diagram of fixed bed co-pyrolysis reactor is shown in Figure 14.3. Fixed bed reactors are used to test the parameters involved in the feedstocks of pyrolysis. On the contrary, only a few studies about fixed bed pyrolysis with different feedstock masses are available in the literature (Marshall et al. 2014; D. Chen et al. 2014).

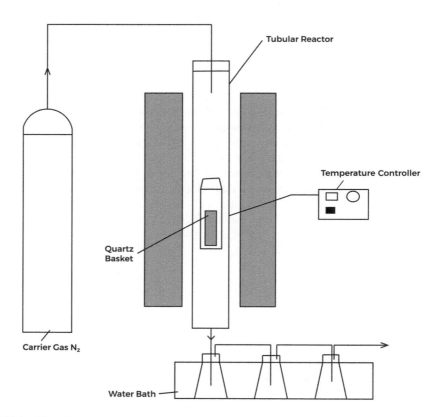

FIGURE 14.3 Fixed bed co-pyrolysis reactor.

Source: Pandey et al. 2019.

14.7.3 Rotary Kiln Reactor

The efficiency problem of fixed bed reactors is addressed by the rotary kiln reactors. The inclined kiln in the rotary kiln enables slow mixing of the feedstocks. Heating rates of rotary kilns are quite low, with their magnitudes not being higher than 10 °C/min. Residence time of the reactor can be as large as 1 h. There are several reasons for lower heating rates in rotary kiln reactors. First, heat

Co-pyrolysis of Biomass with Polymer Waste 197

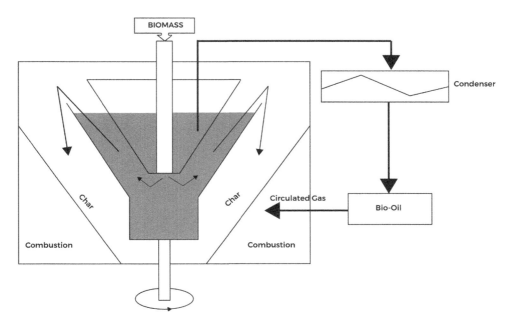

FIGURE 14.4 Rotary kiln reactor.
Source: Pandey et al. 2019.

transfer takes place only through the reactor's walls. The surface area of these walls is often low. Also, particles are often coarse, thereby reducing the overall heating rate. The detailed schematic diagram of Rotary kiln reactor is shown in Figure 14.4.

However, rotary kiln reactors have considerable advantages over other types of reactors. They provide better mixing of feedstocks and can be operated over a range of residence time. The rotary kiln reactor can also take heterogeneous materials, which means that extensive preprocessing of feedstock materials is not necessary (D. Chen et al. 2014). Rotary kilns are usually operated by means of external heating. The design, construction, and operating parameters of these reactors have been extensively explored in literature (Rutgers 1965; Shui Qing Li et al. 2005). Some studies have explored the impact of kiln geometry and feedstock material characteristics on the residence time (S. Q. Li et al. 2002). However, further research is needed on the impact of processing parameters on heat transfer coefficients.

14.7.4 Fluidized Bed Reactor

The problem of low heating rates in rotary kiln reactors is solved by the fluidized bed reactor. Hence, it is used to evaluate the correlation between residence time and pyrolysis temperature on pyrolysis by-products and composition (Mastral et al. 2002, 2003). The reactor is also used to study the secondary reactions and cracking of tars in the environment where there is high residence time. The application of fluidized bed reactors is only limited to academic studies and has not extended to industry. A reason for this is that the bed material separation from biomass is complicated and difficult to accomplish on an industrial scale. However, unlike rotary kiln reactors, fluidized bed reactors are designed for uniform feedstocks. For polymer feedstocks, fluidized bed reactors provide significant advantages over other configurations as polymers are resistant to heat transfer since they have high viscosity and low thermal conductivity. Since fluidized bed reactors can provide high heating rates, they balance the undesirable effects of polymer feedstocks. The line diagram of Fluidized bed reactor is shown in Figure 14.5.

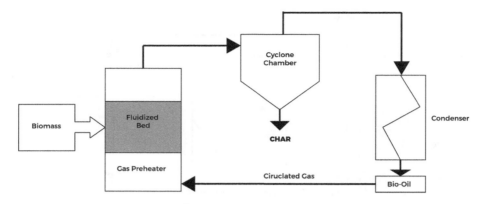

FIGURE 14.5 Fluidized bed reactor.

Source: Pandey et al. 2019.

14.7.5 Tubular Reactor

Another family of externally heated reactors is tubular reactors. They have tubular sections in the reactor. The feedstock is moved inside the reactor through several driving modes. Tubular reactors are hybrids in a sense, as they can fall under any of the categories discussed earlier. Aguado et al. (Aguado et al. 2002) used a continuous screw pyrolyzer, which was a modified tubular reactor. Other types of tubular reactors, such as a rectilinear reactors and tubes with inner reactors, are also reported in the literature (Marculescu et al. 2007; Walendziewski 2002). The detailed schematic representation of tubular reactor is shown in Figure 14.6.

There are several advantages to a tubular reactor. These configurations provide continuous coke and gas removal. Heat transfer per unit volume is high in these reactors. They provide a convenient avenue for syngas reformation. When the heat transfer coefficient is measured beforehand, the design of the tubular reactor becomes simple. The screw tube reactor (a type of tubular reactor) is characterized by very low operation and construction costs. Screw tube reactors are advantageous as their residence time can be altered by altering the screw speed. The speed range is generally 0.5–25 rpm for screw tube reactors. However, they have rigid pretreatment requirements due to their small passage channels. Their operating temperatures also vary with the feedstocks (Aguado et al. 2002). Apart from

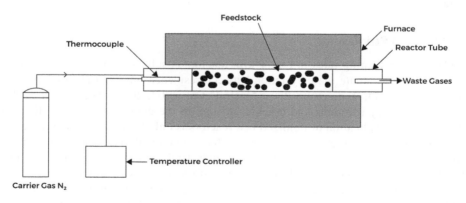

FIGURE 14.6 Tubular reactor.

Source: Pandey et al. 2019.

Co-pyrolysis of Biomass with Polymer Waste 199

these types of reactors, there are multistage reactors as well. Ohmukai et al. (Ohmukai, Hasegawa, and Mae 2008) used a separate stage to reform tar obtained in the first stage. Multistage reactors can be adopted for various applications and are useful to separate various products at different stages.

14.7.6 Plasma Pyrolysis Reactor

The type of heating that occurs in plasma and microwave pyrolysis processes is very different compared to conventional modes of heating, in these processes a volumetric heating occurs (Hrabovsky et al. 2006; Macquarrie, Clark, and Fitzpatrick 2012). Plasma pyrolysis is characterized by very high pyrolysis temperatures (over 1000 °C) in the absence of air using plasma torches. The end products include gases such as CO and H_2 and a vitrified matrix. Plasma pyrolysis is highly efficient, with the freedom to control processing conditions like temperature and syngas composition. A plasma reactor is incredibly safe. The products of the plasma reactor are friendly for material recycling and are safe for both the environment and public health. However, plasma pyrolysis also requires a separate secondary source of energy. This concern must be addressed in future studies. The schematic representation of plasma co-pyrolysis reactor is shown in Figure 14.7.

14.7.7 Microwave Reactor

Microwave pyrolysis is also growing in popularity among researchers. Microwave pyrolysis works on the phenomenon of microwave dielectric heating. The advantages of microwave pyrolysis include efficient volumetric heating from the core of the feedstock to the outer surface, easy control, and easy maintenance in terms of pyrolysis temperature (Baghurst and Mingos 1992). Despite these advantages, finer feedstock particles are necessary to reach the required heating rate. For the

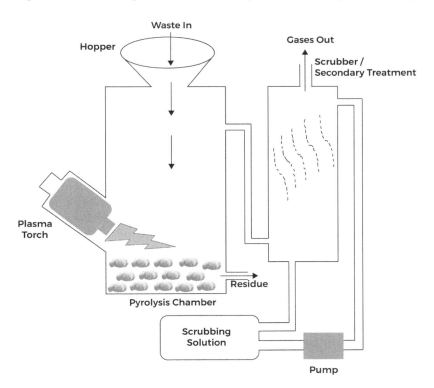

FIGURE 14.7 Plasma co-pyrolysis reactor.

Source: Pandey et al. 2019.

required temperature control, dielectric data of the feedstock in the microwave frequency range is necessary. This data is unavailable for most feedstocks. In a microwave reactor, solid-laden vapors must be wiped out to eliminate secondary reactions of vapor. Additionally, vapors have to be driven out of the reactor to reduce the amount of water-soluble polar molecules (Yin 2012).

14.8 BY-PRODUCTS OBTAINED FROM CO-PYROLYSIS PROCESS

Further, the biomass pyrolysis with plastic also produces two other by-products apart from the liquid: char and gaseous products. According to the experimental study, it is found that the char produced from co-pyrolysis is in the range of 12–13 wt%. Also, the gas generated during co-pyrolysis process possesses lowest calorific value as compared to natural gases. However, the calorific value can be increased by adding some catalyst and increasing the plastic ratio in the biomass. Some details regarding the by-products of the co-pyrolysis process are presented in the following sections.

14.8.1 Gaseous Product

The ideal operating conditions for enhancing the transformation of biomass to gas were higher temperatures and longer residence time. Non-condensable gas is produced only in small quantities by pyrolysis because the process variables needed to increase bio-oil generation is different from those needed for gas. Typically, the gas generated by pyrolysis is between 13 and 25 wt% of the biomass that was consumed (Nunes, Causer, and Ciolkosz 2020). Efforts were made to compile information from several research on the pyrolysis of used tires, and the result found that gaseous product can be produced in amounts between 2.5 and 28.5 wt%. According to the author, a higher gas yield is typically produced under circumstances that include a high temperature for reaction and a slow heating rate (Williams 2013). The composition of the feedstock used will determine the gas composition that results from the co-pyrolysis process. For the most part, the gas generated from wood pyrolysis comprised of CO, CO_2, CH_4, H_2, C_3H_6, C_2H_4, and trace of other hydrocarbons (Mohan, Pittman, and Steele 2006). Finally, the conclusion that the higher biomass content in the waste combination greatly reduced the gas heating value, mostly because less hydrocarbons were formed.

14.8.2 Co-pyrolysis Char

The solid products produced by co-pyrolysis (plastic and biomass wastes) have higher calorific values as compared to some types of coals (Paradela et al. 2009). Similarly, an another study found that char produced by pine wood pyrolysis had lower heating value as compared to char produced by co-pyrolysis. This is owing to the more oxygen matter of char produced by pyrolysis of single pine biomass. The presence of lignin content from the biomass may explain the variation in the composition of the char with polymer and biomass mixture. The author also found that chars have a relatively low sulfur concentration, making them appealing for usage in incineration. As a result, these products are ideal for use as a fuel, for example, in co-combustion in boiler with coal. This char produced can be utilized as an adsorbent to eliminate heavy metals (Brebu et al. 2010).

14.9 CONCLUSION

The quantity and quality of pyrolysis oil can be considerably improved using the co-pyrolysis technology without the use of any catalysts or addition of solvents. As a result, this procedure can be regarded as a quick, inexpensive, and efficient way to get high-quality pyrolysis oil. Additionally, this method helps to raise the heating value of the gas and char that are produced as by-products of co-pyrolysis. The accessibility of plastics which are used as additive materials in co-pyrolysis is crucial to the long-term viability of this process. Co-pyrolysis process is discovered to be a potential solution in conversion of biomass to generate pyrolysis oil from an economic perspective.

Co-pyrolysis has enormous development potential in many nations because to the accessibility and abundance of biomass wastes worldwide. Additionally, this approach makes it simple to manage the amount of biomass wastes. By creating pyrolysis oil from biomass wastes, it may be possible to decrease the necessity for landfills, lower the price for treatment of waste, and address various environmental issues.

REFERENCES

Abnisa, Faisal, and Wan Mohd Ashri Wan Daud. 2014. "A Review on Co-Pyrolysis of Biomass: An Optional Technique to Obtain a High-Grade Pyrolysis Oil." *Energy Conversion and Management.* doi:10.1016/j.enconman.2014.07.007.

Abnisa, Faisal, and Wan Mohd Ashri Wan Daud. 2015. "Optimization of Fuel Recovery through the Stepwise Co-Pyrolysis of Palm Shell and Scrap Tire." *Energy Conversion and Management.* doi:10.1016/j.enconman.2015.04.030.

Acikgoz, C., O. Onay, and O. M. Kockar. 2004. "Fast Pyrolysis of Linseed: Product Yields and Compositions." *Journal of Analytical and Applied Pyrolysis.* doi:10.1016/S0165-2370(03)00124-4.

Aguado, J., D. P. Serrano, J. M. Escola, and E. Garagorri. 2002. "Catalytic Conversion of Low-Density Polyethylene Using a Continuous Screw Kiln Reactor." *Catalysis Today.* doi:10.1016/S0920-5861(02)00077-9.

Ansari, Khursheed B., Saeikh Zaffar Hassan, Rohidas Bhoi, and Ejaz Ahmad. 2021. "Co-Pyrolysis of Biomass and Plastic Wastes: A Review on Reactants Synergy, Catalyst Impact, Process Parameter, Hydrocarbon Fuel Potential, COVID-19." *Journal of Environmental Chemical Engineering.* doi:10.1016/j.jece.2021.106436.

Aylón, E., A. Fernández-Colino, M. V. Navarro, R. Murillor, T. García, and A. M. Mastral. 2008. "Waste Tire Pyrolysis: Comparison Between Fixed Bed Reactor and Moving Bed Reactor." *Industrial and Engineering Chemistry Research.* doi:10.1021/ie071573o.

Azargohar, Ramin, Kathlene L. Jacobson, Erin E. Powell, and Ajay K. Dalai. 2013. "Evaluation of Properties of Fast Pyrolysis Products Obtained, from Canadian Waste Biomass." *Journal of Analytical and Applied Pyrolysis.* doi:10.1016/j.jaap.2013.06.016.

Baghurst, David R., and D. Michael P. Mingos. 1992. "Superheating Effects Associated with Microwave Dielectric Heating." *Journal of the Chemical Society, Chemical Communications.* doi:10.1039/C39920000674.

Bohn, Mark S., and Charles B. Benham. 1984. "Biomass Pyrolysis with an Entrained Flow Reactor." *Industrial and Engineering Chemistry Process Design and Development.* doi:10.1021/i200025a030.

Brebu, Mihai, Suat Ucar, Cornelia Vasile, and Jale Yanik. 2010. "Co-Pyrolysis of Pine Cone with Synthetic Polymers." *Fuel.* doi:10.1016/j.fuel.2010.01.029.

Campuzano, Felipe, Robert C. Brown, and Juan Daniel Martínez. 2019. "Auger Reactors for Pyrolysis of Biomass and Wastes." *Renewable and Sustainable Energy Reviews.* doi:10.1016/j.rser.2018.12.014.

Chen, Dezhen, Lijie Yin, Huan Wang, and Pinjing He. 2014. "Pyrolysis Technologies for Municipal Solid Waste: A Review." *Waste Management.* doi:10.1016/j.wasman.2014.08.004.

Chen, Jiaxin, Biying Zhang, Lingli Luo, Fan Zhang, Yanglei Yi, Yuanyuan Shan, Bianfang Liu, Yuan Zhou, Xin Wang, and Xin Lü. 2021. "A Review on Recycling Techniques for Bioethanol Production from Lignocellulosic Biomass." *Renewable and Sustainable Energy Reviews.* doi:10.1016/j.rser.2021.111370.

Crombie, Kyle, and Ondřej Mašek. 2014. "Investigating the Potential for a Self-Sustaining Slow Pyrolysis System Under Varying Operating Conditions." *Bioresource Technology.* doi:10.1016/j.biortech.2014.03.134.

Cruz, Diana C. 2012. "Production of Bio-Coal and Activated Carbon from Biomass." https://ir.lib.uwo.ca/etd://ir.lib.uwo.ca/etd/1044.

Debdoubi, Abderrahmane, A. El Amarti, E. Colacio, M. J. Blesa, and L. H. Hajjaj. 2006. "The Effect of Heating Rate on Yields and Compositions of Oil Products from Esparto Pyrolysis." *International Journal of Energy Research.* doi:10.1002/er.1215.

Demirbas, Ayhan. 2004. "Pyrolysis of Municipal Plastic Wastes for Recovery of Gasoline-Range Hydrocarbons." *Journal of Analytical and Applied Pyrolysis.* 72 (1). doi:10.1016/j.jaap.2004.03.001.

Ekinci, Ekrem, Murat Citiroglu, Ersan Putun, Gordon D. Love, Christopher J. Lafferty, and Colin E. Snapet. 1992. "Effect of Lignite Addition and Steam on the Pyrolysis of Turkish Oil Shales." *Fuel.* doi:10.1016/0016-2361(92)90227-F.

El Harfi, K., A. Mokhlisse, and M. Ben Chanâa. 1999. "Effect of Water Vapor on the Pyrolysis of the Moroccan (Tarfaya) Oil Shale." *Journal of Analytical and Applied Pyrolysis.* doi:10.1016/S0165-2370(98)00108-9.

Encinar, J. M., F. J. Beltrán, A. Bernalte, A. Ramiro, and J. F. González. 1996. "Pyrolysis of Two Agricultural Residues: Olive and Grape Bagasse. Influence of Particle Size and Temperature." *Biomass and Bioenergy*. doi:10.1016/S0961-9534(96)00029-3.

Engamba Esso, Samy Berthold, Longfei Xu, Hengda Han, Zhe Xiong, Melvina Fudia Kamara, Jun Xu, Long Jiang, et al. 2022. "Effects of Interactions between Organic Solid Waste Components on the Formation of Heavy Components in Oil during Pyrolysis." *Fuel Processing Technology* 225. doi:10.1016/j.fuproc.2021.107041.

Fahmi, R., A. V. Bridgwater, I. Donnison, N. Yates, and J. M. Jones. 2008. "The Effect of Lignin and Inorganic Species in Biomass on Pyrolysis Oil Yields, Quality and Stability." *Fuel* 87 (7). doi:10.1016/j.fuel.2007.07.026.

Fei, Jinxia, Jie Zhang, Fuchen Wang, and Jie Wang. 2012. "Synergistic Effects on Co-Pyrolysis of Lignite and High-Sulfur Swelling Coal." *Journal of Analytical and Applied Pyrolysis* 95. doi:10.1016/j.jaap.2012.01.006.

Garforth, Arthur A., Salmiaton Ali, Jesús Hernández-Martínez, and Aaron Akah. 2004. "Feedstock Recycling of Polymer Wastes." *Current Opinion in Solid State and Materials Science*. doi:10.1016/j.cossms.2005.04.003.

Gibbins-Matham, Jon, and Rafael Kandiyoti. 1988. "Coal Pyrolysis Yields from Fast and Slow Heating in a Wire-Mesh Apparatus with a Gas Sweep." *Energy and Fuels*. doi:10.1021/ef00010a017.

Haykiri-Acma, H. 2006. "The Role of Particle Size in the Non-Isothermal Pyrolysis of Hazelnut Shell." *Journal of Analytical and Applied Pyrolysis*. doi:10.1016/j.jaap.2005.06.002.

Hrabovsky, M., M. Konrad, V. Kopecky, M. Hlina, T. Kavka, O. Chumak, G. Van Oost, E. Beeckman, and Benjamin Defoort. 2006. "Pyrolysis of Wood in Arc Plasma for Syngas Production." *High Temperature Material Processes*. doi:10.1615/HighTempMatProc.v10.i4.70.

Jakab, E., G. Várhegyi, and O. Faix. 2000. "Thermal Decomposition of Polypropylene in the Presence of Wood-Derived Materials." *Journal of Analytical and Applied Pyrolysis*. doi:10.1016/S0165-2370(00)00101-7.

Jiang, Guodong, and Liping Wei. 2018. "Analysis of Pyrolysis Kinetic Model for Processing of Thermogravimetric Analysis Data." *Phase Change Materials and Their Applications*. doi:10.5772/intechopen.79226.

Kumar, R., V. Strezov, H. Weldekidan, J. He, S. Singh, T. Kan, and B. Dastjerdi. 2020. "Lignocellulose Biomass Pyrolysis for Bio-Oil Production: A Review of Biomass Pre-Treatment Methods for Production of Drop-in Fuels." *Renewable and Sustainable Energy Reviews*. doi:10.1016/j.rser.2020.109763.

Li, S. Q., L. B. Ma, W. Wan, and Q. Yao. 2005. "A Mathematical Model of Heat Transfer in a Rotary Kiln Thermo-Reactor." *Chemical Engineering and Technology*. doi:10.1002/ceat.200500241.

Li, S. Q., J. H. Yan, R. D. Li, Y. Chi, and K. F. Cen. 2002. "Axial Transport and Residence Time of MSW in Rotary Kilns – Part I. Experimental." *Powder Technology*. doi:10.1016/S0032-5910(02)00014-1.

Macquarrie, Duncan J., James H. Clark, and Emma Fitzpatrick. 2012. "The Microwave Pyrolysis of Biomass." *Biofuels, Bioproducts and Biorefining*. doi:10.1002/bbb.1344.

Marculescu, C., G. Antonini, A. Badea, and T. Apostol. 2007. "Pilot Installation for the Thermo-Chemical Characterisation of Solid Wastes." *Waste Management*. doi:10.1016/j.wasman.2006.02.011.

Marshall, A. J., Ping F. Wu, Sang Hun Mun, and Charles Lalonde. 2014. "Commercial Application of Pyrolysis Technology in Agriculture." *American Society of Agricultural and Biological Engineers Annual International Meeting 2014, ASABE 2014*. doi:10.13031/aim.20141909089.

Martínez, Juan D., Alberto Veses, Ana M. Mastral, Ramón Murillo, Maria V. Navarro, Neus Puy, Anna Artigues, Jordi Bartrolí, and Tomás García. 2014. "Co-Pyrolysis of Biomass with Waste Tyres: Upgrading of Liquid Bio-Fuel." *Fuel Processing Technology*. doi:10.1016/j.fuproc.2013.11.015.

Mastral, F. J., E. Esperanza, C. Berrueco, M. Juste, and J. Ceamanos. 2003. "Fluidized Bed Thermal Degradation Products of HDPE in an Inert Atmosphere and in Air-Nitrogen Mixtures." *Journal of Analytical and Applied Pyrolysis*. doi:10.1016/S0165-2370(02)00068-2.

Mastral, F. J., E. Esperanza, P. Garcíía, and M. Juste. 2002. "Pyrolysis of High-Density Polyethylene in a Fluidised Bed Reactor: Influence of the Temperature and Residence Time." *Journal of Analytical and Applied Pyrolysis*. doi:10.1016/S0165-2370(01)00137-1.

Minkova, V., M. Razvigorova, E. Bjornbom, R. Zanzi, T. Budinova, and N. Petrov. 2001. "Effect of Water Vapour and Biomass Nature on the Yield and Quality of the Pyrolysis Products from Biomass." *Fuel Processing Technology*. doi:10.1016/S0378-3820(00)00153-3.

Mohan, Dinesh, Charles U. Pittman, and Philip H. Steele. 2006. "Pyrolysis of Wood/Biomass for Bio-Oil: A Critical Review." *Energy and Fuels*. doi:10.1021/ef0502397.

Mwanza, Bupe G., and Charles Mbohwa. 2017. "Drivers to Sustainable Plastic Solid Waste Recycling: A Review." *Procedia Manufacturing* 8. doi:10.1016/j.promfg.2017.02.083.

Nanduri, Arvind, Shreesh S. Kulkarni, and Patrick L. Mills. 2021. "Experimental Techniques to Gain Mechanistic Insight into Fast Pyrolysis of Lignocellulosic Biomass: A State-of-the-Art Review." *Renewable and Sustainable Energy Reviews*. doi:10.1016/j.rser.2021.111262.

Nunes, L. J. R., T. P. Causer, and D. Ciolkosz. 2020. "Biomass for Energy: A Review on Supply Chain Management Models." *Renewable and Sustainable Energy Reviews*. doi:10.1016/j.rser.2019.109658.

Ohmukai, Yoshikage, Isao Hasegawa, and Kazuhiro Mae. 2008. "Pyrolysis of the Mixture of Biomass and Plastics in Countercurrent Flow Reactor Part I: Experimental Analysis and Modeling of Kinetics." *Fuel*. doi:10.1016/j.fuel.2008.04.005.

Olukcu, Nuray, Jale Yanik, Mehmet Saglam, and Mithat Yuksel. 2002. "Liquefaction of Beypazari Oil Shale by Pyrolysis." *Journal of Analytical and Applied Pyrolysis*. doi:10.1016/S0165-2370(01)00168-1.

Önal, Eylem, Başak Burcu Uzun, and Ayşe Eren Pütün. 2014. "Bio-Oil Production via Co-Pyrolysis of Almond Shell as Biomass and High Density Polyethylene." *Energy Conversion and Management*. doi:10.1016/j.enconman.2013.11.022.

Pandey, Ashok, Christian Larroche, Claude Gilles Dussap, Edgard Gnansounou, Samir Kumar Khanal, and Steven Ricke. 2019. *Biomass, Biofuels, Biochemicals: Biofuels: Alternative Feedstocks and Conversion Processes for the Production of Liquid and Gaseous Biofuels*. doi:10.1016/C2018-0-00957-3. Academic Press, 2nd edition, Elsevier, Cambridge, United States.

Paradela, Filipe, Filomena Pinto, Ibrahim Gulyurtlu, Isabel Cabrita, and Nuno Lapa. 2009. "Study of the Co-Pyrolysis of Biomass and Plastic Wastes." *Clean Technologies and Environmental Policy* 11 (1). doi:10.1007/s10098-008-0176-1.

"Plastic Waste Worldwide – Statistics & Facts | Statista." 2022. Accessed May 27. www.statista.com/topics/5401/global-plastic-waste/.

Pütün, Ayşe E., Nurgül Özbay, Esin Apaydin Varol, Başak B. Uzun, and Funda Ateş. 2007. "Rapid and Slow Pyrolysis of Pistachio Shell: Effect of Pyrolysis Conditions on the Product Yields and Characterization of the Liquid Product." *International Journal of Energy Research*. doi:10.1002/er.1263.

Rutgers, R. 1965. "Longitudinal Mixing of Granular Material Flowing Through a Rotating Cylinder. Part II. Experimental." *Chemical Engineering Science*. doi:10.1016/0009-2509(65)80112-9.

Salvilla, John Nikko V., Bjorn Ivan G. Ofrasio, Analiza P. Rollon, Ferdinand G. Manegdeg, Ralf Ruffel M. Abarca, and Mark Daniel G. de Luna. 2020. "Synergistic Co-Pyrolysis of Polyolefin Plastics with Wood and Agricultural Wastes for Biofuel Production." *Applied Energy* 279. doi:10.1016/j.apenergy.2020.115668.

Scott, Donald S., Piotr Majerski, Jan Piskorz, and Desmond Radlein. 1999. "Second Look at Fast Pyrolysis of Biomass – the RTI Process." *Journal of Analytical and Applied Pyrolysis*. doi:10.1016/S0165-2370(99)00006-6.

Şensöz, S., D. Angin, and S. Yorgun. 2000. "Influence of Particle Size on the Pyrolysis of Rapeseed (Brassica Napus L.): Fuel Properties of Bio-Oil." *Biomass and Bioenergy*. doi:10.1016/S0961-9534(00)00041-6.

Singh, Maninderjit, Shakirudeen A. Salaudeen, Brandon H. Gilroyed, Sultan M. Al-Salem, and Animesh Dutta. 2021. "A Review on Co-Pyrolysis of Biomass with Plastics and Tires: Recent Progress, Catalyst Development, and Scaling Up Potential." *Biomass Conversion and Biorefinery*. doi:10.1007/s13399-021-01818-x.

Song, Hu, Sun Xuexing, Xiong Youhui, Xiang Jun, Li Min, and Li Peisheng. 2002. "Percolation Research on Fractal Structure of Coal Char." *Energy and Fuels*. doi:10.1021/ef020003s.

Strezov, V., B. Moghtaderi, and J. A. Lucas. 2003. "Thermal Study of Decomposition of Selected Biomass Samples." *Journal of Thermal Analysis and Calorimetry*. doi:10.1023/A:1025003306775.

Sukiran, Mohamad Azri, Nor Kartini Abu Bakar, and Chow Mee Chin. 2009. "Optimization of Pyrolysis of Oil Palm Empty Fruit Bunches." *Journal of Oil Palm Research*.

Tsai, W. T., M. K. Lee, and Y. M. Chang. 2007. "Fast Pyrolysis of Rice Husk: Product Yields and Compositions." *Bioresource Technology*. doi:10.1016/j.biortech.2005.12.005.

Walendziewski, Jerzy. 2002. "Engine Fuel Derived from Waste Plastics by Thermal Treatment." *Fuel*. doi:10.1016/S0016-2361(01)00118-1.

Williams, Paul T. 2013. "Pyrolysis of Waste Tyres: A Review." *Waste Management* 33 (8). doi:10.1016/j.wasman.2013.05.003.

Yin, Chungen. 2012. "Microwave-Assisted Pyrolysis of Biomass for Liquid Biofuels Production." *Bioresource Technology*. doi:10.1016/j.biortech.2012.06.016.

Zhang, Kai, Huahong Shi, Jinping Peng, Yinghui Wang, Xiong Xiong, Chenxi Wu, and Paul K. S. Lam. 2018. "Microplastic Pollution in China's Inland Water Systems: A Review of Findings, Methods, Characteristics, Effects, and Management." *Science of the Total Environment*. doi:10.1016/j.scitotenv.2018.02.300.

15 Thermal Degradation Behaviors and Kinetics of Pyrolysis

Uthayakumar Azhagu and Anand Ramanathan

CONTENTS

15.1 Introduction 205
15.2 Thermogravimetric Analysis 206
15.3 Thermal Degradation Behaviors 206
15.4 Kinetic Study 207
     15.4.1 Arrhenius Equation 207
15.5 Model-free Method 209
     15.5.1 Friedman Method.. 210
     15.5.2 Flynn–Wall–Ozawa Method 212
     15.5.3 Kissinger–Akahira–Sunose Method 213
15.6 Model-fitting Method... 213
     15.6.1 Coats–Redfern Method 215
15.7 Activation Energy of Pyrolysis Process... 216
15.8 Identification of Pyrolysis Reaction Mechanism 216
References... 218

15.1 INTRODUCTION

Population blast, fast-growing economy, living standard improvement, and technological development create huge energy demand and uncontrolled waste generation (Kumari et al. 2019). Overutilization of fossil fuels and their by-products is continuous and unavoidable in developing and developed countries (Kumar, Panda, and Singh 2011). Waste recycling is the best solution for energy demand and waste disposal. Pyrolysis is one of the sustainable tertiary thermochemical recycling processes to generate fuel from carbonous waste. Pyrolysis is the oxygen-free endothermic thermal degradation process that converts waste into pyro-oil, pyro-gas, and carbon-rich char (Williams and Williams 1997). The thermochemical reaction behind the pyrolysis process is complex and it is difficult to predict the exact reaction. Pyrolysis of waste has a sequence of more than thousands of chemical decomposition reactions. Kinetic methods for thermal degradation analysis must have single reactions. But pyrolysis encompasses several reactions, hence kinetic methods cannot give correct results. The isoconversional kinetic method can handle these multiple reactions and provide a meaningful result from thermogravimetric analysis data (Das and Tiwari 2017). Isoconversional methods are able to obtain activation energy without the solid-state reaction model assumption. Arrhenius rate law helps to assume thermal degradation solid-state reaction rate constant ($K(T)$). Reaction kinetic analysis of the waste degradation process helps to map the pyrolysis process for designing the optimum reactor and foundation for scale-up. Thermogravimetric analysis mass loss data with respect to time and temperature lead to the evaluation of conversion (x). Differential and integral methods exist to compute the kinetic parameters from conversion (x) and Arrhenius rate law. Thermal degradation reaction helps to convert higher molecular weight complex waste into lower molecular weight elements. It may be condensable and non-condensable, depending on feedstock composition, operating temperature, and heating rate. Thermogravimetry is the analysis tool to measure the mass change with respect to temperature change over time. In particular, pyrolysis chemical study needs dynamic thermogravimetry analysis (TGA) for thermal decomposition behavior with various heating rates and different medium gas supplies (Deng, Zhang, and Wang 2008). The accuracy of the pyrolysis reaction kinetics solely depends on the kinetic triplets such as reaction model ($f(x)$), activation energy (E), and pre-exponential factor (A). This section gives an overall idea about thermal degradation kinetic study of medical waste composition with various reaction models and approaches.

15.2 THERMOGRAVIMETRIC ANALYSIS

In dynamic TGA, mass loss with respect to temperature can draw a curve and help to predict the reaction stages in a single or more than one stage. However, TGA can only deal with mass loss by evaporation; not only chemical change but also physical phase change gives mass loss. The endothermic or exothermic action will be identified through differential thermal analysis (DTA). Medical waste has a wide variety of composition, and degradation also has more than one stage. For example, PVC and PET degradation happen in two steps: PP and PE degradation in a single step (Dubdub and Al-Yaari 2020).

For evaluating degradation kinetics according to ASTM E1641, of the pyrolysis process, non-isothermal TGA of carbonaceous waste needs to be done at an inert (N_2, CO_2, He) atmosphere and several heating rate such as 5, 10, 15, 20, 25, 30, 35, and 40 °C/min in the temperature range of 25–900 °C. According to ASTM E1877, dynamic TGA also provides information about thermal stability and lifetime of carbonaceous wastes under oxygen or oxygen-free environment.

15.3 THERMAL DEGRADATION BEHAVIORS

Carbonaceous waste thermal degradation happen in many routes which is not exactly predictable. Typical carbonaceous waste thermal degradation pathways are shown in Figure 15.1. The reaction rate of the thermal degradation process is assumed as a function of temperature and conversion only:

$$\frac{dx}{dt} = F(T, x) \qquad (15.1)$$

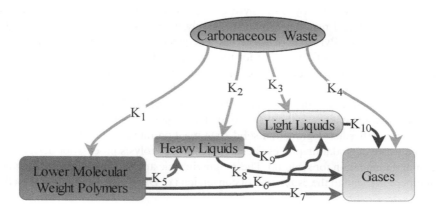

FIGURE 15.1 Carbonaceous waste thermal degradation pathways.

Thermal Degradation Behaviors and Kinetics of Pyrolysis

Take $k(T)$ as a function of temperature and $f(x)$ as a function of conversion. Hence, Equation 15.1 becomes as follows:

$$\frac{dx}{dt} = k(T)f(x) \tag{15.2}$$

x is known as the degree of conversion, which is expressed as follows:

$$x = \left(\frac{(w_0 - w)}{w_0 - w_\infty}\right)$$

Here w, w_0, and w_∞ represent time instant, initial mass, and final mass of the reactants, respectively.

Then the activation energy can be obtained from the following three different methodologies. In isothermal measurements, process time and temperature are taken as the variable parameters. Plot is drawn between time and temperature. In integral and incremental approach with linear heating rate, operating temperature and heating rate are taken as the variables. Plot is drawn between temperature and heating rate.

- *For the differential method:* reaction rate versus temperature (Straka, Bičáková, and Šupová 2017)

The conversion rate (x) depends on time (t) and temperature. Spain is doing much research regarding plastic pyrolysis and gasification. Solid-state reaction processes are too complicated and unknown phenomena. Therefore, chemical kinetic study with aid of dynamic TGA data is most recommended by thermochemical field researchers (Mannocci et al. 2020). With more than two heating rates, TGA data helps to identify the kinetic parameters via an isoconversional method.

The thermogravimetric analysis provides information about the decomposition of medical waste in an inert (helium/nitrogen) atmosphere. It clearly shows the major weight loss zones and conversion rates. Carbonaceous waste degradation data has a small weight loss at below 150 °C because of water/moisture content evaporation. In the range of 250–450 °C, major weight loss occurs in a single or two stages (Deng, Zhang, and Wang 2008). This is because of the volatile release from waste.

15.4 KINETIC STUDY

The kinetic study offers knowledge about the rate of chemical reaction, reaction mechanism, and all of the elements which have an impact on the chemical reaction. Thermochemical conversion kinetics of carbonaceous waste will be delineated by integral and differential methods to estimate inconsistent kinetic parameters. Thermochemical conversion kinetics of carbonaceous waste will be delineated inconsistent kinetic parameters by integral and differential methods. In pyrolysis kinetic study, activation energy, pre-exponential factor, and reaction mechanism are known as kinetic triplets. Pyrolysis kinetics is not exactly reaction kinetics but rather an approximation obtained from experimental dynamic TGA curves. Most appropriate kinetic triplets for pyrolysis of carbonaceous waste are estimated by solid-state reaction model-fitting and free methodologies. Various approaches of pyrolysis chemical kinetic study is clearly visualized in figure 15.2.

15.4.1 Arrhenius Equation

Swedish chemist Svante Arrhenius developed one physical chemistry relationship between the Boltzmann distribution law and activation energy concept, which is expressed by the following equation:

$$\mathbf{K(T)} = \mathbf{A}e^{-\frac{E_a}{RT}} \tag{15.3}$$

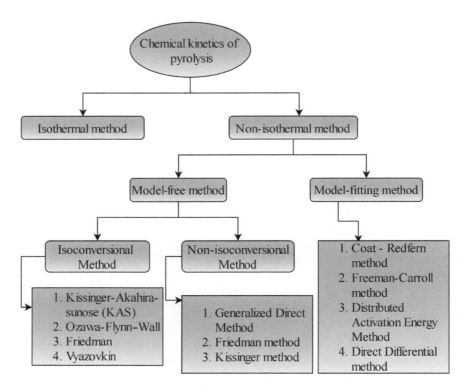

FIGURE 15.2 Kinetic study.

In Equation 15.3, A denotes the pre-exponential factor which deals with the molecule-to-molecule collisions. The frequency of the molecular collisions has adequate energy to stimulate a chemical reaction. The pre-exponential factor can be expressed as follows:

$$A = \rho Z \tag{15.4}$$

Here ρ is the steric factor (depends on the orientation of molecules) and Z is the frequency of collisions.

In the Arrhenius equation, RT denotes the average kinetic energy. E_a/RT is the ratio between activation energy and average kinetic energy. A negative sign before E_a/RT denotes a high rate of reaction and has a small ratio (E_a/RT). As a result of low activation energy, reactions speed and rate constant are high, hence this ratio is an exponent. In Arrhenius's expression, rate constant $K(T)$ is a dependent variable and temperature (T) is an independent variable. Equation 15.3 can be expressed in a non-exponential form handy to correlate graphically. A simple way to determine activation energy and pre-exponential factor is as follows.

Take natural logarithm on both sides of Equation 15.3 for removal of the exponential term:

$$\operatorname{Ln} K(T) = \ln\left(A\, e^{-\frac{E_a}{RT}}\right) \tag{15.5}$$

$$\operatorname{Ln} K(T) = \ln A - \frac{E_a}{RT} \tag{15.6}$$

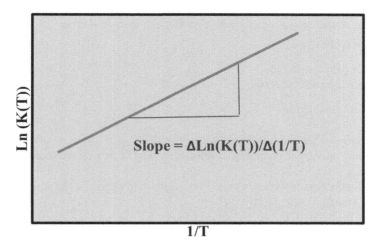

FIGURE 15.3 Arrhenius plot.

$$\text{Ln K}(T) = \ln A - \frac{E_a}{R}\left(\frac{1}{T}\right) \tag{15.7}$$

If the particular thermal degradation reaction rate constant and corresponding temperature are known, then one can plot a curve $1/T$ versus $\ln(K(T))$. The equation can fit with a straight line equation (Wu et al. 2006).

$$y = mx + C \tag{15.8}$$

From the plot, the straight line slope (*m*) can be derived, which is equal to $(-E_a/R)$. Arrhenius plot. and slope identification is shown in Figure 15.3 for better understanding.

Hence activation energy:

$$E_a = -m \times R \tag{15.9}$$

where *R* is the universal gas constant.

This is a simple affordable way to estimate the thermochemical reaction activation energy and pre-exponential factor from the $1/T$ versus $\ln(K(T))$ graph.

15.5 MODEL-FREE METHOD

The model-free approach is one of the isoconversional methods to evaluate the activation energy without any solid-state reaction model assumptions. But reaction model assumption is essential for pre-exponential factor determination. In isoconversional method, the reaction rate is a function of temperature because here conversion rate remains constant. For each conversion rate (*x*), activation energy (E_a) can be calculated directly without any assumption regarding the reaction model $f(x)$. Isoconversional methods are traditional model-free methods for calculating activation energy only by using TGA without assumptions and giving acceptable results. Model-free method can be further categorize into differential and integral approach, which is presented in Figure 15.4. The differential isoconversional methods provide authentic values of kinetic parameters, whereas integral isoconversional methods give average values only.

The general expression for thermochemical conversion rate is as follows:

$$\frac{dx}{dt} = A e^{-\frac{E_a}{RT}} f(x) \tag{15.10}$$

where

$\dfrac{dx}{dt}$ = Conversion rate

x = Degree of conversion

$f(x)$ = Solid-state reaction function

$A\,e^{-\frac{E_a}{RT}}$ – Arrhenius equation

The model-free approach has the following assumptions:
- The activation energy is changing continuously and it is not constant throughout the reaction.
- For the particular conversion (isoconversion), activation energy is the function of the temperature program.
- The reaction mechanism or type is not mandatory for evaluating activation energy.

The model-free method is commonly categorized as a single-value and multiple-value model-free methods (Table 15.1).

15.5.1 Friedman Method

Friedman method is one of the straightforward differential isoconversional approach to evaluate the activation energy (E_a) as a function degree of reaction (x). Friedman kinetic expression for thermal degradation process is obtained from Equation 15.10. Reaction rate (dx/dt) can be replaced with conversion based on the temperature change (Larraín, Carrier, and Radovic 2017):

$$\frac{dx}{dt} = \frac{dx}{dT} \cdot \frac{dT}{dt} = \beta \frac{dx}{dT} \qquad (15.11)$$

Here $\dfrac{dT}{dt} = \beta$ is known as heating rate and ramp rate.

By the effect of Equation 15.11, Equation 15.10 becomes

$$\beta \frac{dx}{dT} = A\,e^{-\frac{E_a}{RT}} f(x) \qquad (15.12)$$

Take natural logarithm on both sides for removal of an exponential term in Equation 15.12. Now equation becomes

---

TABLE 15.1
Model-free Method Classification

Single-value Model-free Methods	Multiple-value Model-free Methods
For the entire process, obtain only one value of activation energy E_a and pre-exponential factor A independent of the degree of conversion x	For each degree of conversion, obtain different activation energy E_a and pre-exponential factor A
• ASTM E2070(A) for isothermal data. • ASTM E2890 • Dynamic Arrhenius for failure temperature (ASTM E2070D) • ASTM E1641 • Isothermal Arrhenius for time-to-event • ASTM E698	• Friedman, • Ozawa-Flynn-Wall (OFW), • Kissinger-Akahira-Sunose (KAS), • Vyazovkin,

Thermal Degradation Behaviors and Kinetics of Pyrolysis

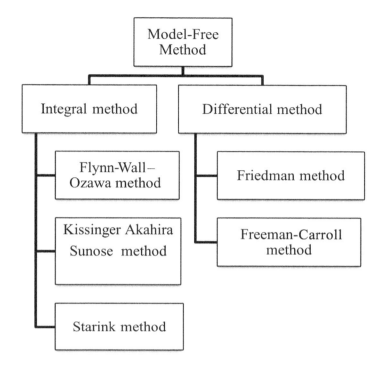

FIGURE 15.4 Classification of model-free method.

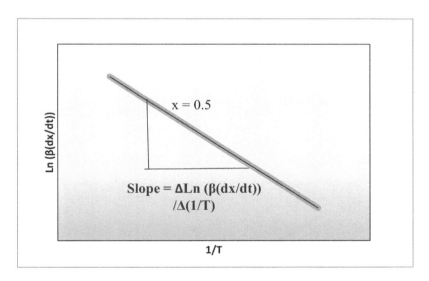

FIGURE 15.5 Friedman plot.

$$\ln\left[\beta\left(\frac{dx}{dT}\right)\right] = \ln\left[Af(x)\right] - \left(\frac{E_a}{RT}\right) \tag{15.13}$$

The degree of conversion will be in the range of 0–1, where heating rate (β) remains constant. For the determination of activation energy through the Friedman method, Equation 15.13 is matched with the straight line in Equation 15.8. Hence the magnitude of the slope $\left(\dfrac{E_a}{R}\right)$ is obtained from the plot between $\frac{1}{T}$ and $\ln\left[\beta\left(\frac{dx}{dT}\right)\right]$ for the constant degree of conversion and heating rate, which is demonstrated in Figure 15.5.

Conversion concerning temperature obtained from different heating rates can give different kinetic triplets. Friedman's method is a simple and straightforward method for the evaluation of kinetic triplets, but it does not give the most appreciable activation energy value. The more secure alternative to the model-free differential method is model-free integral methods, which will be discussed in the upcoming section.

15.5.2 Flynn–Wall–Ozawa Method

Flynn–Wall–Ozawa (FWO) method is an integral model-free method to evaluate kinetic triplets reliably. This integral method of thermal degradation kinetic expression is assumed as temperature dependent and follows the Arrhenius equation. FWO method assesses the temperature counterpart to the degree of conversion from TGA data for different heating rates. Equation 15.12 can be rewritten as follows:

$$\beta \frac{dx}{f(x)} = A e^{-\frac{E_a}{RT}} dT \tag{15.14}$$

Take integration on both sides of Equation 15.14:

$$\beta \int_0^1 \frac{dx}{f(x)} = A \int_{T_0}^T e^{-\frac{E_a}{RT}} dT \tag{15.15}$$

$$\int_0^1 \frac{dx}{f(x)} = g(x) \tag{15.16}$$

Kinetic expression for the FWO is shown in Equation 15.17:

$$\ln \beta = \ln\left(\frac{AE_a}{Rg(x)}\right) - 2.315 - 0.4567\left(\frac{E_a}{RT}\right) \tag{15.17}$$

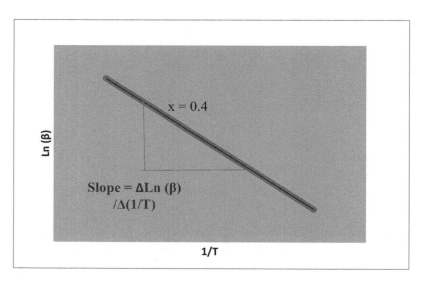

FIGURE 15.6 FWO plot.

Thermal Degradation Behaviors and Kinetics of Pyrolysis

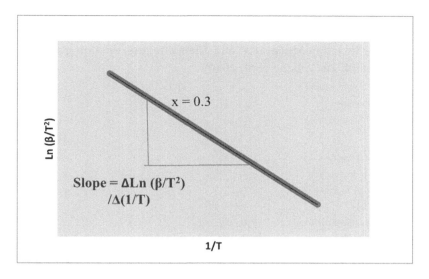

FIGURE 15.7 KAS model plot.

Equation 15.17 is compared with the straight line equation (Equation 15.8). To obtain the slope (−0.4567 E_a/R) for each degree of conversion × graph was drawn between $1/T$ and $\ln \beta$ Which is shown in Figure 15.6. In each degree of conversion, activation energy varies and denotes multistep reaction mechanisms. Many thermochemical reaction activation energy varies over time and it can be effectively studied by the FWO method.

However, the method is predicted to fail if reactions of widely different types (and hence having very different activation energies) are occurring simultaneously. Competitive reactions that have different products also render the method inapplicable. In addition, the Flynn–Wall–Ozawa method is less precise than Friedman's method.

15.5.3 Kissinger–Akahira–Sunose Method

Homer E. Kissinger introduced one integral model-free method for kinetic study in 1957. It is the most standard approach for determining kinetic parameters from thermal analysis data because it is fast and user-friendly. Based on the Kissinger method, another modernized kinetic method has been established, which is known as the Kissinger–Akahira–Sunose (KAS) method. This model-free integral method also determines the activation energy directly without any assumption of thermal degradation mechanisms. The KAS method is derived from Equation 15.18, which will be integrated at initial condition degree of conversion 0 and initial temperature T_0.

$$\ln \frac{^2}{T^2} = \ln \left[\frac{AR}{g(x)} \right] - \frac{E_a}{RT} \tag{15.18}$$

Equation 15.18 is compared with the straight line equation (Equation 15.8). To obtain the slope $(-E_a/R)$ for each degree of conversion × graph was drawn between $1/T$ and $\ln \beta/T^2$. In KAS method slope determination through plot is displayed in Figure 15.7. The E_a can be determined by the gradient of the curve which is represented by $-E_a/R$ through multiplication with the universal gas constant R.

15.6 MODEL-FITTING METHOD

Model-fitting methods are generally defined as the fitting of thermal analysis data to the predefined theoretical kinetic models, which are mathematical functions that reveal the association between the degree of conversion and reaction rate and can be correlated with the reaction mechanism. Model fitting method further branched into integral and deferential approach which is shown in figure 15.8.

TABLE 15.2
Different Solid-state Reaction Models f(x) and Their Integral Function g(x) (Aboulkas, El Harfi, and El Bouadili 2010; Jiang et al. 2010)

Notations	Reaction Model	f(x)	$g(x)=\int_0^x \frac{dx}{f(x)}$
Accelerating model: reaction order			
F_n	Order of reaction ($n \neq 1$)	$(1-x)^n$	$\frac{-(1-x)^{(1-n)}}{(1-n)} + \frac{1}{(1-n)}$
F_1	First-order reaction	$(1-x)$	$-\ln(1-x)$
F_2	Second-order reaction	$(1-x)^2$	$(1-x)^{-1} - 1$
F_3	Third-order reaction	$(1-x)^3$	$0.5(1-x)^{-2} - 0.5$
Dimensional diffusion models			
D_1	One-dimensional diffusion	$0.5x^{-1}$	x^2
D_2	Two-dimensional diffusion	$(-\ln(1-x))^{-1}$	$-(1-x)\ln(1-x) + x$
D_3	Three-dimensional diffusion	$\frac{3}{2}(1-x)^{\frac{2}{3}}\left(1-(1-x)^{\frac{1}{3}}\right)^{-1}$	$\left(1-(1-x)^{\frac{1}{3}}\right)^2$
Nucleation models			
P_n	Power law n	$nx^{(n-1)/n}$	$x^{\frac{1}{n}}$
P_2	Power law 2	$2x^{\frac{1}{2}}$	$x^{\frac{1}{2}}$
P_3	Power law 3	$3x^{\frac{2}{3}}$	$x^{\frac{1}{3}}$
P_4	Power law 4	$4x^{\frac{3}{4}}$	$x^{\frac{1}{4}}$
$P_{2/3}$	Power law 2/3	$\frac{2}{3}x^{\frac{-1}{2}}$	$x^{\frac{3}{2}}$
A_n	Avrami–Erofeev – nuclei growth	$n(1-x)\left[-\ln(1-x)\right]^{(n-1)/n}$	$\left[-\ln(1-x)\right]^{1/n}$
A_2	Avrami–Erofeev – two dimensional	$2(1-x)\left[-\ln(1-x)\right]^{1/2}$	$\left[-\ln(1-x)\right]^{1/2}$
A_3	Avrami–Erofeev – three dimensional	$3(1-x)\left[-\ln(1-x)\right]^{2/3}$	$\left[-\ln(1-x)\right]^{1/3}$
A_4	Avrami–Erofeev – four dimensional	$4(1-x)\left[-\ln(1-x)\right]^{3/4}$	$\left[-\ln(1-x)\right]^{1/4}$
Geometrical contraction models			
R_1	Contracting disk	x	$(1-x)^0$
R_3	Contracting volume	$3(1-x)^{\frac{2}{3}}$	$1-(1-x)^{\frac{1}{3}}$
R_2	Contracting area	$2(1-x)^{\frac{1}{2}}$	$1-(1-x)^{\frac{1}{2}}$
Random scission models			
L_2	Random scission	$2(x^{\frac{1}{2}} - x)$	$\left(\frac{4}{\sqrt{x}} - 2\right)$

Reaction models are used to describe the pyrolysis process reactions approximately. The acceleration model, deceleration model, and sigmoidal model are the common practice reaction models for waste pyrolysis kinetics study (Mishra et al. 2020)

In accelerating models, conversion rate continuously increases till the complete degradation. Power-law model is the best example of the acceleration model. The general expression for the acceleration model is shown in Equation 15.19.

Thermal Degradation Behaviors and Kinetics of Pyrolysis

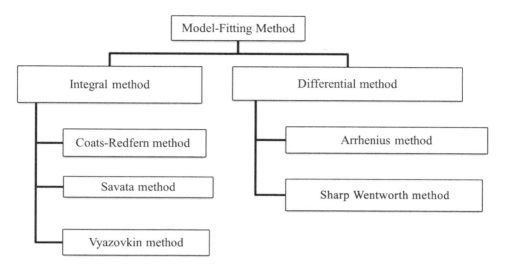

FIGURE 15.8 Classification of model-fitting method.

$$f(x) = nx^{(n-1)/n} \quad (15.19)$$

Here n is a constant.

In decelerating models, conversion rate is maximum at the beginning of the conversion and then it will decrease continuously for whole thermal degradation. The reaction-order model and diffusion model are the common deceleration model. The general expression for the deceleration model is shown in Equation 15.20.

$$f(x) = (1-x)^n \quad (15.20)$$

Here n is the order of the reaction.

In the sigmoidal model, degradation behavior is not a uniform trend like previous models. Reaction rate may reach maximum at the mid of conversion. The Avrami–Erofeev model is the example for sigmoidal model shown in Equation 15.21.

$$f(x) = n(1-x)\left[-\ln(1-x)\right]^{(n-1)/n} \quad (15.21)$$

Sestak and Berggren have introduced a common empirical expression for all the aforementioned three category models (Gibson et al. 2021).

$$f(x) = x^l (1-x)^m \left[-\ln(1-x)\right]^n \quad (15.22)$$

Equation 15.22 depends on l, m, n, and x. Using this single empirical relation can be derived many reaction models in a combination of two or three models. Equation 15.22 can be extended to Prout–Tompkins model by assuming the value of "l = 1 and m = 1." Hence $f(x) = x(1 − x)$.

15.6.1 COATS–REDFERN METHOD

The Coats–Redfern (C-R) method is one of the extensively used integral model-fitting methods to predict the kinetic triplets (activation energy, pre-exponential factor, and reaction mechanism) at a time. The mathematical expression for the C-R method is shown in Equation 15.23:

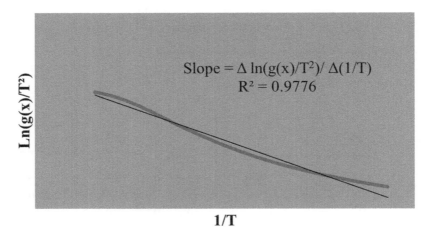

FIGURE 15.9 C-R method plot.

$$\ln\left(\frac{g(x)}{T^2}\right) = \ln\frac{AR}{^2E} - \frac{E}{RT} \quad (15.23)$$

For complex substances, thermal degradation behavior can be analyzed by the C-R integral method with single ramp rate TGA data. The most suitable model can be picked on a trial and error basis by the linear fitting of all solid-state reaction kinetic models and the linear regression coefficient R^2 of each solid-state reaction model. After identifying the best fit model, the regression value is successfully employed to evaluate the pre-exponential factor and activation energy of thermal degradation process. In the C-R method, Equation 15.23 is matched with straight line equation (Equation 15.8) for obtaining slope from the graph.

The graph is plotted between $1/T$ and $g(x)/T^2$ and the slope is equated with $(-E/R)$, which is graphically described in Figure 15.9. The pre-exponential factor can be obtained from the intercept of this graph. $g(\alpha)$ can be diverse according to the assumed models and reaction mechanisms. Solid-state degradation reactions fall into five groups: reaction-order models, dimensional diffusion models, nucleation models, random scission models, and geometrical contraction models, which are detailed in Table 15.2. Linear regression R^2 value should be in the range of 0.95–1 and could be considered the most appropriate mechanism model. Still, this is not a durable and fast rule. The R^2 value alone is not enough to ensure selected reaction mechanism of the pyrolysis process was best fitted or not.

15.7 ACTIVATION ENERGY OF PYROLYSIS PROCESS

The minimum energy required to form a product from reactants is known as activation energy. Activation energy includes the energy required to overcome internal molecular attraction and get close enough to react and obtain final products. Generally, high-temperature process activation energy is much lower than the low-temperature process, because high-temperature molecules are in an excited state. Catalyst addition is also one of the best ways to minimize the activation energy. Table 15.3 displays the earlier works in pyrolysis kinetics of carbonaceous waste.

15.8 IDENTIFICATION OF PYROLYSIS REACTION MECHANISM

The most common tool to identify a pyrolysis reaction model of carbonaceous is the C-R method involved in elucidating its macroscopic decomposition mechanism. Among various reaction models,

TABLE 15.3
Earlier Works in Pyrolysis Kinetics of Carbonaceous Waste

Sample	Method	Reaction Model Used	Activation Energy (kJ/mol)	Pre-exponential Factor (min⁻¹)	R^2	Reference
Wood	C-R method – 10 K/min	Second-order model	92.04	1.8×10^7	0.99	(Sobek and Werle 2020)
Banana leaves	C-R method – 30 K/min	Order of reaction = 1.62	50.01	8545.8	0.978	(Singh et al. 2020)
Polypropylene	Reaction model obtained by C-R and confirmed by Criado and FR	Contacting cylinder	187	NA	0.997	(Aboulkas, El harfi, and El Bouadili 2010)
LDPE	Reaction model obtained by C-R and confirmed by Criado and KAS	Contracting area	214	NA	0.998	(Aboulkas, El harfi, and El Bouadili 2010)
HDPE	Reaction model obtained by C-R and confirmed by Criado and FR	Contracting area	248	NA	0.998	(Aboulkas, El harfi, and El Bouadili 2010)
PS	Augis and Bennetis method (model-free method)	–	102–202	1.1×10^8 – 7.6×10^{14}	NA	(Nisar et al. 2019)
PET (soft drinks bottles)	KAS	–	322.3	NA	NA	(Saha and Ghoshal 2005)
PVC – Stage 1 ($x \leq 0.6$)	Activation energy calculated by FWO method and reaction model identified through C-R method	Contracting-area model (R_2)	141.54	NA	NA	(Liu et al. 2020)
PVC – Stage 2 ($x > 0.6$)	Activation energy calculated by FWO method and reaction model identified through C-R method	Fifth-order model (F_5)	235.37	NA	NA	(Liu et al. 2020)
Rubber waste	Iteration method	Modified power law model	97.77	1.37×10^{10}	N/A	(Chen et al. 2019)
Municipal solid waste (MSW) – average	Various studies	N/A	64.9	2.8×10^6	0.9486	(Sorum, Gronli, and Hustad 2001)

NA – not available.

we chose a reaction model that allowed the best fit for experimental reduced time plots. Solid-state reactions (SSRs) induced by heating occur in a broad range of processes such as the desolvation and dehydration of pharmaceutical crystals or thermal decomposition of solids, e.g., pyrolysis. To elucidate the mechanisms of chemical changes occurring when heating the solid reactants and to characterize the factors that control this process, various solid-state kinetic models have been developed. The great challenge, when these models are implemented, is to accurately determine the so-called kinetic triplet, i.e., the kinetic model, activation energy, and frequency factor, for a complete kinetic description of the overall reaction. The kinetic analysis requires two main steps: (i) the collection of experimental thermal analysis data; and (ii) the appropriate computational techniques to calculate the kinetic triplet.

Particularly, we are interested in using model-fitting method for analyzing single-step isothermal solid-state reactions, which usually take place during the drying process of pharmaceutical hydrates/solvates. The most crucial step when performing model-fitting is to identify the "best" candidate model for a particular solid-state reaction and then the complete kinetic triplet is calculated. This is done by fitting the integral reaction rate of each model explored with a straight line. The highest determination coefficient of these fits is the criterion that model-fitting relies on to finally choose the best candidate model. However, the implementation of model-fitting method on real experimental data often leads to several indistinguishable best candidate models that fit the kinetic data for a single solid-state reaction model equally well. Therefore, the major drawback of model-fitting is that a subjective selection becomes unavoidable and thus the procedure of choosing the best candidate model is not automatic and possibly prone to user errors.

REFERENCES

Aboulkas, A., K. El Harfi, and A. El Bouadili. 2010. "Thermal Degradation Behaviors of Polyethylene and Polypropylene. Part I: Pyrolysis Kinetics and Mechanisms." *Energy Conversion and Management* 51 (7): 1363–9. https://doi.org/10.1016/j.enconman.2009.12.017.

Chen, Ruiyu, Quanwei Li, Yang Zhang, Xiaokang Xu, and Dongdong Zhang. 2019. "Pyrolysis Kinetics and Mechanism of Typical Industrial Non-Tyre Rubber Wastes by Peak-Differentiating Analysis and Multi Kinetics Methods." *Fuel* 235 (January): 1224–37. https://doi.org/10.1016/J.FUEL.2018.08.121.

Das, Pallab, and Pankaj Tiwari. 2017. "Thermal Degradation Kinetics of Plastics and Model Selection." *Thermochimica Acta* 654 (May): 191–202. https://doi.org/10.1016/j.tca.2017.06.001.

Deng, Na, Yu Feng Zhang, and Yan Wang. 2008. "Thermogravimetric Analysis and Kinetic Study on Pyrolysis of Representative Medical Waste Composition." *Waste Management* 28 (9): 1572–80. https://doi.org/10.1016/J.WASMAN.2007.05.024.

Dubdub, Ibrahim, and Mohammed Al-Yaari. 2020. "Pyrolysis of Mixed Plastic Waste: I. Kinetic Study." *Materials* 13 (21): 1–15. https://doi.org/10.3390/ma13214912.

Gibson, Rebecca L., Mark J. H. Simmons, E. Hugh Stitt, John West, Sam K. Wilkinson, and Robert W. Gallen. 2021. "Kinetic Modelling of Thermal Processes Using a Modified Sestak-Berggren Equation." *Chemical Engineering Journal* 408 (March): 127318. https://doi.org/10.1016/J.CEJ.2020.127318.

Jiang, Haiyun, Jigang Wang, Shenqing Wu, Baosheng Wang, and Zhangzhong Wang. 2010. "Pyrolysis Kinetics of Phenol–Formaldehyde Resin by Non-Isothermal Thermogravimetry." *Carbon* 48 (2): 352–8. https://doi.org/10.1016/J.CARBON.2009.09.036.

Kumar, Sachin, Achyut K. Panda, and R. K. Singh. 2011. "A Review on Tertiary Recycling of High-Density Polyethylene to Fuel." *Resources, Conservation and Recycling* 55 (11): 893–910. https://doi.org/10.1016/j.resconrec.2011.05.005.

Kumari, Kanchan, Sunil Kumar, Vineel Rajagopal, Ankur Khare, and Rakesh Kumar. 2019. "Emission from Open Burning of Municipal Solid Waste in India." *Environmental Technology (United Kingdom)* 40 (17): 2201–14. https://doi.org/10.1080/09593330.2017.1351489.

Larraín, T., M. Carrier, and L. R. Radovic. 2017. "Structure-Reactivity Relationship in Pyrolysis of Plastics: A Comparison with Natural Polymers." *Journal of Analytical and Applied Pyrolysis* 126 (January): 346–56. https://doi.org/10.1016/j.jaap.2017.05.011.

Liu, Haoran, Changjian Wang, Jiaqing Zhang, Weiping Zhao, and Minghao Fan. 2020. "Pyrolysis Kinetics and Thermodynamics of Typical Plastic Waste." *Energy and Fuels* 34 (2): 2385–90. https://doi.org/10.1021/acs.energyfuels.9b04152.

Mannocci, Alice, Ornella di Bella, Domenico Barbato, Fulvio Castellani, Giuseppe La Torre, Maria De Giusti, and Angela Del Cimmuto. 2020. "Assessing Knowledge, Attitude, and Practice of Healthcare Personnel Regarding Biomedical Waste Management: A Systematic Review of Available Tools." *Waste Management and Research* 38 (7): 717–25. https://doi.org/10.1177/0734242X20922590.

Mishra, Asmita, Usha Kumari, Venkata Yasaswy Turlapati, Hammad Siddiqi, and B. C. Meikap. 2020. "Extensive Thermogravimetric and Thermo-Kinetic Study of Waste Motor Oil Based on Iso-Conversional Methods." *Energy Conversion and Management* 221 (October): 113194. https://doi.org/10.1016/J.ENCONMAN.2020.113194.

Nisar, Jan, Ghulam Ali, Afzal Shah, Munawar Iqbal, Rafaqat Ali Khan, Sirajuddin, Farooq Anwar, Raqeeb Ullah, and Mohammad Salim Akhter. 2019. "Fuel Production from Waste Polystyrene via Pyrolysis: Kinetics and Products Distribution." *Waste Management* 88: 236–47. https://doi.org/10.1016/j.wasman.2019.03.035.

Saha, B., and A. K. Ghoshal. 2005. "Thermal Degradation Kinetics of Poly(Ethylene Terephthalate) from Waste Soft Drinks Bottles." *Chemical Engineering Journal* 111 (1): 39–43. https://doi.org/10.1016/J.CEJ.2005.04.018.

Singh, Rajnish Kumar, Deeksha Pandey, Trilok Patil, and Ashish N. Sawarkar. 2020. "Pyrolysis of Banana Leaves Biomass: Physico-Chemical Characterization, Thermal Decomposition Behavior, Kinetic and Thermodynamic Analyses." *Bioresource Technology* 310 (August): 123464. https://doi.org/10.1016/J.BIORTECH.2020.123464.

Sobek, Szymon, and Sebastian Werle. 2020. "Kinetic Modelling of Waste Wood Devolatilization During Pyrolysis Based on Thermogravimetric Data and Solar Pyrolysis Reactor Performance." *Fuel* 261 (February): 116459. https://doi.org/10.1016/J.FUEL.2019.116459.

Sorum, L., M. G. Gronli, and J. E. Hustad. 2001. "Pyrolysis Characteristics and Kinetics of Municipal Solid Wastes." *Fuel* 80 (9): 1217–27. https://doi.org/10.1016/S0016-2361(00)00218-0.

Straka, Pavel, Olga Bičáková, and Monika Šupová. 2017. "Thermal Conversion of Polyolefins/Polystyrene Ternary Mixtures: Kinetics and Pyrolysis on a Laboratory and Commercial Scales." *Journal of Analytical and Applied Pyrolysis* 128: 196–207. https://doi.org/10.1016/j.jaap.2017.10.010.

Williams, Elizabeth A., and Paul T. Williams. 1997. "Analysis of Products Derived from the Fast Pyrolysis of Plastic Waste." *Journal of Analytical and Applied Pyrolysis* 40–1: 347–63. https://doi.org/10.1016/S0165-2370(97)00048-X.

Wu, R. M., D. J. Lee, C. Y. Chang, and J. L. Shie. 2006. "Fitting TGA Data of Oil Sludge Pyrolysis and Oxidation by Applying a Model Free Approximation of the Arrhenius Parameters." *Journal of Analytical and Applied Pyrolysis* 76 (1–2): 132–7. https://doi.org/10.1016/J.JAAP.2005.10.001.

16 Techniques for Biodiesel Production from Wastes

Gopi R. and Anand Ramanathan

CONTENTS

16.1 Introduction 221
16.2 Feedstocks and Pretreatment for Biodiesel 222
16.3 Processes for Biodiesel Production..222
16.4 Catalysts Used for Biodiesel Production 223
    16.4.1 Homogeneous Catalyst...223
    16.4.2 Heterogeneous Catalyst..224
        16.4.2.1 Waste Shells 224
        16.4.2.2 Bone Waste 225
        16.4.2.3 Agro-waste 225
        16.4.2.4 Other Biowastes 226
16.5 Microwave-assisted Biodiesel Production 226
16.6 Ultrasonic-assisted Biodiesel Production 226
16.7 Membrane Technology for Biodiesel Production 227
16.8 Posttreatments in Biodiesel Production 227
16.9 Future Perspectives 228
16.10 Conclusion 228
References 228

16.1 INTRODUCTION

Concerns about the depletion of fossil fuels and the harmful impacts of their use on people and the environment have risen dramatically in recent decades. Carbon monoxide, nitric oxide, and carbon dioxide are released as a result of the continuous usage of fossil fuels. These gases are the only cause of significant climate change and pollution, both of which have detrimental consequences for humans and the environment (Yoro and Daramola 2020). According to a research conducted in 2015 by the International Council on Clean Transportation (ICCT) in partnership with two other institutions, vehicle exhaust is a major contributor of outdoor air pollution, resulting in almost 3.8 lakh deaths worldwide in 2015, including 74,000 deaths in India (Gaur and Goyal 2022).

    When the fuel is burned, the nature composition of hydrocarbon is the primary source of CO emissions. Fossil fuels are created over millions of years by geological processes on buried animals and plants. Overconsumption of fossil fuels is unavoidable in order to fulfill the rising fuel demand for vehicles, power plants, industries, and even household appliances (Verma et al. 2021). As a result of the strong demand, gasoline prices have risen significantly, necessitating the development of both economical and environmentally beneficial fuel. When it becomes necessary to avoid causing harm to the environment, the primary focus shifts to natural resources (Antony Casmir Jayaseelan et al. 2021). The extensive biofuel research has resulted in the development of non-harmful, naturally dependable, sustainable, and cost-effective sources such as various edible sources from corn, sugarcane, sunflower, rapeseed, and other crops. Similarly, there are other crops that are edible and non-edible, which has the potential to convert it into biodiesel. But usage of wastes than crops can limit the agricultural costs and it can improve the waste-to-energy process (Brindhadevi et al. 2021).

16.2 FEEDSTOCKS AND PRETREATMENT FOR BIODIESEL

The bio-oil is extracted from crops with high sugar and starch content (Muthuraman, Murugappan, and Soundharajan 2021). Microalgae are utilized to make biofuel alongside agricultural crops. Natural biofuel sources abound, and each has been thoroughly investigated. As the world population continues to rise, it will create imbalance and shortage of the biofuels produced from the edible sources in near future due to the necessity for food (Voloshin et al. 2016). Furthermore, it necessitates the use of another natural resource in order to be grown. It was found that production of biodiesel using the edible oil is greater than the cost of diesel four times and it is commercially unviable. To address the disadvantages of biofuels produced from edible sources, wastes from the industry and agriculture can be used for biofuel production and it is proved to be beneficial (Odude et al. 2019). Non-edible sources have attracted a lot of attention, as they would not only facilitate the management of waste gathered from many industries, but also lessen the pressure on bulk crop production. The valuable land area will be spared and landfills would be avoided. The waste from various sources can be used and some of the examples are waste tires–generated oil, waste chicken fat oil, waste cooking oil, rubber seed oils, mahua, Karanja, polanga, *Jatropha*, and neem. These bio-oils have been employed with minor changes, such as the addition of diesel in various quantities. Continuous research to improve the quality and effectiveness of the system has been ongoing (Mihajlovski et al. 2021).

Considering the various non-edible sources, waste cooking oil, waste animal fat, and waste tires are found to be superior because of their mass availability and it is available all over the season. It was reported that billions of new tires are produced each year with more than half of the old tires being discarded. Tire manufacturing involves a number of complicated and powerful chemical processes, rendering them non-biodegradable in the environment (Sadeghinezhad et al. 2014). Furthermore, fryer oils have been squandered at large hotels and even in households after they have been consumed. Chicken fat is another prevalent waste item after the separation of consumable chicken parts (Haghighi et al. 2021). The chicken fat which cannot be consumed is discarded as waste and it is left out. To increase the appropriateness of the oil extracted from these wastes, chemical processes such as fermentation and transesterification are used (Su et al. 2020). It was found that cetane number and viscosity of the biofuels produced from wastes are low. The fuel produced from wastes is more advantageous because of its readiness to the availability and it is source of renewable energy, which can reduce the fossil fuels dependence and it can be low cost, it can reduce the area of landfills by generating income to the concerned persons with less resource cost, and it has effect of producing carbon neutral (Enweremadu and Mbarawa 2009; Gnanaprakasam et al. 2013).

16.3 PROCESSES FOR BIODIESEL PRODUCTION

The best way for making biodiesel is to transesterify vegetable oils with alcohol. There are two types of transesterifications: (a) with a catalyst, and (b) without a catalyst. The rate and production of biodiesel are improved by using several types of catalysts. Excess alcohol moves the equilibrium to the product side of the transesterification process, which is reversible and is shown in Figure 16.1 (Talebian-Kiakalaieh, Amin, and Mazaheri 2013).

Methanol, ethanol, propanol, and butanol are just a few of the alcohols that may be utilized in this process. Because of its low cost and physical and chemical benefits, such as being polar and having the shortest alcohol chain, methanol is a more viable option: 1 mole of triglyceride reacts with 3 moles of methanol to generate 3 moles of methyl ester (biodiesel) and 1 mole of glycerol in this process (Ling et al. 2019).

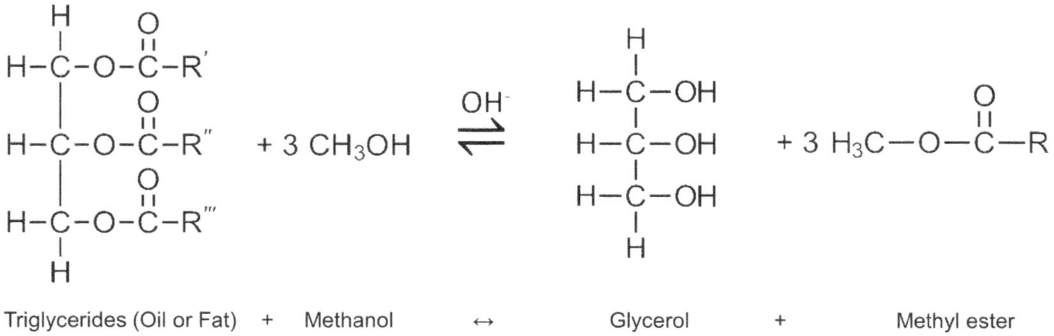

FIGURE 16.1 Transesterification reaction for the biodiesel production (Degfie, Mamo, and Mekonnen 2019).

In general, the transesterification reaction has a number of important factors that have a big impact on the ultimate conversion and yield. Reaction temperature, free fatty acid concentration in the oil, water content in the oil, type of catalyst, amount of catalyst, reaction duration, molar ratio of alcohol to oil, type or chemical stream of alcohol, usage of cosolvent, and mixing intensity are the most essential factors.

16.4 CATALYSTS USED FOR BIODIESEL PRODUCTION

Catalysts are essential for speeding up the otherwise slow transesterification reaction and creating a large amount of biodiesel. Chemical and biological catalysts are the most common types of catalysts utilized in this process. Acidic and basic catalysts are chemical catalysts, whereas enzymes are biological catalysts. Because of their great selectivity, flexibility to employ any type of feedstock, and gentle reaction conditions, enzyme catalysts are regarded the most effective catalysts in transesterification (Parida, Singh, and Pradhan 2022). However, the expensive cost of the catalyst, the extremely slow reaction rate, and the deactivation of enzymes are key flaws in the enzymes that overshadow their benefits and restrict their application (Mandari and Kumar 2021).

16.4.1 Homogeneous Catalyst

Homogeneous catalysis is the process of catalyzing a series of reactions with a chemical that is in the same phase as the reaction system. Due to advantages like less reaction time for complete reaction and easy in handling, most often homogeneous catalyst is used in the synthesis of biodiesel. This category includes both acid and base catalysts. Homogeneous catalysts are usually dissolved in a solvent that is the same phase as all of the reactants.

Alkali metal-based hydroxides, such as potassium or sodium hydroxide; alkali metal-based oxides, such as potassium and sodium methoxides; and carbonates are examples of homogeneous base catalysts. Base catalysts have a strong transesterification activity. Metallic hydroxides are commonly employed as catalysts because they are less expensive; however, they have lesser activity than alkoxides. It has been observed that a process catalyzed by a base is 4000 times quicker than one catalyzed by an acid. If the oil has higher limits of FFA, then it has certain disadvantages in the conversion of biodiesel and it can lead to formation of soap and settle in large volumes (Helwani et al. 2009). The process can still be catalyzed using an alkali catalyst if the FFAs are less than 5%, but more catalyst is needed to balance for the catalyst drop to soap. For biodiesel synthesis employing a homogeneous catalyst, most studies indicate an FFA level of less than 2 wt% (Rizwanul Fattah et al. 2020).

16.4.2 Heterogeneous Catalyst

Heterogeneous catalysts, which are usually solids, exist in a separate phase from the reaction mixture. Several studies have been proposed to investigate the behaviors of a range of heterogeneous materials in order to alleviate the various problems connected with the use of homogeneous bases and liquid acids. Heterogeneous catalysts are non-corrosive, reusable, and less sensitive to FFA content. In this context, acidic and basic kinds, as well as homogeneous catalysts, are present. Apart from the use of heterogeneous catalysts that aid in both the esterification and transesterification processes, undesirable saponification reactions can be avoided (Talha and Sulaiman 2016).

Synthesis of heterogeneous catalyst from wastes of biomass have received a lot of interest and it adds value to trash while also ensuring long-term catalyst production. In transesterification, the function of catalysts produced from various wastes like animal wastes bones, shells of various eggs, biomass ashes, and catalyst synthesized with carbon support or base has been discussed and it is illustrated in Figure 16.2. Also, Table 16.1 presents the various heterogeneous catalysts prepared using wastes for biodiesel production.

16.4.2.1 Waste Shells

The most common base heterogeneous catalyst which is often preferred for the production of biodiesel and which has low free fatty acids is calcium oxide (CaO), due to its easy solubility, strong basicity, and low cost. CaO is often made from limestone, which is a non-renewable resource that is expensive to convert to CaO. The dependency of the limestone for the production of CaO can be decreased by utilizing the waste shells which is rich in calcium. A considerable number of discarded shells is found all over the world as a result of excessive consumption of chicken eggs and mollusk flesh. The shells majorly constitute calcium carbonate ($CaCO_3$) for 95%, which may be calcined at high temperature to produce CaO. The high calcination temperature encourages $CaCO_3$ breakdown, which starts at the surface and moves inward. As a result, the surface porosity rises, resulting in an active CaO catalyst (Jayakumar et al. 2021). To improve the catalytic activity of the heterogeneous catalyst, several research has been undertaken: hydration–dehydration methods, doping of alkali metal, conversion of nanocatalyst particles, and chemical treatments (Borah et al. 2019; Rahman et al. 2019).

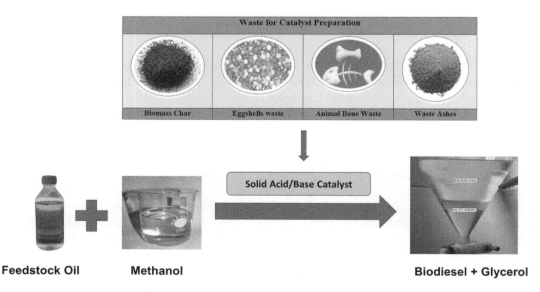

FIGURE 16.2 Biodiesel production using various types of solid heterogeneous catalysts synthesized from different sources of wastes.

TABLE 16.1
Heterogeneous Catalyst Prepared Using Waste Sources for Biodiesel Production

Reference	Feedstock	Waste Material Source	Calcination Conditions for Catalyst Time (h)	Calcination Conditions for Catalyst Temperature (°C)	Biodiesel Reaction Conditions Reaction Time (h)	Biodiesel Reaction Conditions Reaction Temperature (°C)	Biodiesel Reaction Conditions Catalyst Concentration (wt%)	Biodiesel Reaction Conditions Alcohol-to-molar Ratio	Reusability	Yield (%)
(Ur Rahman et al. 2021)	Waste frying oil	Chicken eggshells	4	900	2.5	65	4	15:1	–	93.27
(Sahu 2021)	Waste cooking oil	Rice straw waste		500	2.5	65	3.5	15:1	8	97.3
(Lin et al. 2020)	Waste cooking oil	Recycled waste oyster shell	2	1000	3	65	6	9:1	–	87.3
(Mahmood Khan et al. 2020)	Waste cooking oil	Ostrich (*Struthio camelus*) bone	4	900	4	60	5	15:1	4	90.56
(Kirubakaran 2018)	Chicken fat	Chicken eggshell	4	900	5	57.5	8.5	13:1	5	90.41
(Gohain, Devi, and Deka 2017)	Waste cooking oil	*Musa balbisiana*	4	700	3	60	2	6:1	5	100

16.4.2.2 Bone Waste

The bone waste has calcium which enables them to be used as heterogeneous catalyst. Ghanei et al. used waste bones as catalyst by impregnating the calcined bones for 8 h at 600 °C with CaO for the biodiesel generation from canola oil. It was reported that using impregnated catalyst of 5 wt% with 12:1 methanol-to-molar ratio at 60 °C for 300 min produced a maximum yield of 95.18%, due to its increased catalytic activity which in turn further improved the reaction rate (Ghanei et al. 2016). The calcined waste chicken bones at 900 °C for 4 h are used to make cost-effective solid base catalyst. According to the experiments, the transesterification of soybean oil with catalyst of 5 wt% produced fatty acid methyl ester of 89.33%. The optimized experimental conditions are 65 °C for 240 min with 15:1 methanol molar ratio. To boost process efficiency, the manufactured catalyst might be utilized four times. Due to its catalytic reusability, it has the potential to be replaced for homogeneous catalyst with improved effectiveness (Farooq and Ramli 2015).

16.4.2.3 Agro-waste

Agro-waste may be used to produce cleaner products and many research have been conducted to investigate the potential value of agro-waste into heterogeneous catalyst. The heterogeneous base catalyst produced with waste eggshell with supporting material as rice husk ash (RHA) has been experimented. The biodiesel was produced with palm oil and 30% of RHA$_{800}$ catalyst under the following conditions of catalyst: 7 wt%, methanol-to-molar ratio of 9:1 for 240 min under 65 °C which yielded 91.5%. The catalyst basicity and surface area specificity were found to be 8.5 mmol/g and 11.75 m²/g, respectively. The reusability of catalyst was found to be high and it produced 80% yield even after eight trials of transesterification, and it was reported that using waste ash as catalyst can be promoted due to its increased activity (Chen et al. 2015). Buasri et al. used synthesized low-cost heterogeneous catalyst with discarded waste eggshells from ducks and hens for the synthesis of biodiesel from palm oil. The calcined eggshell waste catalyst (under 900 °C) has high activity, and it has greater yield for the conditions of methanol-to-molar ratio of 9:1 at 65 °C for 240 min with catalyst load of 20 wt% (Buasri et al. 2013).

16.4.2.4 Other Biowastes

Xie et al. used turtle shell to synthesize heterogeneous catalyst with higher performance rate and the catalyst was synthesized using an incomplete carbonization-KF impregnation–activation approach. Using bioimpregnation method with activated KF of 25% at 300 °C, the specific area of the shell was improved from 15.43 m²/g to 63.43 m²/g, thermal conversion method was done after this approach. In the preparation of catalyst, methanol-to-molar ratio of 9:1 was used to heat shells at 70 °C. The catalyst was used for transesterification of rapeseed oil with 3 wt% and it has yielded 97.5% for 180 min. It was reported that the synthesized catalysts have grater catalytic life, high basicity, slender pore size distribution, larger particle size distribution, and it comes under ecological limit with greater permanence in organic solvents (Xie et al. 2009).

16.5 MICROWAVE-ASSISTED BIODIESEL PRODUCTION

Microwave irradiation is electromagnetic irradiation with wavelengths of 0.01–1 m and frequencies 0.3–300 GHz. The wavelengths between 0.01 and 0.25 m are often used for radar broadcasts, whereas the remaining wavelengths are used for telecommunications. The frequency of 2.45 GHz, which corresponds to a wavelength of 12.25 cm, is used in all microwave reactors for chemical synthesis and all home microwave ovens. This is to prevent interfering with cellular phone and telecommunications frequencies (Motasemi and Ani 2012). Because it sends energy directly to the reactant, microwave irradiation is a well-established means of speeding up and improving chemical processes. As a result, heat transmission is more efficient than traditional heating, and the reaction may be finished in less time. As a result, microwave irradiation is one of the most effective ways for shortening reaction times and increasing biodiesel yields. When compared to traditional heating, it speeds up the reaction and simplifies the separation process. Microwave-assisted biodiesel manufacturing has received a thorough examination in this area.

Microwave-assisted esterification (MAE) uses vegetable oil refinery waste, including acid oil as an affordable feedstock for manufacturing biodiesel. The optimum reaction conditions for MAE are a 1:10 molar ratio of free fatty acid to methanol, a reaction period of 60 min, and a catalyst concentration of 3%, with a conversion yield of 95.79%. Density and color brightness rise as the conversion yield of biodiesel rises, whereas viscosity and refractive index fall. The physicochemical and thermal characteristics of biodiesel generated by MAE and magnetic stirrer esterification (MSE) techniques are not significantly different. Meanwhile, the MAE approach uses nearly four times less energy than the MSE method (Alishahi, Golmakani, and Niakousari 2021).

Thushari produced solid catalyst from wastes of palm empty fruit bunch (PEFB), coconut meal residue (CMR), and coconut coir husk (CH) waste. Transesterification of the waste palm oil with synthesized catalyst was studied. When employing microwave heating for biodiesel synthesis from waste palm oil (WPO) and manufactured catalysts, results demonstrate that biodiesel output rises with increasing reaction duration and methanol loading. Even though the catalysts' activity diminished after reuse, they are still of interest since waste biomass from PEFB, CMR, and CH may be utilized for catalyst preparation and microwave-assisted WPO biodiesel manufacture (Thushari and Babel 2018).

16.6 ULTRASONIC-ASSISTED BIODIESEL PRODUCTION

The esterification of free fatty acids to free fatty acid methyl ester and the transesterification of triglycerides to free fatty acid methyl ester are the two main processes in biodiesel manufacturing. In ultrasonic chemical reaction, the catalyst and reactants are homogenized to improve the catalyst surface activation providing more emulsification of the reactants improving the reaction rate. The most mixing method used for homogeneous transesterification in the production of biodiesel is the ultrasonic assistance which improves the reaction rate and mass transfer without affecting thermodynamic equilibrium of the reaction, thus allowing the chemical reaction in shortest time of the same reaction (Bashir et al. 2022).

Fravardin et al. studied the continuous production of biodiesel using the waste cooking oil by optimizing the parameters controlling the yield with ultrasonic power. According to the findings, increasing the residence duration increases the methyl esters production; however, increasing the time up to 90 s reduces the methyl esters yield. When the catalyst concentration is increased by 1%, the methyl esters yield increases by 7%, and when the catalyst concentration is increased by 1.25%, the methyl esters yield reduces by 7%. When the methanol-to-oil ratio is increased to 6:1, the yield increases, and when the methanol-to-oil ratio is increased to 8:1, the methyl esters yield follows a rising trend with a minor slope. The results concluded that using ultrasonic power, the energy consumption for the production of biodiesel can be reduced with optimum conditions (Farvardin et al. 2019).

16.7 MEMBRANE TECHNOLOGY FOR BIODIESEL PRODUCTION

Membrane reactors provide a unique strategy for concurrently transesterifying oils and purifying the glycerol layer without the need of water or other refining processes. These reactors can use novel catalysts, which allows for more efficient transesterification processes and shorter reaction times, as well as greater yields, improved selectivity, and less waste by-products (Xu, Gao, and Xiao 2015). They also prevent unreacted oil from polluting the finished biodiesel, resulting in high-quality biodiesel. The ability of membrane reactors to decrease glycerol from crude biodiesel at an early stage has been investigated utilizing various membrane materials and structures. Dube et al. were the first to effectively employ a membrane reactor in the manufacture and purification of biodiesel (Dubé, Tremblay, and Liu 2007).

Moyo et al. experimented the waste cooking oil as feedstock which was pretreated with sulfated zirconia and the experiments were optimized with various parameters, including circulation rate of membrane reactors. The greater the catalyst-to-WCO ratio, the higher the free fatty acids (FFA) concentration, according to the data. At a temperature of 61 °C and a circulation flow rate of 26 mL/min, a maximum biodiesel production of 92.6% was produced across a TiO_2/Al_2O_3 membrane employing a KOH catalyst concentration of 1.3 wt%. At 58.5 °C, circulation flow rate of 18.78 mL/min, and catalyst concentration of 1.24 wt%, a biodiesel yield of 94.03 mol% was obtained after membrane optimization (Moyo et al. 2021).

16.8 POSTTREATMENTS IN BIODIESEL PRODUCTION

Various requirements are followed to ensure the quality of biodiesel. The chemical composition and qualities of biodiesel as an alternative fuel are described in quality standards such as ASTM D6751 and EN14214 (Gopi et al. 2022). Biodiesel should be devoid of contaminants that might harm engine components by forming incrustations and cause corrosion when deposited in nozzles. After transesterification process, the end products have impurities due to additional alcohol, glycerol, catalyst residues, unsaponifiable elements, feedstock content, and water (Veljković, Banković-Ilić, and Stamenković 2015). To ensure biodiesel purity, such contaminants are eliminated. Methanol reduces density, viscosity, and flash point, as well as corroding zinc and aluminum components. Presence of glycerol can form gum and it can get settled at the bottom of the fuel tank which can affect the injector by clogging and reducing the durability of the engine. Water in biodiesel may create a slew of issues, including methyl ester hydrolysis, and microbiological development that leads to filter obstruction, fuel tube corrosion, and a loss in combustion heat. Dry washing, wet washing, distillation, and membrane extraction are some of the posttreatment processes for removing contaminants from biodiesel. In recent years, wet and dry washing procedures have been the most widely employed methods in the business (Fonseca et al. 2019).

To remove contaminants from biodiesel, wet washing is a frequent approach. Wet washing has a proven track record of efficiency. If the feedstock has high FFA and water content, it can lead to problems such as polluted discharges, emulsion products, and low yield of end product (Bateni, Saraeian, and Able 2017). In wet washing, a particular amount of water is continuously agitated to avoid the emulsion formation and it will be repeated till all the containments left and it is noticed by colorless water (Farid et al. 2020). The fundamental goal of dry washing techniques is to create a waterless purifying procedure that is both environmentally beneficial and water efficient. Waterless agents such as absorbents and acid resin are used to remove impurities from crude biodiesel. The crude biodiesel is taken in stirred vessel and the agents are usually utilized as suspended materials or as a fixed column bed through which the biodiesel flows. When utilized in a suspended state, the agents are recovered following filtering treatment. Ion exchange, filtration, adsorption, filtration, and soap/glycerol interaction are used to remove contaminants from crude biodiesel during dry washing (Chozhavendhan et al. 2020). One or more mechanisms will be more efficient depending on the process circumstances and agent properties. Because waterless washing agents are capable of removing entire methanol and glycerol, it must be eliminated with the use of gravitational separation. To boost the cleanliness of biodiesel, it is washed with distilled water in a stirred tank. The ester is then decanted and outgassed from the aqueous phase (Alba-Rubio et al. 2012).

16.9 FUTURE PERSPECTIVES

Biodiesel production is still in laboratory level with some reactors. The continuous flow production is mostly preferred than the batch-type reactors because of its advantages. There are many continuous flow reactors such as oscillatory flow reactor, microchannel reactor, reactive distillation, slit channel reactor, centrifugal contactor separator, and laminar flow reactor and separator. Among them, microreactor has more advantages and it is considered due to its advantages such as high heat transfer, diffusion rate, and larger surface-to-volume ratio. But still it needs some transformation into commercial scale. The assisted techniques with microreactor can improve the production rate and quality of the biodiesel. The stacked microreactor is still in modeling level, which has the scope for commercialization. The use of waste to produce commercial heterogeneous catalyst has improved effective resource utilization. Waste cooking oil, animal waste, and biomass waste have great potential to be used as feedstock for the biodiesel production.

16.10 CONCLUSION

The biodiesel production has been growing rapidly in recent days and it is found to be one of the predominant methods to reduce the renewable sources. Still the search of feedstocks is ongoing and waste to energy also is considered for the biodiesel production. The conversion of wastes into feedstocks can reduce the dependence of the harvesting product. The waste biomass which can be transformed into oils has major play in the production. The production of catalysts necessary for biodiesel production has been discussed. The heterogeneous catalyst can be synthesized from the waste, which can be recycled and reused for multiple times. The production of biodiesel majorly depends upon the reactors where it gets transesterified, the continuous reactors with assisted ultrasonic and microwave can reduce the energy consumption of the process. The biodiesel produced needs to be treated for their quality enhancement, so the reactors equipped with continuous post-treatment will have more advantages. Still research is going on in the improvement of the reactors and it can lead to commercialization of biodiesel.

REFERENCES

Alba-Rubio, A. C., M. L. Alonso Castillo, M. C. G. Albuquerque, R. Mariscal, C. L. Cavalcante, and M. López Granados. 2012. "A New and Efficient Procedure for Removing Calcium Soaps in Biodiesel Obtained Using CaO as a Heterogeneous Catalyst." *Fuel*. https://doi.org/10.1016/j.fuel.2011.12.024.

Alishahi, Afsaneh, Mohammad-Taghi Golmakani, and Mehrdad Niakousari. 2021. "Feasibility Study of Microwave-Assisted Biodiesel Production from Vegetable Oil Refinery Waste." *European Journal of Lipid Science and Technology* 123 (9): 2000377.

Antony Casmir Jayaseelan, G., A. Anderson, Sekar Manigandan, Ashraf Elfasakhany, and Veeman Dhinakaran. 2021. "Effect of Engine Parameters, Combustion and Emission Characteristics of Diesel Engine with Dual Fuel Operation." *Fuel* 302 (June): 121152. https://doi.org/10.1016/j.fuel.2021.121152.

Bashir, Muhammad Aamir, Sarah Wu, Jun Zhu, Anilkumar Krosuri, Muhammad Usman Khan, and Robinson Junior Ndeddy Aka. 2022. "Recent Development of Advanced Processing Technologies for Biodiesel Production: A Critical Review." *Fuel Processing Technology* 227 (December 2021): 107120. https://doi.org/10.1016/j.fuproc.2021.107120.

Bateni, Hamed, Alireza Saraeian, and Chad Able. 2017. "A Comprehensive Review on Biodiesel Purification and Upgrading." *Biofuel Research Journal*. https://doi.org/10.18331/BRJ2017.4.3.5.

Borah, Manash Jyoti, Ankur Das, Velentina Das, Nilutpal Bhuyan, and Dhanapati Deka. 2019. "Transesterification of Waste Cooking Oil for Biodiesel Production Catalyzed by Zn Substituted Waste Egg Shell Derived CaO Nanocatalyst." *Fuel* 242 (January): 345–54. https://doi.org/10.1016/j.fuel.2019.01.060.

Brindhadevi, Kathirvel, Rajasree Shanmuganathan, Arivalagan Pugazhendhi, P. Gunasekar, and S. Manigandan. 2021. "Biohydrogen Production Using Horizontal and Vertical Continuous Stirred Tank Reactor- a Numerical Optimization." *International Journal of Hydrogen Energy* 46 (20): 11305–12. https://doi.org/10.1016/j.ijhydene.2020.06.155.

Buasri, Achanai, Nattawut Chaiyut, Vorrada Loryuenyong, Chaiwat Wongweang, and Saranpong Khamsrisuk. 2013. "Application of Eggshell Wastes as a Heterogeneous Catalyst for Biodiesel Production." *Sustainable Energy* 1 (2): 7–13. https://doi.org/10.12691/rse-1-2-1.

Chen, Guan Yi, Rui Shan, Jia Fu Shi, and Bei Bei Yan. 2015. "Transesterification of Palm Oil to Biodiesel Using Rice Husk Ash-Based Catalysts." *Fuel Processing Technology* 133: 8–13. https://doi.org/10.1016/j.fuproc.2015.01.005.

Chozhavendhan, S., M. Vijay Pradhap Singh, B. Fransila, R. Praveen Kumar, and G. Karthiga Devi. 2020. "A Review on Influencing Parameters of Biodiesel Production and Purification Processes." *Current Research in Green and Sustainable Chemistry*. https://doi.org/10.1016/j.crgsc.2020.04.002.

Degfie, Tadesse Anbessie, Tadios Tesfaye Mamo, and Yedilfana Setarge Mekonnen. 2019. "Optimized Biodiesel Production from Waste Cooking Oil (WCO) Using Calcium Oxide (CaO) Nano-Catalyst." *Scientific Reports* 9 (1): 1–8. https://doi.org/10.1038/s41598-019-55403-4.

Dubé, M. A., A. Y. Tremblay, and J. Liu. 2007. "Biodiesel Production Using a Membrane Reactor." *Bioresource Technology* 98 (3): 639–47. https://doi.org/10.1016/j.biortech.2006.02.019.

Enweremadu, C. C., and M. M. Mbarawa. 2009. "Technical Aspects of Production and Analysis of Biodiesel from Used Cooking Oil-A Review." *Renewable and Sustainable Energy Reviews* 13 (9): 2205–24. https://doi.org/10.1016/j.rser.2009.06.007.

Farid, Mohammed Abdillah Ahmad, Ahmad Muhaimin Roslan, Mohd Ali Hassan, Muhamad Yusuf Hasan, Mohd Ridzuan Othman, and Yoshihito Shirai. 2020. "Net Energy and Techno-Economic Assessment of Biodiesel Production from Waste Cooking Oil Using a Semi-Industrial Plant: A Malaysia Perspective." *Sustainable Energy Technologies and Assessments*. https://doi.org/10.1016/j.seta.2020.100700.

Farooq, Muhammad, and Anita Ramli. 2015. "Biodiesel Production from Low FFA Waste Cooking Oil Using Heterogeneous Catalyst Derived from Chicken Bones." *Renewable Energy* 76: 362–8. https://doi.org/10.1016/j.renene.2014.11.042.

Farvardin, Mahrokh, Bahram Hosseinzadeh Samani, Sajad Rostami, Ahmad Abbaszadeh-Mayvan, Gholamhassan Najafi, and Ebrahim Fayyazi. 2019. "Enhancement of Biodiesel Production from Waste Cooking Oil: Ultrasonic-Hydrodynamic Combined Cavitation System." *Energy Sources, Part A: Recovery, Utilization and Environmental Effects* 1–15. https://doi.org/10.1080/15567036.2019.1657524.

Fonseca, Jhessica Marchini, Joel Gustavo Teleken, Vitor de Cinque Almeida, and Camila da Silva. 2019. "Biodiesel from Waste Frying Oils: Methods of Production and Purification." *Energy Conversion and Management*. https://doi.org/10.1016/j.enconman.2019.01.061.

Gaur, Roopesh Kanwar, and Rahul Goyal. 2022. "Materials Today : Proceedings A Review : Effect on Performance and Emission Characteristics of Waste Cooking Oil Biodiesel-Diesel Blends on IC Engine." *Materials Today: Proceedings*. https://doi.org/10.1016/j.matpr.2022.04.447.

Ghanei, R., R. Khalili Dermani, Y. Salehi, and M. Mohammadi. 2016. "Waste Animal Bone as Support for CaO Impregnation in Catalytic Biodiesel Production from Vegetable Oil." *Waste and Biomass Valorization* 7 (3): 527–32. https://doi.org/10.1007/s12649-015-9473-1.

Gnanaprakasam, A., V. M. Sivakumar, A. Surendhar, M. Thirumarimurugan, and T. Kannadasan. 2013. "Recent Strategy of Biodiesel Production from Waste Cooking Oil and Process Influencing Parameters: A Review." *Journal of Energy* 1–10. https://doi.org/10.1155/2013/926392.

Gohain, Minakshi, Anuchaya Devi, and Dhanapati Deka. 2017. "Musa Balbisiana Colla Peel as Highly Effective Renewable Heterogeneous Base Catalyst for Biodiesel Production." *Industrial Crops and Products* 109: 8–18.

Gopi, R., Vinoth Thangarasu, Angkayarkan Vinayakaselvi M, and Anand Ramanathan. 2022. "A Critical Review of Recent Advancements in Continuous Flow Reactors and Prominent Integrated Microreactors for Biodiesel Production." *Renewable and Sustainable Energy Reviews* 154 (October 2021): 111869. https://doi.org/10.1016/j.rser.2021.111869.

Haghighi, Seyyedeh Faezeh Mirab, Payam Parvasi, Seyyed Mohammad Jokar, and Angelo Basile. 2021. "Investigating the Effects of Ultrasonic Frequency and Membrane Technology on Biodiesel Production from Chicken Waste." *Energies* 14 (8). https://doi.org/10.3390/en14082133.

Helwani, Z., M. R. Othman, N. Aziz, J. Kim, and W. J.N. Fernando. 2009. "Solid Heterogeneous Catalysts for Transesterification of Triglycerides with Methanol: A Review." *Applied Catalysis A: General* 363 (1–2): 1–10. https://doi.org/10.1016/j.apcata.2009.05.021.

Jayakumar, Mani, Natchimuthu Karmegam, Marttin Paulraj Gundupalli, Kaleab Bizuneh Gebeyehu, Belete Tessema Asfaw, Soon Woong Chang, Balasubramani Ravindran, and Mukesh Kumar Awasthi. 2021. "Heterogeneous Base Catalysts: Synthesis and Application for Biodiesel Production – A Review." *Bioresource Technology* 331 (March): 125054. https://doi.org/10.1016/j.biortech.2021.125054.

Kirubakaran, M. 2018. "Eggshell as Heterogeneous Catalyst for Synthesis of Biodiesel from High Free Fatty Acid Chicken Fat and Its Working Characteristics on a CI Engine." *Journal of Environmental Chemical Engineering* 6 (4): 4490–503.

Lin, Yuan-Chung, Kassian T. T. Amesho, Chin-En Chen, Pei-Cheng Cheng, and Feng-Chih Chou. 2020. "A Cleaner Process for Green Biodiesel Synthesis from Waste Cooking Oil Using Recycled Waste Oyster Shells as a Sustainable Base Heterogeneous Catalyst under the Microwave Heating System." *Sustainable Chemistry and Pharmacy* 17: 100310. https://doi.org/10.1016/j.scp.2020.100310.

Ling, Jasmine Si Jie, Yie Hua Tan, Nabisab Mujawar Mubarak, Jibrail Kansedo, Agus Saptoro, and Cirilo Nolasco-Hipolito. 2019. "A Review of Heterogeneous Calcium Oxide Based Catalyst from Waste for Biodiesel Synthesis." *SN Applied Sciences*. https://doi.org/10.1007/s42452-019-0843-3.

Mahmood Khan, Haris, Tanveer Iqbal, Chaudhry Haider Ali, Ansar Javaid, and Izzat Iqbal Cheema. 2020. "Sustainable Biodiesel Production from Waste Cooking Oil Utilizing Waste Ostrich (Struthio Camelus) Bones Derived Heterogeneous Catalyst." *Fuel*. https://doi.org/10.1016/j.fuel.2020.118091.

Mandari, Venkatesh, and Santhosh Kumar. 2021. "Biodiesel Production Using Homogeneous, Heterogeneous, and Enzyme Catalysts via Transesterification and Esterification Reactions : A Critical Review." *BioEnergy Research*. https://doi.org/10.1007/s12155-021-10333-w.

Mihajlovski, Katarina, Aneta Buntić, Marija Milić, Mirjana Rajilić-Stojanović, and Suzana Dimitrijević-Branković. 2021. "From Agricultural Waste to Biofuel: Enzymatic Potential of a Bacterial Isolate Streptomyces Fulvissimus CKS7 for Bioethanol Production." *Waste and Biomass Valorization* 12 (1): 165–74. https://doi.org/10.1007/s12649-020-00960-3.

Motasemi, F., and F. N. Ani. 2012. "A Review on Microwave-Assisted Production of Biodiesel." *Renewable and Sustainable Energy Reviews*. https://doi.org/10.1016/j.rser.2012.03.069.

Moyo, L. B., S. E. Iyuke, R. F. Muvhiiwa, G. S. Simate, and N. Hlabangana. 2021. "Application of Response Surface Methodology for Optimization of Biodiesel Production Parameters from Waste Cooking Oil Using a Membrane Reactor." *South African Journal of Chemical Engineering* 35 (July 2019): 1–7. https://doi.org/10.1016/j.sajce.2020.10.002.

Muthuraman, R. M., A. Murugappan, and B. Soundharajan. 2021. "Highly Effective Removal of Presence of Toxic Metal Concentrations in the Wastewater Using Microalgae and Pre-Treatment Processing." *Applied Nanoscience*. https://doi.org/10.1007/s13204-021-01795-7.

Odude, Victoria O., Ayo J. Adesina, Oluwaseyi O. Oyetunde, Omowumi O. Adeyemi, Niyi B. Ishola, Anietie Okon Etim, and Eriola Betiku. 2019. "Application of Agricultural Waste-Based Catalysts to Transesterification of Esterified Palm Kernel Oil into Biodiesel: A Case of Banana Fruit Peel Versus Cocoa Pod Husk." *Waste and Biomass Valorization* 10 (4): 877–88. https://doi.org/10.1007/s12649-017-0152-2.

Parida, Soumya, Monika Singh, and Subhalaxmi Pradhan. 2022. "Biomass Wastes: A Potential Catalyst Source for Biodiesel Production." *Bioresource Technology Reports*. https://doi.org/10.1016/j.biteb.2022.101081.

Rahman, Wasi Ur, Anam Fatima, Abdul Hakeem Anwer, Moina Athar, Mohammad Zain Khan, Naseem Ahmad Khan, and Gopinath Halder. 2019. "Biodiesel Synthesis from Eucalyptus Oil by Utilizing Waste Egg Shell Derived Calcium Based Metal Oxide Catalyst." *Process Safety and Environmental Protection* 122: 313–19. https://doi.org/10.1016/j.psep.2018.12.015.

Rizwanul Fattah, I. M., H. C. Ong, T. M. I. Mahlia, M. Mofijur, A. S. Silitonga, S. M. Ashrafur Rahman, and Arslan Ahmad. 2020. "State of the Art of Catalysts for Biodiesel Production." *Frontiers in Energy Research* 8 (June): 1–17. https://doi.org/10.3389/fenrg.2020.00101.

Sadeghinezhad, E., S. N. Kazi, Foad Sadeghinejad, A. Badarudin, Mohammad Mehrali, Rad Sadri, and Mohammad Reza Safaei. 2014. "A Comprehensive Literature Review of Bio-Fuel Performance in Internal Combustion Engine and Relevant Costs Involvement." *Renewable and Sustainable Energy Reviews* 30: 29–44. https://doi.org/10.1016/j.rser.2013.09.022.

Sahu, Omprakash. 2021. "Characterisation and Utilization of Heterogeneous Catalyst from Waste Rice-Straw for Biodiesel Conversion." *Fuel* 287: 119543. https://doi.org/10.1016/j.fuel.2020.119543.

Su, Mingxue, Wenzhi Li, Qiaozhi Ma, and Bowen Zhu. 2020. "Production of Jet Fuel Intermediates from Biomass Platform Compounds via Aldol Condensation Reaction Over Iron-Modified MCM-41 Lewis Acid Zeolite." *Journal of Bioresources and Bioproducts* 5 (4): 256–65. https://doi.org/10.1016/j.jobab.2020.10.004.

Talebian-Kiakalaieh, Amin, Nor Aishah Saidina Amin, and Hossein Mazaheri. 2013. "A Review on Novel Processes of Biodiesel Production from Waste Cooking Oil." *Applied Energy* 104: 683–710. https://doi.org/10.1016/j.apenergy.2012.11.061.

Talha, Nur Syakirah, and Sarina Sulaiman. 2016. "Overview of Catalysts in Biodiesel Production." *ARPN Journal of Engineering and Applied Sciences* 11.

Thushari, Indika, and Sandhya Babel. 2018. "Preparation of Solid Acid Catalysts from Waste Biomass and Their Application for Microwave-Assisted Biodiesel Production from Waste Palm Oil." *Waste Management and Research* 36 (8): 719–28. https://doi.org/10.1177/0734242X18789821.

Ur Rahman, Wasi, Syed Mohd Yahya, Zahid A. Khan, Naseem Ahmad Khan, Gopinath Halder, and Sumit H. Dhawane. 2021. "Valorization of Waste Chicken Egg Shells towards Synthesis of Heterogeneous Catalyst for Biodiesel Production: Optimization and Statistical Analysis." *Environmental Technology and Innovation* 22: 101460. https://doi.org/10.1016/j.eti.2021.101460.

Veljković, Vlada B., Ivana B. Banković-Ilić, and Olivera S. Stamenković. 2015. "Purification of Crude Biodiesel Obtained by Heterogeneously-Catalyzed Transesterification." *Renewable and Sustainable Energy Reviews*. https://doi.org/10.1016/j.rser.2015.04.097.

Verma, Tikendra Nath, Pankaj Shrivastava, Upendra Rajak, Gaurav Dwivedi, Siddharth Jain, Ali Zare, Anoop Kumar Shukla, and Puneet Verma. 2021. "A Comprehensive Review of the Influence of Physicochemical Properties of Biodiesel on Combustion Characteristics, Engine Performance and Emissions." *Journal of Traffic and Transportation Engineering (English Edition)* 8 (4): 510–33. https://doi.org/10.1016/j.jtte.2021.04.006.

Voloshin, Roman A., Margarita V. Rodionova, Sergey K. Zharmukhamedov, T. Nejat Veziroglu, and Suleyman I. Allakhverdiev. 2016. "Review: Biofuel Production from Plant and Algal Biomass." *International Journal of Hydrogen Energy* 41 (39): 17257–73. https://doi.org/10.1016/j.ijhydene.2016.07.084.

Xie, Jie, Xinsheng Zheng, Aiqin Dong, Zhidong Xiao, and Jinhua Zhang. 2009. "Biont Shell Catalyst for Biodiesel Production." *Green Chemistry* 11 (3): 355–36. https://doi.org/10.1039/b812139g.

Xu, Wei, Lijing Gao, and Guomin Xiao. 2015. "Biodiesel Production Optimization Using Monolithic Catalyst in a Fixed-Bed Membrane Reactor." *Fuel* 159: 484–90. https://doi.org/10.1016/j.fuel.2015.07.017.

Yoro, Kelvin O., and Michael O. Daramola. 2020. *CO2 Emission Sources, Greenhouse Gases, and the Global Warming Effect: Advances in Carbon Capture*. Elsevier Inc. https://doi.org/10.1016/b978-0-12-819657-1.00001-3.

17 Environmental Impact of Municipal Waste Energy Recovery Plant

Mane Yogesh G. and Anand Ramanathan

CONTENTS

17.1 Introduction 233
17.2 Category of Waste.. 234
    17.2.1 Municipal Waste 234
    17.2.2 Solid Waste.. 234
17.3 Problems due to Municipal Waste 235
17.4 Statewise Waste Generation in India (Table 17.1)... 235
17.5 Global Warming 236
17.6 Basic Techniques of Energy Recovery 236
17.7 Parameters Affecting Energy Recovery 237
17.8 Methods of Energy Recovery 237
    17.8.1 Landfilling.. 237
    17.8.2 Bioreactor Landfill 237
    17.8.3 Landfill Gas Recovery 237
    17.8.4 Incinerator 238
    17.8.5 Stoker-type Incinerator 238
    17.8.6 Fluidized Bed–type Incinerator 238
    17.8.7 Gasification 238
    17.8.8 Plasma Gasification... 238
    17.8.9 Gasification Melting Furnace 238
    17.8.10 Pyrolysis.. 239
    17.8.11 Microbial... 239
17.9 Impact 239
References... 240

17.1 INTRODUCTION

The growth in population and new advancements in industry result in a considerable increase in waste creation, and the lack of sustainable waste management strategies and their execution is the primary source of environmental damage/problems such as soil, water, and air pollution (Kaur et al. 2021). It also contaminates plants and animals. Waste dump in landfills cause soil pollution, decay of fertility of soil, and contamination of groundwater, which directly affect the agriculture. It involves occupying some ground space and is not environment-friendly to maintain a landfill. Chemical waste affects aquatic life and burning of municipal waste causes air pollution, which affects global warming and greenhouse gas effect, and that is why another essential method and perfect alternative are required (Amulen et al. 2022). Municipal solid waste (MSW) is recognized as an energy source. If waste management systems prioritize operations for prevention, reuse, and recycling, energy recovery from municipal solid waste contributes significantly to the shift to a circular economy. In recent years, technologies have been created that aid in the generation of large amounts of decentralized energy as well as the reduction of trash for safe disposal. The ministry is encouraging all possible technological possibilities for establishing energy recovery projects.

17.2 CATEGORY OF WASTE

17.2.1 Municipal Waste

Municipal wastes generally comes from household and commercial activities. It does not include wastes from construction or other non-decomposable components. Open dumping grounds of these can be very dangerous for us humans, as they can contaminate the waterbodies. Human health is threatened as this can cause life-threatening diseases (e.g., Chagas disease, onchocerciasis). This municipal waste also generates some amount of plastic waste. Municipal waste may be taken to a transfer station for consolidation before it is taken to a landfill or an incinerator. It may be sent to a permitted processing facility.

17.2.2 Solid Waste

It is classified into three types:

(i) Biodegradable trash (food and kitchen waste)
(ii) Non-biodegradable and inert waste
(iii) Reusing trash

It is believed that 17% of rubbish is recyclable, and this figure has been rising over time. Biodegradable garbage fluctuates between 55% and 60% on an annual basis, according to statistics from a few cities.

Biodegradable waste includes fruits, vegetables, flowers, poultry, agricultural, and gardening waste like tree cuttings, grain trash, etc.

Non-biodegradable waste includes plastic, thermocol, paper, bottle cap, glass, metal, etc. It mainly contains recycle, non-recycle, and hazardous waste (Figure 17.1).

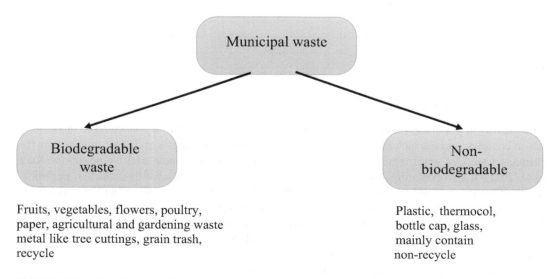

FIGURE 17.1 Classification of wastes.

17.3 PROBLEMS DUE TO MUNICIPAL WASTE

Due to population and economic expansion, as well as changes in production and consumption patterns, municipal solid waste (MSW) generation has expanded during the past few decades on a global scale. In 2015, the world's population created 1.9 Gt/year of MSW, and by 2050, that number is expected to be closer to 3.5 Gt/yr. Statistically, the rapid pace of plastic consumption worldwide has led to the creation of increasing amounts of garbage, which increases the amount of waste that mus t be treated and/or disposed of. This is because plastic items have a very limited usable life – about 40% of them have a life span of less than 1 month per year. Of course, Many of the waste goods made from plastic today are not adequately recovered, recycled, or disposed of, which contributes to the growing buildup of plastic debris in rivers and oceans that are easily accessible to animals for consumption. Hazardous hospital waste is waste that is contaminated with hospital chemicals. These substances include mercury, which is used in thermometers and other measuring devices, as well as phenols and formaldehyde, which are employed as sanitizers. The majority of hospitals in India lack the necessary disposal infrastructure to properly dispose of this hazardous waste. The main industries in the industrial sector that produce hazardous waste are those that deal with metal, chemicals, paper, pesticides, dyes, refining, and rubber. Direct contact with toxic substances in hazardous waste, such as mercury and cyanide, can be fatal. These are significant issues brought on by municipalities. Marine and aquatic life are likewise affected by the issue of municipal trash. Corals, weeds, pink dolphins, and other distinctive aquatic and marine life become extinct as a result of this.

17.4 STATEWISE WASTE GENERATION IN INDIA (TABLE 17.1)

TABLE 17.1
Statewise Waste Generation

States	Total Solid Waste Generation (000 TPA)	Total Liquid Waste Generation (MLD)
Uttar Pradesh	97,728	6,450
Maharashtra	35,548	8,348
Gujarat	37,582	4,608
Tamil Nadu	23,146	3,799
Punjab	13,372	1,638
Madhya Pradesh	44,055	2,516
West Bengal	34,836	3,687
Karnataka	23,623	3,681
Andhra Pradesh	22,594	2,389
Rajasthan	46,577	2,303
Bihar	42,827	1,579
Telangana	14,905	1,864
Assam	15,982	463
Odisha	12,078	866
Haryana	11,752	1,376
Jharkhand	14,626	1,015
Uttarakhand	4,440	499
Chhattisgarh	11,645	723
Kerala	3,990	887
Delhi	5,895	3,331
Jammu and Kashmir	5,532	275

(Continued)

**TABLE 17.1 *(Continued)*
Statewise Waste Generation**

States	Total Solid Waste Generation (000 TPA)	Total Liquid Waste Generation (MLD)
Himachal Pradesh	3,978	58
Tripura	1,293	82
Puducherry	4,292	148
Meghalaya	1,150	48
Chandigarh	4,233	191
Manipur	525	48
Nagaland	167	50
Arunachal Pradesh	445	14
Daman and Diu	4,115	12
Goa	190	41
Andaman and Nicobar Islands	4,195	20
Mizoram	128	72
Sikkim	230	21
Dadra and Nagar Haveli	4,138	25
Lakshadweep	4,103	0
Ladakh	142	
Total	**556,057**	**53,127**

17.5 GLOBAL WARMING

Municipal solid trash has increased significantly due to economic growth and urbanization (MSW). In addition, for a number of decades, effective waste management has created environmental problems. These include greenhouse gas emissions that result in significant global warming and other climate-related changes (Yong et al. 2019). As we see in old days or even nowadays, in some areas people burn the waste which contain all types of municipal waste; mostly in dumping areas, it is the main cause of air pollution and global warming (Yadav et al. 2020).

17.6 BASIC TECHNIQUES OF ENERGY RECOVERY

One of the two ways used for obtaining energy from the organic element may be from biodegradable or non-biodegradable:

(i) The thermal breakdown of organic molecules for getting heat energy, gas, or fuel oil is known as thermochemical conversion.
(ii) In a biochemical conversion, organic matter is broken down by microorganisms using enzymes to produce methane gas or alcohol.

Waste with a high percentage of organic, non-biodegradable components and low moisture can be converted using thermochemical methods. In this category, pyrolysis/gasification and incineration are the two most widely used technological solutions. Biochemical conversion techniques are recommended for wastes that have a high content of organic biodegradable (putrescible) matter and a high moisture/water content that promotes microbial activity. The most frequent technical solution in this area is anaerobic digestion, often known as biomethanation.

17.7 PARAMETERS AFFECTING ENERGY RECOVERY

The amount of garbage and the physical and chemical qualities (quality) of a waste are the two most important factors that influence the feasibility of recovering energy from waste (including MSW). Component size, density, and moisture content are all physical parameters to consider. As the components are small in size, the waste decompose fast. The percentage of biodegradable organic material and moisture in high-density garbage is high. Paper, plastics, and other flammables are abundant in low-density garbage. The parameter values listed here are merely the best case scenarios for implementing a specific waste treatment method, and they might or might not be applicable to the wastes produced, gathered, and transported to the waste treatment facility (Yong et al. 2019). To make the waste more suitable for the recommended course of action, it must be appropriately separated, processed, and mixed on site with the appropriate additives before treatment. Before proceeding, this must be assessed and assured. Straw, paper, and other high carbon wastes, for instance, can be added if the C/N (carbon/nitrogen) ratio is low; if it is high, high nitrogen wastes can be added (Yadav et al. 2020)

17.8 METHODS OF ENERGY RECOVERY

17.8.1 Landfilling

A landfill is a piece of land where rubbish is dumped. The goal of landfilling is to prevent garbage from coming into touch with the environment, particularly groundwater. In India, open, uncontrolled, and poorly managed dumping is common and causes serious environmental damage. In cities and towns, 60–90% of MSW is immediately disposed of on land in an inadequate way. The daily cover measures are insufficient, increasing the likelihood of leakage. The daily cover measures are insufficient, increasing the likelihood of leakage. Due to this, local governments are forced to restrict the usage of even well-known safeguards and practices. Although there is a lack of land for waste disposal in major cities like Delhi, Mumbai, Kolkata, and Chennai, it appears that landfilling will remain the most popular waste disposal method in India in the years to come. During that time, certain improvements will be needed to ensure sanitary landfilling (Nandan et al. 2017)

17.8.2 Bioreactor Landfill

The bioreactor landfill is a new advancement in landfill technology. Bioreactor landfills are built, planned, and operated to maximize moisture content and anaerobic biodegradation rates. Leachate recirculation is the main feature that separates bioreactor landfills from regular landfills. The bioreactor option is the outcome of designing and constructing a new generation of ecologically sound landfills; it ensures environmental security while allowing and promoting quick stabilization of easily and moderately decomposable organic waste components (Nandan et al. 2017).

17.8.3 Landfill Gas Recovery

The waste that is placed in a landfill is subjected to anaerobic conditions over time, and its organic portion is gradually volatilized and degraded in accordance with a process similar to that discussed in the section before. The gas wells are available at the time of purchase. Site selection for gas recovery from existing landfills before moving on with gas recovery initiatives determine whether the current open dumps or landfills are significant enough to warrant consideration. Sites with more than a million tons of trash already there, the majority of which should be no older than 10 years, are typically regarded as suitable for energy recovery. Over the course of several years, such locations should create enough gas to fund a lucrative gas recovery operation (Nandan et al. 2017).

17.8.4 INCINERATOR

Direct burning of trash produces heat energy, inert gases, and ash at 8000 °C and higher when oxygen (air) is present. The amount of moisture and inert materials that contribute to heat loss, the size and shape of the constituents, the ignition temperature, the density and composition of the wastes, the design of the combustion system (fixed bed/fluidized bed), and other factors all have an impact on the net energy yield. While incineration is an essential technique of waste disposal, it is also connected with various harmful outputs that are of concern to the environment, though to varied degrees. Fortunately, they may be efficiently managed by installing appropriate pollution control systems as well as proper furnace architecture and combustion process control (Lam et al. 2012).

17.8.5 STOKER-TYPE INCINERATOR

A set of grates is referred to as a "stoker." Sliding grates cause the input waste to burn off more and more as it flows downstream. The "dry zone," "combustion zone," and "burn-out zone" are the three divisions of the combustion chamber. With enough time in the dry zone, even trash with a high moisture content may be successfully combusted. It is one of the causes of the widespread use of stoker-type incinerators for MSW. The design and operational parameters of this three-stage combustion chamber must be modified based on the quantity and quality of waste (Escamilla-García et al. 2020).

17.8.6 FLUIDIZED BED–TYPE INCINERATOR

Sand is placed at the bottom of the combustion chamber of a fluidized bed incinerator, where air is forced in from the bottom, turning the sand into fluid. The fluidized bed's rubbish begins to burn once the sand layer has heated up. Because of the sand's powerful heating capacity, MSW may be quickly dried and burned even when additional waste with a lot of moisture is added. Instead of burning heterogeneous MSW, this kind of incinerator is more suited for burning homogeneous materials like sludge (Liu et al. 2020)

17.8.7 GASIFICATION

Gasification is the process of turning a carbon-rich solid material, such as coal or biomass, into a gas. The feedstock is heated to extremely high temperatures because it is a thermochemical process. Gases that can undergo chemical reactions and produce synthesis gas, or syngas, are released. Electricity can be produced using syngas, a mixture of hydrogen and carbon monoxide (Rajasekhar et al. 2015).

17.8.8 PLASMA GASIFICATION

For generating heat, plasma is required; plasma is in the form of jet and plasma gasification is done in the absence of gasification oxidant but in the presence of some noble gases like air, nitrogen, oxygen, etc. The feedstock is broken down into its individual components of carbon, hydrogen, and oxygen at extremely high temperatures (more than 5000 °C). This produces syngas of exceptionally high quality that is free of tars, hydrocarbons, and methane, which swiftly combines to form gaseous hydrogen and carbon monoxide (Rajasekhar et al. 2015).

17.8.9 GASIFICATION MELTING FURNACE

By immediately melting bottom ash in the furnace, the gasification melting furnace creates molten slag. Molten slag has a greater potential for use as a building material than bottom ash since it is denser. Gasification melting is the process of thermally decomposing waste while utilizing less oxygen, or indirectly heating waste to create pyrolysis (partial combustion) at a temperature lower than that of combustion. In fluidized bed and kiln-type furnaces, this technique is applied in each furnace that is independent of the melting furnace. The second stage of pyrolysis involves the complete combustion of the pyrolysis gas at extremely high temperatures, which melts the particles (1200–1300 °C) of the gas. Kiln-style furnaces are becoming more and more rare.

17.8.10 Pyrolysis

The use of renewable energy sources has become an essential component of long-term growth. When paired with cutting-edge technologies like pyrolysis, municipal solid waste (MSW) has a lot of potential as a sustainable energy source. The assessment also contains information on the environmental implications of pyrolysis, as well as problems connected with its usage and a set of recommended suggestions for overcoming them (Hasan et al. 2021). The most often used method for pyrolysis of MSW is rotational pyrolysis, which offers adequate heat transmission with relatively low energy consumption. Temperature is the most important parameter in MSW pyrolysis research. Emission control systems should be installed in pyrolysis facilities to make MSW pyrolysis environment-friendly. The riser of this reactor employs a fast bed, a unique hydrodynamic technique. It guarantees constant mixing and temperature management. The surface gas velocity in the circulating type is significantly higher than in the bubbling kind. This high gas velocity, along with excellent mixing, allows the circulating type to handle a large volume of feedstock (Chen et al. 2014).

17.8.11 Microbial

MFC is a technology for providing the modern world with safe, clean, minimal carbon dioxide emissions, highly efficient energy generation, and wastewater treatment. There were several sources of water pollution, each with a negative impact on freshwater and natural water. Surface water is exposed to cathode electrodes, whereas wastewater is exposed to anode electrodes (Lachos-Perez et al. 2017). Numerous bacteria in MFC chambers can transport protons and electrons between electrodes. Firmicutes, proteobacteria, acidobacteria, fungi, and algae were the five main microbe groups that showed electrical production for their respiration in the MFC chamber. Biofilm is created by microorganisms that colonize electrode surfaces and transmit electrons more effectively than utilizing insoluble electron acceptors. Exoelectrogens can transmit electrons from electrodes in a variety of methods, such as direct electron transfer, long-distance electron transport via conductive pili, soluble electron-shuttling molecules, and short-distance electron transfer via redox active proteins. Microbes come into contact with the electrode surface by direct electron transfer. For the transportation of electrons to the anode surface, neither a source nor an intermediate exist. Long-range electron transport via conductive pili is a superior option.

17.9 IMPACT

Thermal and biochemical processes including incineration, pyrolysis, gasification, waste-generated fuel, and biomethanation can be used to turn LFG produced from landfilled MSW into energy. An integrated approach to managing MSW, lowering landfill LFG emissions, and creating a sizable non-conventional energy source is the landfill gas recovery system (LFGRS) (Srivastava et al. 2020).

The sewage energy recovery potential might cut the sewage treatment plants' energy requirement significantly. As a result, sludge might be used to recover energy (Siddiqi et al. 2020).

Thermal or anaerobic digestion may assist to tackle various issues, such as decreasing energy consumption and reducing sludge bulk. It would also reduce the expense of transportation and landfilling (Singh et al. 2020)

REFERENCES

Amulen, Judith, Hillary Kasedde, Jonathan Serugunda, and Joseph D. Lwanyaga. 2022. "The potential of energy recovery from municipal solid waste in Kampala City, Uganda by incineration." *Energy Conversion and Management*, 14, 100204.

Chen, Dezhen, Lijie Yin, Huan Wang, and Pinjing He. 2014. "Pyrolysis technologies for municipal solid waste: A review." *Waste Management*, 34, no. 12, 2466–86.

Escamilla-García, P. E., Camarillo-López, R. H., Carrasco-Hernández, R., Fernández-Rodríguez, E., and Legal-Hernández, J. M. 2020. "Technical and economic analysis of energy generation from waste incineration in Mexico." *Energy Strategy Reviews*, 31, 100542.

Hasan, M. M., M. G. Rasul, M. M. K. Khan, Nanjappa Ashwath, and M. I. Jahirul. 2021. "Energy recovery from municipal solid waste using pyrolysis technology: A review on current status and developments." *Renewable and Sustainable Energy Reviews*, 145, 111073.

Kaur, Prabhjot, Gagan Jyot Kaur, Winny Routray, Jamshid Rahimi, Gopu Raveendran Nair, and Ashutosh Singh. 2021. "Recent advances in utilization of municipal solid waste for production of bioproducts: A bibliometric analysis." *Case Studies in Chemical and Environmental Engineering*, 4, 100164.

Lachos-Perez, D., A. B. Brown, A. Mudhoo, J. Martinez, M. T. Timko, M. A. Rostagno, and T. Forster-Carneiro. 2017. "Applications of subcritical and supercritical water conditions for extraction, hydrolysis, gasification, and carbonization of biomass: A critical review." *Biofuel Research Journal*, 4, no. 2, 611–26.

Lam, Su Shiung, and Howard A. Chase. 2012. "A review on waste to energy processes using microwave pyrolysis." *Energies*, 5, no. 10, 4209–32.

Liu, Chen, Toru Nishiyama, Katsuya Kawamoto, and So Sasaki. 2020. "Waste-to-energy incineration-CCET guideline series on intermediate municipal solid waste treatment technologies." https://wedocs.unep.org/handle/20.500.11822/32795?show=full.

Nandan, Abhishek, Bikarama Prasad Yadav, Soumyadeep Baksi, and Debajyoti Bose. 2017. "Recent scenario of solid waste management in India." *World Scientific News*, 66, 56–74.

Rajasekhar, M., N. Venkat Rao, G. Chinna Rao, G. Priyadarshini, and N. Jeevan Kumar. 2015. "Energy generation from municipal solid waste by innovative technologies–plasma gasification." *Procedia Materials Science*, 10, 513–18.

Siddiqi, Afreen, Masahiko Haraguchi, and Venkatesh Narayanamurti. 2020. "Urban waste to energy recovery assessment simulations for developing countries." *World Development*, 131, 104949.

Singh, Vipin, Harish C. Phuleria, and Munish K. Chandel. 2020. "Estimation of energy recovery potential of sewage sludge in India: Waste to watt approach." *Journal of Cleaner Production*, 276, 122538.

Srivastava, Abhishek N., and Sumedha Chakma. 2020. "Quantification of landfill gas generation and energy recovery estimation from the municipal solid waste landfill sites of Delhi, India." *Energy Sources, Part A: Recovery, Utilization, and Environmental Effects*, 1–14.

Yadav, Shailendra Kumar, Kanagaraj Rajagopal, A. K. Priya, and Gyan Deep Sharma. 2020. "Smart waste management and energy extraction from waste in Indian smart cities–a review." *Contaminants and Clean Technologies*, 321–30.

Yong, Zi Jun, Mohammed J. K. Bashir, Choon Aun Ng, Sumathi Sethupathi, Jun Wei Lim, and Pau Loke Show. 2019. "Sustainable waste-to-energy development in Malaysia: Appraisal of environmental, financial, and public issues related with energy recovery from municipal solid waste." *Processes*, 7, no. 10, 676.

18 A Comprehensive Review on the Modeling of Biomass Gasification Process for Hydrogen-rich Syngas Generation

Kalil Basha Jeelan Basha, Sathishkumar Balasubramani, and Vedharaj Sivasankaralingam

CONTENTS

18.1 Introduction 241
18.2 Modeling of the Biomass Gasification Process 243
    18.2.1 Equilibrium Modeling..243
        18.2.1.1 Stoichiometric Approach 244
        18.2.1.2 Non-stoichiometric Approach...245
    18.2.2 Kinetic Modeling 246
    18.2.3 CFD Modeling Approach 246
    18.2.4 ASPEN Plus Modeling...247
    18.2.5 Machine Learning Approaches...248
        18.2.5.1 Support Vector Machine Approach 248
        18.2.5.2 Artificial Neural Network Approach 248
18.3 Optimization of the Biomass Gasification Process 250
    18.3.1 Effect of Gasification Temperature 250
    18.3.2 Effect of Gasification Pressure 250
    18.3.3 Effect of the Bed Material and Catalyst...251
    18.3.4 Effect of Gasifying Agent 251
    18.3.5 Effect of Biomass Particle Size 251
    18.3.6 Effect of Moisture Content...252
    18.3.7 Effect of Equivalence Ratio 252
18.4 Summary 252
References..255

18.1 INTRODUCTION

For the past few decades, 80–82% of the total global energy supply was produced from fossil fuels like coal, natural gas, and oil and the remaining part was produced through renewable energy sources like solar, wind, hydro, nuclear, and biomass (IEA 2021). Using fossil fuels for electricity generation increases global carbon dioxide (CO_2) emissions, intensifying global warming potential. Many developed and developing nations are devising policies to curtail CO_2 emissions and reduce the increase in global earth temperature (Vedharaj 2021). Recently, there has been a growing interest in increasing the contribution of renewable energy sources to the total global energy supply. In particular, India has planned to achieve 50% of its energy requirements from renewable energy sources by 2030 (National Statement at COP26 Summit 2021).

Biomass is one of the most promising fossil-fuel alternatives as it can be used in the existing facilities with slight modifications. It contributes nearly 50–55% of the total global energy supply from various renewable sources (Popp et al. 2021). Biomass includes plant materials, agricultural residues, and organic wastes from industries and households. Biomass is readily available, widespread, and has lower environmental concerns when compared to fossil fuels; however, its energy density is lower (Krigstin, Levin, and Wetzel 2012). Further, the storage, transportation, and handling of solid biomass is cumbersome and incurs additional costs (Whittaker and Shield 2018). Thus, it is necessary to convert biomass to liquid or gaseous fuels to realize the benefits of biomass with improved fuel properties and lower costs. Recently, biomass biofuels (ethanol, bio-oil, and biodiesel) have been widely adopted in the transportation sector. Further, energy production from organic waste sources has been witnessed as a promising solution for waste disposal; thus, the waste-to-energy conversion technologies are also growing faster (Parashar et al. 2020).

The most widely adopted methods for converting solid biomass to liquid or gaseous biofuels are biochemical or thermochemical processes. Microorganisms are employed to break down the complex hydrocarbons from solid biomass into simple hydrocarbons in biochemical processes (Chen and Wang 2017), whereas heat is employed in thermochemical processes (García et al. 2017). Anaerobic digestion and fermentation are the two prominent biochemical processes employed to produce liquid and gaseous biofuels. The major components in biomass include lignin, hemicellulose and cellulose, which are highly branched complex hydrocarbons (Tovar-Facio, Cansino-Loeza, and Ponce-Ortega 2022). Generally, the biochemical processes convert only the cellulose and hemicellulose content of the biomass to produce biofuels, whereas the lignin part is extracted separately and utilized in the thermochemical processes. The higher cost involved in the growth of microorganisms, maintaining proper pH levels for the sustainability of microorganism and slower reaction rate, makes biochemical processes less favorable compared to thermochemical processes.

Thermochemical processes are economical and can process a wide variety of feedstock with slight process modification in the existing systems. Pyrolysis and gasification are the two basic types of thermochemical processes (Ren et al. 2020). Pyrolysis is carried out at low temperatures (500–600 °C) to maximize liquid fuel yield (Varma, Shankar, and Mondal 2018), whereas gasification is carried out at higher temperatures (800–1000 °C) to maximize gaseous fuel yield. Biomass gasification is a prominent thermochemical process widely employed due to its capacity to handle a wide range of biomass resources. Steam, air, or oxygen can be used as an oxidizing agent for gasification (Safarian, Unnthorsson, and Richter 2021; Habibollahzade, Ahmadi, and Rosen 2021). The different zones in the gasification process are depicted in Figure 18.1 and it consists of drying (100–150 °C), devolatilization (200–500 °C), oxidation (800–1200 °C), and reduction (650–900 °C) processes.

Gasification is a partial thermal oxidation process that yields a high proportion of gaseous by-products as well as small traces of char, ash, tars, and oils. Gasifiers employed in the gasification processes are broadly classified as the fixed bed (updraft, downdraft, and cross-draft), fluidized bed (bubbling and circulating), and entrained flow reactors. Many research works on the gasifier design and its types were carried out for nearly a century (Basu 2010).

Syngas (H_2 and CO) is the primary product obtained from the gasification process and is used in various applications such as the generation of low-carbon fuels, production of valuable chemicals, and direct use in cogeneration of heat and electricity. The quality of produced syngas depends on various factors such as feedstock quality, gasifier design parameters, and operating conditions (Yaghoubi et al. 2018). Thus, it is necessary to optimize the gasifier design and operating parameters to achieve hydrogen-rich syngas. Experimental determination of the optimum design and operating condition requires higher costs and is time-consuming. Thus, many researchers are involved in developing a suitable numerical model to predict the performance of the gasifiers. The most widely

A Review on the Modeling of Biomass Gasification Process

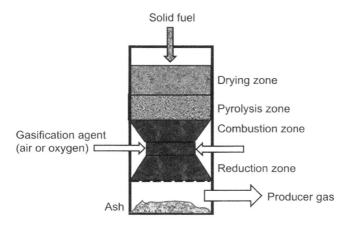

FIGURE 18.1 The different zones in the gasification process (Pang 2016).

adopted modeling approach for the gasification process includes the equilibrium model, kinetic model, CFD model, Aspen plus model, and machine learning model. This chapter reviews the different models adopted for the biomass gasification process. Here, an overview of the different modeling approaches is detailed. Further, the studies on the influence of different design and operating parameters on the gasifier performance with the developed models are elaborated.

18.2 MODELING OF THE BIOMASS GASIFICATION PROCESS

The critical operating factors influencing the gasification process include the gasifying agent, feedstock flow rate, reactor pressure, reactor temperature, and equivalence ratio. The aforementioned factors significantly impact the product gas composition and the performance of the gasifier. Experimentally optimizing these operating factors for a particular gasifier design is time-demanding and expensive. Under these circumstances, mathematical modeling is an effective tool for studying gasifier behavior and optimizing the various parameters. Biomass gasification models can be classified into two groups based on the distinct mechanisms: kinetic and equilibrium models, which are detailed in the following sections.

18.2.1 Equilibrium Modeling

The thermodynamic equilibrium model is a basic mathematical tool used to conduct an initial feasibility analysis in the gasification process (George, Arun, and Muraleedharan 2016). The calculation involved in the equilibrium models does not account for the gasifier design parameters (Puig-Arnavat, Bruno, and Coronas 2010). Due to its simplicity and high prediction accuracy, the equilibrium modeling was adopted extensively to model the gasifier and optimize its performance. The equilibrium modeling is divided into two categories: stoichiometric and non-stoichiometric. The equilibrium in the stoichiometric method is calculated by considering the equilibrium constant of the individual reactions (George, Arun, and Muraleedharan 2016). However, the equilibrium in the non-stoichiometric method is determined by minimizing the total Gibbs free energy (Huang and Ramaswamy 2009). Thus, a reaction mechanism for the gasification process is required for the stoichiometric method, whereas the species details are sufficient for the non-stoichiometric method.

A simple equilibrium model for biomass gasification by considering one-dimensional, perfect gas behavior and well-stirred reactor conditions has been developed. It was found that the simulated data slightly deviated from the experimental results due to the model's simplicity (Ruggiero and Manfrida 1999). A novel three-zone equilibrium and kinetic-free model has been devised for the biomass gasifier. The sub-zones considered in this model include drying and pyrolysis, oxidation and reduction zones. Each zone was solved separately, and the individual solutions were coupled to get the overall solution for the model. The proposed model can handle different raw materials with accurate prediction levels (Ratnadhariya and Channiwala 2009).

The global reduction reaction equilibrium model for biomass gasification with a downdraft gasifier has been studied. The model accurately predicts the equilibrium constant and compositions in the reaction zone. Further, it was reported that the reactants concentration and initial temperature conditions determine the prediction accuracy of equilibrium model (Sharma 2008). A robust model was developed to simulate the biomass gasification plant. The thermodynamic equilibrium model was modified to adopt for real applications, and the equations were solved in the Engineering Equation Solver (EES). Here, only partial chemical equilibrium was achieved. It was mentioned that the developed model could predict the performance of the different gasifiers with various biomass types (Puig-Arnavat, Bruno, and Coronas 2012). A detailed review of the stoichiometric and non-stoichiometric modeling approaches is provided in the following sections.

18.2.1.1 Stoichiometric Approach

The stoichiometric method needs a specific reaction mechanism which consists of all the species and chemical reactions. Consider one mole of biomass feedstock is processed in a gasifier with x moles of steam and y moles of air and the following equation represents the reaction:

$$CH_aO_bN_c + xH_2O + y(O_2 + 3.76N_2) \rightarrow c_1C + c_2H_2 + c_3CO + c_4H_2O + c_5CO_2 + c_6CH_4 + c_7N_2$$

where

$CH_aO_bN_c$ = chemical representation of biomass
a, b, and c = mole ratios (H/C, O/C, and N/C, respectively)
x and y = coefficient of steam and air
$c_1, c_2, \ldots, c_7$ = stoichiometric coefficient

The following reaction mechanism is well-established and represents the major reactions that occur during the gasification process. The equilibrium modeling (stoichiometric approach) is based on these reactions:

$R1: CO_2 + C \rightarrow 2CO$

$R2: C + H_2O \rightarrow H_2 + CO$

$R3: C + 2H_2 \rightarrow CH_4$

$R4: CO + H_2O \rightarrow CO_2 + H_2$

The stoichiometric coefficients of the product gases can be determined by formulating seven equations. Four equations can be formulated by balancing the C, H, O, and N atoms. The remaining three equations can be formulated from the equilibrium constant relation for R1, R2, and R3 reactions:

$$K1 = \left(\frac{y_{CO}^2}{y_{CO_2}}\right) * P$$

$$K2 = \left(\frac{y_{CO} \, y_{H_2}}{y_{H_2O}}\right) * P$$

$$K3 = \left(\frac{y_{CH_4}}{y_{H_2}^2}\right) * P$$

where

    y_i = mole fraction for species i of H_2, CO, CO_2, and H_2O
    P = operating pressure
    K_1, K_2, and K_3 = reaction rate

A biomass gasification model with a stoichiometric approach was formulated. It has been reported that a combination of minimum O/C and maximum H/C ratio can achieve hydrogen yield above 85 g/kg of dry biomass without ash (George, Arun, and Muraleedharan 2016). A stoichiometric equilibrium model with and without accounting char for the downdraft gasifier has been developed. It was concluded that the equilibrium model that accounts for char is more suitable and can relate to the commercial gasifiers (Huang and Ramaswamy 2009).

Prediction of the downdraft gasifier performance using an equilibrium model with various biomass materials has been conducted. The influence of the moisture content and gasification temperature on the calorific value of the producer gas was analyzed. The calorific values from the model prediction are matched well with the experimental data (Zainal et al. 2001). The downdraft biomass air gasifier was modeled with the stoichiometric thermodynamic equilibrium method to predict the syngas compositions. The influence of the moisture content in the feedstock, equivalence ratio, and air-to-fuel ratio on syngas composition has been investigated. A good agreement between the simulated and experimental syngas composition was observed (Akyurek et al. 2019). The downdraft gasifier with the stoichiometric equilibrium model was simulated to obtain producer gas composition and lower heating values. The simulated values agreed with the experimental values from the various research works (Azzone, Morini, and Pinelli 2012).

18.2.1.2 Non-stoichiometric Approach

This model does not require a detailed reaction mechanism, as with the stoichiometric approach. The feedstock composition and species involved in the process are sufficient input for this model. This model is targeted to minimize the Gibbs free energy in the system. The stable equilibrium is achieved in the reaction system when the system is maintained at minimum Gibbs free energy. The Gibbs free energy for the gasification product that consists of N species is represented by

$$G_{total} = \sum_{i=1}^{N} n_i \Delta G^0_{f,i} + \sum_{i=1}^{N} n_i RT \ln\left(\frac{n_i}{\sum n_i}\right)$$

where $\Delta G^0_{f,i}$ = Gibbs free energy of formation for species i at the standard pressure condition of 1 bar. The composition of the product gases can be obtained by performing an optimization problem to minimize the Gibbs free energy.

The performance of the circulating fluidized bed coal gasifier with a non-stoichiometric thermodynamic equilibrium model was predicted. The modeled gasifier also applies for biomass gasification and includes 5 elements and 44 species. It has been noted that the operating temperature range to obtain an H_2-rich gas is 1100–1300 K at atmospheric pressure and air ratio of 0.15–0.25 (X. Li et al. 2001; X. T. Li et al. 2004). The bi-equilibrium model with the non-stoichiometric method was proposed to predict the heating value and syngas composition. The bi-equilibrium model results are more accurate than the results from the model with a single equilibrium method (Biagini, Barontini, and Tognotti 2016). The air-blown gasification process, which accounts for the non-stoichiometric equilibrium model under adiabatic and isothermal conditions, was studied. It has been reported that the moisture content and temperature are the influencing parameters for limiting the formation of tar and carbon residues in a gasifier (Biagini 2016). ASPEN Plus was used to simulate wood gasification in a fluidized bed gasifier with the Gibbs free energy minimization method. The authors concluded that there is a limiting value for preheating and oxygen enrichment, beyond which the gasifier performance declines (Mathieu and Dubuisson 2002).

18.2.2 KINETIC MODELING

The equilibrium model can only predict the distribution of end-reaction product and not the distribution of intermediate product. As a result, this model cannot be utilized to analyze or construct reactors. Thermodynamic equilibrium models also fail to consider for the multistep chemical and physical events occurring inside a gasifier, resulting in incorrect estimation of some species. Kinetic models, on the other hand, were developed to address these issues and have proven capable of simulating a variety of gasifier-operating conditions. The kinetic model can simulate reaction conditions at various times and locations, making it useful for reactor design and optimizing operation parameters. A kinetic model is utilized to predict the yield and product composition from a gasifier after a set amount of time. Both the kinetics of gasification reactions and the hydrodynamics of the gasifier reactor are considered in the kinetic model. There are three types of kinetic models: semi kinetic, total kinetic, and combined kinetic and equilibrium (CRF) approach.

The semi-kinetic approach evaluates kinetically regulated temperature and concentrations for all the other regions while assuming local equilibrium for some gasifier regions. In the semi-kinetic model, the volatiles liberated during the pyrolysis stage enter into a homogeneous gas-phase reaction module where the equilibrium constant was evaluated from the equilibrium approach. Further, the equilibrated gases, along with steam/air, enter into the chemical kinetic module, where the producer gas composition and the rate of reaction are evaluated. The semi-kinetic approach employs few rate equations and is best suited when the gases are closer to equilibrium. The volatile and char gasification kinetics, local temperature, and composition could be modeled simultaneously in total kinetic models. The total kinetic model is employed to model non-ideal reactors with reliable rate equations and fluid flow models.

The combined kinetic and equilibrium technique kinetically simulates the reduction zone. The rate equations considered during reduction reaction are usually expressed as a simple Arrhenius kind reversible reaction with the equilibrium constants included in the rate expressions. The char reactivity factor (CRF) is an adjustable parameter used to alter the rate expressions for char reactions. These models are used to simulate gasifiers in which the chemical composition of the produced gas differs from that of the equilibrium mixture.

The multiscale kinetic model for syngas production in updraft gasifiers has been developed. The linkage of the kinetics and transport phenomena in this multiphase, multicomponent, and multiscale system contributes to the difficulty of solid fuel gasification processes. The complete mathematical model provides information on the chemical and physical processes of a gasifier (Corbetta et al. 2015). A simple steady-state kinetic model was established for biomass steam gasification in a dual fluidized bed reactor. Steam gasification of biomass yields a higher H_2 concentration in the producer gas when compared to gasification process with a combination of oxygen and steam. The created model accurately predicted H_2 and CH_4, but not CO and CO_2 (Hejazi, Grace, and Mahecha-Botero 2019). A kinetic rate model was selected for the reduction zone of a gasifier. The kinetic studies of reduction reactions are evaluated under isothermal conditions over the temperature range of 1000–1300 K. The fourth-order Runge–Kutta approach was used to solve the kinetic rate equations of reduction processes. As a result, under both isothermal and non-isothermal conditions, the compositions of H_2, CO, CO_2, H_2O, CH_4, and char were obtained (Hameed et al. 2014).

18.2.3 CFD MODELING APPROACH

A CFD modeling for biomass gasification solves the governing equation such as conservation of species, mass, momentum, and energy equation for the particular domain specified in the model with appropriate boundary conditions under a definite kinetic reaction rate. The operations like secondary reactions in pyrolysis, char oxidation, and biomass vaporization are considered separate sub-models in a CFD model. A simple Navier–Stokes equation and Reynolds-averaged Navier–Stokes (k–ε) equation were adopted to model the flow and turbulence inside the gasifier.

The Eulerian–Lagrangian approach is commonly utilized with fluidized bed gasifiers. The information on feedstock material, gasification agent, and the operating parameters should also be specified in the model. Only limited studies were conducted on biomass gasification using the CFD modeling. As selecting an appropriate chemical reaction mechanism for the CFD modeling is difficult and the difficulty in estimating the actual reaction rates, there is an inconsistency in predicting the composition of the syngas (Kumar and Paul 2019).

A two-dimensional CFD model was developed for a downdraft gasifier to estimate the end gas composition by considering the volatile breakup. This model accounted for all gasifier zones, i.e., drying, pyrolysis, oxidation, and reduction. Also, a corrected chemical reaction mechanism along with the model was proposed. As the equivalence ratio increases from 0.35 to 0.6, the syngas composition (H_2 and CO) reduces gradually. The model with corrected chemical reactions showed the best fit with the experimental data (Kumar and Paul 2019). The syngas composition obtained from the two-stage downdraft gasifier working under different fluids was predicted using a 3D CFD model created in ANSYS Fluent. A probability density function is accounted in interpreting the chemical kinetics in this model. The probability density function and the non-premixed combustion approach were used to find the effect of various gasification reagents and also offer the quick prediction of syngas components (Yepes Maya et al. 2021). The unsteady-state 3D CFD model was developed for the circulating fluidized bed gasifier to simulate the producer gas composition from the biomass gasifier (Liu et al. 2014). The hydrodynamic performance of the circulating fluidized bed gasifier has been assessed using an accurate 3D Eulerian–Eulerian multiphase model. The solid spatial distribution in the circulated fluidized bed gasifier has been quantitatively estimated for the first time (Yu et al. 2018).

A new Computational fluid dynamics–discrete element method (CFD-DEM) model was incorporated to simulate the gasification process of biomass in a fluidized bed reactor. The concept is based on an Eulerian–Lagrangian approach, in which the Eulerian approach is adopted to solve the gas phase, and the discrete element method is utilized to solve the particle phase. It was reported that the H_2 and CO_2 proportion increased with the steam-to-biomass ratio, whereas CO concentrations reduced. It is due to the higher reaction rate for water-gas shift reaction with increased steam-to-biomass ratio. Furthermore, CH_4 composition is slightly reduced due to the methane–steam reforming reaction (Ku, Li, and Løvås 2015). CFD-DEM model was used to investigate biomass gasification in a 3D bubbling fluidized bed reactor. It was concluded that increase in the equivalence ratio, gasification temperature, and steam-to-biomass ratio increase the fluid force, impact force, velocity, and dispersion coefficients of both sand and biomass particles. Increasing the temperature increases the heterogeneous rate of chemical reaction, and increasing the equivalence ratio improves the reaction of char oxidation (Yang et al. 2019).

18.2.4 ASPEN Plus Modeling

In earlier days, ASPEN Plus tool was applied to simulate coal conversion and methanol synthesis (Knudsen, Bailey, and Fabiano 1982). The ASPEN Plus simulator was recently utilized to simulate biomass gasification (Tavares et al. 2020). The simulation of gasification process using ASPEN Plus helps to determine the composition of the generated gas (Eikeland, Thapa, and Halvorsen 2015). ASPEN Plus was used to model a double fluidized bed rice husk gasifier. Under various operating conditions such as equivalence ratio, velocity of fluidization, the oxygen concentration in the fluidizing gas, moisture content, and bed height, the models could accurately predict temperature of reactor, heating value of gas, and composition of producer gas (Mansaray et al. 2000).

ASPEN Plus was utilized to model the gasification process of a fluidized bed reactor with the steam and air as an gasifying agent. Drying, devolatilization–pyrolysis, and gasification–combustion were the three steps in their gasification model. The immediate drying was modeled using the RSTOIC module. The pyrolysis/devolatilization section of the model was modeled using the RYIELD module. The combustion and gasification reactions were modeled using the RGIBBS reactor module (Mitta et al. 2006). A gasification model using an ASPEN Plus was created for estimating the steady-state behavior of an air fluidized bed gasifier. The RYIELD, RGIBBS, and RCSTR modules were used to model the feed decomposition, volatile combustion, and char gasification, respectively. The findings revealed that carbon conversion efficiency and H_2 production were increased with temperature (Nikoo and Mahinpey 2008).

18.2.5 Machine Learning Approaches

Machine learning approaches can model biomass gasification and predict the end products and their properties. The prediction ability of the machine learning methods is always higher than the conventional modeling methods. The research on modeling gasification using machine learning methods increased tremendously in the past two decades (Ascher, Watson, and You 2022). A detailed report on the adaptation of machine learning methods for modeling biomass pyrolysis and gasification is presented in the literature (Ascher, Watson, and You 2022). In this section, only the important machine learning method have been discussed.

18.2.5.1 Support Vector Machine Approach

The two different classifiers, namely, multiclass random forests and binary least square support vector machine, were proposed to determine the calorific value and end gas composition obtained from the biomass gasification in a downdraft gasifier. The developed models are cross-validated with the samples of 5237 data and achieved the accuracy level of 89% and 96% for the multiclass random forests and binary least square support vector methods, respectively (Mutlu and Yucel 2018). The least square support vector machine modeling has been implemented to optimize and reduce the tar in the biomass gasification process, showing a greater efficiency than the traditional modeling (D. Li and Guo 2015). It was observed that the performance of a multilayer perceptron method is greater than the performance of SVM in the case of predicting the composition of syngas and heating value of biomass gasification process with a larger data set. The lack of SVM compared with multilayer perception is the consideration of a large data set (Elmaz, Yücel, and Mutlu 2020). It has been reported that the performance of the SVM is comparatively better than other machine learning methods only when the data set is minimum (Ascher, Watson, and You 2022).

18.2.5.2 Artificial Neural Network Approach

An artificial neural network model can predict the highly non-linear functions. The architecture of ANN is comprised of three layers, i.e., input, hidden, and output layers, which are connected in series. The available input and output data are fed into the model as input and output layers. The relation between input and output parameters were processed through the hidden layer. The performance of the ANN model hardly depends on the number of neurons in the each hidden layer and the number of hidden layers. The feed-forward back propagation algorithm is widely adopted due to the fast processing of the data and simple in nature. The Levenberg–Marquardt approach is used as a training function, and mean square error is used as a performance function to train the neural network. The elemental composition like carbon content, hydrogen content, oxygen content, ash content, moisture content, operating parameters like gasifier pressure, gasifier temperature, bed material, and other gasifier parameters are considered as input parameters in the ANN model. The product gas yield, energy efficiency and composition like CO, CO_2, H_2, H_2O, and CH_4 are considered as output parameters. The architecture of the ANN model for the gasification process of biomass is shown in Figure 18.2 which represents the architecture with the possible input and output parameters.

In 2006, a novel equilibrium model was proposed to simulate a biomass gasifier. Authors have mentioned that a non-linear relation existed between the gasifier operational variables, temperature, and fuel composition that were modeled through ANN (D. Brown, Fuchino, and Maréchal 2006; D. W. M. Brown, Fuchino, and Maréchal 2007). The biomass gasification modeling for the fluidized bed gasifiers was carried out to estimate the yield and producer gas composition (CO, CO_2, CH4,

A Review on the Modeling of Biomass Gasification Process 249

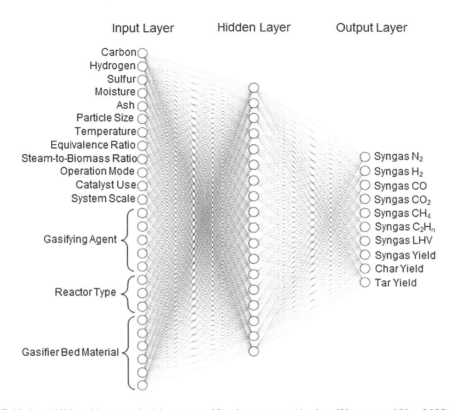

FIGURE 18.2 ANN architecture for biomass gasification process (Ascher, Watson, and You 2022).

and H₂) using the ANN tool. There were two ANN designs developed: one for a bubbling-type fluidized bed gasifier and one for a circulating-type fluidized bed gasifier. The input layer of the model consists of eight neurons, including ash content, moisture content, carbon elements, hydrogen elements, oxygen elements, gasifier temperature, the steam-to-dry biomass ratio and equivalence ratio. The hidden layer is formed with two neurons, and the one output layer is accounted in the model. The models are trained with a back propagation algorithm using the experimental results from the previous literature. The developed models achieved a high correlation coefficient (greater than 0.98) with minimum RMSE values. In the case of the circulating fluidized bed model, the equivalence ratio has a high influence (37.6%) than the other input parameters on the output of producer gas yield. The authors suggested that these models are best suited to control and optimize gasification (Puig-Arnavat et al. 2013).

An ANN model was developed for a fixed bed gasifier to predict producer gas concentrations. It has been observed that the prediction capability of the model is better with larger training data (Baruah, Baruah, and Hazarika 2017). The biomass gasification process in the fluidized bed gasifier was simulated using the ANN approach, and the effect of gasifier length on the H₂ yield, heating value, and efficiency has been studied. The hydrogen yield and efficiency are improved as the gasifier length increases due to a significant improvement in the reaction rate (Y. Li et al. 2018). An ANN model combined with the thermodynamic equilibrium model was proposed for a downdraft gasifier. The developed model was utilized to predict the net power output of the system. The compositions from the proximity analysis, elemental analysis, and operating parameters are considered as an input parameters to the developed model. Among all input parameters, the gasification temperature significantly influenced the power output. It was reported that the correlation coefficient (R^2) value is greater than 0.999 (Safarian et al. 2020).

The composition and yield of product gas (CO, CO_2, H_2, and CH_4) from a bubbling fluidized bed gasifier were predicted using an ANN model and the bed material was also taken as an input parameter. The model was trained with Levenberg–Marquardt back propagation algorithm. The correlation coefficient (R^2) was more than 0.95, and the mean squared error was less than 0.0017 for the developed model. It was concluded that the ANN modeling has the potential to design and control the process in bubbling fluidized bed gasifiers under various operating scenarios (Serrano, Golpour, and Sánchez-Delgado 2020).

18.3 OPTIMIZATION OF THE BIOMASS GASIFICATION PROCESS

Researchers have developed and utilized numerous numerical models to optimize the design and operating parameters of the gasifier. The performance of the gasifier can be ascertained by considering the carbon conversion efficiency, composition of the end gas, total gas yield, gas heating value, and hydrogen production rate. The gasifier performance depends on the gasifier pressure, temperature, bed material, catalyst, gasifying agent, equivalence ratio, particle size of biomass, and moisture content present in biomass. The influence of these parameters on the hydrogen gas yield has been detailed in the following section. The effect of gasifier operating conditions cannot be generalized as it varies with gasifier types. A large amount of tar and char were formed in the fixed bed gasifier due to irregular and low heat and mass transfer behavior between the gasifying medium and the solid feedstocks (Warnecke 2000). Fluidized bed gasifiers improved mixing and gas–solid contact, resulting in a faster reaction rate and higher conversion efficiency. Table 18.1 shows the influence of the design and operating parameters on the performance of the gasification.

18.3.1 Effect of Gasification Temperature

The gasification temperature is an important influencing parameter on the composition of producer gas in the case of ANN model output (Bahadar et al. 2022). Hydrogen concentration is decreased with increased gasifier temperature for a downdraft biomass gasifier with air as a gasification medium (Akyurek et al. 2019). The influence of temperature on hydrogen production showed a reverse trend for fluidized bed gasifiers. The steam gasification process in a fluidized bed reactor was studied. It was found that improving the gasification temperature in reactor from 600 °C to 800 °C resulted in reduced heavy tars and char percentage while boosting the overall gas production, H_2 and CO yields (Rapagnà and Latif 1997). The studies were performed on the gasification processes in a fluidized bed gasifier with EFB (empty fruit bunch) and air at temperatures 700–1000 °C. The amount of H_2, CO, and CH_4 yield improved, whereas the yield of CO_2 decreased as the gasification temperature increased. The concentrations of char and tar were also decreased with a reduction in lower heating value (LHV) of producer gas for higher temperatures (Mohammed et al. 2011). It was found that the amount of gas produced, the composition of hydrogen in the gas, and the carbon conversion efficiency increased with gasification temperature (Lv et al. 2004). The effect of steam gasification with a tri-metallic catalyst in a fluidized fixed bed reactor was studied. The concentrations of H_2 and CO_2 improved as the gasification temperature varied from 750 °C to 900 °C, whereas the amounts of CO and CH_4 decreased. Higher gasification temperature promotes product endothermic reaction and favors exothermic reaction of reactants. As a result, the endothermic hydrocarbon reforming reaction was improved as the temperature was increased, resulting in higher H_2 concentration (J. Li et al. 2009).

18.3.2 Effect of Gasification Pressure

The pressurized gasification systems improved gasification efficiency with less tar (Situmorang et al. 2020). Supercritical water gasification using rice husk under different pressure and temperature range was carried out. The authors reported that the pressure influence is more significant at high temperatures (Basu and Mettanant 2009). The composition of syngas in a pressurized dual fluidized bed gasifier was investigated. The H_2 and CO composition decreased with increased pressure. The composition of the CO_2 and CH_4 showed a reverse trend, i.e., increased with an increase in the pressure (Feng et al. 2013). A similar trend of gas composition with pressure variation was observed. Though these trends in gas compositions were noted, the overall gas yield is increased with an increase in pressure of gasifier (Song et al. 2013; Berrueco et al. 2014). The influence of the operating pressure in gasifier on the LHV of gas, syngas yield, and energy efficiency was assessed. It was concluded that the high gasifier pressure increases the energy efficiency and affects the LHV of gas and syngas yield (Motta et al. 2018).

18.3.3 Effect of the Bed Material and Catalyst

The most common bed particle is sand, which has high mechanical properties and has been widely used in many commercial gasifiers. In addition to the bed materials, catalysts were added to enhance the chemical reactions in the gasifier, increase the yield, and improve the quality of product gases. The commonly used catalysts are nickel, magnesium, limestone, orthosilicate, and other materials which can meet the resistance and mechanical activity requirements (Shen, Gao, and Xiao 2008). The CCE, H_2 composition, and total gas production were improved, while the CO_2 gas composition decreased when calcium oxide was used as a sorbent. Calcium oxide served as both a catalyst and a sorbent, as the hydrocarbons and tars were altered in the absence of CaO, increasing H_2 production (Mahishi and Goswami 2007). The H_2 and CO yield improved as the gasification temperature and catalyst-to-biomass ratio increased; however, the CH_4 and CO_2 composition in end gas reduced slightly. Further, using a calcined dolomite catalyst also accelerated the reaction of tar, lowering the tar yield (Wongsiriamnuay, Kannang, and Tippayawong 2013).

18.3.4 Effect of Gasifying Agent

The amount of tar in the produced gas and the quality could be controlled with a gasifying agent. The gasifier performance of the municipal solid wastes was analyzed using an equilibrium model. The gasifying agents adopted in their study includes air, H_2, and steam. The H_2 and CO mole fractions are higher than CO_2 and H_2O in producer gas, using H_2 gas as a gasifying agent. The yield of H_2 and H_2O is higher compared to CO and CO_2 yield with steam as a gasifying agent (Xu, Jin, and Cheng 2017). The effect of gasifying agents, namely, oxygen, air, air with enriched oxygen and steam, on H_2 content in product gas, the calorific value, and energy efficiency were studied. It was observed that the yield of H_2 is high in the gasification temperature range of 1000–1400 K with the steam as a gasifying agent compared to other agents. The reason is that the nitrogen proportion is zero in the steam gasification process, which causes the shift in the reaction chemical equilibrium, leading to a high yield of H_2 (Shayan, Zare, and Mirzaee 2018). The steam gasification was more effective in boosting hydrogen output than air gasification. For oxygen/steam gasification, higher concentrations of H_2 and CO were recorded (Lv et al. 2004). The CCE and gas output increased when air–steam was employed as the gasifying agent instead of air and the concentration of H_2 and CO in the end producer gas were higher (Wongsiriamnuay, Kannang, and Tippayawong 2013). It was reported that the overall gas and hydrogen yields increased when the steam-to-biomass ratio was increased from 0 to 1.33. However, the overall gas and hydrogen yields started to decrease when the steam-to-biomass ratio increased from 1.33 to 2.67 (J. Li et al. 2009). The H_2 gas concentration varied drastically when the steam-to-biomass ratio increased from 0 to 1. However, H_2 gas concentration and LHV decreased with a further increase in steam-to-biomass ratio beyond 1 (Chang et al. 2011).

18.3.5 Effect of Biomass Particle Size

The effect of particle size distribution in the dual fluidized bed gasifier was investigated. It was found that the limiting particle diameter is based on superficial gas velocity in the gasifier (Wilk and

Hofbauer 2013). The study on the performance of the gasification process is carried out by varying the size of the biomass particles between 0.3 and 1.0 mm. With increasing particle size, the total yield of the end gas reduced while yields of the tar and char increased. For particle sizes of 0.3 and 0.5 mm diameter, hydrogen yields remained nearly constant and then decreased with an increase in the particle size. The highest LHV and the best-quality product gas composition were obtained with particle size 0.3–0.5 mm (Mohammed et al. 2011). The researchers also discovered that smaller particles yield the higher H_2, CH_4, and CO and less CO_2 than the larger particles (J. Li et al. 2009; Lv et al. 2004; Luo et al. 2009). Further, the smaller particles showed higher LHV of gas, yield of end gas, and the best CCE. However, the percentage of char and tar increased for large-diameter particles (Luo et al. 2009).

18.3.6 Effect of Moisture Content

Many researchers studied the influence of moisture content on the performance characteristics of a gasifier. When considering the dry gas composition, the increase in H_2 molar fraction is associated with an increase in the moisture content (dry basis) of the biomass (Sharma 2008; Ratnadhariya and Channiwala 2009). Many researchers encountered the similar trend while increasing the moisture content (Zainal et al. 2001; Akyurek et al. 2019). Further, increasing the moisture in biomass reduces the heating value of end gas and cold gas efficiency (Ghassemi and Shahsavan-Markadeh 2014). It was found that the product gas calorific value reduced with increased moisture content and gasification temperature. Gasification efficiency attains the maximum value at a 20% moisture content level and the efficiency starts to decline with a further increase in moisture in biomass (Zainal et al. 2001). The similar trend was noted for the calorific value of the gas, wherein the maximum value is obtained at a 10% moisture content level (Sharma 2008).

18.3.7 Effect of Equivalence Ratio

The equivalence ratio (ER) is the ratio of the operating oxygen level to the oxygen level required for the stoichiometric condition. The impact of equivalence ratio on the gasifier performance has been widely researched for many different feedstocks and gasifiers in the past few decades. The increase in equivalence ratio led to a decrement in the H_2 gas molar fraction (Sharma 2008; Akyurek et al. 2019). This may be due to hydrogen oxidation with the available excess oxygen. A similar trend is also noticed (Ghassemi and Shahsavan-Markadeh 2014) and it is also observed that the temperature increases with an increase in the equivalence ratio due to the exothermic reaction. The gasification efficiency and calorific value were high for biomass at a lower equivalence ratio (Sharma 2008). At an equivalence ratio of 0.25, the total gas output and the H_2 concentration were higher and then it started to decline as the ER was increased to 0.35 (Mohammed et al. 2011). A slight reduction in the char and tar percentage was noticed when the equivalence ratio was increased from 0.14 to 0.42. However, the total gas production, LHV, and H_2 gas proportion increased and achieved a maximum at an equivalence ratio of 0.21, beyond which a marginal decline was reported. The yields of CH_4 and CO gas decreased as the ER value increased, but the yield of CO_2 gas increased. As the equivalence ratio increases, the gas quality deteriorates due to the increase in oxidation processes, which increases CO_2 production and decreases combustible gas production (Skoulou et al. 2008).

18.4 SUMMARY

A detailed review of the different modeling approaches such as equilibrium, kinetic, ASPEN plus, CFD, and machine learning adopted for the gasification process has been presented. The equilibrium modeling approach considered chemical equilibrium conditions with equilibrium constants or minimization of Gibbs free energy to determine the product gas compositions. This approach is not capable of estimating the local gas composition with time. Kinetic models are employed

TABLE 18.1
Influence of Design and Operating Parameters on the Performance of the Gasification Process

Reference	Gasifier Type	Gasification Medium	Feedstock	Influencing Parameter	Influence	Inference
(Rapagnà and Latif 1997)	Fluidized bed reactor	Steam	Ground almond shells	Temperature	Increase from 600 °C to 800 °C	H_2 and CO increases
				Feedstock particle size	Increase from 300 μm to 1 mm	H_2 decreases gradually
(Mohammed et al. 2011)	Fluidized bed reactor	Air	Empty fruit bunch	Temperature	Increase from 700 °C to 1000 °C	H_2 increases progressively from 10.27 vol% to 38.02 vol%. Total gas yield increases significantly.
				Feedstock particle size	Increase from 0.3 mm to 1 mm	Total gas yield decreases
				Equivalence ratio	Less than 0.3 mm	Gas yield increases to 74.79 wt%
					Increase from 0.15 to 0.35	Gas yield increases from 70.75 wt% to 86.46 wt%
(Lv et al. 2004)	Fluidized bed reactor	Air/steam	Pine sawdust	Temperature	Increase from 700 °C to 900 °C	H_2 content increases
						Gas yield, carbon conversion efficiency, and steam deposition increase
				Feedstock particle size	Increase from 0.2 mm to 0.9 mm	Gas yield, carbon conversion efficiency, steam deposition and LHV of gas increases
				Equivalence ratio	Increase from 0.19 to 0.27	Gas yield, carbon conversion efficiency, team deposition, and LHV of gas first increase and then decrease
(J. Li et al. 2009)	Fixed bed reactor	Steam	Palm oil wastes	Catalyst	No catalyst	Hydrogen yield is 39.75 (g H_2/kg biomass)
					Calcined dolomite	Hydrogen yield is 52.98 (g H_2/kg biomass)
					Nano-NiLaFe/g-Al_2O_3 catalyst	Hydrogen yield is 101.78 (g H_2/kg biomass)
				Feedstock particle size	Decrease from 5 to 0.15 mm	H_2 gas yield is increased LHV of gas decreases
(Feng et al. 2013)	Interconnected fluidized bed reactor	Air/steam	Sawdust from Jiangsu Province	Temperature	Increase from 650 °C to 950 °C *750 is opt	Syngas composition increases around 60–90 mol%
						Carbon conversion of feedstock decreases significantly
				Pressure	Increase from 0.1 MPa to 0.6 MPa	Syngas composition decreases around from 87 mol% to 75 mol%

(Continued)

TABLE 18.1 (Continued)
Influence of Design and Operating Parameters on the Performance of the Gasification Process

Reference	Gasifier Type	Gasification Medium	Feedstock	Influencing Parameter	Influence	Inference
(Song et al. 2013)	Interconnected fluidized bed reactor	Air/steam	—	Pressure	Increase from 10 bar to 20 bar	Syngas composition increases from 0.737 kg/kg to 0.756 kg/kg
(Berrueco et al. 2014)	Fluidized bed reactor	Steam	Norwegian spruce and Norwegian forest residues	Temperature	Increase from 750 °C to 850 °C	Gas yield and char gasification reaction enhance significantly
				Bed material	Dolomite	H_2 yield is high for both biomass with increase in temperature Cracking reactions are dominant
					Sand	CO and CO_2 increase with increase in the temperature Tar repolymerization reactions are dominant
(Shen, Gao, and Xiao 2008)	Fluidized bed reactor	Air/steam	Straw	Temperature	Increase from 650 °C to 900 °C	H_2 increases gradually up to 800 °C and then decreases. CO increases gradually. CO_2 and CH_4 decease gradually
(Mahishi and Goswami 2007)	Simple batch-type reactor	Steam	Southern pine bark	Temperature	Increase from 500 °C to 700 °C	H_2, CO, and CO_2 increase gradually and CH_4 deceases
(Wongsiriamnuay, Kannang, and Tippayawong 2013)	Fluidized bed reactor	Air/steam	Bamboo	Temperature and air/steam	Increase from 400 °C to 600 °C	CO_2 increases gradually H_2, CO, and CH_4 decrease with increase in temperature for both air and steam gasification agents
(Wilk and Hofbauer 2013)	Dual fluidized bed gasifier	Steam	Sawdust and pellets	Particle size	0–1 mm (0–60%)	H_2 and CO_2 decrease, and CO, CH_4, and C_2H_4 increase
(Luo et al. 2009)	Fixed bed reactor	Steam	Pine sawdust	Temperature	Increase from 600 °C to 900 °C	H_2 and CO_2 increase from 25.2% to 51.5%, and CO decreases by 51%
				Steam-to-biomass ratio	Increase from 0 to 2.8	H_2 increases up to S/B = 2.1. CO and CH_4 decrease. Carbon conversion efficiency increases up to S/B = 1.43
(Skoulou et al. 2008)	Fixed bed reactor	Air	Olive kernels and olive tree cuttings	Temperature	Increase from 750 °C to 950 °C	H_2 and CO increase, and CO_2 decreases. CH4 increases and then decreases
			Olive kernels	Equivalence ratio	Increase from 0.14 to 0.42	H_2 and CH_4 increase and then decreases. CO_2 increases. CO decreases
			Olive tree cuttings			H_2 increases and then decreases CO_2 increases. CO and CH_4 decrease

to determine the gas composition at various locations and timescales by solving the appropriate reaction rate equations. Different standard modules in ASPEN plus are adopted to represent the gasifier, and the output of these modules is interlinked to obtain the gasifier performance. The hydrodynamic and chemical kinetic behavior of the fluid and solid particles inside the gasifier are modeled using commercial CFD software. Though these simulations are computationally intensive, they can accurately predict the temperature and gas distribution inside the gasifier. An extensive set of experimental data are used to train the machine learning model. The prediction accuracy of this model is higher than all other models. The developed numerical models can be used to optimize the gasifier design and operating conditions, such as the gasifier pressure, temperature, bed material, catalyst, gasifying agent, equivalence ratio, biomass particle size, and biomass moisture content. A review of the effect of these parameters on the hydrogen gas yield has been detailed.

REFERENCES

Akyurek, Z., A. Akyuz, M. Y. Naz, S. A. Sulaiman, B. C. Lütfüoğlu, and A. Gungor. 2019. "Numerical Simulation of Stoichiometric Thermodynamic Equilibrium Model of a Downdraft Biomass Air Gasifier." *Solid Fuel Chemistry* 53 (6). https://doi.org/10.3103/S0361521919070012.

Ascher, Simon, William Sloan, Ian Watson, and Siming You. 2022. "A Comprehensive Artificial Neural Network Model for Gasification Process Prediction." *Applied Energy* 320 (August): 119289. https://doi.org/10.1016/J.APENERGY.2022.119289.

Ascher, Simon, Ian Watson, and Siming You. 2022. "Machine Learning Methods for Modelling the Gasification and Pyrolysis of Biomass and Waste." *Renewable and Sustainable Energy Reviews*. https://doi.org/10.1016/j.rser.2021.111902.

Azzone, Emanuele, Mirko Morini, and Michele Pinelli. 2012. "Development of an Equilibrium Model for the Simulation of Thermochemical Gasification and Application to Agricultural Residues." *Renewable Energy* 46 (October): 248–54. https://doi.org/10.1016/J.RENENE.2012.03.017.

Bahadar, Ali, Ramesh Kanthasamy, Hani Hussain Sait, Mohammed Zwawi, Mohammed Algarni, Bamidele Victor Ayodele, Chin Kui Cheng, and Lim Jun Wei. 2022. "Elucidating the Effect of Process Parameters on the Production of Hydrogen-Rich Syngas by Biomass and Coal Co-Gasification Techniques: A Multi-Criteria Modeling Approach." *Chemosphere* 287. https://doi.org/10.1016/j.chemosphere.2021.132052.

Baruah, Dipal, D. C. Baruah, and M. K. Hazarika. 2017. "Artificial Neural Network Based Modeling of Biomass Gasification in Fixed Bed Downdraft Gasifiers." *Biomass and Bioenergy* 98. https://doi.org/10.1016/j.biombioe.2017.01.029.

Basu, Prabir. 2010. "Design of Biomass Gasifiers." *Biomass Gasification Design Handbook*, (January): 167–228. https://doi.org/10.1016/B978-0-12-374988-8.00006-4.

Basu, Prabir, and Vichuda Mettanant. 2009. "Biomass Gasification in Supercritical Water – A Review." *International Journal of Chemical Reactor Engineering* 7 (1). https://doi.org/10.2202/1542-6580.1919.

Berrueco, C., D. Montané, B. Matas Güell, and G. del Alamo. 2014. "Effect of Temperature and Dolomite on Tar Formation During Gasification of Torrefied Biomass in a Pressurized Fluidized Bed." *Energy* 66 (March): 849–59. https://doi.org/10.1016/J.ENERGY.2013.12.035.

Biagini, Enrico. 2016. "Study of the Equilibrium of Air-Blown Gasification of Biomass to Coal Evolution Fuels." *Energy Conversion and Management* 128. https://doi.org/10.1016/j.enconman.2016.09.068.

Biagini, Enrico, Federica Barontini, and Leonardo Tognotti. 2016. "Development of a Bi-Equilibrium Model for Biomass Gasification in a Downdraft Bed Reactor." *Bioresource Technology* 201. https://doi.org/10.1016/j.biortech.2015.11.057.

Brown, David W.M., Tetsuo Fuchino, and François M.A. Maréchal. 2006. "Solid Fuel Decomposition Modelling for the Design of Biomass Gasification Systems." *Computer Aided Chemical Engineering* 21 (C). https://doi.org/10.1016/S1570-7946(06)80286-5.

Brown, David W.M., Tetsuo Fuchino, and François M.A. Maréchal. 2007. "Stoichiometric Equilibrium Modelling of Biomass Gasification: Validation of Artificial Neural Network Temperature Difference Parameter Regressions." *Journal of Chemical Engineering of Japan* 40 (3). https://doi.org/10.1252/jcej.40.244.

Chang, Alex C.C., Hsin Fu Chang, Fon Jou Lin, Kuo Hsin Lin, and Chi Hung Chen. 2011. "Biomass Gasification for Hydrogen Production." *International Journal of Hydrogen Energy* 36: 14252–60. https://doi.org/10.1016/j.ijhydene.2011.05.105.

Chen, Hongzhang, and Lan Wang. 2017. "Introduction." *Technologies for Biochemical Conversion of Biomass* (January): 1–10. https://doi.org/10.1016/B978-0-12-802417-1.00001-6.

Corbetta, Michele, Andrea Bassani, Flavio Manenti, Carlo Pirola, Enrico Maggio, Alberto Pettinau, Paolo Deiana, Sauro Pierucci, and Eliseo Ranzi. 2015. "Multi-Scale Kinetic Modeling and Experimental Investigation of Syngas Production from Coal Gasification in Updraft Gasifiers." *Energy and Fuels* 29 (6). https://doi.org/10.1021/acs.energyfuels.5b00648.

Eikeland, Marianne S., Rajan K. Thapa, and Britt M. Halvorsen. 2015. "Aspen Plus Simulation of Biomass Gasification with Known Reaction Kinetic." Proceedings of the 56th Conference on Simulation and Modelling (SIMS 56), October, 7–9, Linköping University, Sweden, vol. 119. https://doi.org/10.3384/ecp15119149.

Elmaz, Furkan, Özgün Yücel, and Ali Yener Mutlu. 2020. "Predictive Modeling of Biomass Gasification with Machine Learning-Based Regression Methods." *Energy* 191. https://doi.org/10.1016/j.energy.2019.116541.

Feng, Fei, Guohui Song, Laihong Shen, Jun Xiao, and Lei Zhang. 2013. "Simulation of Bio-Syngas Production from Biomass Gasification via Pressurized Interconnected Fluidized Beds." *Nongye Jixie Xuebao/Transactions of the Chinese Society for Agricultural Machinery* 44 (3). https://doi.org/10.6041/j.issn.1000-1298.2013.03.024.

García, Roberto, Consuelo Pizarro, Antonio G. Lavín, and Julio L. Bueno. 2017. "Biomass Sources for Thermal Conversion. Techno-Economical Overview." *Fuel* 195 (May): 182–9. https://doi.org/10.1016/J.FUEL.2017.01.063.

George, Joel, P. Arun, and C. Muraleedharan. 2016. "Stoichiometric Equilibrium Model Based Assessment of Hydrogen Generation through Biomass Gasification." *Procedia Technology* 25. https://doi.org/10.1016/j.protcy.2016.08.194.

Ghassemi, Hojat, and Rasoul Shahsavan-Markadeh. 2014. "Effects of Various Operational Parameters on Biomass Gasification Process; a Modified Equilibrium Model." *Energy Conversion and Management* 79 (March): 18–24. https://doi.org/10.1016/J.ENCONMAN.2013.12.007.

Habibollahzade, Ali, Pouria Ahmadi, and Marc A. Rosen. 2021. "Biomass Gasification Using Various Gasification Agents: Optimum Feedstock Selection, Detailed Numerical Analyses and Tri-Objective Grey Wolf Optimization." *Journal of Cleaner Production* 284 (February): 124718. https://doi.org/10.1016/J.JCLEPRO.2020.124718.

Hameed, Samreen, Naveed Ramzan, Zaka Ur Rahman, Muhammad Zafar, and Sheema Riaz. 2014. "Kinetic Modeling of Reduction Zone in Biomass Gasification." *Energy Conversion and Management* 78. https://doi.org/10.1016/j.enconman.2013.10.049.

Hejazi, Bijan, John R. Grace, and Andrés Mahecha-Botero. 2019. "Kinetic Modeling of Lime-Enhanced Biomass Steam Gasification in a Dual Fluidized Bed Reactor." *Industrial and Engineering Chemistry Research* 58 (29). https://doi.org/10.1021/acs.iecr.9b01241.

Huang, Hua Jiang, and Shri Ramaswamy. 2009. "Modeling Biomass Gasification Using Thermodynamic Equilibrium Approach." *Applied Biochemistry and Biotechnology* 154. https://doi.org/10.1007/s12010-008-8483-x.

IEA. 2021. *Key World Energy Statistics 2021*. IEA.

Knudsen, R. A., T. Bailey, and L. A. Fabiano. 1982. "Experience with Aspen While Simulating a New Methanol Plant." AIChE Symposium Series. https://searchworks.stanford.edu/view/1022263.

Krigstin, S., R. Levin, and S. Wetzel. 2012. "Bioenergy for the Urban Environment." *Metropolitan Sustainability: Understanding and Improving the Urban Environment* (January): 556–84. https://doi.org/10.1533/9780857096463.3.556.

Ku, Xiaoke, Tian Li, and Terese Løvås. 2015. "CFD-DEM Simulation of Biomass Gasification with Steam in a Fluidized Bed Reactor." *Chemical Engineering Science* 122. https://doi.org/10.1016/j.ces.2014.08.045.

Kumar, Umesh, and Manosh C. Paul. 2019. "CFD Modelling of Biomass Gasification with a Volatile Break-Up Approach." *Chemical Engineering Science*. https://doi.org/10.1016/j.ces.2018.09.038.

Li, Dazhong, and Fang Guo. 2015. "A LSSVM Modeling Method Based on GRA for the Removal Process of Biomass Gasification Tar." *Taiyangneng Xuebao/Acta Energiae Solaris Sinica* 36 (7).

Li, Jianfen, Yanfang Yin, Xuanming Zhang, Jianjun Liu, and Rong Yan. 2009. "Hydrogen-Rich Gas Production by Steam Gasification of Palm Oil Wastes Over Supported Tri-Metallic Catalyst." *International Journal of Hydrogen Energy* 34 (22). https://doi.org/10.1016/j.ijhydene.2009.09.030.

Li, X. T., J. R. Grace, A. P. Watkinson, C. J. Lim, and A. Ergüdenler. 2001. "Equilibrium Modeling of Gasification: A Free Energy Minimization Approach and Its Application to a Circulating Fluidized Bed Coal Gasifier." *Fuel* 80 (2): 195–207. https://doi.org/10.1016/S0016-2361(00)00074-0.

Li, X. T., J. R. Grace, C. J. Lim, A. P. Watkinson, H. P. Chen, and J. R. Kim. 2004. "Biomass Gasification in a Circulating Fluidized Bed." *Biomass and Bioenergy* 26 (2): 171–93. https://doi.org/10.1016/S0961-9534(03)00084-9.

Li, Yingfang, Li Yan, Bo Yang, Wei Gao, and Mohammad Reza Farahani. 2018. "Simulation of Biomass Gasification in a Fluidized Bed by Artificial Neural Network (ANN)." *Energy Sources, Part A: Recovery, Utilization and Environmental Effects* 40 (5). https://doi.org/10.1080/15567036.2016.1270372.

Liu, Hui, Ali Elkamel, Ali Lohi, and Mazda Biglari. 2014. "Effect of Char Combustion Product Distribution Coefficient on the CFD Modeling of Biomass Gasification in a Circulating Fluidized Bed." *Industrial and Engineering Chemistry Research* 53 (13). https://doi.org/10.1021/ie404239u.

Luo, Siyi, Bo Xiao, Xianjun Guo, Zhiquan Hu, Shiming Liu, and Maoyun He. 2009. "Hydrogen-Rich Gas from Catalytic Steam Gasification of Biomass in a Fixed Bed Reactor: Influence of Particle Size on Gasification Performance." *International Journal of Hydrogen Energy* 34 (3). https://doi.org/10.1016/j.ijhydene.2008.10.088.

Lv, P. M., Z. H. Xiong, J. Chang, C. Z. Wu, Y. Chen, and J. X. Zhu. 2004. "An Experimental Study on Biomass Air-Steam Gasification in a Fluidized Bed." *Bioresource Technology* 95 (1). https://doi.org/10.1016/j.biortech.2004.02.003.

Mahishi, Madhukar R., and D. Y. Goswami. 2007. "An Experimental Study of Hydrogen Production by Gasification of Biomass in the Presence of a CO2 Sorbent." *International Journal of Hydrogen Energy* 32 (14). https://doi.org/10.1016/j.ijhydene.2007.03.030.

Mansaray, K. G., A. M. Al-Taweel, A. E. Ghaly, F. Hamdullahpur, and V. I. Ugursal. 2000. "Mathematical Modeling of a Fluidized Bed Rice Husk Gasifier: Part I – Model Development." *Energy Sources* 22 (1). https://doi.org/10.1080/00908310050014243.

Mathieu, Philippe, and Raphael Dubuisson. 2002. "Performance Analysis of a Biomass Gasifier." *Energy Conversion and Management* 43. https://doi.org/10.1016/S0196-8904(02)00015-8.

Mitta, Narendar R., Sergio Ferrer-Nadal, Aleksandar M. Lazovic, José F. Parales, Enric Velo, and Luis Puigjaner. 2006. "Modelling and Simulation of a Tyre Gasification Plant for Synthesis Gas Production." *Computer Aided Chemical Engineering* 21 (C). https://doi.org/10.1016/S1570-7946(06)80304-4.

Mohammed, M. A. A., A. Salmiaton, W. A. K. G. Wan Azlina, M. S. Mohammad Amran, and A. Fakhru'L-Razi. 2011. "Air Gasification of Empty Fruit Bunch for Hydrogen-Rich Gas Production in a Fluidized-Bed Reactor." *Energy Conversion and Management* 52 (2). https://doi.org/10.1016/j.enconman.2010.10.023.

Motta, Ingrid Lopes, Nahieh Toscano Miranda, Rubens Maciel Filho, and Maria Regina Wolf Maciel. 2018. "Biomass Gasification in Fluidized Beds: A Review of Biomass Moisture Content and Operating Pressure Effects." *Renewable and Sustainable Energy Reviews*. https://doi.org/10.1016/j.rser.2018.06.042.

Mutlu, Ali Yener, and Ozgun Yucel. 2018. "An Artificial Intelligence Based Approach to Predicting Syngas Composition for Downdraft Biomass Gasification." *Energy* 165. https://doi.org/10.1016/j.energy.2018.09.131.

National Statement at COP26 Summit. 2021. *National Statement by Prime Minister Shri Narendra Modi at COP26 Summit in Glasgow*, Press Release. National Statement at COP26 Summit, November.

Nikoo, Mehrdokht B., and Nader Mahinpey. 2008. "Simulation of Biomass Gasification in Fluidized Bed Reactor Using ASPEN Plus." *Biomass and Bioenergy* 32 (12). https://doi.org/10.1016/j.biombioe.2008.02.020.

Pang, Shusheng. 2016. "Fuel Flexible Gas Production: Biomass, Coal and Bio-Solid Wastes." *Fuel Flexible Energy Generation: Solid, Liquid and Gaseous Fuels* (January): 241–69. https://doi.org/10.1016/B978-1-78242-378-2.00009-2.

Parashar, C. K., P. Das, S. Samanta, A. Ganguly, and P. K. Chatterjee. 2020. "Municipal Solid Wastes – A Promising Sustainable Source of Energy: A Review on Different Waste-to-Energy Conversion Technologies." *Energy Recovery Processes from Wastes*. https://doi.org/10.1007/978-981-32-9228-4_13.

Popp, József, Sándor Kovács, Judit Oláh, Zoltán Divéki, and Ervin Balázs. 2021. "Bioeconomy: Biomass and Biomass-Based Energy Supply and Demand." *New Biotechnology* 60 (January): 76–84. https://doi.org/10.1016/J.NBT.2020.10.004.

Puig-Arnavat, Maria, Joan Carles Bruno, and Alberto Coronas. 2010. "Review and Analysis of Biomass Gasification Models." *Renewable and Sustainable Energy Reviews*. https://doi.org/10.1016/j.rser.2010.07.030.

Puig-Arnavat, Maria, J. Alfredo Hernández, Joan Carles Bruno, and Alberto Coronas. 2013. "Artificial Neural Network Models for Biomass Gasification in Fluidized Bed Gasifiers." *Biomass and Bioenergy* 49. https://doi.org/10.1016/j.biombioe.2012.12.012.

Puig-Arnavat, Maria, Juan Carlos Bruno, and Alberto Coronas. 2012. "Modified Thermodynamic Equilibrium Model for Biomass Gasification: A Study of the Influence of Operating Conditions." *Energy and Fuels* 26 (2). https://doi.org/10.1021/ef2019462.

Rapagnà, Sergio, and Ajmal Latif. 1997. "Steam Gasification of Almond Shells in a Fluidised Bed Reactor: The Influence of Temperature and Particle Size on Product Yield and Distribution." *Biomass and Bioenergy* 12 (4). https://doi.org/10.1016/S0961-9534(96)00079-7.

Ratnadhariya, J. K., and S. A. Channiwala. 2009. "Three Zone Equilibrium and Kinetic Free Modeling of Biomass Gasifier – a Novel Approach." *Renewable Energy* 34 (4). https://doi.org/10.1016/j.renene.2008.08.001.

Ren, Jie, Yi Ling Liu, Xiao Yan Zhao, and Jing Pei Cao. 2020. "Biomass Thermochemical Conversion: A Review on Tar Elimination from Biomass Catalytic Gasification." *Journal of the Energy Institute* 93 (3): 1083–98. https://doi.org/10.1016/J.JOEI.2019.10.003.

Ruggiero, M., and G. Manfrida. 1999. "An Equilibriljm Model for Biomass Gasiflcatlon Processes." *Renewable Energy* 16.

Safarian, Sahar, Seyed Mohammad Ebrahimi Saryazdi, Runar Unnthorsson, and Christiaan Richter. 2020. "Artificial Neural Network Integrated with Thermodynamic Equilibrium Modeling of Downdraft Biomass Gasification-Power Production Plant." *Energy* 213 (December): 118800. https://doi.org/10.1016/J.ENERGY.2020.118800.

Safarian, Sahar, Runar Unnthorsson, and Christiaan Richter. 2021. "Hydrogen Production via Biomass Gasification: Simulation and Performance Analysis under Different Gasifying Agents." *Biofuels*. https://doi.org/10.1080/17597269.2021.1894781.

Serrano, Daniel, Iman Golpour, and Sergio Sánchez-Delgado. 2020. "Predicting the Effect of Bed Materials in Bubbling Fluidized Bed Gasification Using Artificial Neural Networks (ANNs) Modeling Approach." *Fuel* 266. https://doi.org/10.1016/j.fuel.2020.117021.

Sharma, Avdhesh Kr. 2008. "Equilibrium Modeling of Global Reduction Reactions for a Downdraft (Biomass) Gasifier." *Energy Conversion and Management* 49 (4). https://doi.org/10.1016/j.enconman.2007.06.025.

Shayan, E., V. Zare, and I. Mirzaee. 2018. "Hydrogen Production from Biomass Gasification; a Theoretical Comparison of Using Different Gasification Agents." *Energy Conversion and Management* 159 (March): 30–41. https://doi.org/10.1016/J.ENCONMAN.2017.12.096.

Shen, Laihong, Yang Gao, and Jun Xiao. 2008. "Simulation of Hydrogen Production from Biomass Gasification in Interconnected Fluidized Beds." *Biomass and Bioenergy* 32 (2). https://doi.org/10.1016/j.biombioe.2007.08.002.

Situmorang, Yohanes Andre, Zhongkai Zhao, Akihiro Yoshida, Abuliti Abudula, and Guoqing Guan. 2020. "Small-Scale Biomass Gasification Systems for Power Generation (<200 kW Class): A Review." *Renewable and Sustainable Energy Reviews* 117 (January): 109486. https://doi.org/10.1016/J.RSER.2019.109486.

Skoulou, V., A. Zabaniotou, G. Stavropoulos, and G. Sakelaropoulos. 2008. "Syngas Production from Olive Tree Cuttings and Olive Kernels in a Downdraft Fixed-Bed Gasifier." *International Journal of Hydrogen Energy* 33 (4). https://doi.org/10.1016/j.ijhydene.2007.12.051.

Song, Guohui, Lulu Chen, Jun Xiao, and Laihong Shen. 2013. "Exergy Evaluation of Biomass Steam Gasification via Interconnected Fluidized Beds." *International Journal of Energy Research* 37 (14). https://doi.org/10.1002/er.2987.

Tavares, Raquel, Eliseu Monteiro, Fouzi Tabet, and Abel Rouboa. 2020. "Numerical Investigation of Optimum Operating Conditions for Syngas and Hydrogen Production from Biomass Gasification Using Aspen Plus." *Renewable Energy* 146. https://doi.org/10.1016/j.renene.2019.07.051.

Tovar-Facio, Javier, Brenda Cansino-Loeza, and José María Ponce-Ortega. 2022. "Management of Renewable Energy Sources." *Sustainable Design for Renewable Processes* (January): 3–31. https://doi.org/10.1016/B978-0-12-824324-4.00004-4.

Varma, Anil Kumar, Ravi Shankar, and Prasenjit Mondal. 2018. "A Review on Pyrolysis of Biomass and the Impacts of Operating Conditions on Product Yield, Quality, and Upgradation." *Recent Advancements in Biofuels and Bioenergy Utilization*. https://doi.org/10.1007/978-981-13-1307-3_10.

Vedharaj, S. 2021. "Advanced Ignition System to Extend the Lean Limit Operation of Spark-Ignited (SI) Engines – A Review." *Energy, Environment, and Sustainability*. https://doi.org/10.1007/978-981-16-1513-9_10.

Warnecke, Ragnar. 2000. "Gasification of Biomass: Comparison of Fixed Bed and Fluidized Bed Gasifier." *Biomass and Bioenergy* 18 (6). https://doi.org/10.1016/S0961-9534(00)00009-X.

Whittaker, Carly, and Ian Shield. 2018. "Biomass Harvesting, Processing, Storage, and Transport." *Greenhouse Gas Balances of Bioenergy Systems* (January): 97–106. https://doi.org/10.1016/B978-0-08-101036-5.00007-0.

Wilk, V., and H. Hofbauer. 2013. "Influence of Fuel Particle Size on Gasification in a Dual Fluidized Bed Steam Gasifier." *Fuel Processing Technology* 115 (November): 139–51. https://doi.org/10.1016/J.FUPROC.2013.04.013.

Wongsiriamnuay, Thanasit, Nattakarn Kannang, and Nakorn Tippayawong. 2013. "Effect of Operating Conditions on Catalytic Gasification of Bamboo in a Fluidized Bed." *International Journal of Chemical Engineering*. https://doi.org/10.1155/2013/297941.

Xu, Pengcheng, Yong Jin, and Yi Cheng. 2017. "Thermodynamic Analysis of the Gasification of Municipal Solid Waste." *Engineering* 3 (3): 416–22. https://doi.org/10.1016/J.ENG.2017.03.004.

Yaghoubi, Ebrahim, Qingang Xiong, Mohammad Hossein Doranehgard, Mehdi Mihandoust Yeganeh, Gholamreza Shahriari, and Mehdi Bidabadi. 2018. "The Effect of Different Operational Parameters on Hydrogen Rich Syngas Production from Biomass Gasification in a Dual Fluidized Bed Gasifier." *Chemical Engineering and Processing – Process Intensification* 126 (April): 210–21. https://doi.org/10.1016/J.CEP.2018.03.005.

Yang, Shiliang, Hua Wang, Yonggang Wei, Jianhang Hu, and Jia Wei Chew. 2019. "Particle-Scale Modeling of Biomass Gasification in the Three-Dimensional Bubbling Fluidized Bed." *Energy Conversion and Management* 196. https://doi.org/10.1016/j.enconman.2019.05.105.

Yepes Maya, Diego Mauricio, Electo Eduardo Silva Lora, Rubenildo Vieira Andrade, Albert Ratner, and Juan Daniel Martínez Angel. 2021. "Biomass Gasification Using Mixtures of Air, Saturated Steam, and Oxygen in a Two-Stage Downdraft Gasifier. Assessment Using a CFD Modeling Approach." *Renewable Energy* 177. https://doi.org/10.1016/j.renene.2021.06.051.

Yu, Xi, Paula H. Blanco, Yassir Makkawi, and Anthony v. Bridgwater. 2018. "CFD and Experimental Studies on a Circulating Fluidised Bed Reactor for Biomass Gasification." *Chemical Engineering and Processing – Process Intensification* 130 (August): 284–95. https://doi.org/10.1016/J.CEP.2018.06.018.

Zainal, Z. A., R. Ali, C. H. Lean, and K. N. Seetharamu. 2001. "Prediction of Performance of a Downdraft Gasifier Using Equilibrium Modeling for Different Biomass Materials." *Energy Conversion and Management* 42 (12). https://doi.org/10.1016/S0196-8904(00)00078-9.

19 A Comprehensive Review of Bio-catalyst Synthesis, Characterization, and Feedstock Selection for Biodiesel Synthesis Using Different Methods

Babu Dharmalingam, Malinee Sriariyanun, Anand Ramanathan, Santhoshkumar A., Selvakumar Ramalingam, Deepakkumar R., and Kasturi Bhattacharya

CONTENTS

19.1 Introduction to Energy Scenario in India and around the World 261
19.2 Feedstock Selection and Biodiesel Production through Different Techniques 262
    19.2.1 First-, Second-, and Third-generation Feedstock for Biodiesel Production 262
    19.2.2 Heterogeneous Bio-catalyst in Biodiesel Production 263
    19.2.3 Conventional Biodiesel Production Technique 264
    19.2.4 Microwave-assisted Transesterification Process 265
    19.2.5 Ultrasonic-assisted Transesterification Process 266
19.3 Conclusion 267
Acknowledgments 267
References 267

19.1 INTRODUCTION TO ENERGY SCENARIO IN INDIA AND AROUND THE WORLD

According to the International Outlet 2016, world energy consumption can be extended by 48% from 2012 to 2040, due to population growth and economic development. By 2040, world energy demand will reach 1094 million barrels per day (World Oil Outlook 2016). Petroleum diesel demand faces two challenges: a scarcity of sources and the environmental pollution. These challenges are the driving force to look for a new alternative or a long-lasting substitute for diesel. In the recent century, biodiesel production increased due to an increase in energy demand, consumption, and dwindling of fossil fuels at a much higher rate (Olkiewicz et al. 2016). Biodiesel consumption and production have increased in many countries, including the United States, Austria, France, Italy, Brazil, China, Germany, Japan, India, and Malaysia. Africa has a large number of feedstock like grains, dry beans, canola, groundnuts, sunflower, soybeans, barley, etc., but there are only fewer bioenergy production plants. Africa, Europe, and the United States consumed 1 billion, 4 million, and 2.1 billion liters of biodiesel in 2007, 2013, and 2015, respectively. Transportation in Europe and the United States consume a high amount of diesel compared to biodiesel, which is only 2% and 0.5%, respectively. In Argentina, biodiesel demand in 2010 was 700 million liters and has the most developed vegetable oil industry. Brazil is leading for ethanol produced from sugarcane and biodiesel mostly produced from cotton seeds, soybeans, and palm (Biodiesel Magazine 2017).

India is a developing country with a faster rate of development than other developing countries; hence, the industrial and transportation sectors require a lot of energy. Because of the rising prices and dwindling supply of crude oil, there is a demand for renewable and sustainable alternative fuels. The Indian Government has encouraged companies to invest in biofuels. In 2003, the Indian Government has issued national biofuel policies to encourage alternative fuel generation, plant cultivation, and the use of biofuel instead of petroleum diesel. To avoid future food versus fuel conflicts, technical advancements are leading to the utilization of non-edible feedstock. In 2009, the Ministry of New and Renewable Energy issued a National Biofuel Policy to promote biofuel in the energy and transportation sectors. Two projects in Maharashtra and Andhra Pradesh, funded by the Indian Government, are testing the efficiency of biodiesel produced from 85 different plants in diesel engines in collaboration with 21 other countries (Nikul et al. 2012). The above literature studies have clearly mentioned energy scarcity, fossil fuel depletion, and environmental pollution due to the combustion of fossil fuels in diesel engines. Hence, the upcoming section will discuss feedstock selection, heterogeneous catalyst and bio-catalyst synthesis and characterization, and various biodiesel production techniques, including microwave-assisted and ultrasonic-assisted transesterification process.

19.2 FEEDSTOCK SELECTION AND BIODIESEL PRODUCTION THROUGH DIFFERENT TECHNIQUES

19.2.1 First-, Second-, and Third-generation Feedstock for Biodiesel Production

The region's climatic conditions, local soil characteristics, topographical location, and farming procedures have influenced the cost and availability of raw materials for biodiesel synthesis. Globally, around 350 crops have been recognized as possible biodiesel production feedstock. Among 350 crops, the primary and most important source of biodiesel production is edible oil (Nikul et al. 2012). About 91% of this biodiesel is made from edible oil sources. Numerous edible oils have contributed to the production of biodiesel, including sunflower oil 3%, palm oil 1%, rapeseed oil 85%, and other remaining sources 2%. Biodiesel was derived from edible oils, for instance, canola oil, coconut oil, groundnut oil, palm oil, rapeseed oil, and soybean oil. Identification of both conventional and unconventional sources for biodiesel production such as groundnut oil, rice bran oil, coconut oil, sunflower oil, soybean oil, safflower oil, palm oil, rapeseed oil, *Jatropha* oil, cotton seed oil, and Karanja. Lard oil, tallow, fish oil, algae, fungi, and microalgae are among non-traditional sources. The biodiesel was produced through the transesterification process from sunflower oil using a heterogeneous catalyst (Singh et al. 2021).

Earlier research found that using vegetable oil in biodiesel production resulted in environmental issues such as deforestation and soil erosion. In addition, the value of so produced vegetable oil would hike, putting biodiesel's economic sustainability at threat. Furthermore, it causes a society-wide food versus fuel problem, as observed by Baskar and Aiswarya (2016). As a result of the higher price of feedstock, vegetable oil–based biodiesel costs around 1.5 times as much as the regular diesel. Feedstock should be inexpensive and accessible. As a result, choosing edible oil for biodiesel synthesis as a feedstock is a quick decision (Demirbas et al. 2016). In biodiesel production, the usage of edible oils may result in food versus fuel conflict. Due to this, the inclusion of damaging components and non-edible oils are not recommended for consumption by human beings (Jahirul et al. 2015). Moreover, biodiesel derived from edible plants currently does not compete economically with petroleum-based fuels due to their higher cost, but with the recent spike in petroleum charges and uncertainty, there stands increased attention to non-edible oil-based biodiesel for engines.

Second-generation feedstock is a possible option for biodiesel generation because of its cost and availability (Boutesteijn et al. 2017). The second-generation feedstock is Karanja oil, Mahua oil, *Jatropha* oil, *Pongamia pinnata* oil, and rubber seed oil which are some of the non-edible oil crops that can be used in biodiesel production. Planting non-edible oil plants in wasteland throughout the world can help to minimize deforestation while also being efficient and environment-friendly. Biodiesel production from *Papaver somniferum* seed oil (PSO) is a low-cost biodiesel feedstock; it also does not contend with other food crops. The maximum biodiesel yield is produced by the trans-esterification of the oil process, which is affected by a variety of constraints such as molar ratio, reaction duration, stirrer speed, catalyst concentration, and temperature. PSO-derived biodiesel might be a next-generation feedstock that fits contemporary market demand for transportation of fuel while simultaneously addressing environmental concerns. The physicochemical properties and fatty acid composition make *Jatropha* oil a prominent second-generation feedstock for biodiesel production (Kurczyński et al. 2021).

Third-generation microalgae biofuel is viewed as a feasible and long-term solution for future energy requirements. This concise summary reviews the recent studies and advances in the field of microalgae biofuel production, including production planning, growth strategy, harvesting, and conversion technologies, as well as economic viability (Lindorfer et al. 2014). It can be farmed all year and generates more oil than any other crop; microalgae are a potential third-generation biodiesel feedstock. Improvements in algal biodiesel in the case of oil extraction approach and its difficulties, plant biodiesel synthesis, as well as fuel properties are being explored. The United States produces 10 million tons of soybean-based WFO each year, whereas China produces 4.5 million tons. In Middle East European nations, 0.7–10 million tons of sunflower-based WFO are generated per year. Due to its global availability and low cost relative to edible and non-edible oils, waste frying oil might be regarded as an essential feedstock for biodiesel generation. Using WFO as a feedstock for biodiesel synthesis could help to reduce water pollution, blockage in drainage systems, and the disposal of additional WFO created by houses and restaurants. Furthermore, European countries prohibited feeding waste frying oil mixture to animals because it could reappear in the food chain as a hazardous component through animal meat. As a result, WFO should be employed in a way that does not endanger humans. As a result, using WFO as a biodiesel feedstock lowers biodiesel costs and also avoids unlawful dumping (Findlater and Kandlikar 2011).

19.2.2 Heterogeneous Bio-catalyst in Biodiesel Production

Many researchers have used homogeneous catalysts to make biodiesel, but they had several problems, including biodiesel separation from glycerol, soap formation, and the requirement for water for washing. For biodiesel synthesis, a heterogeneous catalyst was chosen to overcome the issues mentioned earlier. It was simple in nature and did not have any separation or purification issues Babu et al. (2019). A homogeneous catalyst produces more biodiesel yield than a heterogeneous catalyst due to its higher reaction rate. However, the disadvantages of utilizing homogeneous catalysts include soap generation, increased water consumption for purification, and the inability to use the catalyst a second time during the transesterification reaction. Heterogeneous catalysts had advantages over homogeneous catalysts in biodiesel synthesis, such as being free of saponification and hydrolysis reactions (Kolakoti et al. 2020). A new catalyst was developed in order to generate biodiesel employing Ni-Ca hydroxyapatite solid acid catalyst impregnated with $Ni(NO_3)_2·6H_2O$. The heterogeneous catalyst has produced a maximum biodiesel yield of 60% at a catalyst concentration of 1.5 wt%, a molar ratio of 1:6, 3 h at 60 °C. Waste rohu fishbone was turned into a low-cost catalyst that may be recycled and reused, as noticed by Chakraborty and Das (2011). The fishbone was calcined for 2 h at 997.42 °C. The H-CAT catalyst was used to produce biodiesel from soybean oil. The maximum yield of 97% was achieved at a catalyst concentration of 1 wt%, a molar ratio of 1:6, 5 h and a reaction temperature of 70 °C. The fishbone catalyst may be reused up to six times.

Biodiesel was produced from palm oil using a KOH impregnated with a bentonite catalyst. With a catalytic load of 4%, a mass ratio of 4:1, and a reaction temperature of 60 °C, the highest biodiesel production of 93% was achieved (Goharimanesh et al. 2016). Xie and Huang (2020) synthesized a composite as an efficient and recyclable biocatalyst for biodiesel production from soybean oil. The Fe_3O_4-POLY (GMA-*co*-MAA) composites were characterized by FTIR, XRD, SEM, TEM, XPS, and nitrogen absorption–desorption techniques. The study revealed that the lipase was anchored on the magnetic composites with an 88.7% binding efficiency and a 67.3% activity recovery. Furthermore, the transesterification of soybean oil showed excellent catalytic activity. The maximum biodiesel yield of 92% was obtained for magnetic biocatalyst at 40 °C with a three-step methanol injection. Zulfiqar et al. (2021) developed a novel nanocatalyst (lipase-PDA-TiO_2-NPs) for enzymatic transesterification of *Jatropha curcas* seed oil. The TiO_2-NPs were prepared by using a hydrothermal method and were modified by the polydopamine polymer. RSM was applied to optimize the biodiesel production from JCSO using lipase-PDA-TiO_2-NPs. The optimum biodiesel yield of 92% was achieved by the transesterification process for 30 h at a temperature of 37 °C with a biocatalyst concentration of 10 wt% and a molar ratio of 1:6.

Takeno et al. (2021) extracted a biocatalyst from silver croaker's stone. Silver croaker is a fish, which is widely consumed in Brazil and the stone is a residue from the fish's head. It was found that the stone is composed of calcium carbonate and the catalyst is obtained by calcination of calcium oxide. Also, the parameters were optimized by RSM-based Box–Behnken method. The study unveiled that the optimum yield of 97% was achieved at 67 °C, 5.3 wt%, 175 min, and 11:1 molar ratio. Farooq et al. (2015) synthesized a cost-effective solid-base catalyst (calcined at 900 °C for 4 h) from chicken bones. According to the study, a maximum biodiesel yield of 89% was obtained through the transesterification process at 5 wt% catalyst loading and 15:1 molar ratio. The catalyst can be utilized four times in biodiesel production. Therefore, the biocatalyst catalyst has been claimed to be an effective and a viable replacement for homogeneous catalysts. The effectiveness of calcined animal bone treated with potassium hydroxide catalytic was investigated in biodiesel production by Nisar et al. (2017). Biodiesel was produced from *J. curcas* oil through a transesterification reaction. The study reported that a maximum yield of 96.1% was obtained at 6 wt% catalyst loading, a molar ratio of 9:1, reaction temperature of 70 °C, and reaction time of 3 h.

19.2.3 Conventional Biodiesel Production Technique

The transesterification process, also known as the alcoholized process, involves triglyceride reacting with methanol in the presence of a catalyst, yielding an ester and a glycerol by-product. The effect of process parameters on biodiesel synthesis from WFO via the transesterification process was investigated by Kulkarni and Dalai (2006). Maximum biodiesel yield of 89–90% is achieved with a methanol-to-oil ratio of 7:1 to 8:1, reaction temperatures of 40–50 °C, and a catalyst concentration of 0.75 wt%. Fan et al. (2011) used a transesterification process to extract biodiesel from cottonseed oil. Sodium hydroxide was used as a catalyst to make biodiesel from crude cottonseed oil. The extreme yield of 97% was produced using the reaction condition of 12 molar ratio of 1:6, reaction temperature of 50 °C, catalyst concentration of 1.3 wt%, and stirring speed of 600 rpm. Canacki and Van Gerpen (2001) explored biodiesel synthesis from soybean oil using a transesterification process with a molar ratio of 1:30, a reaction temperature of 60 °C, a catalyst concentration of 5 wt%, and the maximum yield of 98.4%.

Vinoth Arul Raj et al. (2021) produced biodiesel using calcium oxide solid nanocatalyst obtained from eggshells and *Nannochloropsis salina*. RSM and ANN were used to investigate process variable optimization for biodiesel production. Under ideal process conditions, nanocatalyst amount of 3 wt%, oil-to-methanol ratio 1:6, reaction temperature of 60 °C, and reaction time of 55 min, the maximal biodiesel yield was found to be 86.1% using nanocatalyst CaO. Babatunde et al. (2022)

obtained the maximum biodiesel conversion of 72% through the conventional transesterification process, from beef tallow under 7 wt% catalyst loading, 9:1 molar ratio, the temperature of 60 °C, and reaction time of 96 min. Mohadesi et al. (2019) optimized the parameters for biodiesel generation from waste cooking oil using the Box–Behnken approach, at a 9.4:1 molar ratio of oil to alcohol, 1.16 wt% KOH catalyst loading, and a reaction temperature of 2.4 °C, the best yield of 98.26% was produced. Abdullah et al. (2017) used a KOH catalyst to explore biodiesel generation from palm oil sludge via transesterification and esterification processes. The transesterification method yielded a maximum of 93% at a reaction temperature of 60 °C, a period of 60 min, and a catalyst concentration of 1.5 wt% KOH.

Anjana et al. (2016) used potassium iodide impregnated with calcium oxide as a catalyst in a biodiesel synthesis from *Pongamia*. In 2 h at 65 °C reaction temperature, a biodiesel yield of 95.7% was reached with a molar ratio of 1:12 and a catalyst concentration of 4 wt%. Evangelista et al. (2016) found that employing potassium fluoride (KF) loaded with 15% aluminum oxide (Al_2O_3) as a catalyst resulted in a maximum biodiesel yield of 97.7%. Kaur and Ali (2011) used lithium-ion impregnated with a calcium oxide catalyst to make biodiesel from Karanja oil. At a molar ratio of 12:1, a reaction temperature of 65 °C, a period of 2 h, and a catalyst loading of 5%, the highest yield was attained. Hamze et al. (2015) used a design of experiment to optimize the process parameters for biodiesel synthesis. KOH was used as a catalyst to make biodiesel from leftover frying oil. At 65 °C, 1 wt% KOH concentration and a molar ratio of 7.5:1, an optimal yield of 99.3% was obtained. Efavi et al. (2018) used a NaOH catalyst to extract biodiesel from *Citrullus vulgaris* seed at a molar ratio of 1:5, at a temperature of 60 °C, a reaction duration of 90, 120, and 150 min, and at various catalyst concentrations (0.13, 0.15, and 0.18 wt% NaOH). Biodiesel yields of 70%, 53%, and 49% were obtained at catalyst concentrations of 0.13, 0.15, and 0.18 wt%, respectively.

19.2.4 Microwave-assisted Transesterification Process

Potassium fluoride (KF) modified as hydrotalcite has been utilized as a solid-base catalyst in the microwave-assisted transesterification process. The catalyst was made by a simple and quick reaction between KF and HT, and it has been characterized by utilizing X-ray diffraction, surface profile analysis with gas sorption investigations, SEM, FTIR, and solid basicity determination methods. Microwave-assisted transesterification process gives better output compared to the reflux method. In the MW-aided transformation system, the materials performed well as a solid-base catalyst and had outstanding catalytic behavior (Efavi et al. 2018). Another study revealed that biodiesel was synthesized from cottonseed cooking oil through microwave-assisted transesterification utilizing a potassium hydroxide and calcium oxide as a catalyst. It was discovered that, when compared to the standard method, the microwave method highly increases the reaction, resulting in a substantial decrease in reaction time (Ahmad et al. 2019). Biodiesel is produced through catalyzing the transesterification of oils employing various homogeneous catalysts, the latter being non-recoverable and promoting the solubility of the generated glycerol into the otherwise immiscible ester phase, which must be purified. When heterogeneous catalysts are used, the requirement for product purification is avoided, with the added benefit of catalyst recovery. However, extreme reaction conditions (higher temperatures and pressures, longer reaction durations) are required to get more biodiesel (Sharma et al. 2019).

The esterification and transesterification reactions of WCO having higher free fatty acid (less than 1% by weight) were studied. The benefits of combining the esterification–transesterification technique with microwave heating were examined. Microwave esterification significantly reduces reaction time. The reaction time for the esterification and transesterification was lowered from 199 min to 79 and 9 min, respectively (Rocha et al. 2019). The microwave-assisted transesterification process is followed by hot spots, which are regions of enhanced reaction temperature. A blend of methanol and sodium hydroxide or potassium hydroxide may result in fast warming during microwave heating and hence an efficient reaction yield (Hassan and Smith 2020). Microwave irradiation was used to investigate the influence of concurrent separation and esterification of dried microalgae biomass to biofuel. In situ transesterification has shown to be a rapid and straightforward approach for manufacturing biodiesel from dried microalgal biomass (Sajjadi et al. 2014). The synergies between the manufactured catalysts and microwave heater have significantly reduced the transesterification reaction's resident time, indicating its efficiency and low-energy expenditure in the PWL transesterification process. Furthermore, the catalyst may be isolated from the reaction products, as well as its reusability, which contributes to WLB production's cost-effectiveness (Patil et al. 2011). The microwave-assisted esterification of papaya oils was investigated by utilizing a fixed microwave power and constant magnetic swirling. The results showed that molar proportions, heat, and catalyst concentration all had a substantial effect on POME productivity. The biodiesel yield was 99.9% based on the optimal conditions, whereas the actual experimental yield was 99.3% (Lawan et al. 2020).

19.2.5 Ultrasonic-assisted Transesterification Process

Pig-tallow oil (PTO) contains a huge fatty acid content and can be used as feedstock for biodiesel production. Biodiesel is produced from pig tallow using ultrasonic-aided dual-step transesterification process. FTIR, XRD, TEM, and particle size distribution were used to characterize copper oxide nanoparticles made from a novel *Cinnamomum tamala* (Nayak and Vyas 2019). Hoseini et al. (2019) extracted the biodiesel through ultrasonic technology. The dual-step esterification process, combined with ultrasonication, is an effective and time-conserving approach for immediate biodiesel production from unprocessed *Jatropha* seeds (Suresh et al. 2021). Eliane et al. (Tan et al. 2019) produced biodiesel from fish processing residue. The material was extracted from the fish residue using thermal treatment, and the biodiesel was synthesized by warming and swirling with the help of ultrasonic waves. Bismuth silicate (BS) catalysts may be effectively produced biodiesel using the ultrasonication method. As the ultrasonic irradiation period increases, the area of surface, total acidity, pore volume, and bandgap energies of the produced catalysts grow to a maximum at 30 min and then decrease (de Medeiros et al. 2019).

In research to improve the oil recovery from *Maesopsis eminii* beans using ultrasonic-assisted solvent extraction (UASE); RSM was utilized as an experimental design. The exploratory oil refining and transesterification resulted in a biodiesel yield of 84.6%, indicating its high conversion rate and demonstrating its potential as a feedstock for green diesel generation (Mahmoud 2019). Hamed et al. (Joven et al. 2020) investigated ultrasonic-assisted palm oil transesterification using oxide catalysts such as Cao, SrO, and BaO. When an ultrasonic processor was utilized, the yield of all catalysts increased significantly. In the ultrasound process, the barium oxide catalyst generated up to 95% of the biodiesel, whereas the SrO catalyst provided a little lower yield. Adewale et al. (2015) investigated the utilization of ultrasonic treatment for its position to give optimal blending while also providing enough enzyme activity for the enzyme esterification reaction of waste tallow. The reaction rate constants did not alter significantly between ultrasonic amplitudes of 45% and 50%. According to Adewale et al. (2015), the ultrasonic considerably enhanced the process by decreasing the response time to less than 49 min and the catalyst concentration to 3 wt% to obtain a biodiesel yield of 96%.

Ultrasonic treatment has been shown to be beneficial in increasing the mass transfer rate during the reaction by enhancing the emulsification of the reactants. Several approaches to improving the biodiesel synthesis process have been developed (Salamatinia et al. 2010). The investigation of the viability of biodiesel production through safflower seed oil with the help of an ultrasonic method produced biodiesel (Badday et al. 2012). The transesterification reaction was carried out using an ultrasonic processor (Topsonic Model, UP400, Iran). The processor, sonotrode, and PC controller were all part of the setup. They discovered that increasing the ultrasound power from 160 W to 400 W boosted performance by 3.83%. Babak et al. (Hosseinzadeh Samani et al. 2020) used low-frequency ultrasonic irradiation to explore the intensification of the biodiesel production process. They developed a quadratic model with a confidence level of 97%. Almasi et al. (2019) investigated the possibility of producing biodiesel via a single rapeseed genotype (TERI (OE) R-983). To amplify the reaction, an ultrasonic approach was used. According to the findings of this study, the conversion of biodiesel was 87.175% under optimized conditions.

Mostafaei et al. (2015) investigated and optimized the consequences of using varied kinds of ultrasonic variables on the reaction efficiency for successive biofuel synthesis. This investigation's optimized parameters resulted in a reaction yield of 92% and energy consumption of 103 W. As ultrasonic energy increases, the reaction yield increases, but as the energy increases, the reaction efficiency decreases. Furthermore, they found that ultrasonic-assisted biodiesel production is more efficient than mechanical stirring production, requiring less time and energy. Adewale et al. (2016) investigated the effects of ultrasonic parameters (magnitude, cycle, and pulse) on the kinetics of discarded lard-biocatalyzed lipase in the manufacture of biodiesel. At 40% ultrasound amplitude and above, there was a rapid increase in forwarding reaction rate constants and a decrease in the reversed rate of reaction constants. Maneechakr et al. (2015) researched a new sulfonated carbonaceous catalyst by hydrothermally carbonizing cyclodextrin, hydroxyethyl sulfonic acid, and citric acid in a single step.

19.3 CONCLUSION

This study provides a comprehensive overview of international and national energy scenario, various feedstock and selection, heterogeneous and biocatalyst synthesis and characterization in biodiesel production, and various biodiesel production techniques, including conventional, microwave-assisted transesterification, and ultrasonic-assisted transesterification process. Earlier research found that using vegetable oil in biodiesel production resulted in environmental issues such as deforestation and soil erosion. In addition, the value of so produced vegetable oil would increase, putting biodiesel's economic sustainability at threat. Furthermore, it causes a society-wide food versus fuel problem. Therefore, the second- and third-generation feedstocks such as microalgae and waste cooking oil and animal fats are recommended for biodiesel production due to their easy availability and low cost. In addition, homogeneous catalysts could produce higher biodiesel yield but have the drawbacks like water washing, glycerol separation and purification, and never using the same catalyst for another experiment. However, heterogeneous biocatalysts are free from these problems. The conventional transesterification process consumes more energy for converting triglycerides into biodiesel apart from more reaction time and temperature, whereas microwave- and ultrasonic-assisted transesterification reactions have converted triglycerides into biodiesel within a short period along with lower reaction temperature and time with minimum catalyst loading. Hence, this study concludes that the third-generation feedstock, along with microwave- and ultrasonic-assisted biocatalyst, would be recommended for biodiesel production.

ACKNOWLEDGMENTS

The authors would like to thank the King Mongkut's University of Technology North Bangkok (Grant Contract No. KMUTNB-FF-65–37, KMUTNB-Post-65–09, KMUTNB-Post-65–05) for financial support during this work.

REFERENCES

Abdullah, R., N. Rahmawati Sianipar, D. Ariyani and I. Fatyasari Nata. (2017). Conversion of palm oil sludge to biodiesel using alum and KOH as catalysts. *Sustainable Environment Research*, 27, 291–5.

Adewale, P., M.J. Dumont and M. Ngadi. (2015). Enzyme-catalyzed synthesis and kinetics of ultrasonic-assisted biodiesel production from waste tallow. *Ultrasonics Sonochemistry*, 27, 1–9.

Adewale, P., M.J. Dumont and M. Ngadi. (2016). Enzyme-catalyzed synthesis and kinetics of ultrasonic assisted methanolysis of waste lard for biodiesel production. *Chemical Engineering Journal*, 284, 158–65.

Ahmad, T., M. Danish, P. Kale, B. Geremew, S.B. Adeloju, M. Nizami and M. Ayoub. (2019). Optimization of process variables for biodiesel production by transesterification of flaxseed oil and produced biodiesel characterizations. *Renewable Energy*, 139, 1272–80.

Almasi, S., B. Ghobadian, G.H. Najafi, T. Yusaf, M.D. Soufi and S.S. Hoseini. (2019). Optimization of an ultrasonic-assisted biodiesel production process from one genotype of rapeseed (Teri (OE) R-983) as a novel feedstock using response surface methodology. *Energies*, 12(14).

Anjana, P.A., S. Niji, S. Meera, K.M. Begam, N. Anantharaman, R. Anand and D. Babu. (2016). Studies on biodiesel production from pongamia oil using heterogeneous catalyst and its effect on diesel engine performance and emission characteristics. *Biofuels*, 7, 377–87.

Arul Raj, J. Vinoth, R. Praveen Kumar, B. Vijayakumar, Edgard Gnansounou, B. Bharathiraja. (2021). Modelling and process optimization for biodiesel production from Nannochloropsis salina using artificial neural network. *Bioresource Technology*, 329, 124872.

Babatunde, Esther Olubunmi, Fatai Alade Aderibigbe, Ogunjobi Samuel Ogbeide Ebhodaghe and Tokunbo Oladapo. (2022). Optimization and kinetic study of biodiesel production from beef tallow using calcium oxide as a heterogeneous and recyclable catalyst. *Energy Conversion and Management*, 14, 100221.

Babu, D. and R. Anand. (2019). Biodiesel-diesel-alcohol blends as an alternative fuel for DICI engines. *Advanced Biofuels: Applications, Technologies, and Environmental Sustainability*, 338–65, Woodhead Publishing Series in Energy.

Badday, A.S., A.Z. Abdullah, K.T. Lee, and M.S. Khayoon. (2012). Intensification of biodiesel production via ultrasonic-assisted process: A critical review on fundamentals and recent development. *Renewable and Sustainable Energy Reviews*, 16(7), 4574–87.

Baskar, G. and R. Aiswarya. (2016). Trends in catalytic production of biodiesel from various feedstocks. *Renewable and Sustainable Energy Reviews*, 57, 496–504.

Biodiesel Magazine. (2017). Latest news and data about biodiesel production. *Biodiesel Magazine*.

Boutesteijn, C., D. Drabik and T.J. Venus. (2017). The interaction between EU biofuel policy and first- and second-generation biodiesel production. *Industrial Crops and Products*, 106, 124–9, doi: 10.1016/j.indcrop.2016.09.067.

Canakci, M. and J. Van Gerpen. (2001). Biodiesel production via acid catalyst. *Transactions of the American Society of Agricultural Engineers*, 42, 1203–10.

Chakraborty, R. and S.K. Das. (2011). Optimization of biodiesel synthesis from waste frying soyabean oil using fish scale-supported Ni catalyst. *Industrial and Engineering Chemistry Research*, 51, 404–14.

de Medeiros, E.F., B.M. Vieira, C.M.P. de Pereira, W.C. Nadaleti, M.S. Quadro and R. Andreazza. (2019). Production of biodiesel using oil obtained from fish processing residue by conventional methods assisted by ultrasonic waves: Heating and stirring. *Renewable Energy*, 143, 1357–65.

Demirbas, A., A. Bafail, W. Ahmad and M. Sheikh. (2016). Biodiesel production from non-edible plant oils. *Energy Exploration and Exploitation*, 34(2), 290–318.

Efavi, J.K., D. Kanbogtah, V. Apalangya, E.E.K. Nyankson Tiburu, D. Dodoo-Arhin, B. Onwona-Agyeman and A. Yaya. (2018). The effect of NaOH catalyst concentration and extraction time on the yield and properties of Citrullus vulgaris seed oil as a potential biodiesel feed stock. *South African Journal of Chemical Engineering*, 25, 98–102.

Evangelista, J.P., A.D. Gondim and L.D. Souza. (2016). Alumina supported potassium components as heterogeneous catalysts for biodiesel production: A review. *Renewable and Sustainable Energy Reviews*, 59, 887–94.

Fan, X., Xi Wang and F. Chen. (2011). Biodiesel production from crude cottonseed oil: An optimization process using response surface methodology. *The Open Fuels and Energy Science Journal*, 4, 1–8.

Farooq, M., A. Ramli and A. Naeem. (2015). Biodiesel production from low FFA waste cooking oil using heterogeneous catalyst derived from chicken bones. *Renewable Energy*, 76, 362–68.

Findlater, K.M. and M. Kandlikar. (2011). Land use and second-generation biofuel feedstocks: The unconsidered impacts of Jatropha biodiesel in Rajasthan, India. *Energy Policy*, 39(6), 3404–13.

Goharimanesh, M., A. Lashkaripour and A.A. Akbari. (2016). Optimization of biodiesel production using multi-objective genetic algorithm. *Journal of Applied Science and Engineering*, 19(2), 117–24.

Jahirul, M.I. et al. (2015). Physio-chemical assessment of beauty leaf (*Calophyllum inophyllum*) as second-generation biodiesel feedstock. *Energy Reports*, 1, 204–15.

Joven, J.M.O. et al. (2020). Optimized ultrasonic-assisted oil extraction and biodiesel production from the seeds of *Maesopsis eminii*. *Industrial Crops and Products*, 155, July, 112772.

Hamze, H., M. Akiaa and F. Yazdani. (2015). Optimization of biodiesel production from the waste cooking oil using response surface methodology. *Process Safety and Environmental Protection*, 94, 1–10.

Hassan, A.A. and J.D. Smith. (2020). Investigation of microwave-assisted transesterification reactor of waste cooking oil. *Renewable Energy*, 162, 1735–46.

Hoseini, S.S., G. Najafi, B. Ghobadian, R. Mamat, M.T. Ebadi and T. Yusaf. (2019). Characterization of biodiesel production (ultrasonic-assisted) from evening-primroses (*Oenothera lamarckiana*) as novel feedstock and its effect on CI engine parameters. *Renewable Energy*, 130, 50–60.

Hosseinzadeh Samani, B., M. Ansari Samani, A. Shirneshan, E. Fayyazi, G. Najafi and S. Rostami. (2020). Evaluation of an enhanced ultrasonic-assisted biodiesel synthesized using safflower oil in a diesel power generator. *Biofuels*, 11(4), 523–32.

Kaur, M. and A. Ali. (2011). Lithium ion impregnated calcium oxide as nano catalyst for the biodiesel production from karanja and jatropha oil. *Renewable Energy*, 36, 2866–71.

Kolakoti, A., P. Jha, P.R. Mosa, M. Mahapatro and T.G. Kotaru. (2020). Optimization and modelling of mahua oil biodiesel using RSM and genetic algorithm techniques. *Mathematical Modelling of Engineering Problems*, 6(2), 134–46.

Kulkarni, M.G. and A.K. Dalai. (2006). Waste cooking oils an economical source for biodiesel. *Industrial and Engineering Chemistry Research*, 45, 2901–13.

Kurczyński, D., P. Łagowski and G. Wcisło. (2021). Experimental study into the effect of the second-generation BBuE biofuel use on the diesel engine parameters and exhaust composition. *Fuel*, 284, August, doi: 10.1016/j.fuel.2020.118982.

Lawan, I., Z.N. Garba, W. Zhou, M. Zhang, and Z. Yuan. (2020). Synergies between the microwave reactor and CaO/zeolite catalyst in waste lard biodiesel production. *Renewable Energy*, 145, 2550–60.

Lindorfer, J., K. Fazeni and H. Steinmüller. (2014). Life cycle analysis and soil organic carbon balance as methods for assessing the ecological sustainability of 2nd generation biofuel feedstock. *Sustainable Energy Technologies and Assessments*, 5, 95–105, doi: 10.1016/j.seta.2013.12.003.

Mahmoud, H.R. (2019). Bismuth silicate ($Bi_4Si_3O_{12}$ and Bi_2SiO_5) prepared by ultrasonic-assisted hydrothermal method as novel catalysts for biodiesel production via oleic acid esterification with methanol. *Fuel*, 256, August, 115979.

Maneechakr, P., J. Samerjit and S. Karnjanakom. (2015). Ultrasonic-assisted biodiesel production from waste cooking oil over novel sulfonic functionalized carbon spheres derived from cyclodextrin via one-step: A way to produce biodiesel at short reaction time. *RSC Advances*, 5(68), 55252–61.

Mohadesi, M., B. Aghel, M. Maleki and A. Ansari. (2019). Production of biodiesel from waste cooking oil using a homogeneous catalyst: Study of semi-industrial pilot of micro reactor. *Renewable Energy*, 136, 677–82.

Mostafaei, M., B. Ghobadian, M. Barzegar and A. Banakar. (2015). Optimization of ultrasonic assisted continuous production of biodiesel using response surface methodology. *Ultrasonics Sonochemistry*, 27, 54–61.

Nayak, M.G. and A.P. Vyas. (2019). Optimization of microwave-assisted biodiesel production from Papaya oil using response surface methodology. *Renewable Energy*, 138, 18–28.

Nikul, K. Patel, Padamanabhi S. Nagar, and Shailesh N. Shah. (2012). *Identification of Non-Edible Seeds as Potential Feedstock for the Production and Application of Bio-Diesel*. Vadodara: The MS University of Baroda.

Nisar, Jan, Rameez Razaq, Muhammad Farooq, Munawar Iqbal, Rafaqat Ali Khan, Murtaza Sayed, Afzal Shah, and Inayat Rahman. (2017). Enhanced biodiesel production from Jatropha oil using calcined waste animal bones as catalyst. *Renewable Energy*, 101, 111–19.

Olkiewicz, M., C.M. Torres, l. Jiménez, J. Font, and C. Bengoa. (2016). Scale-up and economic analysis of biodiesel production from municipal primary sewage sludge. *Bioresource Technology*, 214, 122–31.

Patil, P.D. et al. (2011). Optimization of microwave-assisted transesterification of dry algal biomass using response surface methodology. *Bioresource Technology*, 102(2), 1399–405.

Rocha, P.D., L.S. Oliveira and A.S. Franca. (2019). Sulfonated activated carbon from corn cobs as heterogeneous catalysts for biodiesel production using microwave-assisted transesterification. *Renewable Energy*, 143, 1710–16.

Sajjadi, B., A.R. Abdul Aziz and S. Ibrahim. (2014). Investigation, modelling and reviewing the effective parameters in microwave-assisted transesterification. *Renewable and Sustainable Energy Reviews*, 37, 762–77.

Salamatinia, B., H. Mootabadi, S. Bhatia and A.Z. Abdullah. (2010). Optimization of ultrasonic-assisted heterogeneous biodiesel production from palm oil: A response surface methodology approach. *Fuel Process Technology*, 91(5), 441–8.

Sharma, A., P. Kodgire, and S.S. Kachhwaha. (2019). Biodiesel production from waste cotton-seed cooking oil using microwave-assisted transesterification: Optimization and kinetic modeling. *Renewable and Sustainable Energy Reviews*, 116, August, 109394.

Singh, D. et al. (2021). A comprehensive review of physicochemical properties, production process, performance and emissions characteristics of 2nd generation biodiesel feedstock: *Jatropha curcas*. *Fuel*, 285, August, 119110.

Suresh, T., N. Sivarajasekar and K. Balasubramani. (2021). Enhanced ultrasonic assisted biodiesel production from meat industry waste (pig tallow) using green copper oxide nanocatalyst: Comparison of response surface and neural network modelling. *Renewable Energy*, 164, 897–907.

Takeno, Mitsuo L., Iasmin M. Mendonça, Silma de S. Barros, Paulo J. de Sousa Maia, Wanison A.G. Pessoa Jr., Mayane P. Souza, Elzalina R. Soares, Rosane dos S. Binda, Fabio L. Calderaro, Ingrity S.C. S, Claudia C. Silva, Lizandro Manzato, Stefan Iglauer and Flavio A. de Freita. (2021). A novel CaO-based catalyst obtained from silver croaker (*Plagioscion squamosissimus*) stone for biodiesel synthesis: Waste valorization and process optimization. *Renewable Energy*, 172, 1035–45.

Tan, S.X. et al. (2019). Two-step catalytic reactive extraction and transesterification process via ultrasonic irradiation for biodiesel production from solid Jatropha oil seeds. *Chemical Engineering and Processing: Process Intensification*, 146, 107687.

World Oil Outlook. (2016). Organisation of the Petroleum Exporting Countries, 10th edition. Vienna, Austria: OPEC, October.

Xie, Wenlei and Mengyun Huang. (2020). Fabrication of immobilized Candida rugosa lipase on magnetic Fe_3O_4-poly(glycidyl methacrylate-co-methacrylic acid) composite as an efficient and recyclable biocatalyst for enzymatic production of biodiesel. *Renewable Energy*, 158, 474–86.

Zulfiqar, Anam, Muhammad Waseem Mumtaz, Hamid Mukhtar, Jawayria Najeeb, Ahmad Irfan, Sadia Akram, Tooba Touqeer and Ghulam Nabi. (2021). Lipase-PDA-TiO2 NPs: An emphatic nano-biocatalyst for optimized biodiesel production from *Jatropha curcas* oil. *Renewable Energy*, 169, 1026–37.

20 A Techno-economic Analysis of Green Hydrogen Production from Agricultural Residues and Municipal Solid Waste through Biomass-steam Gasification Process

Pon Pavithiran C.K., P. Raman, and D. Sakthivadivel

CONTENTS

20.1	Introduction	272
20.2	Current Status of Hydrogen Demand	272
20.3	Key Hydrogen Production Technologies	273
20.4	Green Hydrogen and Its Significance	273
20.5	Gasification Processes	274
	20.5.1 Background	274
	20.5.2 Mechanism of Gasification Process	274
	20.5.2.1 Drying Zone	275
	20.5.2.2 Pyrolysis Zone	275
	20.5.2.3 Combustion Zone	275
	20.5.2.4 Reduction Zone	275
	20.5.3 Principles of Steam Gasification	275
20.6	Literature Survey on Steam-injected Gasification	276
	20.6.1 National-level Research Studies on Hydrogen Production through Biomass Gasification	276
	20.6.2 International Research Studies on Hydrogen Production through Biomass Gasification	277
20.7	Potential Feed for Green Hydrogen Production	277
	20.7.1 Agricultural Residual Waste	278
	20.7.2 Municipal Solid Waste	279
20.8	Methodology of Hydrogen Separation	279
	20.8.1 Pressure Swing Adsorption Technology	279
	20.8.2 Membrane System	280
	20.8.3 Cryogenic Separation	280
20.9	Standard Methods for Techno-economic Analysis	280
20.10	Techno-economic Analysis of Green Hydrogen Production	281
	20.10.1 Technical Analysis	281

DOI: 10.1201/9781003334415-20

        20.10.2 Economic Analysis 281
            20.10.2.1 Capital Expenditure 281
            20.10.2.2 Operational Expenditure...282
        20.10.3 Hydrogen Production Cost 283
20.11 Green Hydrogen Production through Electrolysis..283
20.12 Conclusion 283
Funding 284
References..285

20.1 INTRODUCTION

Hydrogen is considered green energy that can be used as a feasible alternative to fast depleting traditional fossil fuels. Currently, India consumes about 6.7 million metric tons per annum of gray hydrogen (CSO 2019), which is about 8.5% of the global hydrogen demand and comes mainly from the industrial sector. The study conducted by the Energy and Resources Institute (TERI) has predicted that the hydrogen demand is likely to increase (by 5% and 7%) in 2050 and 2060, respectively, concurrent with an increase in population and industrial development sectors. These predicted advancements in green hydrogen production from renewables will result in lowering costs by more than 50% by 2030 (Hall et al., 2020). Among the two primary methods of water splitting process and biomass process used for green hydrogen production, generating hydrogen from biomass feedstock material is considered to be superior. The thermochemical conversion technique of steam biomass gasification is the most economical and clean energy production method when compared to other hydrogen generation processes. The gasifier is powered with various low-density biomass fuels like agro-residues, municipal solid waste (MSW), aquatic plants and water weeds, etc., due to their abundant availability and low carbon emission characteristics.

The key to developing this technology is to find the optimal cost-effective solutions in dealing the problems associated with the technical and environmental aspects. Thus, the techno-economic analysis can help in the carefully deliberated selection of appropriate research and development paths in complex areas. An in-depth study of available literature on biomass-gasified hydrogen generation reveals that the hydrogen production rate in the range of 12.75 to 9.6 €/kg has been considered suitable for the techno-economic approach (Rajabi et al. 2016). In a study conducted in the Unites States by Mohamed et al., it was found that the cost of hydrogen production using empty fruit bunch from oil palm was US$2.11/kg with the operating effect of 33%. Lv et al. (2008) evaluated hydrogen production with downdraft biomass oxygen gasification and CO-shift at atmospheric pressure; their study achieved US$1.69/kg of H_2 production. Taking many aspects into consideration, Sara et al. concluded that when green hydrogen purification cost was taken into account, the production cost was likely to be only nominally higher. Hydrogen production can be made cost-effective with low biomass cost, and by using a system of bigger size and relatively higher efficiency.

In the following section, the basic knowledge on green hydrogen and its significance along with the various production technologies are discussed, followed by a comparative techno-economic analysis between the two major green hydrogen generation techniques of biomass–steam gasification integrated pressure swing adsorption and electrolysis technology powered with coal, solar energy, wind energy, and biomass. Detailed descriptions on the sequential processes of green hydrogen generation are also provided.

20.2 CURRENT STATUS OF HYDROGEN DEMAND

Hydrogen has recently been recognized as the "future fuel," but it has yet to establish itself as a key participant in the energy system. Historically, the deployment of new energy technologies has not been smooth. New technological developments are usually directed toward meeting/addressing increasing industrial demands and growing interest in ensuring the widespread/successful

A Techno-economic Analysis of Green Hydrogen Production

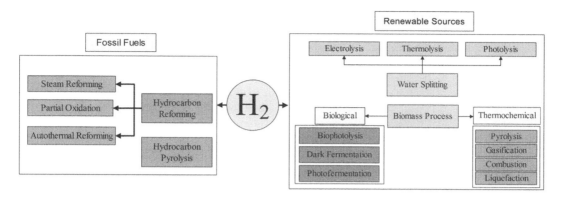

FIGURE 20.1 Classifications of various hydrogen production methods.

implementation of renewable energy systems. It is envisaged that the demand for hydrogen will increase nearly fivefold from 6 Mt in 2020 to 28 Mt in 2050. TERI has stated that the increase in hydrogen demand is largely from industries and fertilizers, refineries, power, transport, and steel sectors. Added to this are the demands made by the steel, transportation, shipping, and aviation sectors. The demand for green hydrogen is therefore likely to increase even further to 40 Mt in 2060. This increasing demand trend is due to the favorable lower emission and cost-effective characteristics of hydrogen (Hall et al. 2020).

20.3 KEY HYDROGEN PRODUCTION TECHNOLOGIES

Hydrogen is the only carbon-free fuel which has the characteristics of higher energy content of 141.9 MJ/kg and higher heating value (HHV), which is 60.8%, 63.42%, 66.5%, 68.42%, and 85.9% more than methane, ethane, gasoline, diesel, and methanol fuels, respectively. Thus, hydrogen is considered as an effective alternative energy source to dwindling conventional fossil fuels. The process of hydrogen production can be split into two categories based on the raw materials used: conventional and renewable technologies (Dincer 2012).

In conventional technologies, hydrogen production from the fossil fuels is carried out either in hydrocarbon reforming or hydrocarbon pyrolysis. In the hydrocarbon reforming process, the chemical techniques involved are steam reforming, partial oxidation, and auto thermal reforming. Among the renewable technologies, the production of hydrogen is processed with the two readily available sources of biomass and water as feed. The biomass feed is treated either with the biological methods of bio photolysis, dark fermentation, and photo fermentation or with the thermochemical methods of pyrolysis, gasification, combustion, and liquefaction. Another class of hydrogen production using water as the feed is processed through the water splitting processes of electrolysis, thermolysis, and photo electrolysis (Nikolaidis and Poullikkas 2017). From this, it is evident that the hydrogen produced from the fossil fuels feedstock is called "black/brown hydrogen," whereas that obtained from renewable resources is termed as "green hydrogen." The classifications of various hydrogen production methods are depicted in Figure 20.1.

20.4 GREEN HYDROGEN AND ITS SIGNIFICANCE

Apart from considering the hydrogen generated through renewable sources as green hydrogen, it can also be defined as follows: "Any renewable energy source with an explicit focus on air pollution, energy security and global climate problems" or "Any renewable source or any other net zero carbon energy through carbon capture and sequestration (CCS) and/or emissions offsets."

For the production of green hydrogen from renewable sources, the two methods of biomass processing and water splitting are considered as they result in zero greenhouse and carbon dioxide emissions. In biomass processing, the energy production can be sub classified into two categories of biological and thermochemical processes. The biochemical reaction is carried out with the organic waste resulting in higher methane production through anaerobic digestion. This method faces the limitation of converting all biomass into hydrogen due to its thermodynamic constraints (Niel 2016). Thus, this process is considered as a challenging method for hydrogen production.

On the other hand, the thermochemical biomass conversion processes, including combustion, pyrolysis, liquefaction, and gasification; degrade the biomass into combustible fuels. Among these, combustion converts the biomass into heat and electricity with lesser efficiency and higher levels of pollution, and thus it cannot be considered as the best method for hydrogen production. Liquefaction offers lesser hydrogen production with higher operating difficulties. Hydrogen production is low in slow pyrolysis as this process is carried out for charcoal production whereas fast pyrolysis achieves better hydrogen production if the reactor is maintained under higher temperature with sufficient volatile-phase residence time. Hydrogen generation is feasible in fast pyrolysis if the purpose of the pyrolysis process is to produce charcoal and bio-liquid fuels. In gasification, the biomass undergoes partial oxidization resulting in gas composed of hydrogen, carbon monoxide, carbon dioxide, and methane, along with charcoal as a by-product. The production of green hydrogen can be increased by the feasible method of passing steam into the gasification process. Thus, steam gasification is considered as the best among all methods of green hydrogen production in recent years.

20.5 GASIFICATION PROCESSES

20.5.1 Background

Gasification was developed independently in France and England in 1798. Later, in 1850, syngas generated from gasification was used for supplying electricity in London. Also, during World Wars I and II, gasoline played an important role in meeting the energy demands of warfare. However, with dwindling fuel supply, the utilization of gas from biomass reduced. Alternatively, petroleum fuels began to be increasingly used to meet the required energy demands as they are considered to be more reliable. Currently, biomass gasification is again being favored due to the prevailing energy crisis and towering petroleum costs. It is considered as an effective method for getting energy from many types of organic materials, with the added benefit of using the waste management technique of recovering eco-friendly energy from biomass wastes. Currently, two different designs of gasifiers can be seen in India. The first one (NERIFIER gasification unit) has been installed at Nohar town in Hanungarh district, Rajasthan, by Narvreet Energy Research and Information (NERI) society for burning of agro-residual wastes, sawmill dust, and forest wastes, while the second one is the Tata Energy Research Institute (TERI) gasifier installed at Gaul Pahari campus, New Delhi.

20.5.2 Mechanism of Gasification Process

The thermochemical conversion of solid organic compounds into a gaseous fuel of syngas and a solid fuel of char as a by-product is referred to as biomass gasification. The partial oxidation of the carbon in the feed material occurs in the presence of a gasifying carrier such as air, oxygen, steam, or carbon dioxide. Carbon monoxide, carbon dioxide, hydrogen, and methane are all present in the syngas produced. In the gasification process, undesired gases such as hydrogen sulfide, hydrochloric acid, and inert gases such as nitrogen can also be detected. The generated syngas has its lower heating value in the range of 4–13 MJ/Nm^3 and the presence of undesirable gases mainly depends on the pretreatment of fuels and the operational conditions. On the other hand, the amount of ash produced depends on the biomass processed and the quantity of unconverted organic portion in the char.

Drying, pyrolysis, combustion, and reduction are all common steps in the biomass gasification process. The steps of gasification's mechanistic process are described in the following sections.

20.5.2.1 Drying Zone

The biomass feedstock eliminates its water humidity at 160 °C as a vapor phase in this drying zone. A part of the removed water vapor flows downward to reduce into hydrogen in the reduction zone. The remaining will finish up as moisture within the gas. Chemical reactions are restricted due to lesser operating temperatures in this zone.

20.5.2.2 Pyrolysis Zone

In this zone, the devolatilization occurs on the removal of hydrogen, carbon monoxide, and carbon dioxide in the absence of air/oxygen. This happens with the endothermic reactions where the required heat is supplied from the combustion zone.

20.5.2.3 Combustion Zone

The products of the pyrolysis reactions are partially oxidized with oxygen from the air supply, resulting in carbon monoxide, carbon dioxide, and water. This occurs at the operational temperature of 1200 °C with the heat release of an exothermic reaction, as represented in Equations 20.1–20.3. This zone acts as a source for the overall heat required for the gasification process.

$$C + O_2 \rightarrow CO_2 \tag{20.1}$$

$$C + \frac{1}{2}O_2 \rightarrow CO \tag{20.2}$$

$$H + \frac{1}{2}O_2 \rightarrow H_2O \tag{20.3}$$

20.5.2.4 Reduction Zone

The gaseous fuel emitting from the gasifier passes over the hot charcoal within this region. The chemical reactions in this zone take place in the absence of oxygen. The reaction initiates with the generation of carbon monoxide and is known as Boudourad reaction, as given in Equation 20.4. Following this, water-gas reaction takes place between 600 and 950 °C for the production of hydrogen and carbon monoxide, as represented in Equation 20.5:

$$C + CO_2 \rightarrow 2CO \tag{20.4}$$

$$C + H_2O \rightarrow CO + H_2 \tag{20.5}$$

20.5.3 Principles of Steam Gasification

Steam gasification is the most promising method of hydrogen production as it does not leave behind any carbon footprint in the environment. It also has the advantage of lesser production of char and tar since steam initiates more water-gas reaction along with the elimination of the drawbacks of earlier methods. Additionally, this technology is most suited for biomass having less moisture content of 35%. Researchers have proved that the hydrogen production rate is three times higher than air gasification. Moreover, in addition to being an economical method, this system is also an effective means of green hydrogen production and offers a cleaner product with lesser environmental impacts. Table 20.1 depicts the comparison of the various functions of air-, steam-, and oxygen-fed biomass gasification (Parthasarathy and Narayanan 2014).

TABLE 20.1
Comparative Functions of Air, Steam, and Oxygen-fed Biomass Gasification

Methods	Air Gasification	Steam Gasification	Oxygen Gasification
Gas composition	Hydrogen 15%	Hydrogen 40%	Hydrogen 40%
	Carbon monoxide 20%	Carbon monoxide 25%	Carbon monoxide 40%
	Methane 2%	Methane 8%	Carbon dioxide 20%
	Carbon dioxide 15%	Carbon dioxide 25%	
	Nitrogen 48%	Nitrogen 2%	
Reactor temperature	900–1100 °C	700–1200 °C	1000–1400 °C
Heating value (MJ/Nm^3)	4–6	15–20	10–15
Cost	Low	Medium	Costly

There are some vital process parameters which influence the yield of hydrogen generated in steam gasification:

- Biomass type
- Biomass feed particle size
- Temperature
- Steam-to-biomass ratio
- Addition of catalysts
- Sorbent-to-biomass ratio

20.6 LITERATURE SURVEY ON STEAM-INJECTED GASIFICATION

20.6.1 NATIONAL-LEVEL RESEARCH STUDIES ON HYDROGEN PRODUCTION THROUGH BIOMASS GASIFICATION

S. Dassppa and his research team from the Indian Institute of Science and Technology (IISc), Bangalore, focused on energy and exergy analyses of oxy-steam gasification and compared it with air gasification to optimize the H_2 yield, efficiency, and syngas energy density. They processed this in the first and second law thermodynamic analysis of air and oxy-steam biomass gasification (Sandeep and Dasappa 2014). Prior to this, hydrogen production from biomass using oxygen-blown gasification was modeled by Atmadeep Bhattacharya in 2012 from the Department of Power Engineering, Jadavpur University, Kolkata (Bhattacharya, Bhattacharya, and Datta 2012). His study resulted in 54.4% hydrogen in the product gas stream for 95% oxygen in the gasification agent and the yield of hydrogen from every kilogram of biomass was found to be 102 g. V.M. Jaganathan conducted experimental investigations on oxy-steam gasification of biomass in a downdraft packed bed reactor at the Indian Institute of Technology, Madras. He processed it by restricting the upstream bed temperatures between 120 °C (to avoid steam condensation) and 150 °C (to prevent bulk devolatilization of bed), the intrinsic H_2 yield from biomass was determined over an equivalence ratio (F) range of 3.5 to 1.2. Interestingly, the H_2 yield over this entire range was within 30–40 g/kg of biomass (Jaganathan, Mohan, and Varunkumar 2019). In 2019, an experimental analysis of air–steam gasification was made in a dual-fired downdraft biomass gasifier enabling hydrogen enrichment in the producer gas by the Department of Energy and Environment, TERI School of Advanced Studies, Delhi. The study showed that the maximum amount of 27.24% (by volume) hydrogen was achieved at equivalence number (EN) 1.54 suitable for biohydrogen production. However, equivalence number (EN) 1.5–2.2 is more suitable for power generation applications since maximum higher heating value (HHV) occurs in this range, i.e., 6.33 MJ/Nm^3. The enrichment of producer gas resulted in an increase of HHV by 44% (Ram et al. 2019). Novel research has been carried out on the enhancement of hydrogen production in various laboratories for achieving maximum hydrogen production from biomass gasification.

20.6.2 International Research Studies on Hydrogen Production through Biomass Gasification

There are a number of research studies that have been conducted at a global level on biomass, which is considered as one of the promising renewable energy sources due to its abundant availability. Yanlei Xiang from China examined the production of hydrogen with different gasifying agents, including steam, CO_2, steam + O_2, CO_2 + O_2, steam + CO_2, and steam + CO_2 + O_2. Besides, the effects of equivalence ratios and gasifying agent-to-biomass ratios on syngas production were studied. The results revealed that steam contributed to the yield of H_2, while it inhibited the generation of CO. CO_2 in the gasifying agent promoted the yield of CO. O_2 had little impact on H_2 production, but it significantly reduced the generation of CO and CH_4. The introduction of O_2 in the gasifying agent reduced the lower heating value (LHV) of syngas, while it generally raised the gasification efficiency. Furthermore, the net power generation efficiencies under steam + O_2 gasification, CO_2 + O_2 gasification, and steam + CO_2 + O_2 gasification were 26.3%, 28.3%, and 27.6%, respectively (Xiong et al. 2020). Wei Chen from Texas undertook a study to identify the optimum conditions for producing gases from woody biomass with enhanced heating values. Air, air/steam, and a carbon dioxide/oxygen (CO_2/O_2) mixture were used as gasification media. It was found that peak temperature in the gasifier decreased from 1050 °C to 850 °C as steam-to-fuel ratio (*S/F*) was increased from 0 to 0.45. However, the carbon dioxide (CO_2) and hydrogen (H_2) concentrations increased and the carbon monoxide (CO) percentage decreased with the introduction of larger amounts of steam. For air/steam gasification, the HHV of the producer gas was estimated to be in the range of 2800–3800 kJ/Nm³. The highest HHV was obtained at *S/F* = 0.3 and ER = 2.7 (optimum). The HHV of the gas from air/steam gasification was found to be almost the same as that produced from pure air gasification (2800–4000 kJ/Nm³), except for an increased H_2 yield for air/steam (Chen et al. 2013). The Energy Technologies Department, Italy, dealt with the effects of oxygen and steam equivalence ratios on updraft gasification of biomass and concluded that the oxy-steam gasification of torrefied wood gave the best results in terms of cold gas efficiency and LHV when carried out in the range of 0.23–0.27 for both the equivalent ratios (Cardoen et al. 2015). A research-scale fluidized bed reactor was built in France and used to study the effect of steam/biomass ratio, time duration of experiments, reactor temperature, and biomass particle size on hydrogen yield and tar content in produced syngas during steam gasification of biomass (Fremaux et al. 2015). Steam gasification of polyethylene was conducted using a two-stage gasifier consisting of a fluidized bed gasifier and a tar-cracking reactor filled with active carbon.

The study was carried out in Korea to produce H_2-rich syngas and simultaneously reduce tar to produce a syngas having 55 vol% H_2 on average (Schweitzer et al. 2018). In a study conducted in Germany on the SER (sorption-enhanced reforming) gasification process, a nitrogen-free combustion environment was used for producing high calorific value. In addition, due to low gasification temperatures of 600–750 °C and the use of limestone as bed material, in situ CO_2 capture was possible, leading to a hydrogen-rich and carbon-lean product gas as a result. There are many ongoing research studies being conducted on the influence of air and oxy-steam gasification characteristics over the system performance with focus on factors such as energy efficiency, pure hydrogen production, and so on.

20.7 POTENTIAL FEED FOR GREEN HYDROGEN PRODUCTION

In India, biomass resources are abundantly available in many forms such as agricultural crops, agricultural residues, forestry, agro-industry, and a part of municipal solid wastes. Thermal gasification technologies offer an eco-friendly energy recovery solution to the problem of solid waste management and disposal by transforming the waste into a renewable energy source without actual combustion. This section deals with the variety of biomass available in the form of agricultural residues and municipal solid waste, and explores their possible use as renewable fuels. Also, proximate and ultimate analyses are used for assessing the potential of their utilization in different industries.

20.7.1 Agricultural Residual Waste

In 2015, the Ministry of New and Renewable Energy (MNRE) estimated that the available biomass in India was around 500 million metric tons per annum. Considering the agriculture and forestry waste residues, the anticipated surplus biomass available was computed to be 120–150 million metric tons per annum, which constituted an energy potential of about 18,000 MW. Generally, two types of biomass wastes are available, namely, woody and non-woody, in which woody biomass is characterized by high bulk density, low ash content, low moisture content, and high calorific value. The woody biomass is classified under forest-based residue and agro-industrial residue. The non-woody biomass is characterized by lower bulk density, higher ash content, higher moisture content, and lower calorific value as in agricultural residue, animal waste, and urban and industrial solid wastes. Since India is a rapidly developing country, most agricultural residues are being used as fuel. They have a lower energy value than woody fuels as they are available in abundance.

Agricultural residues are calculated based on the residue-to-crop ratio (RCR). For those residues whose generation is independent of yield, the crop-independent residue (CIR) is considered. This can be expressed in tons of residue generated per hectare per year. The production of farm-level residues is calculated from Equation 20.6.

$$\text{Residue production} = \text{Residue yield} \times \text{Gross cropped area} \quad (20.6)$$

In addition to this, the fraction of the original crop that is actually processed, as well as the amount that is exported, must be taken into account while processing residues, as represented in Equation 20.7.

$$\text{Residue production} = \text{Residue yield} \times (1 - \text{Fraction exported}) \times \text{Fraction processed} \\ \times \text{Gross cropped area} \quad (20.7)$$

It is important to remember that the data on residue generation does not include residue generated by consumers (e.g., the residue production of mango peels includes only those peels produced by mango-processing plants). Based on these relevant equations, the relation between the yield and RCR is obtained and it is evident that it is not the same for all crops. The term "residue" includes both residues produced at harvesting and during processing of the by-product, for ease of understanding (Cardoen et al. 2015).

In the Indian agricultural scenario, the generated agro-residual wastes are classified as follows.

- Leaves, pseudostems, and peels from banana plants
- Stem and leaves from cabbages
- Stalks and husk from chickpeas
- Fronds, husk, shell, meal, and coir pith from coconut
- Stalks, hull, gin trash, and meal from cotton
- Stalks of eggplants
- Stalks, shell, and meal of groundnuts (with shell)
- Stover, cobs, and corn fiber from maize
- Pruning wood, peels, seeds, and meal from mango trees
- Stalks, seedpod, and meal from mustard plants

- Stalks from onions
- Straw, husk, bran, and de-oiled bran from paddy

The list of agricultural residues also includes pearl millet, pigeon pea, potato, sorghum, soyabean, sugarcane, tapioca, tomato, water hyacinth, and wheat. These wastes are termed as lignocellulosic biomass that comprises 35–55% cellulose, 25–40% hemicellulose, and 15–25% lignin with small percentage of proteins and ash. The heating value ranges from 12 MJ/kg to 20 MJ/kg.

20.7.2 Municipal Solid Waste

MSW can be defined as trash which is a non-homogeneous mixture of waste generated by the residential, commercial, and industrial sectors. Residential and commercial MSW include clothing, disposable tableware, yard trimmings, cans, office disposals, and paper and boxes. Industrial and institutional waste includes restaurant trash, paper, classroom wastes, wood pallets, plastics, corrugated boxes, and office papers. In addition, MSW consists of some metals which often need to be recovered prior to gasification. The MSW composition mostly consists of 27% paper, 13.5% yard trimming, 14.6% food waste, 12.8% plastic, 9.1% metal, 9% rubber/textile, 6.2% wood, 4.5% glass, and 3.3% others, as mentioned by the US Environmental Protection Agency in 2013.

The statistical records of the Indian Central Pollution Control Board show that urban areas of the country generated 62 Mt of MSW in 2015, out of which 82% was collected and the remaining 18% was considered as litter. To predict the generation of MSW in the years ahead, the data of the previous years were analyzed by researchers and they forecasted that the MSW was likely to be 165 Mt by 2030, 230 Mt by 2040, and 436 Mt by 2050 (Sharma and Jain 2019). MSW has an approximate heating value of 14.49 MJ/kg, which was accounted for the utilization of energy production.

20.8 METHODOLOGY OF HYDROGEN SEPARATION

For the separation of hydrogen, the feed is passed into the biomass gasifier for the generation of syngas with varying gas compositions of hydrogen, methane, carbon monoxide, carbon dioxide, and nitrogen. The hydrogen in the gas can be enhanced by passing the steam into the gasifier through the steam generator. The produced gas is then passed into the cyclone filter for the removal of tar and dust particles in the gas. The syngas blower is used for pressuring the air to cross the hydrogen separator. As depicted in Figure 20.2, the hydrogen is separated and the remaining gases are used for other thermal applications or electricity generation. Two hydrogen separation methods are currently in practice among the three available methods: pressure swing adsorption, membrane separation, and cryogenic separation (Hadden 2003). A detailed description of all the methods is provided in the following sections.

20.8.1 Pressure Swing Adsorption Technology

The PSA unit extracts the impurities from hydrogen streams using solid adsorbent beds, resulting in the production of high-purity hydrogen gas. This method is based on gas molecules physically adhering to adsorbent material. The force acting between the gas molecules and the adsorbent material is determined by the gas component, type of adsorbent material, partial pressure of gas components, and operating temperature. Here, adsorption is carried out at high pressure, normally in the range of 10–40 bars until the loading is reached after which the adsorbent material must be regenerated. This regeneration is accomplished by lowering the pressure to slightly above atmospheric pressure resulting in a respective decrease in equilibrium loading. As a result, the impurities on the adsorbent material are desorbed, and the adsorbent material is regenerated. The amount of impurities removed from a gas stream within one cycle corresponds to the difference of adsorption to desorption loading.

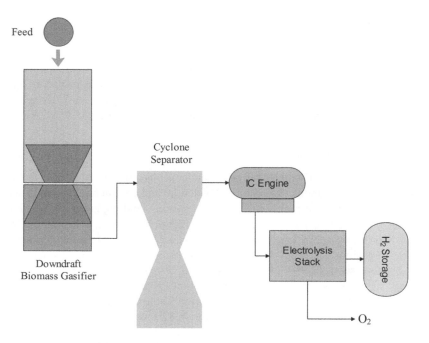

FIGURE 20.2 Process flow diagram of green hydrogen production and separation.

20.8.2 Membrane System

The alternative method of separating the purified hydrogen is through membrane technology. Hydrogen can be separated by either dense-phase metal or porous ceramic membrane. The porous membrane is preferable as it is able to withstand high temperatures and is more stable and more durable, particularly in harsh and hydrothermal environments. This system is considered to be more reliable as it does not have any moving parts. The membrane can produce hydrogen of higher purity. However, a disadvantage is the passing of hydrogen sulfide into the membrane reducing the efficiency of the membrane due to fouling.

20.8.3 Cryogenic Separation

Cryogenic separation involves cooling the gas and condensing some, or all, of the constituents of the gas streams. Separation may include flashing or distillation, depending on the product purity desired. The ability to isolate a range of products from a single feed stream is a benefit of cryogenic systems. The separation of low-boiling olefins from a hydrogen stream is one example. It can offer hydrogen recovery of 95–98%.

20.9 STANDARD METHODS FOR TECHNO-ECONOMIC ANALYSIS

The techno-economic assessment (TEA) is a tool to evaluate a project's cost viability. A sequential method includes the following:

1. An assessment of a given project's economic viability
2. Identification of financial issues over the course of the project
3. Investigation of various accessible technologies in order to get the desired result
4. Comparison of the costs associated with available technologies

Cost analysis, benefit analysis, risk analysis, and finally the techno-economic approach all are part of the techno-economic analysis. There are two kinds of cost assessments: investment cost and operational cost. The cost of investment covers the initial expenditure, as well as planning, consultancy, administration, annual insurance, and infrastructure costs. Similarly, operational cost includes fuel, labor, maintenance, and any other ancillary costs. A number of methods are used to evaluate the cost of a specific project:

- Assessment of static cost-benefit method
- Annuity method
- Net cash flow method
- Net present value method
- Internal rate of return method

20.10 TECHNO-ECONOMIC ANALYSIS OF GREEN HYDROGEN PRODUCTION

Numerous studies have been conducted in order to develop the technology for commercial production of fuels and chemicals as a result of the growing interest in biomass thermochemical conversion. Various literatures have been found on thermochemical conversion of fuels like methanol, ethanol, dimethyl ether, ethylene, ammonia, bio-oil, bio char, hydrogen, and gasoline along with the conversion of biomass into electricity. Limited research studies have been carried out on hydrogen generation and a research gap in downdraft gasification is identified. This study involves the techno-economic analysis of biomass-driven gasification for the generation of green hydrogen through the use of agro-residues and municipal solid waste as the feedstock for energy production. This method is compared with electrolysis, which is an alternative technology for green hydrogen production. The technical and the economic comparison is done and the details are summarized.

20.10.1 TECHNICAL ANALYSIS

From the experimental study, it is observed that the generation of the hydrogen-enriched producer gas through air–steam gasification consists of the components listed in Table 20.2. These values are drafted for the design consisting of gasification reactor, heat exchanger, blower, cyclone filter, bag house filter, cartridge filter, and gas outlet. The technical specifications are considered for a system that can produce 10 kg of hydrogen.

In addition, the hydrogen separation and purification are processed through the pressure swing adsorption technique. The technical specifications of the research study undertaken by Rajabi Sara et al. using the pressure swing adsorption technique are (Rajabi Sara et al. 2016) shown in Table 20.3.

20.10.2 ECONOMIC ANALYSIS

The incurred costs are considered under the two heads of capital expenditure and operating expenses. The former capital expense depreciates with N number of years. The N year is a function of lifetime, maintenance, and long-term usage.

20.10.2.1 Capital Expenditure

From the first assessment, the value of N is taken as 20 years of depreciation. The capital cost is considered to be 7%. Thus, from this, the annual capital cost can be considered as follows:

$$\text{Rs.} / \text{year} = \frac{\text{Capital expenditure in Rs.} \times 0.07}{1-(1+0.07)^{\wedge}(-20)} \tag{20.8}$$

TABLE 20.2
Technical Specifications of Hydrogen-enriched Biomass Gasification System (Kodanda et al. 2020)

Parameters	Values
Fuel consumption rate	150 kg/h
Air rate to gasifier	290 m^3/h
Steam rate to gasifier	44.06 kg
Generated syngas from gasifier	412.5 m^3/h
Syngas temperature	150 °C
Density of syngas	0.95 kg/m^3
Mass of syngas	390 kg/h
Water from fuel	16 kg/h
Water from air	3 kg/h
Water condensed from gas	16 kg/h
Steam converted into gas	46 kg/h
Gas compositions	
Nitrogen	220 kg/h
Carbon monoxide	75 kg/h
Carbon dioxide	125 kg/h
Methane	4 kg/h
Hydrogen	10 kg/h
Water vapor	15 kg/h

TABLE 20.3
Technical Specification of the Pressure Swing Adsorption Technique

Parameters	Values
PSA inlet pressure	7 bar
PSA intercooler compressor Efficiency	62%
PSA intercooler compressor Temperature	40
Intercooler compressor stages	2

20.10.2.2 Operational Expenditure

The operational expenditure comprises the maintenance cost, insurance, taxes, and electrical energy. This system is to be maintained for 7000 annual working hours. The portable purification and separation system is fully automated. The overall system can be scaled up or scaled down based on Equation 20.9:

$$SC = RC \times (SP/RP)^{EXP} \tag{20.9}$$

Here SC is the scaling cost; RC is the reference cost; SP is the scaling parameter; RP is the reference parameter; and EXP is the exponent.

TABLE 20.4
Total Cost of the Plant

	Total Cost (Rs. in Lakhs/Year)
Hardware cost	
Gasifier system	15
Pressure swing adsorption system	9.5
Total equipment cost (TEC)	24.5
Engineering cost	
Engineering and design (13% of TEC)	3.18
Purchasing and construction (14% of TEC)	3.43
Total capital expenditure (TCE)	31.36
Maintenance (2% of TCE)	4.9
Insurance (2% of TCE)	4.9
Biomass (available in abundance)	–
Energy (Rs.5.75/kWh) (TANGEDCO 2017)	4.02
Total operating expenditure	13.82
Total cost	45.18
Hydrogen production (ton/year)	0.70
Hydrogen production cost (Rs./kg)	64.54

20.10.3 Hydrogen Production Cost

The total capital cost of the equipment depends on the material cost, manufacturing cost, labor cost, and miscellaneous cost for 150 kg/h feed capacity gasifier, i.e., 150 kWe is Rs.1,50,00,000 and pressure swing adsorption unit of Rs.95,00,000 (Table 20.4).

20.11 GREEN HYDROGEN PRODUCTION THROUGH ELECTROLYSIS

An economic analysis is done of the electrolytic water splitting process which is considered as one of the best methods for green hydrogen production. In this process, the gasifier system is integrated with the producer gas–driven internal combustion engine (IC Engine). A conceptual view of the hydrogen production system using biomass gasifier and electrolysis is shown in Figure 20.3. The system is equipped with battery energy storage to prevent fluctuations in energy production and to ensure continuous operations.

Researchers from Canada have performed experiments and determined that the electricity requirement of the systems falls within a narrow range of 34.2–34.4 kWh/kg H_2. It can be noted that the by-product of separated oxygen can be sold off based on the processing requirements. The water requirement and energy needed are showcased in Table 20.5 and is compared with the data available in the literature. It is noted that 34.2 kWh of electricity is required for the production of 1 kg of hydrogen using the electrolysis method. A comparative analysis of electricity production through coal, wind energy, PV solar energy and through biomass gasification is also done to identify the best outcome.

20.12 CONCLUSION

This research endeavor offers insights into the techno-economic analysis of various methods adopted on green hydrogen production platforms. The factors governing hydrogen production cost such as feedstock, capital cost, and internal rate of return have been reviewed. Steam gasification

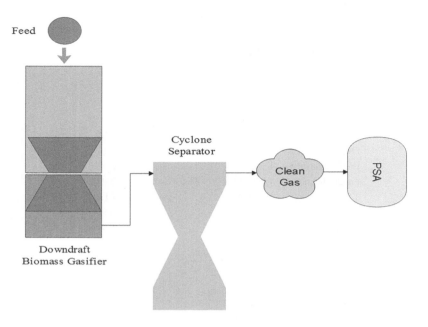

FIGURE 20.3 Process flow diagram of green hydrogen production through electrolysis.

TABLE 20.5
Energy Requirements and Water Consumption Estimates (TANGEDCO 2017)

System	System Requirement	
	Water consumption kg/kg of H_2	Electricity requirement kWh/kg of H_2
Electrolyzer only	8.8	34.2
Ranges from literatures	9.1–15.5	33.1–36.8
Sources	Cost of 1 kWh (Rs.)	Cost of 1 kg of H_2 production
Coal	6.98	240
Biomass gasification	3.65	125
PV solar	4.96	170
Wind	3.87	133

with pressure swing adsorption method (PSA) has been investigated by many researchers due to its viable production cost. The economic study shows that the cost is Rs.64.54/kg of H_2 production using agricultural residual waste and municipal solid waste as the feedstock. This method is considered to be a feasible alternative method of green hydrogen production since it is cheaper than the electrolysis process. In the green hydrogen production method through the electrolysis process, electricity is generated using coal, biomass, solar energy, or wind energy. A comparison of the results shows that it is 73%, 48%, 62%, and 52%, respectively, more expensive than the feasible biomass–steam gasification process.

FUNDING

We would like to extend our sincere gratitude to the Department of Science and Technology – Science Technology Innovation Hub (Project No. DST/SEED/TSP/STI/2020/243), for providing the funding to carry out this research work.

REFERENCES

Bhattacharya, Atmadeep, Abhishek Bhattacharya, and Amitava Datta. 2012. "Modeling of Hydrogen Production Process from Biomass Using Oxygen Blown Gasification." *International Journal of Hydrogen Energy* 37 (24): 18782–90. https://doi.org/10.1016/j.ijhydene.2012.09.131.

Cardoen, Dennis, Piyush Joshi, Ludo Diels, Priyangshu M Sarma, and Deepak Pant. 2015. "Resources, Conservation and Recycling Agriculture Biomass in India: Part 1. Estimation and Characterization." *Resources, Conservation & Recycling* 102: 39–48. https://doi.org/10.1016/j.resconrec.2015.06.003.

Chen, Wei, Siva Sankar Thanapal, Kalyan Annamalai, Robert J. Ansley, and Mustafa Mirik. 2013. "Updraft Gasification of Mesquite Fuel Using Air/Steam and CO_2/O_2 Mixtures." *Energy and Fuels* 27 (12): 7460–9. https://doi.org/10.1021/ef401595t.

CSO. 2019. "Energy Statistics 2019 (Twenty Sixth Issue)." http://mospi.nic.in/sites/default/files/publication_reports/Energy Statistics 2019-finall.pdf.

Dincer, Ibrahim. 2012. "Green Methods for Hydrogen Production." *International Journal of Hydrogen Energy* 37 (2): 1954–71. https://doi.org/10.1016/j.ijhydene.2011.03.173.

Fremaux, Sylvain, Sayyed Mohsen Beheshti, Hojat Ghassemi, and Rasoul Shahsavan-Markadeh. 2015. "An Experimental Study on Hydrogen-Rich Gas Production via Steam Gasification of Biomass in a Research-Scale Fluidized Bed." *Energy Conversion and Management* 91: 427–32. https://doi.org/10.1016/j.enconman.2014.12.048.

Hadden, Raymond. 2003. "Hydrogen Production." *Hydrocarbon Engineering* 8. https://doi.org/10.1002/9781119707875.ch12.

Hall, Will, Thomas Spencer, G Renjith, and Shruti Dayal. 2020. "The Energy And Resources Institute Creating Innovative Solutions for a Sustainable Future A Pathway for Scaling-up Low Carbon Hydrogen across the Economy the Potential Role of Hydrogen in India." https://www.teriin.org/sites/default/files/2021-07/Report_on_The_Potential_Role_of_%20Hydrogen_in_India.pdf.

Jaganathan, V. M., Omex Mohan, and S. Varunkumar. 2019. "Intrinsic Hydrogen Yield from Gasification of Biomass with Oxy-Steam Mixtures." *International Journal of Hydrogen Energy* 44 (33): 17781–91. https://doi.org/10.1016/j.ijhydene.2019.05.095.

Kodanda, Narasimhan, N. K. Ram, Nameirakpam Rajesh, Perumal Raman, Atul Kumar, and Priyanka Kaushal. 2020. "Experimental Study on Performance Analysis of an Internal Combustion Engine Operated on Hydrogen-Enriched Producer Gas from the Air e Steam Gasi Fi Cation." https://ideas.repec.org/a/eee/energy/v205y2020ics0360544220311361.html.

Lv, P., C. Wu, L. Ma, and Z. Yuan. 2008. "A Study on the Economic Efficiency of Hydrogen Production from Biomass Residues in China." *Renewable Energy* 33 (8): 1874–79.

Niel, E. W. J. Van. 2016. "Biological Processes for Hydrogen Production." *Biochemical Engineering and Biotechnology*. https://doi.org/10.1007/10.

Nikolaidis, Pavlos, and Andreas Poullikkas. 2017. "A Comparative Overview of Hydrogen Production Processes." *Renewable and Sustainable Energy Reviews* 67: 597–611. https://doi.org/10.1016/j.rser.2016.09.044.

Parthasarathy, Prakash, and K. Sheeba Narayanan. 2014. "Hydrogen Production from Steam Gasi Fi Cation of Biomass: In Fl Uence of Process Parameters on Hydrogen Yield e A Review." *Renewable Energy* 66: 570–9. https://doi.org/10.1016/j.renene.2013.12.025.

Rajabi Sara, Hamedani, Bocci Enrico, Villarini Mauro, and Di Carlo. 2016. "Techno-Economic Analysis of Hydrogen Production Using Biomass Gasification: A Small Scale Power Plant Study." *Energy Procedia* 101 (September): 806–13. https://doi.org/10.1016/j.egypro.2016.11.102.

Ram, Narasimhan Kodanda, Nameirakpam Rajesh Singh, Perumal Raman, Atul Kumar, and Priyanka Kaushal. 2019. "A Detailed Experimental Analysis of Air–Steam Gasification in a Dual Fired Downdraft Biomass Gasifier Enabling Hydrogen Enrichment in the Producer Gas." *Energy* 187: 1–16. https://doi.org/10.1016/j.energy.2019.115937.

Sandeep, K., and S. Dasappa. 2014. "Oxy – Steam Gasification of Biomass for Hydrogen Rich Syngas Production Using Downdraft Reactor Configuration." *International Journal of Energy Research* (March 2013): 174–88. https://doi.org/10.1002/er.

Schweitzer, Daniel, Friedemann Georg Albrecht, Max Schmid, Marcel Beirow, Reinhold Spörl, Ralph Uwe Dietrich, and Antje Seitz. 2018. "Process Simulation and Techno-Economic Assessment of SER Steam Gasification for Hydrogen Production." *International Journal of Hydrogen Energy* 43 (2): 569–79. https://doi.org/10.1016/j.ijhydene.2017.11.001.

Sharma, Kapil Dev, and Siddharth Jain. 2019. "Overview of Municipal Solid Waste Generation, Composition, and Management in India." *Journal of Environmental Engineering* 145 (3). https://doi.org/10.1061/(ASCE)EE.1943-7870.0001490.

TANGEDCO. 2017. "Category of Consumers * Fully/# Partly Subsidised by the Government I – High Tension Supply II – Low Tension Supply." www.tangedco.gov.in.

Xiong, Shanshan, Jiang He, Zhongqing Yang, Mingnv Guo, Yunfei Yan, and Jingyu Ran. 2020. "Thermodynamic Analysis of CaO Enhanced Steam Gasification Process of Food Waste with High Moisture and Low Moisture." *Energy* 194: 116831. https://doi.org/10.1016/j.energy.2019.116831.

21 The Energy Potential of Brazilian Organic Waste

Luciano Basto Oliveira, Amaro Olímpio Pereira Júnior, Ingrid Roberta de França Soares Alves, Marcelo de Miranda Reis, and Adriana Fiorotti Campos

CONTENTS

21.1 Introduction 287
21.2 Materials and Methods 289
    21.2.1 Biodiesel Estimates 290
    21.2.2 Biogas Estimates 291
    21.2.3 Economic Analysis 292
21.3 Results 294
21.4 Conclusion 295
References 295

21.1 INTRODUCTION

A major challenge facing modern society is addressing the management of solid waste. With the increases in the world's population and the high consumption of materials, the problem of landfill depletion and the pollution generated by the improper disposal of waste has grown. In this scenario, the need to change the paradigm and consumption patterns in the short term has become evident and, taking into account that these technologies cannot be used as an incentive for greater waste production, the opportunity arises to better educate society in the treatment of solid waste, encouraging recycling and obtaining energy and benefits from this.

As more efficient energy conversion technologies are developed, along with the implementation of policies to encourage alternative renewable sources, an increase in the share of biomass in the world energy matrix will occur. In some developed countries, especially in Europe, such policies have been put into practice through tax incentives and the issuance of green certificates, which are related to projects for the generation of electricity from alternative sources to fossil fuels, such as biomass and wind and solar energies (Miranda et al. 2010).

In Brazil, the boost in the use of new technologies for waste treatment tends to materialize with Law No. 12,305/2010, which instituted the National Policy on Solid Waste in 2010. This law establishes the closure of dumps in the country and that only tailings (waste for which there is no technical, environmental, and economic feasibility for use) may be disposed of in landfills, with higher costs than in dumps, in addition to higher and increasing transportation costs. With this, technologies for the energy use of waste are once again evaluated and, due to the high concentration of organic material, the alternative of biogas production gains momentum.

According to the Panorama of Solid Waste in Brazil (2020), most of the municipal solid waste (MSW) collected in the country in 2019 was disposed in sanitary landfills and registered an increase of 10 million tons in a decade, from 33 million tons/year to 43 million tons/year. At the same time, the amount of waste that was sent to inappropriate units (dumps and controlled landfills) also grew, from 25 million tons/year to just over 29 million tons/year. It is worth noting that the composition of municipal solid waste generated in Brazil in 2019 consisted mainly of organic matter (45.3%) (ABRELPE 2020).

Organic residues can be transformed through anaerobic digestion, a process of organic matter degradation carried out by an effective consortium of microorganisms, resulting in fuel gas with methane levels of around 60–70%, and carbon dioxide of 20–30%, in addition to other gases. Digester residues can be used as fertilizer and it is an important factor for Brazilian agriculture, which is responsible for 7% of the world fertilizer consumption only behind China, India, and the United States. Approximately 70% of the fertilizers used in Brazil are imported, reinforcing the dependence on imports to supply Brazilian demand (ANDA 2019).

Biogas also allows for several applications, such as cooking food, power generation in lamps, refrigerators, incubators, industrial ovens, electricity generation, and vehicular consumption, after treatment. Even with fertilizer as a co-product, which is not always simple to dispose of, especially in large cities, whether due to the distance from rural areas or the risk of contamination that the mixture with hazardous waste can cause, biogas has the advantage of being able to be used for some purposes even with no treatment and, when treated, can replace natural gas in all its applications.

An experience developed in Brazil in the 1980s was biogas treatment to achieve vehicular purposes, used by vehicles from the sanitation companies of Rio de Janeiro (CEDAE) and São Paulo (SABESP) and Companhia de Transporte Coletivo (CTC), a public company in the state of Rio de Janeiro responsible for urban buses, which converted its diesel cycle engines to an Otto cycle, an expensive activity that made the vehicles gas captives, since biogas substitutes are gasoline and ethanol, whose prices are higher and would not justify the choice.

In the 1990s, when a natural gas distribution network in large cities was already available, alongside evidence of climate change and damage from local pollution caused by diesel oil burning in cities, the conversion of urban buses to natural gas was proposed, but came up against the lack of supply in medium-sized cities, for which these vehicles are sold after a few years of use, also indicating fuel dependence.

However, new facts in the world of technology make it possible to change this assessment, namely, the Flexible Diesel Natural Gas System (Dual-Fuel) and Ottolization at the Factory.

The Dual-Fuel system consists of adapting existing diesel engines through the installation of a second injection system, to manage vehicle natural gas and air control, without the need to change basic engine components, with ignition performed by diesel pilot injection (BOSCH 2011). This system allows the replacement of diesel oil for a mixture with up to 90% of natural gas (with each liter of diesel being replaced by a cubic meter of gas) and 10% of diesel oil, or its biomass substitutes.

Scania at the São Bernardo do Campo – SP plant began the production of trucks powered by CNG (natural gas), LNG (liquefied gas), and biomethane (obtained from organic waste) in 2020. In addition to European countries, Brazil is the only one to produce this line of vehicles. The CNG and LNG versions are also able to use biomethane in any proportion, termed flex vehicles (Scania 2020).

This alternative aids in reducing the damage caused by diesel oil combustion with respect to atmospheric emissions, both local (especially in metropolises) and regional, as well as global. Substitution by natural gas and biomethane reduces particulate matter emissions, which cause breathing problems, sulfur, which causes acid rain, and gases responsible for the greenhouse effect. In addition, this reduces the import of diesel oil, a derivative for which Brazil is not yet self-sufficient.

Regarding the replacement of diesel oil with treated biogas, to the point of guaranteeing the purity of the natural gas (renewable natural gas – GNR), and biodiesel, the supply potential is estimated and compared with the demand for selected diesel oil sectors.

In this study, a feasibility analysis of the use of biogas and biodiesel in Dual-Fuel systems was carried out, using data obtained from the literature, as well as the concept of energy balance and the cost-benefit index. This choice stems from the consideration that the fleet's exchange speed for gas-powered diesel vehicles will take longer and is more expensive than the installation of the Dual-Fuel system.

This choice is based on the possibility that the biofuel consortium is able to comply with the National Environment Council (Conama) resolution No. 490, of November 2018, which stated that all heavy vehicles must follow the limit of emissions of the European Euro VI guideline until 2032, with intermediate stages in 2023 and 2027, which will not be the subject of this work. When applicable, the economy was verified before the alternative of replacing the fleet in the assessed scenario.

21.2 MATERIALS AND METHODS

Data from the literature and information from government agencies and companies dealing with the issue of waste production and energy aspects were used for the analysis developed in this study. The concept of energy balance was used to compare biofuel (biogas and biodiesel) availability to diesel oil demands. To analyze the results, information related to the cost-benefit index was used, where the technique described subsequently was applied and the results were compared to the price of diesel oil.

Based on data from Solvi (2006), we considered that the consumption of diesel oil in the collection of solid urban waste would have increased by 65%, from the range of 4.5 L/t to 7.5 L/t. Thus, diesel oil consumption was estimated at about 545 ML/year to collect approximately 72.7 Mt of annual waste (ABRELPE 2020), as 6.3 Mt were not even collected, whose emissions, using the factor of 2.6 kg CO_2/L, reach 1.5 Mt CO_2/year. Considering the typical composition of 60% in food waste (Soares et al. 2011) and the methane generation factor of 55 m³/t of organic fraction (OWS 2011), the annual potential of anaerobic digestion of this portion of urban waste adds up to 2.2 Mm³/year, four times the demand.

Another sector that consumes a significant amount of diesel oil and results in organic residues that can be converted into biogas is agriculture. According to the National Energy Balance (EPE 2019a), the consumption of the Agricultural sector was 10% of the national diesel oil, totaling 56.9 Mm³, which corresponds to about 5.7 Mm³/year. As the carbon dioxide emission factor is of 2.6 t CO_2/m³ (in combustion), the emission resulting from this combustion was of 15 Mt CO_2.

As the biomass available in this sector is that left in the field during the harvest, it will be necessary to collect the available material, representing an increase in demands. Typical residual biomass production is higher than that of the commercial product, but most of this must be left in the field, due to agronomic requirements. Typical values are shown in Table 21.1.

Thus, the material that can be collected is equivalent to approximately 21% of the harvested production. Even so, the vehicles that are used must accompany the harvesters throughout the route, since the material removal must be performed equally throughout the terrain.

For this reason, energy consumption will not be equivalent to the collected portion and, on the other hand, will also not exceed the volume currently consumed to transport the harvest. In the

TABLE 21.1
Characteristics of Agricultural Products (Base Year 2017)

Crop	Agricultural Production (Mt)	Residue/ Product (tBbs/t)	Moisture (%)	Collectable Without Causing Damage (%)	Available Material (Mt)
Sugarcane (straw)	759	0.15	15	50	65
Soy	115	1.25	15	30	50
Corn	98	1.82	9,5	40	78
Rice	12.5	1.23	13	40	7
Wheat	4	2.73	7.5	40	5
Manioc	19	0.64	9.6	40	5
Bean	3	1.25	11	40	2
Agriculture	**1010.5**				**212**

Source: EPE (2019c).

most conservative case, this demand would double, which represents the upper limit of agriculture, reaching 11.4 Mm³/year. Thus, the sum of the demands reaches 18.4 12 Mm³ of diesel oil per year.

Considering that the implementation of this proposal is effective, the emission of carbon dioxide to be avoided from the combustion of diesel oil represents only that referring to the currently consumed fuel, which amounts to 13.5 Mt CO_2/year, as the complementary collection of agricultural waste will not be performed with fossil fuel.

In order to avoid doubts about the estimate and the potential, the relationship between the calorific powers of biofuels and diesel oil is considered as 90%. Thus, each biofuel should have more than the volumetric parity to meet demand. Thus, to replace 12 Mm³ of diesel oil, 12,000 Mm³/year of GNR and 1.32 Mm³/year of biodiesel will be required.

21.2.1 BIODIESEL ESTIMATES

Due to the legislation in force in Brazil, 10% of biodiesel consumption is already in place in the country, which means that 0.575 Mm³ of biodiesel is already used in the 5.7 Mm³ of the agricultural sector and 0.5 Mm³ of the sanitation sector. Thus, it would be necessary to increase the offer by 0.75 Mm³ of biodiesel, practically 15% of the current production, which is completely compatible with the existing idle capacity in the country.

Biodiesel production was 5.4 billion liters, an increase of 24.7% in relation to the previous year, while the consumption of diesel B grew only 1.6%. Thus, the increase in the mandatory percentage was the main reason for this growth in biofuel production. On the other hand, the production of diesel A by the national refining park increased by 2.4%. In this context, fossil fuel imports fell 10%, reaching 11.6 billion liters. This value was not higher due to the increases in biodiesel consumption (EPE 2019d).

This additional demand to serve the two current main sectors is lower than diesel imports and slightly higher than the offer already made at auctions (Figure 21.1).

The main raw material available in rural areas, the largest of the markets evaluated in this study and the main input used for biodiesel sold at auctions, is soybean oil. According to EPE (2019d), this input remained as the main raw material for obtaining biodiesel in 2018, with a 69.8% share in the market, which represents 3.7 billion liters, followed by bovine fat, with 13.4 (Figure 21.2).

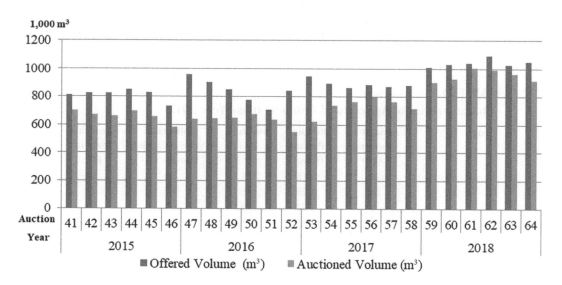

FIGURE 21.1 Volume of biodiesel at auctions: offered × auctioned.

Source: EPE (2019d).

The Energy Potential of Brazilian Organic Waste

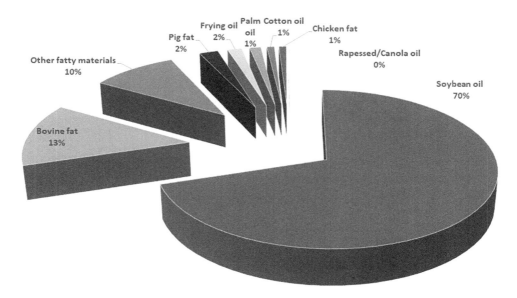

FIGURE 21.2 Composition of the biodiesel input market in Brazil.
Source: EPE (2019d).

Regarding the potential supply of fatty inputs for biodiesel production, the 10-year plans of the Ministry of Agriculture, Livestock and Supply (MAPA), which project the production of oilseeds and the slaughter of animals, of which oil and tallow contents are known, respectively, may be used as reference.

It is also possible to consider fatty urban waste, such as frying oil and the material present in grease boxes. To simplify, as one of the main agricultural crops in Brazil is soy, which is also the main source of biodiesel, a potential of 24 Mm³ of oil for this crop was estimated using data from Table 21.1 and the oil content on the order of 19% of the grain and its typical density of 0.9 kg/m³. As a result, even if all biodiesel came from soybean oil, Brazil would display the capacity to support the proposed expansion and maintain food security.

21.2.2 BIOGAS ESTIMATES

With regard to biogas production, the availability of this material is linked to rural production and urban waste, whose data were obtained from surveys carried out by the Brazilian Institute of Geography and Statistics (IBGE) and the National Sanitation Information System (SNIS), such as the 2017 Census of Agriculture (IBGE 2019) and the 2018 Diagnosis of Water and Sewage Services (SNIS 2019).

The production of methane in the urban environment is on the order of 55 m³/t of organic fraction contained in solid waste (OWS 2011), and 15 L/m³ of treated sewage (Jordão and Pessoa 1995). In rural areas, this factor reaches 200 m³/t (Cortez et al. 2008) for agricultural residues, while in livestock they reach 180 L/day per bovine, 90 L/day per swine, and 5.5 L/day per bird (Nogueira and Lora 2003). Cane vinasse, which is an agro-industrial residue available in sugar and alcohol plants, on the order of 12 times the production of ethanol, has the potential to produce triple its volume in methane. Thus, part of the potential can be accounted for, as displayed in Table 21.2.

In addition to the total amount presented in Table 21.2 which is higher than that required to replace all diesel oil in these sectors, it is important to note that the offers are more than sufficient to meet sectoral demands.

TABLE 21.2
Biomass and Biogas Supply

Material	Heads	Usable Waste (Mt)	Biogas Methane (Mm³/a)
Agriculture		212	42,400
Cane vinasse *			1,100
Confined livestock			4,783
Dairy cattle	11,506.788		756
Pigs	39,346.192		1,295
Birds	1,362,253.509		2,724
Rural subtotal			**48,283**
Solid waste		72.7	3,999
Sewage			64.5
Urban subtotal			**4,064**

Note: *From 33 Mm³ of ethanol produced in 2018 (EPE 2019a).
Source: IBGE (2019); SNIS (2019).

TABLE 21.3
Diesel Oil Prices to Consumers by Region in 2018

Region	Diesel Oil (R$/L)*
Midwest	3.69
North	3.68
South	3.34
Southeast	3.46
Northeast	3.50

Note: * In the case of diesel and biodiesel, the conversion is 1 m³ = 1000 L. Conversion from biomethane is 1 m³ of biomethane = 1 L of diesel.
Source: Based on ANP data (2019).

Considering that commercial biodigesters worldwide produce on average 5000 m³ of methane per day (EUROGAS 2011), this use can guarantee the construction and operation of approximately 26,000 rural digesters. In the urban case, the systems tend to be larger than 8,500 m³/day (OWS 2011), which represents a further 2000 biodigesters, which may correspond to the development of another economic segment still incipient in the country.

21.2.3 ECONOMIC ANALYSIS

As the reference price is that of diesel oil, which varies according to the distance in relation to the refineries that must be transported, Table 21.3 displays the practiced values.

For this reason, the typical price of biodiesel applied herein was that practiced in the last auctions promoted by the National Agency of Petroleum, Natural Gas and Biofuels (ANP), shown in Table 21.4.

The comparison between Tables 21.3 and 21.4 indicates that biodiesel was cheaper than diesel in all regions during most of 2018.

TABLE 21.4
Evolution of Biodiesel Prices at Auctions in 2018

Auction	Mandatory 10% Mix Phase: From March 2018			
	Offered Volume (m³)	Auctioned Volume (m³)	Maximum Reference Price (R$/m³)	Average Price (R$/m³)
59th Auction – ANP No. 01/2018–01 public notice – 07/02/2018	1,013,227	903,225	2,970.48	2,590.87
60th Auction – ANP No. 02/2018 public notice – 05 and 06/04/2018	1,033,422	927,693	2,910.72	2,423.33
61st Auction – ANP No. 03/2018 public notice – 14, 15, and 18/06/2018	1,040,120	1,008,019	2,657.59	5,630.59
62nd Auction – ANP No. 04/2018 public notice – 16, 17, and 20/08/2018	1,093,310	992,574	2,983.75	2,439.31
62nd Auction – ANP No. 05/2018 public notice – 03, 04, and 06/10/2018	1,029,600	964,356	3,161.86	2,814.85

Source: ANP (2019).

Field biogas production requires the collection of agricultural waste, for which the typical cost of harvesting cane straw will be applied, disregarding the values of agricultural losses that this could cause and the opportunity cost of the material in the field (Hassuani et al. 2005), since no damage due to the withdrawal of this portion of R$17.83/t (idem) is noted as a parameter. The advantage of the material being homogeneous and containing almost no contaminants is emphasized, thus not requiring a previous treatment step. In the case of livestock waste and agribusiness, there is no collection cost, and a treatment cost to avoid local environmental damage could have been considered, which is being disregarded in this analysis.

In contrast, urban waste, in addition to being collected and delivered to the treatment site, still must be treated, in the value range of R$15–60/t (EPE 2019b). On the other hand, it is very heterogeneous and requires a sorting system to segregate materials that could clog the biodigester. As its collection and final disposal in an anaerobic environment are required by health authorities, it is a sector that generates methane and whose energy use, by avoiding its emissions, can provide the mitigation of global warming gas emissions (IPCC 2006).

The biodigestion system evaluated for both cases uses a dry stage process, due to both lower investments and the operation and maintenance, equivalent to 10% a year of the total investment (OWS 2011). In order to simplify the assessment, since it would be appropriate to analyze the radius of action of each project, the plants were considered viable in terms of logistics based on the following premises: the cost of rural freight bears the waste harvest, which will accompany the product harvester.

The amounts used for this were R$45.5 million (EPE 2019b) for a sorting and anaerobic digestion plant for urban waste, able to receive 600 t/day and digest 300 t/day of organic material. The same technology was considered for rural cases, disregarding the sorting system, odor filter, liquid effluent treatment, truck scale, maintenance workshop, warehouse, in addition to the facilities for the waste pickers' cooperative, decreasing the investment to R$21.5 million (EPE 2019c).

The methodology for calculating the cost of treated biogas was the leveled cost (Vollebergh 1997; Oliveira and Rosa 2003) annualizing the investment based on a discount rate.

Vollebergh (1997) uses the cost-benefit index (CBI) of the energetics from Barbier et al. (1992). In the Brazilian case, the costs were supplied in a discriminated way, allowing for comparisons of details like necessary investments, and applied to the CBI calculation, as follows:

$$CBI = CI + COM + CT + CC$$

Here:

CI is the annual investment cost in the power plant in $/m^3$, given by

CI = IU*FRU/EG where IU is the total investment cost in the process plant, including interest during construction, in FRU the capital payback factor for the economic useful life of the process plant, expressed by

FRU = $i*(1 + i)^v/((1+i)^v - 1)$

where i is the annual discount rate; v is the useful life in years, and EG is the guaranteed energy from the process plant, in m^3/year.

COM, the annual operating and maintenance cost of the process plant in $/m^3$, is given by

COM = OMU/m^3

where

OMU is the annual operating and maintenance cost of the power plant in $/year

CT is the investment in transport, not applied in this case

CC is the annual fuel cost for the power plant in $/m^3$, given by

CC = CUT/REND where CUT is the fuel unit cost in $/ton and REND is the specific average conversion factor in m^3/ton.

For this, a discount rate of 12% per year, a waste cost of R$30/t, an organic fraction content of 80%, and conversion to methane of 55 m^3/t of organic fraction were applied.

21.3 RESULTS

Based on the survey of biogas generation from solid waste (urban and rural), the costs of treated biogas in each analyzed case were estimated and are displayed in Table 21.5. Rural waste has a higher cost when compared to urban solid waste.

One cubic meter of gas and 0.11 L of biodiesel are required to replace each liter of diesel oil. The cost of this replacement is equivalent to different values, depending on the type of analyzed sector, as shown in Table 21.6.

TABLE 21.5
Costs of Urban and Rural Solid Waste

Material	R$/m³
Urban solid waste	**1.44**
Agricultural solid waste	1.95
Livestock and agribusiness wastes	1.62
Rural waste	**1.79**

Source: The author.

TABLE 21.6
Cost of Mixing Gas and Biodiesel According to Sector

Sector	Gas (R$)	Biodiesel (R$)	Mix (R$)
Sanitation	1.44	0.30	1.74
Agriculture	1.79	0.30	2.09
Agriculture	1.95	0.30	2.25
Livestock and agribusiness	1.62	0.30	1.94

Note: Biodiesel calculated based on the average value of the most recent auction in Table 21.4.

Source: The author.

TABLE 21.7
Solution Cost Considering the Biofuel Kit

Sector	Mistura (R$)	Kit (R$)	Soma (R$)
Sanitation	1.74	0.02	1.76
Agriculture	2.09	0.02	2.11
Agriculture	2.25	0.02	2.27
Livestock and agribusiness	1.94	0.02	1.96

Source: The author.

As the biofuel technology costs R$30,000.00 per city bus (Bosch 2011), which consumes 100 m³/day and has a useful life of 10 years, the cost to be charged for the fuel was estimated in the same way as for the biodigesters. With this, a value of R$0.02/m³ is reached, which must be added to the previous subtotals and then compared to the diesel oil value to assess the attractiveness of the project (Table 21.7).

The results obtained for this scenario indicate that it is possible to replace diesel oil consumption in the sanitation, livestock, and agribusiness sectors with renewable sources and obtain cost reductions. The benefit exceeds R$1/L of substituted diesel oil, which may yield R$1.93/L in the region where it is more expensive in view of the potential of the sanitation sector.

In the other scenario, in which the cost of harvesting is equal to half the previous cost considering that vehicles will only carry 21% of the weight, even if they follow the entire route of the harvesters, this would reduce agriculture costs, increasing the advantage of this substitution and stimulating investments in this alternative to reduce imports, as reported in the National Energy Balance (EPE 2019d).

21.4 CONCLUSION

The survey carried out on the production of biogas and biodiesel from different sources, as well as the methodology of cost analysis and the consumption estimates by the different sectors of the country applied herein, indicates that there is a demand for fuel and availability of residual inputs for this in the sanitation and agriculture sectors.

The new biofuel technology with flexible diesel, the natural gas system (Bosch 2011), allows the replacement of diesel fuel consumption with treated biogas and biodiesel. In this substitution, it is possible to improve the final disposal of urban waste and air quality, reduce the emissions of gases responsible for the greenhouse effect, eliminate the import of diesel oil, and create a biodigester industry with a potential of over 26,000 rural and 2000 urban equipment, or nearly 28,000 biodigesters in the country.

The estimated costs pointed to the viability of the analyzed options, which points to the need for more in-depth assessments on biodigester viability.

REFERENCES

ABRELPE [Associação Brasileira de Empresas de Limpeza Pública e Resíduos Especiais]. 2020. *Panorama dos Resíduos Sólidos no Brasil* [Panorama of Solid Waste in Brazil]. São Paulo: ABRELPE.

ANDA [Associação Nacional para Difusão de Adubos]. 2019. *Principais Indicadores do Setor de Fertilizantes. Pesquisa Setorial – Dados 2019* [Main Indicators of the Fertilizer Sector. Sector Research – 2019]. Available Online http://anda.org.br/wp-content/uploads/2020/02/Principais_Indicadores_2019.pdf. Accessed in February 2020.

ANP [Agência Nacional do Petróleo, Gás Natural e Biocombustíveis]. 2019. *Anuário estatístico brasileiro do petróleo, gás natural e biocombustíveis: 2019* [Brazilian Statistical Yearbook on Petroleum and Natural Gas: 2019]. Rio de Janeiro: ANP.

Barbier, E., Burgess, J. and Pearce, D. 1992. "Technological substitution options for controlling greenhouse gas emissions". In: Dornbusch, R., and Poterba, J.M. (Eds.), *Global Warming: Economic Policy Responses.* Cambridge, MA: MIT Press, pp. 109–60.

Bosch. 2011. *Internal | DS/SA-LA | 1/13/2010.* Robert Bosch Ltd. Available Online https://assets.bosch.com/media/en/global/bosch_group/our_figures/publication_archive/pdf_1/gb2010.pdf.

Cortez, L. A. B., Lora, E. E. S., and Gómez, E. O. 2008. *Biomassa para energia* [Biomass for Energy]. Campinas: Unicamp Ed.

EPE [Empresa de Pesquisa Energética]. 2019a. *Balanço Energético Nacional: Ano base 2018* [National Energy Yearbook 2018]. Rio de Janeiro: EPE. Available Online www.epe.gov.br. Accessed in June 2019.

EPE [Empresa de Pesquisa Energética]. 2019b. *Potencial Energético dos Resíduos Urbanos.* EPE-DEA-IT-007/2019. Rio de Janeiro: EPE.

EPE [Empresa de Pesquisa Energética]. 2019c. *Potencial Energético dos Resíduos Agropecuários* [Energy Potential of Agricultural Waste]. EPE-DEA-IT-006/2019. Rio de Janeiro: EPE.

EPE [Empresa de Pesquisa Energética]. 2019d. *Análise de Conjuntura dos Biocombustíveis Ano 2018* [Conjuncture Analysis of Biofuels Year 2018]. EPE-DPG-SGB-Bios-NT-01-2019-r0. Rio de Janeiro: EPE.

EUROGAS. 2011. *Rede de Biogás da Europa.* Brussels: EUROGAS. Available Online www.european-biogas.eu/eba/index.php. Accessed in June 2019.

Hassuani, S. J., Leal, M. R. L. V., and Macedo, I. C. 2005. "Biomass power generation: Sugar cane bagasse and trash". In: *Série Caminhos para Sustentabilidade.* Piracicaba, Brasil: Programa das Nações Unidas para o Desenvolvimento/Centro de Tecnologia Canavieira – CTC.

IBGE [Instituto Brasileiro de Geografia e Estatística]. 2019. *Censo Agropecuário 2017: Resultados Definitivos 2017* [Agricultural Census: Definitive Results]. Rio de Janeiro: IBGE.

IPCC [Intergovernmental Panel on Climate Change]. 2006. Available Online https://www.ipcc.ch/report/2019-refinement-to-the-2006-ipcc-guidelines-for-national-greenhouse-gas-inventories/.

Jordão, E. P. and Pessoa, C. A. 1995. *Tratamento de Esgotos Domésticos*, 3a edição. Rio de Janeiro, RJ: ABES, 681 p.

Miranda, I. C., Araujo, C. R., and Mothé, C. G. 2010. "Utilização da Biomassa para Fins Energéticos e Inserção na Matriz Energética Mundial: Panorama Atual e Perspectiva Futura [Use of biomass for energy purposes and insertion in the world energy matrix: Current outlook and future perspective]". *Revista Analytica*, 46, pp. 98–110.

Nogueira, L. A. H., and Lora, E. E. S. 2003. *Dendroenergia: Fundamentos e Aplicações* [Dendroenergy: Fundamentals and Applications]. Rio de Janeiro: Editora Interciência.

Oliveira, L. B., and Rosa, L. P. 2003. "Brazilian waste potential: Energy, environmental, social and economic benefits". *Energy Policy*, 31, pp. 1481–91. https://doi.org/10.1016/S0301-4215(02)00204-5.

OWS [Organic Waste Systems]. 2011. Available Online www.ows.be. Accessed in June 2011.

Scania. 2020. *Comunicação Interna.* Available Online www.jornadascania.com.br/nossajornada/perspecr1.php. Accessed in February 2020.

SNIS [Sistema Nacional de Informações Sobre Saneamento]. 2019. *Sistema Nacional de Informações sobre Saneamento: 24º Diagnóstico dos Serviços de Água e Esgotos – 2018* [National Sanitation Information System: 24th Diagnosis of Water and Sewerage Services – 2018]. Brasília: SNS/MDR.

Soares, E., Mahler, C. F., and Schueler, A. S. 2011. "Study of the gravimetric characterisation and heating value of urban solid waste in Rio de Janeiro, Brazil". Proceedings Sardinia 2011, Thirteenth International Waste Management and Landfill Symposium. Cagliari, Italy.

Solvi. 2006. *Annual Report.* Available Online www.solvi.com.br. Accessed in June 2011.

Vollebergh, H. 1997. "Environmental externalities and social optimality in biomass markets: Waste-to-energy in The Netherlands and biofuels in France". *Energy Policy*, 25, pp. 605–21. https://doi.org/10.1016/S0301-4215(97)00052-9

22 Life Cycle Sustainability Assessment of Bioenergy

Literature Review and Case Study

João Gabriel Lassio and Denise Ferreira de Matos

CONTENTS

22.1 Introduction 297
22.2 Materials and Methods 298
    22.2.1 A Brief Overview of Bioenergy 298
    22.2.2 Life Cycle Sustainability Assessment 298
22.3 LCSA Applied to Bioenergy 299
    22.3.1 Publication Evolution and Trends 299
    22.3.2 Dimensions of Sustainability 299
    22.3.3 Sustainability Indicators 300
    22.3.4 Application Methods 302
22.4 Case Study: A Comparison between Soybean and Palm Oil Cultivation for Biodiesel Production in Brazil 302
    22.4.1 A Brief Description of the Brazilian Biofuel Context 302
    22.4.2 Case Study Presentation and Methodology 302
        22.4.2.1 Goal and Scope Definition 303
        22.4.2.2 Sustainability Aspects and Indicators..303
        22.4.2.3 Inventory Analysis 304
        22.4.2.4 Impact Assessment..305
        22.4.2.5 Interpretation...306
22.5 Conclusion 307
References 307

22.1 INTRODUCTION

Since concerns about climate change are in the spotlight, the international energy sector trends and projections indicate a future fueled by low-carbon energy sources. Generating energy from bioenergy instead of fossil fuels has been a promising pathway to keep climate change within acceptable limits. This is because of bioenergy's potential for reducing GHG emissions and its availability and compatibility with several sectors of the worldwide economy (Anoop Singh, Olsen, and Pant 2013; OECD/FAO 2019).

In parallel with this transition toward renewable energy sources, a new core value has emerged in our society: sustainability. This concept has globally brought increased attention to the socio-environmental aspects of energy systems (Maxim 2014). Accordingly, various authors have worked on this topic to place the socio-environmental concerns on the same importance level as the economic ones (Kabayo et al. 2019; Wang et al. 2009; Jin and Sutherland 2018).

Bioenergy is seen as sustainable in itself or, at least, more environmentally sustainable than nonrenewable sources, but its production is not entirely clean (Lijó et al. 2019; Dewulf, Meester, and Alvarenga 2015). Hence, bioenergy requires a robust treatment through specific tools and indicators to deal with its particularities. An excellent example of this is that most of its impacts are dispersed across its supply chain rather than condensed in the power generation phase (Laurent, Espinosa, and Hauschild 2018).

In this context, life cycle sustainability assessment (LCSA) is a suitable technique because it covers issues related to sustainability of the complete products' and services' life cycles. Furthermore, its integrative perspective is consistent with all three sustainability dimensions of the triple bottom line (TBL) model (Elkington 1998) since LCSA can address environmental, social, and economic aspects (Guinée 2016).

This chapter aims to contribute to integrating sustainability into the decision-making process for bioenergy generation. Its first part provides a comprehensive understanding of LCSA and its application in bioenergy. It also guides the methodological choices made to conduct the second part of this chapter, which features a case study comparing the sustainability of soybean and palm oil cultivation for biodiesel production.

22.2 MATERIALS AND METHODS

22.2.1 A Brief Overview of Bioenergy

Bioenergy is derived from biofuels produced directly or indirectly from biomass (FAO 2004; FAO and UNEP 2010). Biofuels are one of the main drivers of decarbonization in various sectors, such as transportation, because they play a crucial role in achieving climate change mitigation targets by reducing GHG emissions (Pathak and Das 2020). However, they are associated with different socio-environmental impacts when compared to fossil-based energy sources (Nieder-Heitmann, Haigh, and Görgens 2019).

Bioenergy implies complex and multifaceted social, environmental, and economic interactions. It includes energy crops cultivation, harvesting, transport activities, biofuels conversion, and end-use of bioenergy (Jin and Sutherland 2018). Therefore, bioenergy requires a comprehensive framework to understand its unclear environmental, social, and economic impacts (Pathak and Das 2020).

Some biofuels are prevalent on the market (bioethanol and biodiesel), while others are more recent or developing. In the case of bioethanol, corn, sugarcane, sugar beet, and wheat are some of the numerous crops that may be used to produce this biofuel. On the other hand, biodiesel's feedstock can be represented mainly by vegetable oils (including recycled), biotechnological sources, and animal fat. As the basis for dedicated energy crops, agricultural production systems use large amounts of resource input (Lora et al. 2011).

22.2.2 Life Cycle Sustainability Assessment

Life cycle assessment (LCA) is a technique for assessing a product's or service's environmental impacts throughout its entire life cycle that has become commonly used to analyze energy systems. LCA is particularly interesting for bioenergy since most of its problems of socio-environmental nature are distributed over its production chains (Laurent, Espinosa, and Hauschild 2018). As a result, this approach better describes the effect of varying feedstocks, enhancing agricultural and industrial processes and evidencing for stakeholders the effect of a specific bioenergy system (Wiloso and Heijungs 2013).

LCA was a step toward sustainability, and LCSA is advancing in the same direction. In this way, the scope of LCA, restricted initially to the environmental dimension, expands to include social and economic issues, resulting in the LCSA. This more holistic approach requires the systematic application and integration of LCA, social life cycle assessment (SLCA), and life cycle cost (LCC) (Kloepffer 2008).

Life Cycle Sustainability Assessment of Bioenergy

BOX 22.1 SLCA AND LCC DEFINITIONS

SLCA is a relatively new technique compared to LCA. It examines the effects of a product's or service's life cycle on stakeholders' well-being (UNEP 2020). However, SLCA's methodological maturing, practice, and agreement are still in their early stages.

LCC is associated with the first reference of the life cycle thinking that occurred at the end of the 1950s when Novick (1959) used a life cycle–based method to analyze military weapon investments. It allows assessing a product's costs over its life cycle and interactions with the economic dimension.

Costa et al. (2019) show that the number of LCSA studies has grown in recent years and point out gaps to be improved in this methodology. For instance, the authors highlight the need for a definition of the system boundary; databases for evaluating the social and economic aspects; a definition of a set of impact categories that allows studies to be compared; comprehensive impact assessment methods; uncertainty analysis methods; and new strategies to communicate LCSA results.

Furthermore, LCSA presents the challenge of integrating results obtained in LCA, SLCA, and LCC and composing a result that expresses the sustainability performance of products or services (Guinée 2016). For this reason, some authors have employed multiple criteria decision-making (MCDM) methods to integrate the results obtained from these three life cycle–based tools (Kalbar and Das 2020).

22.3 LCSA APPLIED TO BIOENERGY

22.3.1 Publication Evolution and Trends

By searching for papers in peer-reviewed scientific journals in the CAPES (Co-ordination and Improvement of Higher-Level Personnel Agency of Brazil) database, we included the keywords "LCSA" and "Life Cycle Sustainability Assessment" to find relevant papers published from 2000. This database comprises Scopus (Elsevier), Web of Science, Materials science and Engineering Database, and Technology Research Database collections.

Table 22.1 shows the current state of development of LCSA through the frequency of papers in scientific journals divided into 5-year periods starting in 2000. We can observe that until 2005, the number of scientific papers was very small: 8 until 2000, 55 in 2005. However, this frequency has grown significantly since 2010. Both the overall increase in publications and the emergence of bioenergy publications since 2010 may have been triggered by relevant publications dedicated to LCSA, such as Kloepffer (2008), Hunkeler et al. (2008), UNEP/SETAC Life Cycle Initiative (2009, 2013), and CALCAS (2009). It is also possible to verify that the relevance of the LCSA studies on liquid biofuels stands out among other kinds of bioenergy.

Although LCSA is not as widely used as LCA, interest in this approach has grown over time. However, the number of sustainability assessment studies is still limited in Brazil. According to Zanghelini et al. (2017), an analysis of the evolution of LCA research topics in the country shows that the biofuels group constituted the central area.

22.3.2 Dimensions of Sustainability

As the sustainability concept usually involves subjective and complicated themes, multiple LCSA interpretations and implementations are possible (Phillis and Andriantiatsaholiniaina 2001; Cinelli, Coles, and Kirwan 2014). Regarding the energy sector, several authors have been working under different points of view to apply LCSA. Amer and Daim (2011) consider the environment, social,

TABLE 22.1
Number of Peer-reviewed Papers and Title of Most Relevant Papers per Period

Period	No. of Papers	Title of the Most Relevant Paper — Generic	Bioenergy
Until 2000	8	"The feasibility of including sustainability in LCA for product development" (Andersson et al. 2000)	–
2001–2005	55	"Applications of LCA to NatureWorks Polylactide (PLA) production" (Vink et al. 2003)	–
2006–2010	176	"LCSA of products" (Kloepffer 2008)	–
2011–2015	798	"Enhancing the practical implementation of LCSA – proposal of a tiered approach" (Neugebauer et al. 2015)	"Prioritization of bioethanol production pathways in China based on LCSA and MCDM" (Ren et al. 2015)
2016–2020	905	"Extending the geopolitical supply risk indicator: application of LCSA to the petrochemical supply chain of polyacrylonitrile-based carbon fibers" (Helbig et al. 2016)	"Inclusive Impact Assessment for the sustainability of vegetable oil-based biodiesel – part II" (Nguyen et al. 2017)

economic, political, and technical dimensions in selecting renewable energy technologies. Santoyo-Castelazo and Azapagic (2014), Kouloumpis and Azapagic (2018), and Traverso et al. (2012) follow the majority of the works of LCSA, which encompass only the environmental, social, and economic dimensions of sustainability.

In the case of bioenergy, this situation does not seem different. By discussing biofuels' sustainability, Lora et al. (2011) propose a framework for sustainability indicators as a tool for performance assessment. On the other hand, Jin and Sutherland (2018) suggest an integrative perspective that considers the environmental, social, and economic dimensions to understand better the potential consequences of establishing a bioenergy system.

Most authors prefer to follow the TBL model (Elkington 1998). In this way, assessing sustainability means that the environmental, social, and economic aspects have to be tuned and checked against each other (Kloepffer 2008). Moreover, this model is consistent with the global plan of action for "people, planet, and prosperity," advocated by the 2030 Agenda for Sustainable Development (UN 2015) and well accepted by the industry (Guinée 2016).

22.3.3 SUSTAINABILITY INDICATORS

One of the first and most relevant works on sustainability indicators focused on energy corresponds to the efforts led by the International Atomic Energy Agency (IAEA) (2005). This international initiative resulted in 30 indicators, classified into environmental, social, and economic dimensions (IAEA 2005; Maxim 2014). Regarding bioenergy, FAO and UNEP (2010) indicate key issues to consider, intending to ensure the bioenergy assessment in further detail with the help of appropriate tools. Both sets of indicators can serve as a starting point in formulating a more complete and widely recognized set of bioenergy indicators following the sustainable development principles.

In scientific literature, the sustainability assessment of energy sources, such as bioenergy, through some indicators, is complicated and time-consuming because it can involve unquantifiable or unavailable information (Doukas, Patlitzianas, and Psarras 2006). For that reason, there is an imbalance in almost all LCSA studies concerning the frequency of issues addressed by each sustainability dimension.

Collotta et al. (2019) conducted an extensive survey on sustainability indicators for biofuels. After examining 60 LCSA studies from 2005 (Figure 22.1), the authors indicate a great

Life Cycle Sustainability Assessment of Bioenergy

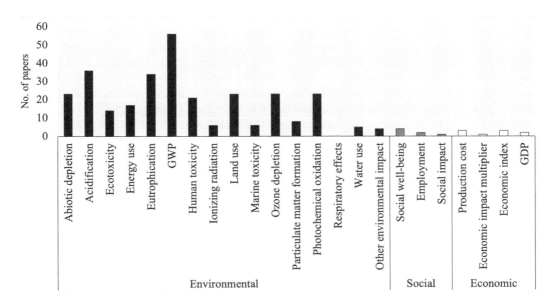

FIGURE 22.1 Environmental, social, and economic indicators considered in the LCSA studies analyzed by Collotta et al. (2019).

Source: Elaborated by the authors based on Collotta et al. (2019).

opportunity to extend the number of sustainability indicators covered, mainly to reflect social and economic aspects. The environmental dimension not only was considered by the majority of studies but also contemplated a larger number of indicators when compared to other sustainability dimensions.

Selecting environmental indicators of most LCSA studies is based on the impact categories obtained in the available life cycle impact assessment (LCIA) methods. Czyrnek-Delêtre et al. (2017) point out that most biofuel LCA studies look only at GHG emissions and/or energy balance. Note that this situation might result in burden-shifting, where a certain biofuel reduces GHG emissions significantly but has unintended consequences on the natural environment. For these authors, a life cycle–based study of biofuels should include at least climate change, acidification, eutrophication, land use, and energy consumption. Complementing this list, Collotta et al. (2019) draw attention to the need to consider water use.

Regarding social aspects, the UNEP/SETAC Life Cycle Initiative Group (2009) provides a set of indicators organized by stakeholder categories. Nevertheless, several researchers report that existing frameworks cannot guide the selection of suitable social indicators (Kühnen and Hahn 2017). Collotta et al. (2019) highlight this challenge faced by the analyzed studies in their survey, which is reflected in the low frequency of studies covering the social dimension and indicators (see Figure 22.1).

For the economic dimension, the lack of a standardized procedure to conduct LCC in an LCSA study implies the difficulty in defining economic indicators. In this context, Hunkeler et al. (2008) provide a comprehensive guide to help practitioners develop their initial understanding and their first application of LCC. Collotta et al. (2019) affirm that considering economic indicators is crucial for biofuels since their costs are the primary factor for their dissemination in the market. Thus, LCC is often employed to analyze the economic feasibility of biofuel production. Other common indicators are economic indices, co-product value, capital expenditures, and operating costs (Sawaengsak et al. 2014; Collotta et al. 2019).

22.3.4 APPLICATION METHODS

Although LCA, LCC, and SLCA are suitable tools for evaluating environmental, economic, and social sustainability, their integration into a common methodological framework is still scarce (Kouloumpis and Azapagic 2018; Costa, Quinteiro, and Dias 2019). This is because of the lack of a standardized procedure to conduct LCSA and the difficulties of dealing with value choices and subjectivity in the weighting step (Kouloumpis and Azapagic 2018; Guinée 2016).

In scientific literature, several authors integrate the results of LCA, LCC, and SLCA using a weighting scheme based on MCDM methods (Costa, Quinteiro, and Dias 2019). Considering that the criteria and weights definition is consistent with a particular decision problem, these methods provide more rational results (Wang et al. 2009). Accordingly, MCDM methods can support the decision-making process for bioenergy generation that encompasses numerous criteria by allowing more clear, reasonable, and efficient choices.

By reviewing MCDM methods for evaluating the sustainability of energy systems, Wang et al. (2009) reported that the analytic hierarchy process (AHP) is the most popular method due mainly to its simplicity. AHP is usually employed to evaluate technologies and scenarios for production and energy supply. Concerning the LCSA application, several authors have used this method to improve the decision-making step (Corona and San Miguel 2019; Roinioti and Koroneos 2019; Ghazvinei et al. 2017).

22.4 CASE STUDY: A COMPARISON BETWEEN SOYBEAN AND PALM OIL CULTIVATION FOR BIODIESEL PRODUCTION IN BRAZIL

22.4.1 A BRIEF DESCRIPTION OF THE BRAZILIAN BIOFUEL CONTEXT

Brazil stands out globally due to its high proportion of renewable energy in its energy supply mix, approximately 45% of the total in 2018. Because of that, its energy sector is one of the least carbon-intensive in the world (Brazil 2019a; IEA 2018). Its transport sector is the largest energy consumer, which accounted for almost a third of the final energy consumption in 2018. Its energy demand is expected to grow annually at an average rate of 2.1% over the next 10 years. The highlight is the increase of bioethanol (from sugarcane) and biodiesel shares (Brazil 2019b).

The current percentage of biofuels in transportation is around 19% for bioethanol and 4% for biodiesel. Biodiesel's share in transport final energy consumption will achieve 7% in 2027 (Brazil 2019b, 2019a). Following this trend, Brazil should retain its position as the third major biodiesel producer worldwide, contributing to more than 50% of the global biodiesel production expansion (OECD/FAO 2019).

Soybean oil is currently the primary feedstock for biodiesel production in Brazil. Almost three-quarters of biodiesel are produced from this vegetable oil, which will maintain its leadership position for the next decennial period (ANP 2018; Brazil 2019b). Palm oil is the closest energetic competitor of biodiesel raw materials among vegetable oils. It presents the second largest production in the international market (with a biodiesel production share of 25%, just behind soybean, with 30%), with more competitive prices and heating value, and higher energy (Rico and Sauer 2015; Lee and Ofori-Boateng 2013; OECD/FAO 2019; Brazil 2019b).

22.4.2 CASE STUDY PRESENTATION AND METHODOLOGY

Considering that the most critical factor affecting biodiesel production's sustainability is the choice of feedstock (Lee and Ofori-Boateng 2013), we carry out a case study comparing the sustainability of soybean with a promising alternative energy crop: palm oil. We believe it can contribute to elaborating strategies for developing sustainable bioenergy generation and provide subsidies for improving the LCSA understanding and usefulness.

Our analysis follows the TBL concept (Elkington 1998), according to which sustainability comprises three dimensions: environmental, social, and economic. In addition, we propose weighting and aggregating the LCSA results to obtain a single score. Both procedures are assisted with the AHP method. Applying this method requires the participation of the stakeholders affected by the system and dimensionless input parameters. With this in mind, we consulted ten researchers who belong to the energy sector and used internal normalization to transform the various LCSA parameters into a dimensionless scale.

As there is no widely agreed methodology for performing LCSA, we sought to follow the principles and framework of the ISO standards for LCA (ISO 2006).

22.4.2.1 Goal and Scope Definition

22.4.2.1.1 The Goal, Functional Unit, and System Boundaries

This LCSA aims to compare the sustainability of soybean and palm oil cultivation for biodiesel production in Brazil. A functional unit of 1.0 MJ of biodiesel produced was chosen as the reference unit for all these energy crops' input and output streams. The system boundaries were defined from "cradle-to-farm gate." As can be seen in Figure 22.2, it includes the four main tasks provided by Korres (2013): seedbed preparation, sowing/planting, field operations, and harvesting.

22.4.2.2 Sustainability Aspects and Indicators

The sustainability indicators are selected based on the issues raised and discussed in the previous sections of this chapter and categorized into the three dimensions of sustainability.

The environmental dimension comprises five indicators based on the problem-oriented (midpoint) approach of ReCiPe 2016 (Huijbregts et al. 2017). They measure the potential impacts on climate change (GWP), terrestrial acidification (TAP), freshwater eutrophication (FEP), terrestrial ecotoxicity (TETP), and freshwater ecotoxicity (FETP). The social dimension is assessed through social welfare, which is represented here by human carcinogenic toxicity (HTPc) and human non-carcinogenic toxicity (HTPnc) by USEtox 2.0 (Rosenbaum et al. 2018), and local employment (LEM), based on Barbier (2010). Finally, regarding the economic dimension, our analysis is focused on two levels. The first level is based on technical-economic aspects of the energy crops represented by land use (LOP), according to ReCiPe 2016 (Huijbregts et al. 2017), and water consumption (WCP), according to AWARE (Boulay et al. 2018). The second level is more comprehensive

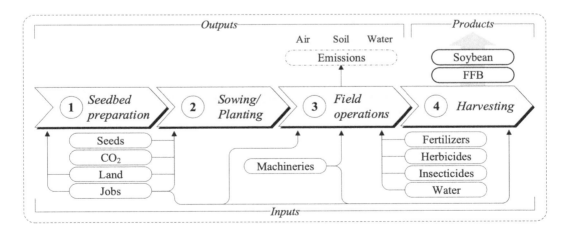

FIGURE 22.2 System boundaries considered in the LCSA of soybean and palm oil cultivation for biodiesel.

Source: Elaborated by the authors.

and related to the contribution to the local economic development (CED), based on IBGE (2017), Nascimento et al. (2018), and Villela et al. (2014).

22.4.2.3 Inventory Analysis

Data used for this analysis were obtained from literature and the Ecoinvent database 3.5. Table 22.2 summarizes materials and energy resources inputs and outputs for producing 1.0 MJ of biodiesel from palm oil and soybean.

TABLE 22.2
LCI of Palm Oil and Soybean Cultivation for the Production of 1.0 MJ of Biodiesel

Inputs/Outputs	Unit	Amount Palm Oil	Amount Soybean	Source
Seedbed preparation (inputs)				
Land use	ha	6.12×10^{-6}	4.04×10^{-5}	Jungbluth and Chudacoff (2007), Villela et al. (2014)
Tillage				
Machine use	ha	–	5.24×10^{-5}	Jungbluth and Chudacoff (2007)
Limestone	kg	–	1.52×10^{-2}	Cavalett and Ortega (2010), de Souza et al. (2010)
Sowing/planting (inputs)				
Seeds	kg	6.65×10^{-6}	2.79×10^{-3}	Cavalett and Ortega (2010), de Souza et al. (2010)
CO_2	kg	1.43×10^{-1}	1.57×10^{-1}	Jungbluth and Chudacoff (2007)
Energy	MJ	1.99	2.34	
Land use change	ha/a	1.65×10^{-4}	2.02×10^{-5}	Cavalett and Ortega (2010), de Souza et al. (2010)
Machine use	ha	–	2.74×10^{-5}	
Field operations (inputs)				
Fertilization				
Fertilizers				
Nitrogen (N)	kg	2.52×10^{-4}	–	Cavalett and Ortega (2010), de Souza et al. (2010)
Phosphate (P_2O_5)	kg	4.92×10^{-4}	1.37×10^{-3}	
Potash (K_2O)	kg	9.04×10^{-4}	2.64×10^{-3}	
Magnesium (MgO)	kg	7.06×10^{-5}	–	
Boron (B)	kg	3.50×10^{-5}	–	
Machine use	ha	4.18×10^{-6}	7.19×10^{-6}	Jungbluth and Chudacoff (2007)
Weed and pest control				
Herbicide	kg	1.53×10^{-5}	1.94×10^{-4}	Cavalett and Ortega (2010), de Souza et al. (2010)
Insecticide	kg	7.34×10^{-6}	1.29×10^{-4}	
Machine use	ha	4.18×10^{-6}	3.42×10^{-5}	Jungbluth and Chudacoff (2007)
Irrigation		n/a	n/a	
(outputs)				
Emissions to air				
N_2O	kg	3.15×10^{-6}	–	Almeida and Moreira (2005)
CO_2	kg	1.26×10^{-4}	–	
NO_x	kg	6.62×10^{-7}	–	Jungbluth and Chudacoff (2007)
Glyphosate	kg	7.65×10^{-7}	9.71×10^{-6}	Villela et al. (2014)
Emissions to water				
Nitrate	kg	2.52×10^{-5}	–	Almeida and Moreira (2005).
Phosphorus (P) (river)	kg	3.64×10^{-5}	1.01×10^{-4}	Jungbluth and Chudacoff (2007), Almeida and Moreira (2005).

TABLE 22.2 *(Continued)*
LCI of Palm Oil and Soybean Cultivation for the Production of 1.0 MJ of Biodiesel

Inputs/Outputs	Unit	Amount Palm Oil	Soybean	Source
P (groundwater)	kg	1.28×10^{-5}	3.55×10^{-5}	
Glyphosate	kg	3.06×10^{-6}	3.88×10^{-5}	Villela et al. (2014)
Emissions to soil				
Glyphosate	kg	1.15×10^{-5}	1.46×10^{-4}	Villela et al. (2014)
Harvesting				
Machine use	ha	1.36×10^{-1}	1.43×10^{-5}	Jungbluth and Chudacoff (2007)

TABLE 22.3
Characterized Impacts from Palm Oil and Soybean Cultivation for the Production of 1.0 MJ of Biodiesel

	Sustainability Indicator	Unit	Energy Crop Soybean	Palm Oil
Environmental				
GWP	Climate change	kg CO_2-equiv.	2.25×10^{-2}	7.73×10^{-1}
TAP	Terrestrial acidification	kg SO_2-equiv.	7.60×10^{-5}	1.51×10^{-2}
FEP	Freshwater eutrophication	kg P-equiv.	1.59×10^{-4}	3.09×10^{-4}
TETP	Terrestrial ecotoxicity	kg 1,4-DCB-equiv.	7.73×10^{-6}	1.67×10^{-4}
FETP	Freshwater ecotoxicity	kg 1,4-DCB-equiv.	3.61×10^{-4}	4.99×10^{-3}
Social				
HTPnc	Human carcinogenic toxicity	Cases	1.49×10^{-11}	1.02×10^{-11}
HTPc	Human non-carcinogenic toxicity	Cases	6.40×10^{-12}	5.79×10^{-11}
LEM	Local employment	Jobs	2.83×10^{-6}	1.22×10^{-6}
Economic				
LOP	Land use	m²a crop land-equiv.	2.53×10^{-1}	4.26
WCP	Water consumption	m³	8.47×10^{-3}	8.28×10^{-1}
CED	Contribution to economic development	%	1.40×10^{-13}	1.21×10^{-12}

Concerning the LEM indicator, we considered 0.07 jobs/ha for soybean cultivation and 0.2 jobs/ha for palm oil cultivation (Barbier 2010). As for the CED indicator, we took into account the economic contribution of soybean cultivation to the GDP of the state of Mato Grosso and the economic contribution of palm oil cultivation to the GDP of the state of Pará, which were estimated at 3.46×10^{-9}%/ha and 1.97×10^{-7}%/ha, respectively (IBGE 2017; Villela et al. 2014; Nascimento, Figueiredo, and Miranda 2018).

22.4.2.4 Impact Assessment

SimaPro 8.9 (PRé Consultants 2014) software was used to carry out LCSA, except for calculating the LEM and CED indicators. The results of soybean and palm oil cultivation for producing 1.0 MJ of biodiesel according to the selected indicators can be seen in Table 22.3.

In short, we first performed an internal normalization by the "division by sum" approach (Hauschild and Huijbregts 2015). Then, we divided the decision-making problem into three hierarchical levels concerning our goal, sustainability dimensions, and indicators. Finally, our AHP model determined the single sustainability score for the energy crops considered by synthesizing their normalized performance against each indicator and the weights obtained for sustainability dimensions and indicators (see Figure 22.3).

22.4.2.5 Interpretation

In this step, we seek to present, evaluate, and compare the results easily and comprehensively. Besides the single sustainability score, it is also necessary to identify each energy crop option's positive and negative aspects for biodiesel production. Thus, we present the results in a graphical form that shows the single score, the quantitative values obtained in each sustainability dimension, and indicator – making the interpretation step more straightforward, understandable, and transparent for all stakeholders (Figure 22.4).

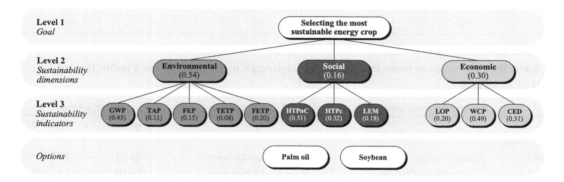

FIGURE 22.3 The hierarchy model for selecting the most sustainable energy crop with the weights of sustainability dimensions and indicators.

Source: Elaborated by the authors.

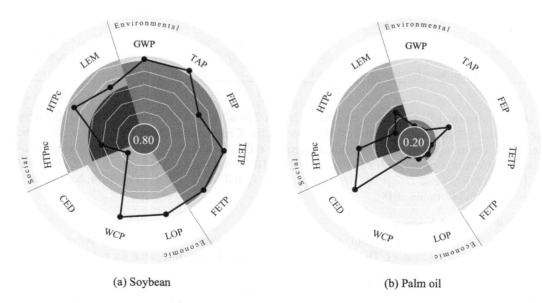

FIGURE 22.4 LCSA results presentation.

Source: Elaborated by the authors.

Figure 22.4 shows that soybean (single score of 0.80) is the most sustainable energy crop for biodiesel production compared to palm oil (single score of 0.20). A polar area diagram illustrates their performances under the sustainability dimensions. Each segment colored in different shades of gray refers to a dimension of sustainability. How far each of these segments is filled from the center of the polar axis depends on the value it represents. Soybean performs better under environmental (0.92), social (0.62), and economic (0.71) dimensions.

When analyzing the energy crops' performances under the sustainability indicators, soybean proves to be the better option as it is a relatively stronger performer on most indicators considered. Palm oil contributes better to economic development and affects low human health due to non-cancerous substances compared to soybean. Note that these results can change whether other practitioners consider different aspects and determine their weights to make our case study valuable for them.

22.5 CONCLUSION

This chapter provided theoretical knowledge of LCSA and responded to the need for practical application of LCSA on different products and services reported by the scientific community. Thus, the capabilities of this methodology have been illustrated by its application to a case study assessing the sustainability of two different energy crops for biodiesel production in Brazil.

LCSA can facilitate the transition toward a more sustainable future by identifying the most sustainable options. Furthermore, presenting the results in a comprehensive way helps to solve the communication problems and identify the trade-offs between different alternatives. Finally, it can be useful for several industry sectors as support for enhancing technologies and products, as well as for policymakers to design future scenarios following the sustainable development principles.

REFERENCES

Almeida, Marcelo, and José Moreira. 2005. "Life Cycle Inventory Analysis of Sugarcane Ethanol: Investigating Renewable Fuels Environmental Sustainability in Brazil." *SAE Technical Papers*. https://doi.org/10.4271/2005-01-3981.

Amer, Muhammad, and Tugrul U. Daim. 2011. "Selection of Renewable Energy Technologies for a Developing County: A Case of Pakistan." *Energy for Sustainable Development* 15 (4): 420–35. https://doi.org/10.1016/j.esd.2011.09.001.

Andersson, Karin, Merete Hogaas Eide, Ulrika Lundqvist, and Berit Mattsson. 2000. "Feasibility of Including Sustainability in LCA for Product Development." *Doktorsavhandlingar Vid Chalmers Tekniska Hogskola*. https://doi.org/10.1016/s0959-6526(98)00028-6.

ANP. 2018. *Anuário Estatístico 2018*. ANP.

Barbier, Edward B. 2010. *A Global Green New Deal: Rethinking the Economic Recovery*. Department of Economics & Finance. https://doi.org/10.1017/CBO9780511844607.

Boulay, Anne, Jane Bare, Lorenzo Benini, Markus Berger, Michael Lathuillière, Alessandro Manzardo, Manuele Margni, et al. 2018. "The WULCA Consensus Characterization Model for Water Scarcity Footprints: Assessing Impacts of Water Consumption Based on Available Water Remaining (AWARE)." *International Journal of Life Cycle Assessment*. https://doi.org/10.1007/s11367-017-1333-8.

Brazil. 2019a. *Brazilian Energy Balance 2018*. https://www.epe.gov.br/en/publications/publications/brazilian-energy-balance.

Brazil. 2019b. *The Ten-Year Energy Expansion Plan 2029*. https://www.mdpi.com/1996-1073/15/24/9286/pdf.

CALCAS. 2009. "D20 Blue Paper on Life Cycle Sustainability Analysis." *Co-Ordination Action for Innovation in Life-Cycle Analysis for Sustainability*. https://theses.hal.science/tel-02869841/document.

Cavalett, Otávio, and Enrique Ortega. 2010. "Integrated Environmental Assessment of Biodiesel Production from Soybean in Brazil." *Journal of Cleaner Production* 18 (1): 55–70. https://doi.org/10.1016/j.jclepro.2009.09.008.

Cinelli, Marco, Stuart Coles, and Kerry Kirwan. 2014. "Analysis of the Potentials of Multi-Criteria Decision Analysis Methods to Conduct Sustainability Assessment." *Ecological Indicators*. https://doi.org/10.1016/j.ecolind.2014.06.011.

Collotta, M., P. Champagne, G. Tomasoni, M. Alberti, L. Busi, and W. Mabee. 2019. "Critical Indicators of Sustainability for Biofuels: An Analysis Through a Life Cycle Sustainability Assessment Perspective." *Renewable and Sustainable Energy Reviews* 115. https://doi.org/10.1016/j.rser.2019.109358.

Corona, Blanca, and Guillermo San Miguel. 2019. "LCSA Applied to an Innovative Configuration of Concentrated Solar Power." *International Journal of Life Cycle Assessment* 24 (8): 1444–60. https://doi.org/10.1007/s11367-018-1568-z.

Costa, D., P. Quinteiro, and A. Dias. 2019. "A Systematic Review of LCSA: Current State, Methodological Challenges, and Implementation Issues." *Science of the Total Environment*. https://doi.org/10.1016/j.scitotenv.2019.05.435.

Czyrnek-Delêtre, Magdalena, Beatrice Smyth, and Jerry Murphy. 2017. "Beyond Carbon and Energy: The Challenge in Setting Guidelines for Life Cycle Assessment of Biofuel Systems." *Renewable Energy*. https://doi.org/10.1016/j.renene.2016.11.043.

Dewulf, Jo, Steven Meester, and Rodrigo Alvarenga. 2015. *Sustainability Assessment of Renewables-Based Products: Methods and Case Studies*. Edited by Jo Dewulf, Steven De Meester, and Rodrigo A. F. Alvarenga. John Wiley & Sons, Ltd. https://doi.org/10.1002/9781118933916.

Doukas, H., K. Patlitzianas, and J. Psarras. 2006. "Supporting Sustainable Electricity Technologies in Greece Using MCDM." *Resources Policy*. www.sciencedirect.com/science/article/pii/S0301420706000341.

Elkington, John. 1998. *Cannibals with Forks: The Triple Bottom Line of Sustainability*. New Society Publishers.

FAO. 2004. *Unified Bioenergy Terminology*, December. www.fao.org/3/b-j4504e.pdf.

FAO and UNEP. 2010. *A Decision Support Tool for Sustainable Bioenergy: Energy*. FAO and UNEP.

Ghazvinei, Pezhman, Masoud Mir, Hossein Darvishi, and Junaidah Ariffin. 2017. "Solid Waste – Management Models." *University Campus Solid Waste Management: Combining Life Cycle Assessment and Analytical Hierarchy Process*. https://doi.org/10.1007/978-3-319-43228-1_3.

Guinée, Jeroen. 2016. "Life Cycle Sustainability Assessment: What Is It and What Are Its Challenges?" In *Taking Stock of Industrial Ecology*, edited by R. Clift and A. Druckman, 45–68. Springer. https://doi.org/10.1007/978-3-319-20571-7_3.

Hauschild, Michael Z., and Mark A. J. Huijbregts. 2015. *The International Journal of Life Cycle Assessment*. Edited by Michael Z. Hauschild and Mark A. J. Huijbregts, vol. 2. Springer. https://doi.org/10.1007/978-94-017-9744-3.

Helbig, Christoph, Eskinder Gemechu, Baptiste Pillain, Steven Young, Andrea Thorenz, Axel Tuma, and Guido Sonnemann. 2016. "Extending the Geopolitical Supply Risk Indicator: Application of Life Cycle Sustainability Assessment to the Petrochemical Supply Chain of Polyacrylonitrile-Based Carbon Fibers." *Journal of Cleaner Production*. https://doi.org/10.1016/j.jclepro.2016.07.214.

Huijbregts, Mark, Zoran Steinmann, Pieter Elshout, Gea Stam, Francesca Verones, Marisa Vieira, Michiel Zijp, Anne Hollander, and Rosalie van Zelm. 2017. "ReCiPe2016: A Harmonised LCIA Method at Midpoint and Endpoint Level." *The International Journal of Life Cycle Assessment* 22 (2): 138–47. https://doi.org/10.1007/s11367-016-1246-y.

Hunkeler, David, Kerstin Lichtenvort, and Gerald Rebitzer. 2008. *Environmental Life Cycle Costing*. Environmental Life Cycle Costing. CRC Press. https://doi.org/10.1201/9781420054736.

IAEA. 2005. *Energy Indicators for Sustainable Development: Guidelines and Methodologies*. International Atomic Energy Agency.

IBGE. 2017. *Censo Agropecuário*. Brazilian Institute of Geography and Statistics.

IEA. 2018. *World Energy Balances 2018. Report*. IEA. https://doi.org/10.1787/3a876031-en.

ISO. 2006. *ISO 14040: Environmental Management – Life Cycle Assessment – Principles and Framework*. International Organization for Standardization.

Jin, Enze, and John W. Sutherland. 2018. "An Integrated Sustainability Model for a Bioenergy System: Forest Residues for Electricity Generation." *Biomass and Bioenergy* 119: 10–21. https://doi.org/10.1016/j.biombioe.2018.09.005.

Jungbluth, N., and Mike Chudacoff. 2007. *Life Cycle Inventories of Bioenergy*. Final Report Ecoinvent.

Kabayo, Jeremiah, Pedro Marques, Rita Garcia, and Fausto Freire. 2019. "Life-Cycle Sustainability Assessment of Key Electricity Generation Systems in Portugal." *Energy* 176 (June): 131–42. https://doi.org/10.1016/j.energy.2019.03.166.

Kalbar, Pradip, and Deepjyoti Das. 2020. "Advancing Life Cycle Sustainability Assessment Using Multiple Criteria Decision Making." *Life Cycle Sustainability Assessment for Decision-Making*. https://doi.org/10.1016/b978-0-12-818355-7.00010-5.

Kloepffer, Walter. 2008. "Life Cycle Sustainability Assessment of Products (with Comments by Helias A. Udo de Haes, p. 95)." *International Journal of Life Cycle Assessment*. https://doi.org/10.1065/lca2008.02.376.

Korres, Nicholas E. 2013. "The Application of LCA on Agricultural Production Systems with Reference to Lignocellulosic Biogas and Bioethanol Production as Transport Fuels." *Green Energy and Technology*. https://doi.org/10.1007/978-1-4471-5364-1_3.

Kouloumpis, Victor, and Adisa Azapagic. 2018. "Integrated Life Cycle Sustainability Assessment Using Fuzzy Inference: A Novel Felicita Model." *Sustainable Production and Consumption* 15 (July): 25–34. https://doi.org/10.1016/j.spc.2018.03.002.

Kühnen, Michael, and Rüdiger Hahn. 2017. "Indicators in Social Life Cycle Assessment: A Review of Frameworks, Theories, and Empirical Experience." *Journal of Industrial Ecology*. https://doi.org/10.1111/jiec.12663.

Laurent, Alexis, Nieves Espinosa, and Michael Z. Hauschild. 2018. "LCA of Energy Systems." In *Life Cycle Assessment*, edited by M. Hauschild, R. Rosenbaum, and S. Olsen. Springer, 633–68. https://doi.org/10.1007/978-3-319-56475-3_26.

Lee, Keat, and Cynthia Ofori-Boateng. 2013. "Sustainability of Biofuel Production from Oil Palm Biomass." *Green Energy and Technology*. https://doi.org/10.1007/978-981-4451-70-3.

Lijó, Lucia, Sara González-García, Daniela Lovarelli, Maria Moreira, Gumersindo Feijoo, and Jacopo Bacenetti. 2019. "Life Cycle Assessment of Renewable Energy Production from Biomass." *Green Energy and Technology*. https://doi.org/10.1007/978-3-319-93740-3_6.

Lora, Electo, José Palacio, Mateus Rocha, Maria Renó, Osvaldo Venturini, and Oscar Olmo. 2011. "Issues to Consider, Existing Tools and Constraints in Biofuels Sustainability Assessments." *Energy*. https://doi.org/10.1016/j.energy.2010.06.012.

Maxim, Alexandru. 2014. "Sustainability Assessment of Electricity Generation Technologies Using Weighted Multi-Criteria Decision Analysis." *Energy Policy* 65 (February): 284–97. https://doi.org/10.1016/j.enpol.2013.09.059.

Nascimento, Alani, Adriano Figueiredo, and Pamela Miranda. 2018. "Dimensão Do PIB Do Agronegócio Na Economia de Mato Grosso." *Ensaios FEE* 38 (4): 903–30.

Neugebauer, Sabrina, Julia Martinez-Blanco, René Scheumann, and Matthias Finkbeiner. 2015. "Enhancing the Practical Implementation of Life Cycle Sustainability Assessment – Proposal of a Tiered Approach." *Journal of Cleaner Production*. https://doi.org/10.1016/j.jclepro.2015.04.053.

Nguyen, Tu Anh, Yasuaki Maeda, Kana Kuroda, and Koji Otsuka. 2017. "Inclusive Impact Assessment for the Sustainability of Vegetable Oil-Based Biodiesel – Part II: Sustainability Assessment of Inedible Vegetable Oil-Based Biodiesel in Ha Long Bay, Vietnam." *Journal of Cleaner Production*. https://doi.org/10.1016/j.jclepro.2017.08.238.

Nieder-Heitmann, M., Kathleen Haigh, and Johann Görgens. 2019. "Life Cycle Assessment and Multi-Criteria Analysis of Sugarcane Biorefinery Scenarios: Finding a Sustainable Solution for the South African Sugar Industry." *Journal of Cleaner Production*. https://doi.org/10.1016/j.jclepro.2019.118039.

Novick, David. 1959. *The Federal Budget as an Indicator of Government Intentions and the Implications of Intentions*. RAND Corporation.

OECD/FAO. 2019. *OECD-FAO Agricultural Outlook 2019–2028*. OECD Publishing. https://doi.org/10.1787/agr_outlook-2019-en.

Pathak, Kankan Kishore, and Sangeeta Das. 2020. "Impact of Bioenergy on Environmental Sustainability." *Biomass Valorization to Bioenergy*. https://doi.org/10.1007/978-981-15-0410-5_10.

Phillis, Y., and L. Andriantiatsaholiniaina. 2001. "Sustainability: An Ill-Defined Concept and Its Assessment Using Fuzzy Logic." *Ecological Economics* 37 (3): 435–56. https://doi.org/10.1016/S0921-8009(00)00290-1.

PRé Consultants. 2014. "SimaPro Database Manual – Methods Library." *Pré*. https://doi.org/10.1017/CBO9781107415324.004.

Ren, Jingzheng, Alessandro Manzardo, Anna Mazzi, Filippo Zuliani, and Antonio Scipioni. 2015. "Prioritization of Bioethanol Production Pathways in China Based on Life Cycle Sustainability Assessment and Multicriteria Decision-Making." *International Journal of Life Cycle Assessment*. https://doi.org/10.1007/s11367-015-0877-8.

Rico, J. A. P., and I. L. Sauer. 2015. "A Review of Brazilian Biodiesel Experiences." *Renewable and Sustainable Energy Reviews* 45: 513–29. https://doi.org/10.1016/j.rser.2015.01.028.

Roinioti, Argiro, and Christopher Koroneos. 2019. "Integrated Life Cycle Sustainability Assessment of the Greek Interconnected Electricity System." *Sustainable Energy Technologies and Assessments* 32 (April): 29–46. https://doi.org/10.1016/j.seta.2019.01.003.

Rosenbaum, Ralph K., Michael Z. Hauschild, Anne-Marie Boulay, Peter Fantke, Alexis Laurent, Montserrat Núñez, and Marisa Vieira. 2018. "Life Cycle Impact Assessment." In *Life Cycle Assessment*, edited by M. Hauschild, R. Rosenbaum, and S. Olsen. Springer, 167–270. https://doi.org/10.1007/978-3-319-56475-3_10.

Santoyo-Castelazo, Edgar, and Adisa Azapagic. 2014. "Sustainability Assessment of Energy Systems: Integrating Environmental, Economic and Social Aspects." *Journal of Cleaner Production* 80 (October): 119–38. https://doi.org/10.1016/j.jclepro.2014.05.061.

Sawaengsak, Wanchat, Thapat Silalertruksa, Athikom Bangviwat, and Shabbir Gheewala. 2014. "Life Cycle Cost of Biodiesel Production from Microalgae in Thailand." *Energy for Sustainable Development*. https://doi.org/10.1016/j.esd.2013.12.003.

Singh, Anoop, Stig Olsen, and Deepak Pant. 2013. "Importance of Life Cycle Assessment of Renewable Energy Sources." *Green Energy and Technology*. https://doi.org/10.1007/978-1-4471-5364-1_1.

Souza, Simone, Sergio Pacca, Márcio Ávila, and José Borges. 2010. "Greenhouse Gas Emissions and Energy Balance of Palm Oil Biofuel." *Renewable Energy*. https://doi.org/10.1016/j.renene.2010.03.028.

Traverso, Marzia, Francesco Asdrubali, Annalisa Francia, and Matthias Finkbeiner. 2012. "Towards Life Cycle Sustainability Assessment: An Implementation to Photovoltaic Modules." *The International Journal of Life Cycle Assessment* 17 (8): 1068–79. https://doi.org/10.1007/s11367-012-0433-8.

UN. 2015. *Transforming Our World: The 2030 Agenda for Sustainable Development United Nations*. A/RES/70/1. United Nations.

UNEP. 2020. *Guidelines for Social Life Cycle Assessment of Products and Organizations 2020*. UNEP.

UNEP/SETAC Life Cycle Initiative. 2009. *Guidelines for Social Life Cycle Assessment of Products: Management*. UNEP.

UNEP/SETAC Life Cycle Initiative. 2013. *The Methodological Sheets for Sub-Categories in Social Life Cycle Assessment (S-LCA): Pre Publication-Version*. UNEP. https://doi.org/10.1007/978-1-4419-8825-6.

Villela, Alberto, D'Alembert Jaccoud, Luiz Rosa, and Marcos Freitas. 2014. "Status and Prospects of Oil Palm in the Brazilian Amazon." *Biomass and Bioenergy*. https://doi.org/10.1016/j.biombioe.2014.05.005.

Vink, Erwin, Karl Rábago, David Glassner, and Patrick Gruber. 2003. "Applications of Life Cycle Assessment to NatureWorks™ Polylactide (PLA) Production." *Polymer Degradation and Stability*. https://doi.org/10.1016/S0141-3910(02)00372-5.

Wang, Jiang-Jiang, You-Yin Jing, Chun-Fa Zhang, and Jun-Hong Zhao. 2009. "Review on Multi-Criteria Decision Analysis Aid in Sustainable Energy Decision-Making." *Renewable and Sustainable Energy Reviews* 13 (9): 2263–78. https://doi.org/10.1016/j.rser.2009.06.021.

Wiloso, Edi Iswanto, and Reinout Heijungs. 2013. "Key Issues in Conducting Life Cycle Assessment of Bio-Based Renewable Energy Sources." In *Green Energy and Technology*, edited by A. Singh, D. Pant, and S. Olsen. Springer, 13–36. https://doi.org/10.1007/978-1-4471-5364-1_2.

Zanghelini, Guilherme, Henrique De Souza Junior, Edivan Cherubini, Luiz Kulay, and Sebastião Soares. 2017. "Análise Da Evolução Dos Temas de Pesquisa Da ACV No Brasil Baseada Na Relação de Co-Words." *LALCA – Revista Latino-Americana Em Avaliação Do Ciclo de Vida*. https://doi.org/10.18225/lalca.v1i1.3071.

23 Environmental, Social, and Economic Aspects of Waste-to-energy Technologies in Brazil

Gasification and Pyrolysis

Suani Teixeira Coelho, Luciano Infiesta, and Vanessa Pecora Garcilasso

CONTENTS

23.1	Introduction	311
23.2	Environmental/Social Aspects	312
23.3	Economic Aspects	314
23.4	Gasification × Incineration Processes for Small- and Medium-size Municipalities	316
23.5	Barriers against MSW WtE Gasification Technologies	318
References		320

23.1 INTRODUCTION

Municipal solid waste (MSW) collection and disposal is indeed a huge challenge in developing countries (DCs), as discussed in Coelho et al (2019). In such regions, most of MSW is disposed in the existing thousands of dumps. For example, in Brazil, about 27% of the municipalities still have dumps, totaling 1493 dumps, with greater representation in the North (247 dumps) and Northeast (844 dumps) regions, which are the less developing ones in the country (ABRELPE 2019).

This has strong and the quite well-known negative impacts that affect the environment and health, among others. Challenges to solve this problem of dumps worldwide were discussed in Coelho et al (2019) and are summarized ahead. These solutions pass through sustainable pathways for MSW energy conversion, the so-called Waste-to-Energy (WtE) technologies, adequately selected from existing and commercialized ones and adapted to the local conditions (economic feasibility, funds availability, adequate legislation, and amount of residues disposed inadequately).

The treatment and disposal of MSW also cover a set of activities and processes that generate direct and indirect jobs. Such activities range from collection, sorting, and recycling, which involve various processes of transforming materials into new products, treatment technologies, energy recovery, and final disposal (DEWHA 2009).

Therefore, it is very important to analyze the synergies between the collection and proper disposal of MSW, as well as the adoption of WtE technologies for energy production. The use of WtE for MSW corresponds to an advantageous process, as it contributes not only to solving the problem of disposal of MSW, but also to increasing the availability of energy, which is a particularly important issue in developing countries that face difficulties in providing access to energy.

This chapter discusses the main aspects related to the environment, social, and economic aspects, based on the extensive assessment performed in Coelho et al (2019). Brazilian experience is also presented as a case study.

DOI: 10.1201/9781003334415-23

23.2 ENVIRONMENTAL/SOCIAL ASPECTS

The adequate final disposal of MSW recorded an index of 59.5% of the annual amount sent to sanitary landfills. Inadequate units – such as dumps and controlled landfills – however, are still present in all regions of the country and receive more than 80,000 tons of waste per day, with an index above 40%, with high potential for environmental pollution and negative impacts on the health (ABRELPE 2019). The packaging of MSW in these inadequate systems can compromise the quality of soil, water, and air, as they are sources of volatile organic compounds, pesticides, solvents, and heavy metals, among others (DEWHA 2009).

The dump is the inappropriate form of final MSW disposal that consists of the discharge of waste on the ground, in the open air, without measures to protect the soil and the environment in general. There is also no control over the types of waste deposited. In such cases, low-hazard waste is deposited along with high-polluting ones, such as industrial and hospital waste.

Inadequate disposal of MSW in dumps presents several environmental and social negative impacts, as discussed previously (Coelho et al 2019). Dumps present impacts on soil and waters (contamination of local and underground waters) and atmosphere (methane emissions from uncontrolled decomposition of organic matter), besides attracting vultures and other birds. Other problems associated with dumps are the risks of fires caused by the biogas generated in the decomposition of waste (which is not capable of being captured) and landslides caused by lack of technical criteria (Ferreira and Rosolen 2012; Coelho et al 2018).

Negative social impacts (Coelho et al 2019) include local poor people (including children) catching garbage with the well-known negative health impacts. It is important to note that, in many cases, people reside in dumps, in unsanitary conditions, without any protection, resulting in a high probability of acquiring diseases.

In this context, considering the need to eliminate the dumps in DCs, WtE technologies appear as an interesting tool. Nowadays, the most common WtE technologies commercialized worldwide include MSW incineration but gasification is also an important option. Such technologies can also be combined with a recycling process and the separation of the organic fraction of the MSW, to be fed into biodigesters (the so-called MBT – mechanical biological treatment) (Coelho et al 2019).

WtE systems are mostly used in developed countries where existing legislation does not allow for the disposal of *in natura* waste in sanitary landfills. The MSW must go through the recycling and recovery processes before the WtE processes, and only the tailings are sent to landfills.

Among the advantages of MSW incineration, there is a reduction in volume for disposal in landfills, the elimination of methane emissions in landfills, and the energy recovered from combustion that can be used for the production of electricity. However, in Brazil, for example, there are still several social, environmental, and economic difficulties.

In addition to energy conversion, gasification process can also be used for fuel synthesis, but this is still a process under development and large-scale plants have faced several difficulties in the recent past (Cardoso 2013). More recently, pilot plants and larger plants for electricity generation are being built, also for MSW (IEA 2020), both in developed countries (Enerkem, for fuel synthesis, in Quebec, Canada) and DCs (Coelho et al 2019), mainly in India (as discussed in other chapters in this book) and Brazil (as mentioned ahead).

WtE processes correspond to a sustainable pathway for MSW disposal, with additional important social benefits. Not only such processes help to eliminate dumps, but they also may transform the unskilled workers existing in DCs (working in unhealthy conditions to separate the waste) into skilled workers, working in recycling cooperatives that may commercialize the recycled products (those with commercial value).

Incineration plants (as well as other WtE plants) are also often objects of concern with a lack of jobs for unskilled people. In many countries, these people usually work in existing dumps, aiming at the recovery of recyclable material for commercialization. Despite being hard work with multiple health impacts, it is seen as the only option in some poor regions. Therefore, any

WtE process is seen as a threat to these workers. However, collectors are fundamental in the garbage collection process. Recycling in Brazil is done at the expense of the work of the collector and not of selective collection. A broad information program is needed to explain that WtE processes need recycling beforehand; therefore, in regions where people do not carry out their own recycling of waste, as in many Brazilian municipalities, the recycling process before WtE processes is mandatory.

This is an important issue for WtE technologies, mainly in the case of gasification processes. In this pathway, differently from incineration process, it is necessary to convert *in natura* MSW into RDF, as discussed ahead; for this, recycling is mandatory and it requires a large number of workers for manual recycling in municipalities where there are no recycling policies (mostly in DCs).

However, in the case of existing dumps, these recycling and biodigestion processes are not possible, and the WtE technologies must be applied to the *in natura* existing MSW (Coelho et al 2019) from the dumps.

Regarding environmental aspects, WtE technologies present the advantages of reducing the amount of MSW to be disposed, with the corresponding environmental advantages of lower local impacts, since only the residues from the process will be disposed in sanitary landfills (Coelho et al 2019).

However, air emissions correspond to an important issue to be carefully discussed. During MSW incineration process (the so-called *mass burning*), there is the formation of dioxins and furans, which are carcinogenic compounds. Such emissions are produced in MSW combustion process due to the presence of oxygen and must be cleaned from exhaust gases before final emissions to atmosphere to comply with environmental standards (FEAM 2012; Infiesta 2015). These sophisticated cleaning systems make incineration process more expensive and only economically feasible for larger plants (municipalities with more than 400–500 t/day of MSW). This is also a reality in other countries; however, incineration plants in these countries, mainly in the European Union, end up becoming economically viable due to the subsidies received, such as non-refundable investments and special tariffs for the electricity generated.

During MSW gasification process, there are almost no formation of dioxins and furans since this is an incomplete combustion in absence of oxygen. Therefore, cleaning systems for MSW gasification process are simpler than in incineration process. On another hand, it is necessary to treat the MSW. This pretreatment must be applied to convert MSW into refuse-derived fuel (RDF). This RDF may be commercialized or fed into gasifier systems. RDF production process from MSW corresponds, basically, to separation, drying, and crushing processes.

The issue of dioxins and furans is also very much related to public perception. Most people fear these carcinogenic emissions, not realizing that nowadays such emissions are not allowed due to existing standards in almost every country. In addition, they do not know that there is almost no such emissions in gasification process, as discussed earlier.

The adoption of technologies for energy use of MSW is an important way to meet the demand for sustainable management of these wastes. The National Solid Waste Policy (PNRS), approved in August 2010 but not yet implemented, is the country's current regulatory framework on MSW management; its fundamental objectives promote environmentally correct ways to manage waste in the 5570 municipalities. However, Brazil still needs to face essential challenges that consist of providing total waste collection, proper processing and disposal, reuse and recycling, and, above all, exploring the potential of generating energy from MSW.

In turn, although the PNRS is a vital starting point for dealing with waste management issues and providing a holistic regulatory framework, the results are still unsatisfactory, which was expected especially in less developed municipalities. The use of WtE systems offers the opportunity to address these issues, as this synergy can help expand the decentralized energy supply, but mainly contribute to reducing the negative impacts of improper disposal. However, clearly, it will require long-term planning and investment, as well as the appropriate regulatory framework that takes regional disparities into account.

23.3 ECONOMIC ASPECTS

The gasification of municipal and/or industrial solid waste seems like a very attractive option, since it provides both electricity production and adequate final disposal for the waste, within an environmentally appropriate process. In this section, economic aspects of a small-scale MSW gasification plant are discussed based on existing results from existing plants and also from the first commercial plant under construction in Brazil (in Boa Esperanza Municipality, Minas Gerais State, Brazil), as Figures 23.1 and 23.2 illustrate.

In brief, MSW gasification process can be divided into three stages:

- An automated line able to process heterogeneous waste into a standard residue derivate fuel (RDF), as illustrated in Figure 23.3

FIGURE 23.1 MSW-WtE gasification process in Boa Esperanza municipality.

Source: Carbogas (published with authorization).

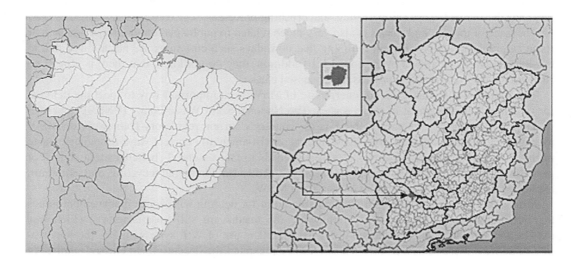

FIGURE 23.2 Geographic location of Boa Esperanza municipality, Minas Gerais States, Brazil.

Source: Authors.

Waste-to-energy Technologies in Brazil

FIGURE 23.3 Conversion process of MSW from the existing dump into RDF, in Boa Esperanza municipality, Minas Gerais, Brazil.
Source: Authors.

- The gasification process, occurring inside the fluidized bed thermochemical reactor
- Applying the syngas produced to generate energy

For this economic discussion, the following process is considered. At the first stage, there is the MSW conversion into RDF through a sequence of automatic operations. In this process, first there is a separation of undesirable and recycling materials, like metals (ferrous and non-ferrous), glasses, stones, and inert components. Then, there is a mechanical and thermal drying system to reduce water content down to adequate levels, ensuring a final fuel with both uniform granulometry and heating value, the so-called RDF. The next stage at the thermochemical reactor, the gasification happens, converting RDF into syngas, a combustible gas with multiple applications. During this gasification phase, the RDF passes through a thermal cracking process, producing the syngas (CO, H_2, CH_4, and CO_2).

This process (technology developed by Carbogas Energia) allows the conversion of RDF-carbon content into syngas. Therefore, ashes are the unique residues from the process, intrinsic to the composition of waste content. The MSW syngas produced has low heating value and, for a typical system, it is in the range of 1100–1500 kcal/Nm³.

At the final step, there is the end-use for the syngas, which usually is electricity production that can occur in two different technological options nowadays commercialized. The first one, an Otto-engine adapted for syngas or a Rankine steam cycle, where syngas can be used as a fuel directly in the engine. Otherwise, in the steam cycle, syngas is burnt in boilers and the steam is fed into turbo-generators.

Looking at the Brazilian scenario, municipalities are responsible to establish the adequate environment-friendly route for the final MSW destination. In this sense, they currently have, for its most part, public or private landfills for disposal and there is a fee for the final disposal. Consequently, the WtE Company is able to take advantage of this revenue.

Thus, the gate fee allows the capitalization for the environmental liability, which it will afterward contribute to the adequate MSW disposal. In addition, the energy produced (thermal or electric) can be sold to the city for supplying public buildings and structures, made available to other customers or could feed the local network. This would be a second revenue.

As an example, a thermochemical power generation plant (TPGP) that treats 275 tons of MSW per day will provide the final destination of 100,375 tons per year. At a fee of around BRL80 per ton, estimated by Via Publica (2012), the annual revenue will be 8,030,000 BRL.

According to average figures from Carbogas Energia, a plant of this size will generate 4.54 MW of exportable electricity, with a selling fee estimated in BRL561.00 (reference value published in 2018 by the Brazilian Ministry of Mines and Energy for units that generate less than 5 MW). This will result in an annual revenue of approximately BRL20,515,000.

Such a module would have a reactor with a thermal capacity of 20 Gcal/h, added to the waste-processing line, the power generation plant, and a wastewater treatment plant. The CAPEX of this enterprise would add up to BRL83,500,000, with an operating cost of BRL5,223,000. Under these premises, considering 80% financing, with interest rate of 6%, taxes on profit of 9.25% and on income of 34%, the project would achieve ROI after approximately 4 years, with internal rates of return (IRR) higher than 25%.

Table 23.1 shows results for different scenarios with TPGP with thermal capacities of 10, 15, and 20 Gcal/h. For larger demands, it is possible to make the composition in parallel of different combinations of these modules.

23.4 GASIFICATION × INCINERATION PROCESSES FOR SMALL- AND MEDIUM-SIZE MUNICIPALITIES

When discussing such WtE technologies, it is important to analyze the following aspects, since all present strong relationship (Coelho et al 2019):

- The environmental aspects, mainly atmospheric emissions such as dioxins and furans
- The amount of MSW to be treated
- The economic feasibility of the WtE process to be used

In recent years, the thermal treatment of waste has been a topic of great interest in the country, while sanitary landfills reach their maximum storage capacity or are distant, increasing the cost of transport and final disposal of the waste. In addition, there is difficulty of licensing new sanitary landfills before the PNRS.

Despite having some advantages, such as reducing the volume of waste destined for landfills and energy recovery, there are controversies regarding incineration, which faces barriers in Brazil. Brazilian society rejects the use of incinerators, mainly for fear of the toxicity characteristics of the exhaust gases. The system has high initial investments and, consequently, the generation cost is also high. Added to these factors is the characteristic of existing incineration systems being large, requiring large amounts of urban solid waste, which makes its implementation in small municipalities difficult. The need for scale can also go against policies that encourage the principles of reducing waste generation and separating it for recycling and other destinations, which can reduce the viability of new projects and units in operation (Leite 2016).

In terms of energy use, the heat from the exhaust gases can be used for thermal purposes or conventional thermoelectric generation in a steam cycle (Rankine), with the possible use of thermal energy (cogeneration). The generation capacity depends directly on the thermodynamic efficiency of the cycle and on the composition of the waste, which defines its calorific value.

The efficiency of conversion into energy, in incineration units, can be compromised, as the priority must be the control of the pollutants generated in the process (COELHO et al. 2019) and the reduction of the volume of waste.

It is well-known that the cleaning systems for incineration systems must be very efficient to comply with environmental standards and so are quite expensive. Due to that, only large-scale systems are economically feasible and are commercialized, corresponding to at least 400–500 metric tons of MSW per day (CENBIO 2013). This means that incineration process can be used only for municipalities of 500,000 inhabitants or more (considering an average of 1 kg per habitant per day of MSW produced).

On the other hand, it is important to note (Coelho et al 2019) that there are several municipalities with much less than 500,000 inhabitants.1This may be a majority in DCs. In Brazil, for instance, more than 95% of the municipalities have less than 100,000 inhabitants and 90% have less than 50,000 inhabitants (Miranda 2014). In addition, in most cases, these are the municipalities with dumps and inadequate MSW disposal.

This means that, in these cases, MSW incineration is not an economic feasible option. In such municipalities, another feasible pathway is MSW gasification (after conversion of RDF produced

TABLE 23.1
Economic Feasibility of Different Sizes of Modules

Summary of Modules: In Reais (BRL)

Modules	Plat Potency Gcal/h	Energy Sell R$/MWh	Capex (R$1.000)	Crude Energy Produced MWh	Exportable Energy MWh	Gate-Fee Income (R$1.000)	Electricity Income R$/Year	Annual Income
1	5	561	R$44.400,00	1,52	1,239	R$2.044,00	R$5.149.357,65	R$7.193.357,65
2	10	561	R$53.600,00	3,01	2,260	R$4.088,00	R$10.216.518,69	R$14.304.518,69
3	15	561	R$71.700,00	4,53	3,399	R$6.132,00	R$15.365.876,33	R$21.497.876,33
4	20	561	R$83.500,00	6,05	4,538	R$8.030,00	R$20.515.233,98	R$28.545.233,98

Modules	Consume MSW per day	MSW Gate fee R$/Ton	Opex	Payback Simple	Payback Discounted	IRR Project	IRR SH	NPV
1	70	80	R$2.301266,14	9,1	13,3	9,5	8,5	R$14.511.477,14
2	140	80	R$3.715.702,42	5,1	6,1	19,1	32,6	R$67.912.015,26
3	210	80	R$4.453.728,55	4,2	4,8	23,2	41,7	R$123.026.782,39
4	275	80	R$5.223.372,82	3,6	4,0	27,0	50,3	R$181.616.942,55

Source: Author's evaluation. Also: (1) Via Pública (2012). (2) www.mme.gov.br/web/guest/todas-as-noticias/-/asset_publisher/pdAS9IcdBICN/content/mme-fixa-novos-valores-anuais-de-referencia-para-sistemas-de-geracao-distribuida/pop_up?_101_INSTANCE_pdAS9IcdBICN_viewMode=print&_101_INSTANCE_pdAS9IcdBICN_languageId=pt_BR

from MSW).2The association of several municipalities to implement a WtE process may be difficult due to logistic issues for MSW transportation (in the case of large municipalities and lack of infrastructure, such as in Brazilian Amazonia) and (in some cases) due to political problems. This is an important market opportunity for such systems, as well as a significant solution for the elimination of existing dumps in DCs. An interesting example is a case study developed for Itanhaem municipality, a medium-size municipality, in Sao Paulo State, Brazil, with 87,000 inhabitants (base year 2010). From the local MSW available, after a recycling process, it would be possible to generate almost 2 MWe (Miranda 2014).

There are in fact some additional challenges for MSW gasification process. The mandatory requirement of RDF conversion before the gasification process increases the total investment needed and even such process is not commercially produced for MSW amounts lower than 50 tons/day (in a fluidized bed gasifier), corresponding to municipalities with at least 50,000 inhabitants.

WtE gasification processes have two demonstration plants in Brazil; one 200 kWe waste gasification plant (1 MWth) in Maua, Sao Paulo State, and the another one, under construction, in Boa Esperansa municipality (Minas Gerais State), in Brazil, both from Carbogas Co,3A joint R&D project FURNAS/ANEEL, under construction by Carbogas Energia Co. which, as mentioned, can run with MSW from small and medium municipalities (with more than 50,000 municipalities).

In addition, another local company (WEG Co)4www.weg.net/institutional/BR/pt/news/produtos-e-solucoes/weg-oferece-ao-mercado-solucao-para-geracao-de-energia-eletrica-com-utilizacao-de-residuos-solidos-urbanos-rsu informed in 2019 that they are launching a MSW gasification system for medium municipalities (with more than 130,000 inhabitants). Their proposal is to supply modules of 2.5 MW for such municipalities and they inform a 45-month payback time.5www.abegas.org.br/arquivos/74019; https://museuweg.net/blog/weg-lanca-solucao-para-a-geracao-de-energia-eletrica-a-partir-do-lixo/ There is no information of any plant in operation in the country yet.

23.5 BARRIERS AGAINST MSW WTE GASIFICATION TECHNOLOGIES

As discussed in detail in Coelho et al 2019, the implementation of WtE gasification technologies would be possible if the different existing barriers would be solved. Identifying such barriers and proposing actions for their managing and mitigation would contribute to promote sustainable and well-designed projects of WtE systems that become truly beneficial for each region.

The existing barriers are related to technical, economic, social, and political aspects, as discussed in Coelho et al 2019 and summarized here:

TECHNICAL BARRIERS

- Lack of knowledge of WtE technologies in many DCs: even considering the advantages of WtE gasification technologies for medium and small municipalities, there is a local lack of knowledge of such processes. Quite often, only the option of municipalities' consortium is taken into account, even considering the difficulties related to logistics and political issues.
- Lack of commercial availability of WtE technologies in many DCs.

ECONOMIC BARRIERS

- High initial investments required and lack of economic feasibility.
- The interests of the electricity sector and their impact of WtE; in several studies, only traditional economic feasibility studies are performed, not considering other benefits, such as the significant synergies of WtE plants and basic sanitation in DCs.

PUBLIC PERCEPTION (SOCIAL BARRIERS)

- The not in my back yard (NIMBY) effect: it is very well-known as the rejection on the part of the community involved near a waste management facility, fearing odors and health impacts, due to the lack of information about the WtE systems. This is especially important considering that WtE gasification technologies present a much lower environmental impacts due to the very low (almost zero) emissions of dioxins and furans, due to the absence of oxygen.6Gasification process is the incomplete combustion and there is not enough oxygen to allow the formation of dioxins and funans.
- Fear of impacts on the large number of waste pickers existing: despite being an extremely unhealthy and hazardous job, there is a perception that WtE technologies will cause a lack of such jobs. Again, there is a lack of adequate knowledge about the need for collecting and adequate recycling of waste before the preparation of the waste for the gasification process, as required.

POLITICAL BARRIERS

- Interests and influence of existing waste managing concessionaires
- Barriers related to the legal and institutional framework for WtE projects
- Lack of implementation of "zero waste" policies, including recycling processes
- Lack of adequate legislation to incentivize WtE projects considering the environmental and social benefits

The heat treatment of MSW (WtE) is quite complex, fragmented in terms of policy and regulation, in addition to having a huge untapped potential, especially in developing countries.

Regional and even international efforts are needed so that the MSW heat treatment market can leverage in Brazil, thus benefiting the waste and energy management sectors, as WtE technologies not only allow for an increase in energy supply in municipalities, but also – and more importantly – contribute to solving the problem of proper disposal of waste.

However, there are still great difficulties in implementing these technologies in developing countries, such as Brazil. Challenges include the need for adequate dissemination of information, local training, as well as the lack of resources for investment and adequate regulation, such as tariffs and mandatory purchase of generated energy, in order to make the process economically viable. In general, the price of electricity generated by WtE systems is not competitive with other renewable energy sources. It would be interesting to analyze the possibility of guaranteeing a fixed subsidized price for the project at the time of implementation.

In addition to the energy use of syngas (gasification product), there are several routes for its conversion into other fuels. Among them, the Fischer–Tropsch synthesis stands out, which allows the production of naphtha, gasoline, and diesel. Other important routes are hydrogen concentration and ammonia and methanol production (Santos and Alencar 2019). These products have in the gasification of biomass a possible low carbon source, and each in its own way is important in the current energy system or is part of possible solutions for the transition to a system based on renewable and sustainable sources.

Many gasification plants in operation in the world make use of energy in cogeneration units for electricity and heat, some are dedicated to heating systems, and there are also plants for the production of fuels by processes such as Fischer–Tropsch, as presented by IEA (2020).

The predominance of cogeneration units, generally small, is due to several factors. The greater stability and technological consolidation of fixed bed units and smaller scales make them preferable, and these can make the installation of complex reaction and fuel production units economically unfeasible.

Another issue is the technical-economic difficulty of the synthesis gas cleaning step. The formation of several unwanted substances in the biomass conversion by gasification, such as dust, ash, tar, ammonia, sulfur compounds, among others is common (Ruiz et al 2013). Synthesis gas cleaning systems, with different configurations available and under development, are responsible for an important part of the costs of a gasification unit (Santos and Alencar 2019), which is considered a bottleneck for the development of technology. The choice of form of use, therefore, may be linked to more robust systems in relation to fuel contamination, prioritizing thermal generation systems or internal combustion engines in relation to the use of gas turbines and conversion into fuels.

More details on biomass gasification systems and their current status in the world can be found in Soares (2012) and Coelho et al. (2019).

REFERENCES

ABRELPE – Associação Brasileira de Empresas de Limpeza Pública e Resíduos Especiais. 2019. *Panorama dos Resíduos Sólidos no Brasil 2018/2019* [Overview of Solid Waste in Brazil 2018/2019]. Accessed February 5, 2020. http://abrelpe.org.br/panorama/.

Cardoso, M. T. 2013. *Da iluminação das cidades do século XIX às biorrefinarias modernas: história técnica e econômica da gaseificação*. Dissertação de Mestrado – Programa de Pós-Graduação em Energia EP/FEA/IEE/IF da Universidade de São Paulo, São Paulo, pp.1–49.

CENBIO – Centro Nacional de Referência em Biomassa/IEE – Instituto de Energia e Ambiente/USP – Universidade de São Paulo. 2013. *Avaliação do Ciclo de Vida (ACV) comparativa entre tecnologias de aproveitamento energético de resíduos sólidos* [Comparative Life Cycle Assessment (LCA) Between Technologies for the Energy Use of Solid Waste]. Activity Report. R&D Project EMAE/ANEEL: 0393-00611.

Coelho, S. T. (Coord.), Garcilasso, V. P., Ferraz Junior, A. D. N., Santos, M. M. and Joppert, C. L. 2018. *Tecnologias de Produção e Uso de Biogás e Biometano* [Technologies for the Production and Use of Biogas and Biomethane]. IEE-USP. ISBN: 978-85-86923-53-1.

Coelho, Suani, Sanches-Pereira, Alessandro, Mani, S. K., Bouille, D. H., Stafford, W. H. L., Recalde, M. Y. and Salvino, A. A. 2019. *Municipal Solid Waste Energy Conversion in Developing Countries: Technologies, Best Practices, Challenges and Policy*, vol. 1, p. 300. Elsevier. ISBN: 978-0-12-813419-1.

DEWHA – Department of the Environment Water, Heritage and the Arts. 2009. *Employment in Waste Management and Recycling*. Accessed July 2020. www.environment.gov.au/system/files/resources/5cc6a848-a93e-4b3f-abf7-fc8891d21405/files/waste-and-recycling-employment.pdf.

FEAM – Fundação Estadual do Meio Ambiente. 2012. *Aproveitamento energético de resíduos sólidos urbanos: Guia de orientação para governos municipais de Minas Gerais* [Energy Use of Urban Solid Waste: Guidance for Municipal Governments of Minas Gerais]. FEAM. Belo Horizonte. Accessed January 29, 2020. https://docplayer.com.br/1858283-Aproveitamento-energetico-de-residuos-solidos-urbanos-guia-de-orientacoes-para-governos-municipais-de-minas-gerais.html.

Ferreira, D. A. and Rosolen, V. 2012. "Disposição de resíduos sólidos e qualidade dos recursos hídricos no município de Uberlândia/MG [Disposal of Solid Waste and Quality of Water Resources in the City of Uberlândia/MG]". *Horizonte Científico* (Uberlândia), vol. VI, pp. 1–21. Accessed April 10, 2020. https://seer.ufu.br/index.php/horizontecientifico/article/view/14758#:~:text=O%20munic%C3%ADpio%20de%20Uberl%C3%A2ndia%20(MG,a%20sa%C3%BAde%20e%20ao%20ambiente.

IEA Bioenergy. 2020. *Gasification of Biomass and Waste – Task 33*. Accessed July 20, 2020. http://task33.ieabioenergy.com/.

Infiesta, L. 2015. *Gaseificação de Resíduos sólidos urbanos RSU no Vale do Paranapanema – Projeto CIVAP* [Gasification of Urban Solid Waste in the Paranapanema Valley – CIVAP Project]. São Paulo – 2015. PECE – Curso de Especialização em Energias Renováveis, Geração Distribuída e Eficiência Energética. USP. Accessed July 15, 2020. https://pt.scribd.com/document/364529223/Gaseificacao-de-Residuos-Solidos-Urbanos-Rsu-No-Vale-Do-Paranapanema-Projeto-Civap-Rev04.

Leite, C. B. 2016. *Tratamento de resíduos sólidos urbanos com aproveitamento energético: Avaliação econômica entre as tecnologias de digestão anaeróbia e incineração*. Dissertação de Mestrado, Programa de Pós-Graduação em Energia, Instituto de Energia e Ambiente, Universidade de São Paulo. Accessed April 11, 2020. https://teses.usp.br/teses/disponiveis/106/106131/tde-28032017-134502/pt-br.php.

Miranda, L. H. T. G. 2014. *Aproveitamento energético de resíduos sólidos urbanos: Estudo de caso no município de Itanhaém-SP* [Energy Use of Urban Solid Waste: A Case Study in the Municipality of Itanhaém-SP]. Trabalho de Monografia. Curso de Especialização em Energias Renováveis, Geração Distribuída e Eficiência Energética do Programa de Educação Continuada da Escola Politécnica (PECE) da Universidade de São Paulo (USP).

Ruiz, J. A., Juárez, M. C., Morales, M. P., Muñoz, P. and Mendívil, M. A. 2013. "Biomass Gasification for Electricity Generation: Review of Current Technology Barriers." *Renewable and Sustainable Energy Reviews*, vol. 18, pp. 174–83. https://doi.org/10.1016/j.rser.2012.10.021.

Santos, R. G. and Alencar, A. C. 2019. "Biomass-derived syngas production via gasification process and its catalytic conversion into fuels by Fischer Tropsch synthesis: A review." *International Journal of Hydrogen Energy*. https://doi.org/10.1016/j.ijhydene.2019.07.133.

Soares, D. H. 2012. *Gaseificação de biomassa de médio e grande porte para geração de eletricidade: Uma análise da situação atual no mundo*. Monografia apresentada no Programa de Educação Continuada (PECE) da Escola Politécnica da USP.

Via Publica. 2012. *Estudo de alternativas de tratamento de resíduos sólidos urbanos: Incinerador mass burn e biodigestor anaeróbio* [Study of Urban Solid Waste Treatment Alternatives: Mass Burn Incinerator and Anaerobic Digester]. Accessed May 8, 2020. https://polis.org.br/publicacoes/estudo-de-alternativas-de-tratamento-de-residuos-solidos-urbanos-incinerador-mass-burn-e-biodigestor-anaerobio/.

24 Life Cycle Assessment of Lubricant Oil Plastic Containers in Brazil
Comparing Disposal Scenarios

Maria Clara Brandt and Alessandra Magrini

CONTENTS

24.1	Introduction	323
24.2	Methodology	324
	24.2.1 Goal and Scope	324
	24.2.2 Life Cycle Inventory	326
	24.2.3 Life Cycle Impact Assessment	326
	24.2.4 Sensitivity Analysis	328
24.3	Results	328
	24.3.1 Environmental Impacts of the Four Scenarios	328
	24.3.2 Sensitivity Analysis Results	331
24.4	Conclusion	333
References		333

24.1 INTRODUCTION

As observed in many emerging countries, Brazil has experienced exponential growth in the number of automobiles in the last decades. The rise in the number of vehicles has led to a higher consumption of maintenance and operational products, such as lubricant oil (EPA 2006).

As lubricant oils are a hazardous waste (ABNIT 2004), their incorrect disposal causes serious environmental impacts, imposing risks to human health and air and water pollution IPCC (2001). According to Willing (2001), one liter of used lubricant oil can contaminate one million liters of water, and its degradation can take 300 years.

After being changed in a vehicle, 30–60 mL of lubricating oil still remains in the "empty" container. Therefore, the container itself is also considered hazardous waste and should be managed separately from other solid wastes (Willing 2001). Lubricant oil plastic containers (LOPCs) are made of high-density polyethylene (HDPE), one of the most consumed plastic resins in Europe and in Brazil (PlasticsEurope 2015; Plastivida 2011).

The implementation of effective and sustainable waste management strategies is of paramount importance (Demertzi et al. 2015). The problem in Brazil, as in other emerging countries, is that cost considerations have led to non-optimal solutions. Therefore, industrial landfills are the most used end-of-life destination for LOPCs, being preferred over other options such as recycling, incineration, or co-processing (Sistema Firjan 2015).

Different assessments and methods for studying and describing environmental and sustainability performance have been developed over the past years. Life cycle assessment (LCA) is the most popular and globally accepted system assessment tool for dealing with waste management, also being a well-established environmental management tool (Kanokkantapong et al. 2009; Finkbeiner et al. 2010).

LCA is a technique used to evaluate environmental impacts during the life cycle of a product, service, or process (Guinée et al. 2011; Lorenz 2014). Some LCA studies deal with integrated waste management systems as a whole, while others, such as the one presented in this chapter, focus on one single waste fraction (Rigamonti, Grosso and Guigliano 2010).

This study may provide input into further investigations on hazardous waste management as well as useful elements to expand the already existing waste management and recycling programs.

In this chapter, life cycle assessment is used as a management tool to evaluate the life cycle of lubricating oil plastic containers, comparing the current practice – recycling – to incineration with energy generation (waste-to-energy), which is a treatment widely used in many countries, as alternative options to industrial landfilling. Due to the difficulties of analyzing the whole country, a case study was carried out in the State of Rio de Janeiro which was chosen for having both a large population and a massive fleet of vehicles (Denatran 2016; IBGE 2016). With 16 million inhabitants, Rio de Janeiro is the third most populated state in Brazil and its number of light vehicles is estimated to be over 6 million units. Its demand for lubricant oil in 2014 was around 118 million tons (Sindicom 2016).

24.2 METHODOLOGY

LCA studies in emerging countries are many times hindered by the lack of reliable information. Their success depends on the amount and quality of the information collected, especially local data, the knowledge of the technology applied, and simplifications and approaches used (Guinée et al. 2002). Case studies are a good option for gathering primary data and qualitative and quantitative information (Becker 1997). Willers and Rodrigues (2014) state that there are many possible applications for LCA studies in Brazil due to its economic profile, and Zanghelini et al. (2016) affirm that the assessment of industrial sectors is fundamental for achieving environmental goals in the country.

The ISO 14040 and 14044 standards provide the principles and frameworks for the conduction of LCA studies (ISO 2006a, 2006b). They distinguish four phases of a life cycle assessment:

- Goal and scope definition
- Inventory analysis, including the inputs and outputs of the system flows
- Impact assessment, including the evaluation of the effects on human health and the environment
- Interpretation of the results

24.2.1 GOAL AND SCOPE

The goal of this study is to evaluate the different impacts associated with the life cycle of LOPCs. This includes different management options for the end-of-life of LOPCs that can be used in the future decision-making process.

A LCA study can have an attributional or consequential approach, according to the *ILCD Handbook*. Attributional modeling is used for existing supply chains, whereas the consequential approach models generic inputs and outputs expected as a consequence of possible decisions made in the system background.

To identify the most appropriate impact assessment modeling and method to comply with the goal and scope of the work, an LCA should be carried out following one of the three different context situations:

- A (micro-level, no large-scale consequences)
- B (macro-level, strategic/large-scale consequences)

The attributional approach was considered more appropriate for the purpose of this analysis as it reflects the real situation of the processes and the life cycle of the analyzed system. Considering that this is a product-based study, it fits situation A (micro-level decision support).

- **Functional Unit**

The functional unit (FU) quantifies the performance of a product system and gives a reference to the input and output flows in a life cycle assessment study. The FU used was 1 ton of high-density polyethylene (HDPE), as it can be applied to the whole chain.

- **Description of Scenarios and System Boundaries**

In order to study the life cycle of LOPCs, the analysis considers their production, distribution in the state of Rio de Janeiro, collection, and final destination, including reverse logistics for recycling and incineration. The scope of this study encompasses the impacts caused and avoided through these different stages.

One producer/actor was chosen for each step of the case study. The production of LOPCs takes place in a factory located in a small town about 250 km from the city of Rio de Janeiro, in the state which is its namesake. Once a week, the oil company sends a lorry (medium weight, 7.5–16 tons) with virgin HDPE pellets and collects the new containers to be filled with oil and distributed. The HDPE pellets are extruded with the pigment, and a material balance is considered in the process. Water is used for cooling the plastic when entering the machine (25 °C) and after molding (15 °C). The oil company is responsible for the supply of 30% of the market in the state of Rio de Janeiro (heavyweight lorries, >28 tons) (Figure 24.1).

After lube oil changes, mostly done at service stations and repair shops, the LOPCs are stored for disposal. Final destination options can be an industrial landfill, a recycling plant, or an incinerator. Using a reverse logistic system, the Instituto Jogue Limpo collects the used containers for recycling.

In this study, different waste treatment scenarios were created following the waste hierarchy. Recycling and incineration were prioritized, followed by industrial landfilling as last option. Even though incineration is not currently applied as a technology for treatment of this kind of waste in Brazil, two scenarios were created using it as a possible option for destination of LOPCs.

The alternative scenarios under study are as follows:

- Scenario 0: Recycling 16%, industrial landfill 84% (current situation)
- Scenario 1: Recycling 50%, industrial landfill 50%
- Scenario 2: Recycling 16%, incineration 16%, industrial landfill 68%
- Scenario 3: Recycling 50%, incineration 50%

The hypotheses assume that the recycling comprises the production of different products (open-loop) and the incineration is used for energy generation, to be supplied to the Brazilian grid.

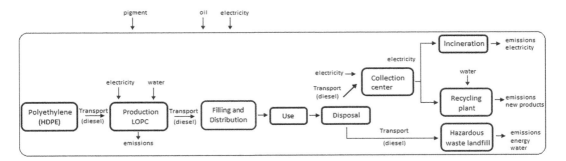

FIGURE 24.1 Life cycle of LOPCs with different disposal options.

The boundaries of the system include only the LOPCs themselves, excluding the caps/bottle tops and the bags in which the used containers are stored for transportation. Data about the filling of the containers with lubricant oil were also not included.

The LOPCs disposal options examined generate several co-products that are substitutes for the products/fuels produced through conventional technology, such as recycled material, electric energy, and energy heat. HDPE substitutes virgin plastic with a ratio of 1:1, assuming that there are neither losses nor degradation in the process of recycling. For the energy, the substitution rate is also of 1:1.

24.2.2 Life Cycle Inventory

In any life cycle study, the life cycle inventory is a crucial stage, because that is when specifications of the data collected are checked, in accordance with the ISO 14040 standards (ISO 2006a).

The data assembly was carried out mainly through direct interviews and primary data gathering. Whenever primary data was not available, information in the literature and the Ecoinvent (Ecoinvent Center 2016) database was used. Inputs and outputs on production of polyethylene and diesel for transport were compiled from an important Brazilian company and the national emissions inventory (IBICT 2016), respectively. Emissions were calculated based on the conversion factors (MMA 2011) presented in Table 24.1. Regarding electricity, the Brazilian medium voltage mix was used (data from Ecoinvent).

Data on the production and distribution of LOPCs was provided by a producer of this plastic packaging and by one of the largest lube oil producers and distributers in Brazil (in order to preserve the companies' identities and data privacy, their names will not be revealed), respectively. The Instituto Jogue Limpo also provided valuable input for the development of the present work, including data on collection of used LOPCs and demand for diesel and electricity. Information regarding water, transportation, energy, and heat was obtained directly in the recycling plant. Table 24.2 presents the inputs and outputs of the LOPC life cycle.

24.2.3 Life Cycle Impact Assessment

The life cycle impact assessment (LCIA) was performed using SimaPro 8.1.1.16 software (PRé-Consultants: Amersfoort, The Netherlands), and the method selected was the ReCiPe 2008 midpoint (Goedkoop et al. 2008). The ReCiPe methodology integrates midpoint and endpoint approaches, both of which are characterization methods with indicators at different levels. The first evaluates the environmental impact at a level in the cause-effect chain, and the latter in the areas of protection (human health, ecosystems, and resources) (European Commission 2010a). SimaPro is one of the leading software programs used for LCA studies, developed by PRè Consultants. ReCiPe 2008 has been adopted worldwide, with uncertainty at the midpoint level results considered to be relatively

TABLE 24.1
Conversion Factors in g/kg of Diesel (MMA 2011)

Type of Emission	Lightweight Lorry, 3.5–7.5 tons	Medium-weight Lorry, 7.5–16 tons	Heavyweight Truck, >28 tons
CO_2	2.7	2.7	2.7
CO	0.42	0.58	1.01
NO_x	0.08	0.11	0.19
PM	0.04	0.054	0.095
NMVOC	2.37	3.25	5.68

PM: particulate matter; NMVOC: non-methane volatile organic compounds.

Assessment of Lubricant Oil Plastic Containers in Brazil

TABLE 24.2
Inventory Data for the Inputs and Outputs of the Life Cycle of Lubricant Oil Plastic Containers (LOPC)

Input			Output		
Material	**Amount**	**Units**	**Material**	**Amount**	**Units**
Production LOPC					
HDPE	1	ton	LOPC	1	ton
Diesel (transport)	63.75	L	Heat	6.14×10^3	MJ
Energy	52,500	kWh	CO_2	172,125	kg
Water (cooling)	2.99	m³	CO	0.023562	kg
			PM	2.244	kg
			NO_x	4.488×10^3	kg
			NMVOC	132,957	kg
Distribution					
LOPC	1	ton	CO_2	6.102×10^3	kg
Diesel (transport)	2.26	L	CO	1.15×10^{-6}	kg
Lubricant oil *	43	ton	PM	1.074×10^{-4}	kg
			NO_x	2.19×10^{-1}	kg
			NMVOC	6.46×10^{-3}	kg
*Collection LOPC (Instituto Jogue Limpo**)*					
LOPC	1	ton	HDPE	1	ton
Diesel (transport)	69.92	L	CO_2	1.88×10^2	kg
Energy	1.080×10^3	kWh	CO	2.58×10^{-2}	kg
Lubricant oil *	1.935	ton	PM	2.46×10^{-3}	kg
			NO_x	4.92×10^{-3}	kg
			NMVOC	1.46×10^{-1}	kg
Recycling LOPC					
HDPE	1	ton	Product	1	ton
Diesel (transport)	0.387	L	Heat	1.83×10^3	MJ
Energy	13,305	kWh	CO_2	1.045×10^4	kg
Water (cooling)	19.7	m³	CO	3.44×10^{-4}	kg
			PM	3.24×10^{-5}	kg
			NO_x	6.47×10^{-5}	kg
			NMVOC	1.93×10^{-3}	kg
Incineration					
HDPE	1	ton	Ash	0.01	ton
Diesel (transport)	21.8	L	Heat	1087×10^3	MJ
Energy	500	kWh	CO_2	58.86	kg
Heating oil	0.008095	ton	CO	1.94×10^{-2}	kg
Natural gas	0.963657	GJ	PM	1.82×10^{-3}	kg
Water	3581	L	NO_x	3.64×10^{-3}	kg
			NMVOC	1.09×10^{-1}	kg
			Water	1098	L
Industrial Landfill					
HDPE	1	ton	CO_2	187.866	kg
Diesel (transport)	69.58	L	CO	3.45×10^{-2}	kg
Energy	0.367	kWh	PM	3.66×10^{-3}	kg
Lubricant oil	1.935	tons	NOx	7.01×10^{-3}	kg
			NMVOC	2.08×10^{-1}	kg
			Heat	32.9	GJ

* Lubricant oil used for calculation of weight for transportation only, considering that 45 mL oil remains in the "empty" bottle. The density of lubricant oil is 0.86 g/cm³ (Willing 2001). HDPE: high-density polyethylene.

** Available online at www.joguelimpo.org.br

low (European Commission 2010b). Furthermore, it is the LCIA method with the highest number of midpoint impact categories (eighteen), enabling a more consistent and complete analysis of the cases being studied.

The impact categories selected were climate change (CC), ozone depletion (OD), human toxicity (HT), photochemical oxidant formation (POF), particulate matter formation (PMF), ionizing radiation (IR), terrestrial acidification (TA), freshwater eutrophication (FE), marine eutrophication (ME), terrestrial ecotoxicity (TET), freshwater ecotoxicity (FET), marine ecotoxicity (MET), agricultural land occupation (ALO), urban land occupation (ULO), natural land transformation (NLT), metal depletion (MD), and fossil depletion (FD).

The results of the LCIA were normalized by applying the European normalization data for the year 2000 encompassed in the ReCiPe midpoint method, which is based on the report of Seleesjik et al. (2008). This step is used in LCAs to show the relative contribution of each environmental impact category to a reference situation, making it easier to compare the different categories.

The normalization factors were as follows: CC: 8.91×10^{-5}; OD: 4.54×10; HT: 1.69×10^{-3}; POF: 1.89×10^{-2}; PMF: 6.71×10^{-2}; IR: 1.60×10^{-4}; TA: 2.91×10^{-2}; FE: 2.41×10; ME: 9.88×10^{-2}; TET: 1.22×10; FET: 9.19×10^{-2}; MET: 1.18×10^{-2}; ALO: 2.21×10^{-4}; ULO: 2.46×10^{-5}; NLT: 6.20×10; MD: 1.40×10^{-3}; FD: 6.14×10^{-4}.

24.2.4 SENSITIVITY ANALYSIS

A sensitivity analysis was conducted to address uncertainties in this LCA study. Two parameters were changed to evaluate their influences on the environmental impact results:

- The amount of lubricant oil that remains in the "empty" LOPCs was changed to both 30 and 60 mL to compare the impacts with those produced by the 45 mL baseline.
- The amount of HDPE that goes to the recycling or incineration plants was changed from 100% to 95% and 90%, as the weighting of the material collected occurs before the separation of possible contaminants, such as other solid wastes that are thrown by mistake in the bins for disposal of used LOPCs.

24.3 RESULTS

24.3.1 ENVIRONMENTAL IMPACTS OF THE FOUR SCENARIOS

Table 24.3 presents the aggregated net values of the selected impact categories for the four proposed scenarios. Positive values represent environmental impacts (costs/burdens to the environment), whereas negative values show environmental benefits.

Figure 24.2 shows this same data graphically, making it easier to analyze the results. It is possible to see, for example, that the highest impact is associated with MET, followed by FET, NLT, and HT for all scenarios.

Scenario 0 is the one with the highest values, followed by Scenario 2. These two scenarios have the greatest amounts of LOPCs being sent to landfills. Scenario 2 shows a reduction of these impacts, which results from raising recycling rates from 16% to 50%. Scenario 3 presents much lower values for the impacts on human toxicity, freshwater toxicity, and marine ecotoxicity than all the other scenarios. Its impact on natural land transformation was higher than the ones previously mentioned, although less significant than the results observed in the other scenarios. This is a scenario in which all LOPCs collected are sent only to recycling and incineration (waste-to-energy).

As mentioned previously, the numbers presented in Table 24.3 and Figure 24.2 are the net values of the impact assessment, which are the sum of the positive and negative values for each impact category evaluated. Impacts can have different values depending on the phase of the life cycle, as some fluxes can cause burdens while others can avoid them. Figures 24.3–24.6 show the separate positive and negative gross values of the impact categories for each of the four scenarios.

Assessment of Lubricant Oil Plastic Containers in Brazil

TABLE 24.3
LOPC Impact Assessment for the Four Scenarios Proposed

Impact Category	Abbreviation	Scenario 0	Scenario 1	Scenario 2	Scenario 3
Climate change	CC	2.61×10^3	9.11×10^2	2.95×10^3	1.98×10^3
Ozone depletion	OD	1.16×10^2	5.90×10^1	1.16×10^2	5.80×10^2
Human toxicity	HT	4.97×10^4	2.87×10^4	4.10×10^4	1.47×10^3
Photochemical oxidant formation	POF	1.04×10^3	3.94×10^2	1.08×10^3	5.20×10^2
Particulate matter formation	PMF	1.48×10^3	6.21×10^2	1.47×10^3	5.96×10^2
Ionizing radiation	IR	5.71×10^2	2.16×10^2	5.91×10^2	2.78×10^2
Terrestrial acidification	TA	1.65×10^3	6.67×10^2	1.62×10^3	5.99×10^2
Freshwater eutrophication	FE	6.93×10^3	3.00×10^3	6.54×10^3	1.81×10^3
Marine eutrophication	ME	7.18×10^2	3.82×10^2	6.25×10^2	9.10×10^1
Terrestrial ecotoxicity	TET	9.22×10^2	4.15×10^2	8.46×10^2	1.75×10^2
Freshwater ecotoxicity	FET	2.37×10^5	1.40×10^5	1.93×10^5	2.19×10^3
Marine ecotoxicity	MET	3.10×10^5	1.84×10^5	2.52×10^5	2.65×10^3
Agricultural land occupation	ALO	3.37×10^2	1.68×10^2	2.99×10^2	5.1×10
Urban land occupation	ULO	1.69×10^2	8.20×10^1	1.55×10^2	4.00×10
Natural land transformation	NLT	1.75×10^5	7.44×10^4	1.69×10^5	5.48×10^4
Metal depletion	MD	2.15×10^3	1.12×10^3	1.87×10^3	2.65×10^2
Fossil depletion	FD	4.65×10^3	2.40×10^3	4.42×10^3	1.67×10^3

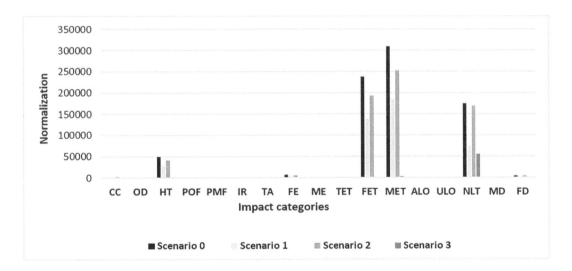

FIGURE 24.2 Graphic comparing the impact assessment of the four scenarios proposed.

In Scenario 0, the most relevant impacts in descending order of importance are related to marine ecotoxicity, freshwater ecotoxicity, natural land transformation, and human toxicity. Positive values at a much lower level can also be observed in relation to fossil depletion, freshwater eutrophication, climate change, particulate matter formation, terrestrial acidification, and mineral depletion. Avoided burdens for ionizing radiation and climate change are also present, but with low intensity, as can be observed in Figure 24.3.

Figure 24.4 demonstrates the impacts for Scenario 1. Those that stand out are marine ecotoxicity, natural land transformation, and freshwater ecotoxicity, in the order of intensity, with human

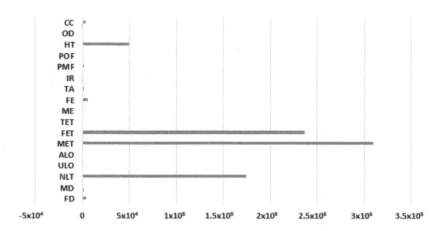

FIGURE 24.3 Graphic showing the positive and negative impacts associated with Scenario 0.

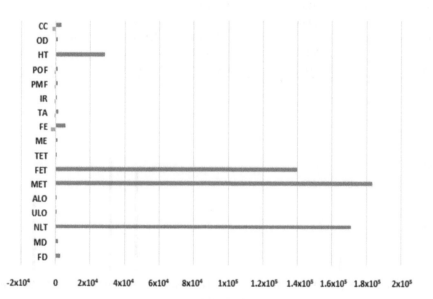

FIGURE 24.4 Graphic showing the positive and negative impacts associated with Scenario 1.

toxicity standing a bit lower. Other impacts that should be mentioned for having positive values are climate change, freshwater eutrophication, metal depletion, and fossil depletion. In this scenario, recycling was raised from 16% to 50%, which explains the negative values of CC, POF, PMF, IR, TA, and FE.

Figure 24.5 shows the same pattern as Figure 24.3, with the highest impacts being associated with MET, FET, and NLT, respectively, followed by HT with a lower value. All other impact categories have positive results much lower than those four. The only burden avoided associated with Scenario 2 is NLT, but with a very low value.

Figure 24.6 reveals a different pattern from the previous ones, as the only impact category that really stands out is natural land transformation, with both high positive and negative values. Scenario 3 is the only one that does not have an industrial landfill as a destination option. In this case, LOPCs are diverted half and half to recycling and incineration (waste-to-energy). The other

Assessment of Lubricant Oil Plastic Containers in Brazil 331

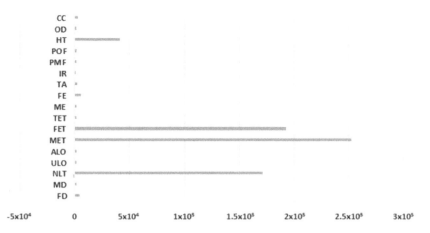

FIGURE 24.5 Graphic showing the positive and negative impacts associated with Scenario 2.

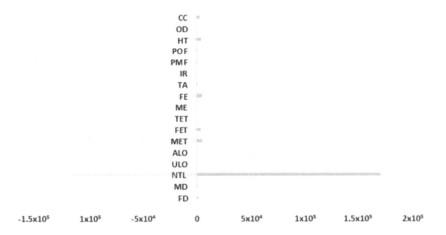

FIGURE 24.6 Graphic showing the positive and negative impacts associated with Scenario 3.

positive impacts observed are associated with CC, HT, FE, FET and MET, and burdens associated with CC, HT, POF, PMF, IR, TA, FE and MET decrease slightly.

Other results may be highlighted besides the net impacts already mentioned and presented in Figure 24.2. It is possible to see environmental impacts caused on climate change, freshwater eutrophication, and fossil resources depletion for all four scenarios. Burdens avoided can be mainly observed in Scenario 1 on climate change and freshwater ecotoxicity and in Scenario 3 on natural land transformation, human toxicity, freshwater eutrophication, freshwater ecotoxicity, marine ecotoxicity, and natural land transformation. These are the scenarios with higher recycling rates, and Scenario 3 is the one with the highest incineration rate.

24.3.2 Sensitivity Analysis Results

When the amount of recycled or incinerated HDPE changed, Scenario 1 showed variations of up to 33% for CC and Scenario 3 showed variations of up to 37% for TE when there was less HDPE to be treated (90%). Most of the impact categories were affected, which demonstrates that this parameter influences the overall results of the study (Table 24.4).

TABLE 24.4
Changes in Parameters Observed on the Sensitivity Analysis for the Variation in Amount of HDPE that Goes to the Recycling or Incineration Facilities

Impact Categories	Abbreviation	Scenario 0	Scenario 1	Scenario 2	Scenario 3
Climate change	CC	2–3%	16–33%	1–2%	8–15%
Ozone depletion	OD	2%	4–9%	0–1%	5–9%
Human toxicity	HT	0%	1–2%	0%	16–31%
Photochemical oxidant formation	POF	1–3%	12–24%	1–3%	9–18%
Particulate matter formation	PMF	1–3%	9–18%	1–3%	9–18%
Ionizing radiation	IR	2–4%	16–31%	2–4%	12–24%
Terrestrial acidification	TA	1–2%	10–20%	2–3%	11–22%
Freshwater eutrophication	FE	1–3%	9–19%	1–3%	15–31%
Marine eutrophication	ME	1%	3–5%	0–1%	10–21%
Terrestrial ecotoxicity	TET	1–2%	8–15%	1–2%	19–37%
Freshwater ecotoxicity	FET	0%	0–1%	0%	10–19%
Marine ecotoxicity	MET	0%	0%	0%	10–20%
Agricultural land occupation	ALO	1%	5–10%	1–2%	17–34%
Urban land occupation	ULO	1–2%	6–12%	1–3%	13–25%
Natural land transformation	NLT	1–3%	11–23%	1–3%	16–31%
Metal depletion	MD	1%	4–8%	2–3%	16–32%
Fossil depletion	FD	1–2%	5–9%	1%	6–13%

TABLE 24.5
Changes in Parameters Observed on the Sensitivity Analysis for the Variation of Lubricant Oil in the "Empty" LOPCs

Impact Categories	Abbreviation	Scenario 0	Scenario 1	Scenario 2	Scenario 3
Climate change	CC	±3%	0–2%	±2%	±0%
Ozone depletion	OD	±19%	±13%	±16%	±0%
Human toxicity	HT	0%	±0%	0%	±0%
Photochemical oxidant formation	POF	±9%	±4%	±5%	±0%
Particulate matter formation	PMF	±8%	±4%	±5%	±0%
Ionizing radiation	IR	±2%	±1%	±1%	±0%
Terrestrial acidification	TA	±12%	±6%	±8%	±0%
Freshwater eutrophication	FE	±4%	±2%	±2%	±0%
Marine eutrophication	ME	±2%	±1%	±1%	±1%
Terrestrial ecotoxicity	TET	±5%	±3%	±3%	±0%
Freshwater ecotoxicity	FET	0–1%	±0%	0%	±0%
Marine ecotoxicity	MET	±0%	±0%	0%	±0%
Agricultural land occupation	ALO	±1%	±0%	0%	±1%
Urban land occupation	ULO	±8%	±5%	±6%	±0%
Natural land transformation	NLT	±9%	±5%	±6%	±1%
Metal depletion	MD	±1%	±1%	±1%	±0%
Fossil depletion	FD	±18%	±12%	±15%	±1%

Table 24.5 shows that the variation of the amount of residual lubricant oil in LOPCs had considerable impact on Scenarios 0, 1, and 2 for the OD, TA, FD, ULO, and NLT categories. Its effect was very low in Scenario 3, showing that this variation has a greater impact on scenarios that include industrial landfills as a destination option for LOPCs.

The sensitivity analysis demonstrates that both the amount of oil remaining in the LOPCs and the quantity of HDPE that goes to recycling or incineration can influence the final impacts caused by the life cycle of LOPCs, showing that more detailed data provides more accurate results, though it should be stressed that this does not change the overall conclusions of this study.

24.4 CONCLUSION

This study shows that fewer impacts are produced during the life cycle of lubricant oil plastic containers if destination to industrial landfills can be avoided. Scenarios with higher recycling rates and incineration as an alternative treatment option for LOPCs have proved to be the ones that caused lower overall environmental impacts.

Despite the fact that waste-to-energy is not a technology currently applied in Brazil for the treatment of LOPCs and that the analysis of its use has uncertainties due to the scarcity of primary data, it shows promising results as a substitute option for the destination of LOPCs, though further studies should be conducted to ascertain its merits.

Using a case study in a Brazilian state, this chapter shows that emerging countries can deal with hazardous waste effectively. However, it takes time, political determination, and adequate legislation, together with a cooperative gesture of the private sector to accomplish desired results.

Furthermore, the implementation of environmental management policies needs inputs provided by the solid technical studies, such as those provided by the LCA methodology, to enable the finding of the best possible solutions.

REFERENCES

Associação Brasileira de Normas Técnicas (ABNT). 2004. *NBR 10004: Classificação de Resíduos Sólidos*. ABNT, Rio de Janeiro, Brazil.

Becker, H.S. 1997. *Métodos de Pesquisa em Ciências Sociais* [Research Methods in Social Sciences]. HUCITEC, São Paulo, Brazil.

Demertzi, M.; Dias, A.C.; Matos, A.; Arroja, L.M. 2015. "Evaluation of different end-of-life management alternatives for used natural cork stoppers through life cycle assessment." *Waste Manag.*, 46, 668–80. https://doi.org/10.1016/j.wasman.2015.09.026

Denatran. *Departamento Nacional de Trânsito*. Available online: www.denatran.gov.br/index.php/estatistica/253-frota-2014 (accessed on 4 April 2016).

Ecoinvent Center. *Ecoinvent Data v3.0. Swiss Centre for Life Cycle Inventories, St. Gallen*. Available online: www.ecoinvent.org (accessed on 15 September 2016).

European Commission. 2010a. *International Reference Life Cycle Data System (ILCD) Handbook – General Guide for Life Cycle Assessment – Detailed Guidance*, 1st ed. Publications Office of the European Union, Joint Research Centre, Institute for Environment and Sustainability, Ispra, Italy, p. 417.

European Commission. 2010b. *International Reference Life Cycle Data System (ILCD) Analysis of Existing Environmental Impact Assessment Methodologies for Use in Life Cycle Assessment*, 1st ed. Publications Office of the European Union, Joint Research Centre, Institute for Environment and Sustainability, Ispra, Italy, p. 417.

Finkbeiner, M.; Schau, E.M.; Lehmann, A.; Traverso, M. 2010. "Towards life cycle sustainability assessment". *Sustainability*, 2, 3309–22. https://doi.org/10.3390/su2103309

Goedkoop, M.; Heijungs, R.; Huijbregts, M.; Schryver, A.D.; Struijs, J.; van Zelm, R. 2008. *ReCiPe 2008 – A Life Cycle Impact Assessment Method Which Comprises Harmonised Category Indicators at the Midpoint and the Endpoint Level*. Available online: www.leidenuniv.nl/cml/ssp/publications/recipe_characterisation.pdf (accessed on 15 April 2016).

Guinée, J.B.; Gorrée, M.; Heijungs, R.; Huppes, G.; Kleijn, R.; de Koning, A.; van Oers, L.; Wegener Sleeswijk, A.; Suh, S.; Udo de Haes, H.A.; et al. 2002. *Handbook on Life Cycle Assessment: Operational Guide to the ISO Standards*, 1st ed. Springer, Dordrecht, The Netherlands, p. 692.

Guinée, J.B.; Heijungs, H.; Huppes, G.; Zamagni, A.; Masoni, P.; Buonamici, R.; Ekvall, T.; Rydberg, T. 2011. "Life cycle assessment: Past, present, and future". *Environ. Sci. Technol.*, 45, 90–6. https://doi.org/10.1021/es101316v

Instituto Brasileiro de Geografia e Estatística (IBGE). 2016. Available online: http://cod.ibge.gov.br/7DM (accessed on 17 October 2016).
Instituto Brasileiro de Informação em Ciência e Tecnologia (IBICT). 2016. *Banco nacional de Inventários de Ciclo de Vida*. Available online: http://sicv.acv.ibict.br/Node/ (accessed on 16 September 2016).
Intergovernmental Panel on Climate Change (IPCC). 2001. *Good Practice Guidance and Uncertainty Management in National Greenhouse Gas Inventories*. Report. Available online: www.ipcc-nggip.iges.or.jp/public/gp/bgp/5_3_Waste_Incineration.pdf (accessed on 10 December 2016).
International Standards Organization. 2006a. *ISO 14040 Environmental Management Life Cycle Assessment Principles and Framework*. International Standards Organization, Brussels, Belgium.
International Standards Organization. 2006b. *ISO 14044. Environmental Management Life Cycle Assessment Requirements and Guidelines*. International Stardard Organization, Brussels, Belgium.
Kanokkantapong, V.; Kiatkittipong, W.; Panyapinyopol, B.; Wongsuchoto, P.; Pavasant, P. 2009. "Used lubricating oil management options based on life cycle thinking". *Resour. Conserv. Recycl.*, 53, 294–9. https://doi.org/10.1016/j.resconrec.2009.01.002
Lorenz, E. 2014. "Life-cycle assessment in US codes and standards". *PCI J.*, 59, 49–54.
Ministério do Meio Ambiente (MMA). 2011. *1_ Inventário Nacional de Emissões Atmosféricas por Veículos Automotores Rodoviários* [1_ National Inventory of Atmospheric Emissions by Road Motor Vehicles]. Relatório Final. Available online: www.mma.gov.br/estruturas/163/_publicacao/163_publicacao27072011055200.pdf (accessed on 17 October 2016).
PlasticsEurope. 2015. *Plastics – The Facts 2014/2015 – An Analysis on European Plastics Production, Demand and Waste Data*. PlasticsEurope, Brussels, Belgium. Available online: http://issuu.com/plasticseuropeebook/docs/final_plastics_the_facts_2014_19122 (accessed on 15 July 2016).
Plastivida. 2011. *Monitoramento dos Índices de Reciclagem Mecânica de Plástico No Brasil (IRmP)*. Available online: www.plastivida.org.br/images/temas/Apresentacao_IRMP2011.pdf (accessed on 25 November 2015).
Rigamonti, L.; Grosso, M.; Giugliano, M. 2010. "Life cycle assessment of sub-units composing a MSW management system". *J. Clean. Prod.*, 18, 1652–62. https://doi.org/10.1016/j.jclepro.2010.06.029
Sindicom. 2016. *Data for the Lubricant Oil Sector*. Available online: www.sindicom.com.br/#conteudo.asp?conteudo=72&id_pai=60&targetElement=leftpart (accessed on 9 December 2016).
Sistema Firjan. *Manual de Gerenciamento de Resíduos: Guia de Procedimento Passo a Passo, 2nd Edição* [Waste Management Manual: Step-by-Step Procedure Guide]. Available online: www.firjan.com.br/lumis/portal/file/fileDownload.jsp?fileId=2C908A8F4EBC426A014ED041F0FB576E&inline=1 (accessed on 10 August 2015).
Sleeswijk, A.W.; van Oers, L.F.; Guinèe, J.B.; Struijs, J.; Huijbregts, M.A. 2008. "Normalization in product life cycle assessment: An LCA of the global and European economic systems in the year 2000". *Sci. Total Environ.*, 390, 227–70.
United States Environmental Protection Agency (EPA). 2006. *Plastic Oil Bottle Recycling*. Final Report. Available online: https://cfpub.epa.gov/ncer_abstracts/index.cfm/fuseaction/display.highlight/abstract/7974/report/F (accessed on 10 December 2016).
Willers, C.D.; Rodrigues, L.B. 2014. "A critical evaluation of Brazilian life cycle assessment studies." *Int. J. Life Cycle Assess.*, 19, 144–52. https://doi.org/10.1007/s11367-013-0608-y
Willing, A. 2001. "Lubricants based on renewable resources – An environmentally compatible alternative to mineral oil products". *Chemosphere*, 43, 89–98. https://doi.org/10.1016/S0045-6535(00)00328-3
Zanghelini, G.M.; de Souza, H.R.A., Jr.; Kulay, L.; Cherubini, E.; Ribeiro, P.T.; Soares, S.R. 2016. "A bibliometric overview of Brazilian LCA research". *Int. J. Life Cycle Assess.*, 21, 1759–75. https://doi.org/10.1007/s11367-016-1132-7

25 Step toward Sustainability
Fuel Production and Hybrid Vehicles

Manoj Eswara Vel S.B., Chandru R., Dhanalakshmi K., Joshua George Stanly, and Anand Ramanathan

CONTENTS

25.1	Municipal Waste Separation	335
25.2	Co-gasification of Solid Waste	336
	25.2.1 Syngas Production and Separation Techniques	336
	25.2.2 H_2 Production and Separation Techniques	336
	25.2.3 Alcohol Production and Separation Techniques	337
25.3	Pyrolysis of Solid Waste	337
	25.3.1 H_2 Production and Separation Techniques	337
	25.3.2 Alcohol Production and Separation Techniques	338
25.4	Alcohol in Petrol Engine	338
	25.4.1 Methanol or Methyl Alcohol	338
	25.4.2 Ethanol or Ethyl Alcohol	338
	25.4.3 Dimethyl Ether	339
	25.4.4 Diethyl Ether	339
25.5	Modifications in the Engine	339
	25.5.1 Possibilities and Advantages	339
	25.5.2 Challenges	340
25.6	Bio-oil in Diesel Engine	341
	25.6.1 Modifications in the Engine	341
	25.6.2 Possibilities and Advantages	341
	25.6.3 Challenges	342
25.7	Hybrid Vehicles	342
	25.7.1 Battery System	342
	25.7.2 Power Controller System	343
	25.7.3 Motor System	343
25.8	Energy Conversion Techniques	343
	25.8.1 Fuel Cell Energy Conversion Technology	343
	25.8.2 Renewable Energy Conversion Techniques	344
25.9	Conclusion	344
References		345

25.1 MUNICIPAL WASTE SEPARATION

The optimum separation procedures must be used to handle municipal solid trash properly. To achieve the maximum recovery and purity rates, the proper waste collection and separation is very much important. The actual garbage is still a mixture of components: plastics, metal, ferrous materials, wood, glass, organic material, textiles, and paper. To avoid garbage being burned or disposed in landfills, these precious materials can be turned into new goods. The municipal solid waste is categorized into six categories: compostable waste, hazardous waste, recyclable waste, bulky waste, combustible waste, and other waste. Based on the variety of municipal solid waste, it is collected and transported to the separation unit and it will be separated by mainly two methods: the first one is automatic separation in which sensors are used to identify and separate a different variety of wastes and the another one is manual separation in which wastes are identified physically to separate into different categories (Chen et al. 2017).

25.2 CO-GASIFICATION OF SOLID WASTE

Since coal has been used for gasification, which first took place in the 1800s and has seen numerous advancements over the past 200 years, to produce synthesis gas and fuel (Chmielniak and Sciazko 2003). The intrinsic qualities of biomass gasification, such as its high moisture, low-energy content, high-sulfur content, hygroscopic nature, and lightweight, present some limitations, making it more important during transit, preparation, and storage for gasification (Kuo and Wu 2016). A promising approach to generate heat, electricity, liquid, and gaseous biofuels utilizing synthesis gas is co-gasification. Two different feedstock can co-gasify more effectively to increase the H_2/CO ratio in the product gas, which is necessary for the production of liquid fuels (Manzanares-Papayanopoulos et al. 2014). The many kind of biomass and waste, together with agricultural products and wastes, sewage sludge, wood waste, and municipal wastes, have been co-gasified ("Natural Gas Substitutes from Coal and Oil (Book) | OSTI.GOV," n.d.). It is thought that the co-gasification of biomass and coal will lead to the production of gaseous fuel. The flammable constituent of biomass quickly decomposes and produces free radicals during the co-gasification of biomass feedstock and coal. These free radicals then react with the organic content of the coal, increasing the rate of conversion. Additionally, hydrogen might bond with coal-derived free radicals as soon as they make them, which might stop recombination events from creating secondary tar molecules that are less reactive. Consequently, hydrogen-enriched fuel gases are identified (Gunasekar and Manigandan 2022).

25.2.1 Syngas Production and Separation Techniques

In a gasifier, unprocessed syngas is generated, which is extremely hot and full of contaminants like particles, soot, and undesired gas components like acid gases. Furthermore, it has high CO levels, which are undesirable for many downstream uses. Syngas is therefore cleaned and conditioned for effective use in a variety of applications. Particulate cleaning is one of the methods in which ash particles can be cleaned from the syngas by using two-stage water wash. Also, wet processing such as solvent adsorption is employed to remove acid gas component, which is mixed from the syngas, namely, monoethanol amine (MEA), diethanol amine (DEA), FLEXSORB (hindered amines), and methyldiethanol amine (MDEA). The syngas constituents are effectively absorbed into the solvent during the physical absorption process (Furimsky 1999).

25.2.2 H_2 Production and Separation Techniques

Syngas produced by commercial gasifiers contains significant levels of CO (30–60%). The water-gas shift process can convert to H_2 with an unshifted syngas having 27–50% H_2 content (J. Wang et al. 2009). This reaction is crucial for increasing the amount of H_2 that may be produced from syngas released by gasifiers. Steam and CO combine in this catalytic reaction to produce H_2 and CO_2 (Garg and Dang 2005). When the synthesis gas generated by the oxygen introduction or the steam during the gasification is supplied to the Fischer–Tropsch reactors, the yield of the synthesis of hydrocarbons improves. Additionally, the pressure swing adsorption method can be used to segregate hydrogen and carbon dioxide after the catalysis water-gas shift reaction, which generates hydrogen from carbon monoxide (Boujjat et al. 2020).

Utilizing a Ni-Cu-Al catalyst, the efficiency of producing hydrogen from glycerol at the temperature of 550–600 °C under the atmospheric pressure was assessed using a continuous flow reactor (Dou et al. 2014). Along with a significant amount of hydrogen in excess water, the formation of negligible methane and carbon monoxide was seen. The generation of hydrogen from glycerol reforming increased with temperature (Dou et al. 2014).

In meeting the high-energy needs of hydrogen while also providing the supply, downstream processes like membrane separation are crucial. The technology is straightforward, less expensive, environmentally benign, and uses less energy (Singla et al. 2022). After the desorption from the surface, gaseous components from the gas sample mixture using the pressure swing adsorptive method yield about 99.9% high-purity hydrogen (Li et al. 2016).

25.2.3 Alcohol Production and Separation Techniques

When the biorefineries strategy is used, the food sector and its associated industries contribute significantly to the organic biomass waste which can be used as feedstock for the generation of alcohols. Wastes and leftovers from dairy, livestock, and other food processing facilities are among the potential biomass resources from the food industry. In addition to these, significant food wastes are also produced in home kitchens, restaurants, as well as those that are gathered as a result of municipal activities. Anaerobic and transesterification technologies are two potential pathways for using organic food waste to produce alcohol. The bioenergy categories include biogas, bioethanol, and biodiesel (Rajkumar and Kurinjimalar 2022).

A technique for extracting a liquid from a combination of many compounds is fractional distillation. For instance, fractional distillation can be used to separate liquid ethanol from an ethanol and aqueous mixture. The fact that the liquids in the mixture have various boiling points makes this approach effective. One liquid evaporates first when the mixture is heated (Berger and McPherson 1979).

25.3 PYROLYSIS OF SOLID WASTE

25.3.1 H_2 Production and Separation Techniques

Fuels produced are in solid, liquid, and gaseous state. Around the world, enormous amounts of waste are produced each year; for instance, the United States produces over 246 million tons (Brody and Avaliani 2021). Only a small portion of the majority of wastes gets eliminated using other methods, like anaerobic digestion and composting. By thermally degrading wastes in the absence of air, pyrolysis creates recyclable by-products such as char, oil, and combustible gases. To create either a solid char, gas, or oil product, the pyrolysis process parameters can be optimized.

The solid biochar can be converted to activated carbon or used as carbon black in addition to being a fuel, char-oil. The oil can be utilized as a feedstock for chemical production, upgraded using catalysts to create premium quality gasoline, added to the stock in oil refineries, or used as fuel additives (Sotoudehnia et al. 2020). Venkateswarlu Chintala et al. studied the feasibility of using solar thermal pyrolysis process of non-edible biomass feedstock and analyzed the three major product yields; the results showed that bio-oil containing alkanes and alkenes group was present in the product and the maximum bio-oil yield achieved was 20%, pyrolytic gas 29%, and biochar yield 51% (Chintala et al. 2017).

Saravanan Sathiya et al. reported that char utilization from rice husk as a catalyst to enhance the hydrogen content can yield 14.24% non-condensable gas at 450 °C as compared with conventional pyrolysis and surface area increases 50% in comparison to initial biochar (Saravana Sathiya Prabhahar et al. 2022). Bleeker et al. performed the steam iron process to produce hydrogen from pyrolysis oil. Their result revealed that at 900 °C, higher amount of hydrogen 1.82 Nm^3 was achieved using fluidized bed reactor (Bleeker, Veringa, and Kersten 2010).

25.3.2 Alcohol Production and Separation Techniques

The best option for switching to liquid fuel is provided by biomass-based sources because of their widespread availability and energy gain. Process engineering trends are being addressed in current fuel research and development to increase the production of bioalcohol during pyrolysis. It has been established that the manufacturing of bioalcohol benefits the biofuels industry. Ethanol (C_2H_5OH), methanol (CH_3OH), butanol (C_4H_9OH), and propanol (C_3H_7OH) are the main bioalcohols generated (Demirbas 2008). The designing of a bioalcohol manufacturing process must take advanced technology and product quality into account. To enhance the quality of the final product, biomass processing also necessitates new chemical transformations. These variables offer stimulating research opportunities in the field of process system engineering (Daoutidis et al. 2013).

It takes a lot of energy to separate the ethanol from this diluted aqueous solution because fermentation comprises only 5–12 wt% ethanol. Furthermore, at 78.15°C, ethanol and water combine to create an azeotrope, making it impossible to separate and dehydrate the ethanol from the solution using a single distillation column. As a result, a two-step procedure is frequently used: first, regular distillation is used to concentrate the solution to around 92.4 wt% ethanol. The resultant ethanol is then thoroughly dehydrated to pure alcohol using azeotropic distillation, liquid–liquid extraction, extractive distillation, adsorption, or methane gasification (Huang et al. 2010).

25.4 ALCOHOL IN PETROL ENGINE

The supply of gasoline and diesel is decreasing every day as a result of the rise in demand in the automotive sector. Nowadays, practically all vehicles run on diesel or gasoline, and because of this, different emissions are produced that have an impact on both the environment and the health of living things. Because of this issue, interest in alternative fuels like alcohol as a fuel for IC engines is growing (Guo, Wang, and Wang 2014). Diesel and gasoline are the two main fuels that were created and became widely used at the same time as IC engines and the automobile industry over the course of the last century. Because of their unique qualities, higher alcohols like propanol and butanol are employed as alternative fuel in the engine. Additionally, the ignition quality of alcohol molecules is improved by a longer carbon chain compared to lesser alcohols like ethanol and methanol. In SI engines, gasoline and alcohol mixtures reduce unburned hydrocarbon, pollutants like carbon monoxide (CO), and the tendency to knock, which is the primary issue with most vehicles (Erdogan 2020).

25.4.1 Methanol or Methyl Alcohol

Methanol has the chemical formula CH_3OH and is sometimes referred to as methyl alcohol and other names. Nowadays, industrial methanol production primarily involves hydrogenating carbon monoxide. The simplest alcohol is methanol, which is made up of a methyl group attached to a hydroxyl group. It is a colorless, flammable liquid that is light and volatile, and its unique odor is comparable to that of ethanol (drinking alcohol). However, methanol is much more hazardous than ethanol.

25.4.2 Ethanol or Ethyl Alcohol

A simple alcohol with the molecular formula C_2H_5OH, ethanol is also known as alcohol, ethyl alcohol, grain alcohol, and drinking alcohol. Its formula, which includes an ethyl group connected to a hydroxyl croup, can be written as $CH_3\_CH_2\_OH$ or $C_2H_5\_OH$; it is frequently abbreviated as EtOH.

A colorless, flammable liquid with a mild distinctive odor, ethanol is volatile and combustible. It is the primary form of alcohol present in alcoholic beverages and is a psychoactive chemical. Ethanol is a naturally occurring substance that is created when sugars are fermented by yeasts or through petrochemical processes. It is most frequently used as a recreational drug. Additionally, it can be used in medicine as an antiseptic and disinfectant.

25.4.3 Dimethyl Ether

Dimethyl ether (DME), commonly referred to as methoxymethane, is an organic molecule whose formula is C_2H_6O and can be written as CH_3OCH_3. The simplest ether is a colorless gas that can be used as an aerosol propellant and a helpful precursor to other organic molecules. Its potential usage as a fuel is currently being explored. It is an ethanol isomer.

25.4.4 Diethyl Ether

It is an inflammable, colorless liquid with a high volatility. It is frequently used as a laboratory solvent and as an engine-starting fluid. Before the development of non-flammable medicines like halothane, it was employed as a general anesthetic. It has been abused recreationally to get people high. According to earlier studies, the low cetane number and high latent heat of vaporization of lesser alcohols like methanol and ethanol when blended in various ratios with diesel cause several difficulties. Because of their high cetane number, improved blend stability, and less hygroscopic nature, higher alcohols like propanol and butanol perform better when blended with diesel than lower alcohols. According to earlier studies, mixing alcohols with gasoline or diesel increases efficiency and lowers gas emissions. Zhang conducted research on the variations in emission characteristics when isopropanol and gasoline are blended in SI engines. He examined the mixing of isopropanol with gasoline in his study, and the findings of that review paper showed that doing so would result in lower emissions of HC and CO. When isopropanol blends with unleaded gasoline in SI engines, Nehare et al. evaluated the exhaust emission characteristics of these engines. The test was conducted in SI engines, and the results showed that the emission of carbon oxide is minimized.

25.5 MODIFICATIONS IN THE ENGINE

The main objective of this chapter is to use what is learned about alcohols' basic combustion chemistry to comprehend how well they work in internal combustion engines. In this part, the effects of alcohol fuel structure on laminar flame speed, homogeneous ignition delay time, and combustion products are predicted using the comprehensive C_1-C_5 alcohol kinetic model. Then, these are connected to phenomena that affect engines, particularly pre-ignition and super-knock, fuel (Cai et al. 2019)

Because it is an oxygenated alternative fuel and a renewable bio-based resource, ethanol has the potential to lower particle emissions from compression–ignition engines. The characteristics and specifications of ethanol fuel that has been blended with diesel and gasoline are also included in this chapter. The elements crucial to these blends' potential commercial use are highlighted in particular. The impact of the fuel on emissions and engine performance (SI and CI engines) as well as material compatibility are also taken into account.

The global petroleum dilemma may have a workable solution in the form of biofuels. Automobiles powered by gasoline and diesel are the main emitters of greenhouse gases (GHG). Researchers from all around the world have looked into a number of alternative energy sources that could slake the population's growing need for energy. Biomass, biogas, primary alcohols, vegetable oils, biodiesel, and other forms of biofuel energy all have been investigated. These alternative energy sources are mainly eco-friendly, but each must be assessed individually for their benefits, drawbacks, and intended uses. While some of these fuels can be utilized right away, others need to be modified in order to achieve qualities that are more similar to those of conventional fuels.

25.5.1 Possibilities and Advantages

Criteria for selecting alternative fuels:

1. It must be affordable and widely available.
2. It has to consume propane and emit zero emissions.

3. It ought to have a high cacao value.
4. Production should be simple and affordable.
5. It should not require any modifications to current IC engines.
6. It ought to boost engine horsepower.
7. There should not be any engine maintenance needed.
8. It needs to be simple to handle and store.

The diesel engine was created by Rudolf Diesel to run on a variety of fuels, including heavy mineral oil, vegetable oils, and coal dust suspended in water. Despite the fact that Diesel's initial engine trials were disastrous failures, by the time he displayed his engine at the World Exhibition in Paris in 1900, it was powered entirely by peanut oil. Diesel had a unique vision. "The diesel engine may be fed with vegetable oils and would aid substantially in the development of renewable based fuels in engines," he claimed in 1911.

25.5.2 CHALLENGES

The volumes needed and the development of the right technology have been defined by the usage of alcohols. In the United States and Brazil, corn and sugarcane, respectively, are fermented to produce ethanol by the synthetic process of ethylene hydration. Prior to low-cost gasoline forcing it off the market in the early part of the century, ethanol was utilized in automobiles. Alcohols have also been employed when there is a lack of oil. Methanol fuel is frequently used by racing automobiles because, among other things, it can produce more power than an equivalent gasoline-fueled engine. Beginning around 1830, methanol was utilized for lighting until being eventually superseded by whale oil. In the 1880s, brighter kerosene took its place in turn. Methanol was still being used in some cooking and heating. Beginning in the 1920s, the emergence of the chemical industry – particularly for the production of plastics – led to a resurgence in the significance of methanol, this time as a chemical intermediate. Unprecedented levels of production and consumption are promised by the widespread use of methanol as a fuel. The oil crisis of the 1970s, which started with papers highlighting the potential for methanol fuels and steadily growing commercial introduction into gasoline blends, was what led to the recent use of methanol as a transportation fuel in the United States. The benefits of alcohols in gasoline blends for increasing octane have recently received more attention, especially with the phaseout of lead in gasoline. In 1923, a facility in Germany started producing methanol from synthesis gas, a blend of hydrogen, carbon monoxide, and carbon dioxide. At first, coal was used to produce synthesis gas, but after World War II, low-cost natural gas and light petroleum distillates have nearly entirely replaced coal. Individual plant capabilities increased throughout this time, rising from 40 tons/day in the 1930s to 2000 tons/day in the 1970s.

1. Even a tiny amount of methanol causes a mixture's vapor pressure to rise noticeably (by around 3 psi), as methanol is polar and gasoline is not. Ethanol has an effect that is only about one-third as strong. Other oxygenates have less of an impact, and combining them with methanol can lessen the impact. As vapor pressure rises, so do evaporative emissions and the likelihood of vapor lock.
2. When a tiny amount of water is present, a methanol–gasoline mixture will separate into two distinct liquid layers. Mixtures of ethanol and gasoline can hold nearly four times as much water. As alcohol concentration, gasoline aromaticity, and temperature rise, the susceptibility to separation decreases. Phase separation is also less likely when CrC9 oxygenates are added, either alone or in combination. Cosolvents are the name for such ingredients. When relatively minimal effects are desired, ethanol can be employed entirely or in part as a cosolvent for a methanol–gasoline mixture.
3. The fuel-to-air ratio of the charge introduced into the combustion chamber affects a variety of operating parameters and emissions. Unless a closed-loop electronic fuel control

system is employed, the oxygen in methanol (and other oxygenated hydrocarbons) alters this ratio when a fuel blend is utilized (relative to neat gasoline) (equivalent to mechanically readjusting the carburetor to a leaner fuel mixture).
4. Enleanment (increased air-to-fuel ratio) often results in lower emissions of carbon monoxide and unburned fuel (hydrocarbons).

25.6 BIO-OIL IN DIESEL ENGINE

The importance of biofuels stems from its ability to replace fossil fuels. Utilizing biofuels has numerous advantages for the environment, economy, and consumers. Fossil fuels can be replaced by bio-oil to produce heat, electricity, and/or chemicals. The conversion of bio-oil to a fuel for transportation is technically possible but requires more work. Making biodiesel from vegetable oil is the best way to use it as fuel. A clean-burning monoalkyl ester-based oxygenated fuel known as "biodiesel" is created from natural, renewable sources including fresh/used vegetable oils and animal fats. The resulting biodiesel's primary features are pretty similar to those of regular diesel. Although biodiesel doesn't contain any petroleum compounds, it may be combined with mineral diesel in any amount to provide a stable fuel that is compatible with traditional diesel (Subramanian 2022).

Before suggesting any alternative fuel to be utilized in existing technologies on a wide scale, there are a number of issues that need to be taken into consideration:

- The extent of the modifications that must be made to the current hardware; for example, if any alternative fuel requires major changes to the current hardware that cost a lot of money, it may be challenging to implement.
- The expense of building the infrastructure needed to process these alternative fuels. The expense of building infrastructure may be a barrier to the development of the energy resource.
- Environmental friendliness in comparison to traditional fuels. The new fuel will not be acceptable as fuel if it is more polluting (Yang, Kumar, and Huhnke 2015).
- Increased user costs related to routine maintenance, equipment wear, and lubricating oil life. The widespread acceptability of this fuel will suffer if the additional costs are too high.

25.6.1 Modifications in the Engine

Without gasoline consumption, there would be no human life. As technology develops, so does the demand for the various types of energy that mankind has been seeking throughout history.

Zhang et al. used mixtures of methyl, isopropyl, and winterized methyl ester of soybean oil with diesel as a fuel to study the combustion characteristics of turbocharged direct injection diesel engines. Except for isopropyl ester, they discovered that all fuel mixtures exhibited comparable combustion characteristics. Ester–diesel blends ignite more quickly than pure diesel fuel. Compared to diesel, Senatore et al. (2000) discovered that methyl ester heat release always occurs earlier with rapeseed oil (Zhang et al. 2020).

25.6.2 Possibilities and Advantages

The global biofuel sector is still in its early stages and is growing quickly. The EU policies on renewable energy and how they interact with state energy policy will determine the future framework for a global biofuel market in Europe. The Commission has thus far made it clear that biomass will be crucial in the future. In light of this, the trade in biofuels appears to be a realistic possibility for Europe. It is possible that as this new fuel market develops, seemingly odd trade flows may come and vanish.

Fuels made from biodiesel have good lubricity, thanks to the oxygen functional groups in them. Additionally, biodiesel fuels have cetane levels that are on par with or even greater than diesel fuels. These many characteristics have a considerable positive impact on particle engine emissions, total hydrocarbon emissions, and carbon monoxide emissions.

25.6.3 CHALLENGES

Legislative and regulatory measures to replace petroleum fuels with a variety of fuels obtained from domestic renewable sources have been prompted by the depletion of crude oil reserves and growing concern over greenhouse gas emissions. In the case of diesel automotive engines, three options are currently regarded as realistic, based on their capacity to replace petroleum-diesel in appreciable amounts without requiring significant engine redesign or modification: biodiesel fuels, hydrotreated oils (both derived from vegetable oils or animal fats), and Fischer–Tropsch (FT) diesel fuels made from lignocellulosic biomass. The only FT fuels that are now commercially available are made from fossil fuels (Das, Sahu, and Panda 2022).

Although their performance is very comparable, biodiesel and its blends have somewhat different attributes from petroleum-diesel. During the refinement of diesel, a significant amount of sulfur- and oxygen-containing compounds are eliminated. The fuel's lubricity is decreased by this removal, but it is made up for by additives. As opposed to ordinary diesel fuel, biodiesel typically has greater cloud and cold filter plugging points, which increases the likelihood that the fuel filter will become clogged.

25.7 HYBRID VEHICLES

In a hybrid electric vehicle (HEV), it will have two or more different forms of power and energy sources to operate the vehicle, hence the term "hybrid" refers to the fusion of many types of technology. Power sources include battery bank, flywheel, battery pack, regenerative braking, fuel cell, ultracapacitor, or internal combustion engine (ICE). The fuel cell/ultracapacitor hybrid power source was created and built by Fathabadi (2018) who discovered that it had a power efficiency of 96.2% at the top speed of 158 km/h, and a range of up to 435 km with a weight of 1880 kg. Longer mileage, lower pollutants, and reduced fuel consumption are all benefits of effective energy management strategies and optimization (Y. Wang et al. 2018). The advantage of HEV is that the secondary source will function as a backup system to the driveline with its maximum range when the primary fuel (diesel or gasoline) storage tank empties while operating the ICE (Thompson et al. 2011). HEVs are further divided into three groups based on the types of energy sources used in the driveline, including parallel, series, and dual HEVs.

25.7.1 BATTERY SYSTEM

The energy storage system that serves as an energy buffer is one of the essential elements of hybrid electric vehicles (HEVs). A smart energy storage system and proper battery-powertrain management are crucial for achieving the best potential pollution and consumption reduction while also increasing vehicle performance. This is accomplished by recovering the braking energy and giving the combustion engine an extra degree of freedom to move the engine's load point. Such lofty objectives place extreme pressure on the battery and battery management to perform. One of the key concerns for applications in vehicles is the energy storage system's specification. Varying hybrid vehicle ideas are logically leading to different needs for the battery system, depending on the sector of use. In light-duty vehicles and tiny cars, for instance, a hybrid powertrain enables the engine to be turned off during particular operating modes. The tiny car's fuel consumption can be greatly decreased by using the Start=Stop operation mode, especially in city traffic. However, a hybrid powertrain with an electric driving capacity enables package delivery services to deliver in confined city

Step toward Sustainability

center locations. HEV uses different battery systems such as lead acid battery, nickel metal hydride battery, lithium-based batteries, and electric double-layer capacitor (Conte 2006).

25.7.2 Power Controller System

According to various driving circumstances, the power controller of the gasoline engine and the electric motor is defined as follows. Estimating the magnitude and duration of the torque under various driving circumstances is the initial step. The second step entails choosing the payoff matrices that were created between the engine and the motor. The third stage involves solving the bimatrix game to arrive at a Nash equilibrium and calculating the contributed power and energy using extensive modeling of both average and extreme drive cycles. Getting a pair of pure or hybrid tactics is the last step. The best techniques influence power sharing to some extent. The motor and gasoline controllers receive the values. The bimatrix game, which featured a four-stage procedure, was used to discover the answers in terms of sharing power and energy storage. There are two stages to the design. The end result should be a controller that causes some desirable emergent behavior in the closed-loop system. Common language used to characterize emergent behavior includes terms like "enhanced safety," "better torque efficiency," "lower environmental impact," etc. (Chin and Jafari 2010).

25.7.3 Motor System

To accomplish the challenging goals for efficiency, power density, and drivetrain cost in HEVs, new motors and generators are greatly required. The motor or generator's specification is determined by its intended use, such as in light-, medium-, or heavy-duty vehicles, off-road vehicles, on-highway vehicles, and locomotives. The machine's performance is primarily influenced by the vehicle duty cycle, the thermal properties, and the cooling system used. In recent years, the traction systems for EVs have evolved due to the invention of power converter topologies for drive control. HEV uses different types of motors based on the requirements and they are switched reluctance motors, brushless DC motors, permanent magnet synchronous motors, and induction motors (Figure 25.1).

25.8 ENERGY CONVERSION TECHNIQUES

25.8.1 Fuel Cell Energy Conversion Technology

A fuel cell is a type of galvanic cell that transforms the chemical reaction energy of an oxidizing agent and a fuel that is continuously supplied into electrical energy. A fuel cell is a converter rather than an energy storage device. The fuel also contains the energy, which is chemically bonded to it. A fuel cell's energy efficiency is typically 40–60%; however, if waste heat is used in a cogeneration

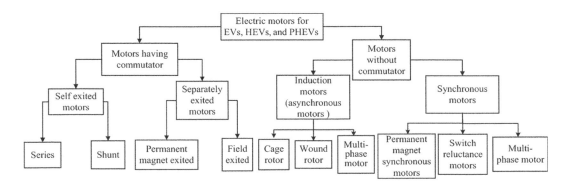

FIGURE 25.1 Classification of electric motors for hybrid vehicles (Singh, Bansal, and Singh 2019).

system, efficiencies of up to 85% can be attained. Although hydrogen–oxygen fuel cells are frequently mentioned, many other fuels, such as methanol, butane, or natural gas, can also be utilized in fuel cells. There are many distinct technological choices that can be classified into different categories based on the electrolyte used, resulting in various operating temperatures and pressures (Fuel Cells and Hydrogen 2 Joint Undertaking (FCH JU) 2018). The main fuel cell technologies used are the solid oxide fuel cell, the alkaline fuel cell, proton-exchange membrane fuel cell (PEMFC), the phosphoric acid, and the molten carbonate fuel cell (Alaswad et al. 2022). Power technologies and fuel cell automobiles are the two main applications of fuel cell technologies at the moment. The majority of greenhouse gas emissions come from the transportation sector, which is in addition to industry and energy generation. Therefore, it is ideal to substitute fossil fuels with various cutting-edge methods. Three distinct strategies for lowering GHG emissions are now being discussed in the mobility sector: combustion engines using synthetic fuels, fuel cell vehicles, and battery-powered cars. Numerous vehicles, including cars, buses, trucks, commercial vehicles, scooters, and bicycles, can be powered by fuel cells. Although some manufacturers have also demonstrated a fuel cell vehicle using methanol, fuel cell vehicles typically use hydrogen to power an electric engine – either directly or by charging a battery. The PEMFC is the most popular fuel cell technology for this application (Liu et al. 2018). The hydrogen is kept in a tank that is under pressure (up to 800 bars). On autos, 700 bar pressure tanks are typically utilized, while on buses, 350 bar pressure tanks are used. The range of vehicles is significantly impacted by the pressure tank used. In the past 10 years, several hydrogen refueling stations have been constructed all over the world. Since a polymer membrane is used to keep the temperature below 100 °C, a motor vehicle fuel cell operates between 60 °C and 80 °C.

25.8.2 Renewable Energy Conversion Techniques

Renewable energy sources are those that do not run out as they are used. In other words, they can be replenished from their respective sources, such as water resources for hydroelectric power. There are five primary categories of renewable energy: geothermal energy, which is heat from deep beneath the ground, wind energy, solar energy from the sun, hydroelectric energy from flowing water, and biomass, which is energy derived from plants and other waste. As opposed to fossil fuel–based energy, which produces waste and additional pollution, renewable energy is sometimes referred to as clean energy (Gopi et al. 2022). These generate fewer greenhouse gases and have a smaller carbon footprint. Recent years have seen a rise in interest in clean energy as various economies and governments seek to reduce their reliance on extremely polluting fossil fuels (Jaiswal et al. 2022). Burning agricultural trash, municipal solid waste, industrial waste, wood pulp liquor, and other waste is the conventional and simplest way to produce energy. Typically, this process is used to create steam, which is subsequently utilized to produce electricity. This process has been used to directly transform solid biomass into a gaseous molecule at high temperatures with a small amount of oxygen; the gases produced may be a combination of CH_4, CO_2, CO, N_2, and H_2. Recently, various techniques are used to convert renewable biomass sources to bioenergy using gasification, pyrolysis, anaerobic digestion, fermentation, extraction, and direct combustion. But conversion of renewable energy biomass source into bioenergy is not simple due to its various process parameters involved such as temperature, catalyst concentration, mixing concentrations, light, biological conditions, etc. (Gururani et al. 2022).

25.9 CONCLUSION

The limitations in the fossil energy leads to the growth of renewable energy research and utilization. Majorly, the energy needed for the transportation is due to its large number of usage creating more environmental risks. So the fuel production using different wastes with advanced techniques can eliminate waste and it can also reduce the dependency of the fossil fuels. In corresponding with fuels produced from various renewable sources, it is also necessary to change the existing design of the engine with updated model to work on different types of fuels. As changes in the existing engine is little complicated, it is improvised with hybrid technology which can power the vehicle in dual mode. Currently, there are many number of hybrid vehicles and battery vehicles in the design and production. The advanced techniques which can convert the energy on board can improve the energy requirement or recharging. The vehicles with fuel cell technology can produce on board energy which can be stored and utilized for the engine applications.

REFERENCES

Alaswad, Abed, A. Palumbo, Michele Dassisti, Mohammad A. Abdelkareem, and Abdul-Ghani Olabi. 2022. "Fuel Cell Technologies, Applications, and State of the Art. A Reference Guide☆." In *Encyclopedia of Smart Materials Olabi*, edited by B. T. Abdul-Ghani, 315–33. Oxford: Elsevier. https://doi.org/https://doi.org/10.1016/B978-0-12-815732-9.00033-4.

Berger, R., and W. McPherson. 1979. "Fractional Distillation." *Journal of the American Oil Chemists' Society* 56 (11). https://doi.org/10.1007/BF02667433.

Bleeker, M. F., H. J. Veringa, and S. R.A. Kersten. 2010. "Pure Hydrogen Production from Pyrolysis Oil Using the Steam-Iron Process: Effects of Temperature and Iron Oxide Conversion in the Reduction." *Industrial and Engineering Chemistry Research* 49 (1). https://doi.org/10.1021/ie900530d.

Boujjat, Houssame, Giberto Mitsuyoshi Yuki Junior, Sylvain Rodat, and Stéphane Abanades. 2020. "Dynamic Simulation and Control of Solar Biomass Gasification for Hydrogen-Rich Syngas Production during Allothermal and Hybrid Solar/Autothermal Operation." *International Journal of Hydrogen Energy* 45 (48). https://doi.org/10.1016/j.ijhydene.2020.01.072.

Brody, Michael, and Simon L. Avaliani. 2021. "Assessment of Health Risks from Environmental Factors. 16 Years of Collaboration Between the United States Environmental Protection Agency (US EPA), Hygiene and Environmental Organizations in the Russian Federation: Results and Reflections." *Gigiena i Sanitariya* 100 (12). https://doi.org/10.47470/0016-9900-2021-100-12-1344-1349.

Cai, Haiming, Jingyong Liu, Wuming Xie, Jiahong Kuo, Musa Buyukada, and Fatih Evrendilek. 2019. "Pyrolytic Kinetics, Reaction Mechanisms and Products of Waste Tea via TG-FTIR and Py-GC/MS." *Energy Conversion and Management* 184 (October 2018): 436–47. https://doi.org/10.1016/j.enconman.2019.01.031.

Chen, Haibin, Yan Yang, Wei Jiang, Mengjie Song, Ying Wang, and Tiantian Xiang. 2017. "Source Separation of Municipal Solid Waste: The Effects of Different Separation Methods and Citizens' Inclination – Case Study of Changsha, China." *Journal of the Air and Waste Management Association* 67 (2). https://doi.org/10.1080/10962247.2016.1222317.

Chin, Hubert H., Ayat A. Jafari, Old Westbury, and New York. 2010. "Design of Power Controller for Hybrid Vehicle Electrical Engineering and Computer Science Department New York Institute of Technology." *New York*, 165–70.

Chintala, Venkateswarlu, Suresh Kumar, Jitendra K. Pandey, Amit K. Sharma, and Sagar Kumar. 2017. "Solar Thermal Pyrolysis of Non-Edible Seeds to Biofuels and Their Feasibility Assessment." *Energy Conversion and Management* 153. https://doi.org/10.1016/j.enconman.2017.10.029.

Chmielniak, Tomasz, and Marek Sciazko. 2003. "Co-Gasification of Biomass and Coal for Methanol Synthesis." *Applied Energy* 74 (3–4). https://doi.org/10.1016/S0306-2619(02)00184-8.

Conte, F. V. 2006. "Battery and Battery Management for Hybrid Electric Vehicles: A Review." *E & i Elektrotechnik Und Informationstechnik* 123 (10): 424–31. https://doi.org/10.1007/s00502-006-0383-6.

Daoutidis, Prodromos, Adam Kelloway, W. Alex Marvin, Srinivas Rangarajan, and Ana I. Torres. 2013. "Process Systems Engineering for Biorefineries: New Research Vistas." *Current Opinion in Chemical Engineering*. https://doi.org/10.1016/j.coche.2013.09.006.

Das, Amar Kumar, Santosh Kumar Sahu, and Achyut Kumar Panda. 2022. "Current Status and Prospects of Alternate Liquid Transportation Fuels in Compression Ignition Engines: A Critical Review." *Renewable and Sustainable Energy Reviews* 161 (June): 112358. https://doi.org/10.1016/J.RSER.2022.112358.

Demirbas, Ayhan. 2008. "Biofuels Sources, Biofuel Policy, Biofuel Economy and Global Biofuel Projections." *Energy Conversion and Management* 49 (8). https://doi.org/10.1016/j.enconman.2008.02.020.

Dou, Binlin, Chao Wang, Yongchen Song, Haisheng Chen, and Yujie Xu. 2014. "Activity of Ni-Cu-Al Based Catalyst for Renewable Hydrogen Production from Steam Reforming of Glycerol." *Energy Conversion and Management* 78. https://doi.org/10.1016/j.enconman.2013.10.067.

Erdogan, Sinan. 2020. "Recycling of Waste Plastics into Pyrolytic Fuels and Their Use in IC Engines." *Sustainable Mobility*, no. 1: 1–23. https://doi.org/10.5772/intechopen.90639.

Fathabadi, Hassan. 2018. "Fuel Cell Hybrid Electric Vehicle (FCHEV): Novel Fuel Cell/SC Hybrid Power Generation System." *Energy Conversion and Management* 156: 192–201. https://doi.org/10.1016/j.enconman.2017.11.001.

Fuel Cells and Hydrogen 2 Joint Undertaking (FCH JU). 2018. *Addendum to the Multi-Annual Work Plan 2014–2020*. FCH.

Furimsky, E. 1999. "Gasification in Petroleum Refinery of 21st Century." *Oil and Gas Science and Technology* 54 (5). https://doi.org/10.2516/ogst:1999051.

Garg, M. O., and G. S. Dang. 2005. "Technology: Residue Upgrading Options: Petroleum Residue Processing R & D Thrust Area & Policy." *Hydrocarbon Asia* 15 (4).

Gopi, R., Vinoth Thangarasu, Angkayarkan Vinayakaselvi M, and Anand Ramanathan. 2022. "A Critical Review of Recent Advancements in Continuous Flow Reactors and Prominent Integrated Microreactors for Biodiesel Production." *Renewable and Sustainable Energy Reviews* 154 (October 2021): 111869. https://doi.org/10.1016/j.rser.2021.111869.

Gunasekar, P., and S. Manigandan. 2022. "Biofuels and Bioenergy." *International Journal of Ambient Energy* 43 (1). https://doi.org/10.1080/01430750.2019.1613264.

Guo, Zuogang, Shurong Wang, and Xiangyu Wang. 2014. "Stability Mechanism Investigation of Emulsion Fuels from Biomass Pyrolysis Oil and Diesel." *Energy* 66: 250–5. https://doi.org/10.1016/j.energy.2014.01.010.

Gururani, Prateek, Pooja Bhatnagar, Bhawna Bisht, Krishna Kumar Jaiswal, Vinod Kumar, Sanjay Kumar, Mikhail S. Vlaskin, Anatoly V. Grigorenko, and Kirill G. Rindin. 2022. "Recent Advances and Viability in Sustainable Thermochemical Conversion of Sludge to Bio-Fuel Production." *Fuel* 316: 123351. https://doi.org/10.1016/j.fuel.2022.123351.

Huang, H. J., S. Ramaswamy, U. W. Tschirner, and B. V. Ramarao. 2010. "Separation and Purification Processes for Lignocellulose-to-Bioalcohol Production." *Bioalcohol Production: Biochemical Conversion of Lignocellulosic Biomass*. https://doi.org/10.1533/9781845699611.3.246.

Jaiswal, Krishna Kumar, Chandrama Roy Chowdhury, Deepti Yadav, Ravikant Verma, Swapnamoy Dutta, Km Smriti Jaiswal, SangmeshB, and Karthik Selva Kumar Karuppasamy. 2022. "Renewable and Sustainable Clean Energy Development and Impact on Social, Economic, and Environmental Health." *Energy Nexus* 7: 100118. https://doi.org/https://doi.org/10.1016/j.nexus.2022.100118.

Kuo, Po Chih, and Wei Wu. 2016. "Thermodynamic Analysis of a Combined Heat and Power System with CO2 Utilization Based on Co-Gasification of Biomass and Coal." *Chemical Engineering Science* 142. https://doi.org/10.1016/j.ces.2015.11.030.

Li, Baojun, Gaohong He, Xiaobin Jiang, Yan Dai, and Xuehua Ruan. 2016. "Pressure Swing Adsorption/Membrane Hybrid Processes for Hydrogen Purification with a High Recovery." *Frontiers of Chemical Science and Engineering* 10 (2). https://doi.org/10.1007/s11705-016-1567-1.

Liu, Feiqi, Fuquan Zhao, Zongwei Liu, and Han Hao. 2018. "The Impact of Fuel Cell Vehicle Deployment on Road Transport Greenhouse Gas Emissions: The China Case." *International Journal of Hydrogen Energy* 43 (50): 22604–21. https://doi.org/https://doi.org/10.1016/j.ijhydene.2018.10.088.

Manzanares-Papayanopoulos, E., J. R. Herrera-Velarde, A. Arriola-Medellín, A. M. Alcaraz-Calderón, J. A. Altamirano-Bedolla, and M. Fernández-Montiel. 2014. "The Co-Gasification of Coal-Biomass Mixtures for Power Generation: A Comparative Study for Solid Fuels Available in Mexico." *Energy Sources, Part A: Recovery, Utilization and Environmental Effects* 36 (1). https://doi.org/10.1080/15567036.2010.536817.

"Natural Gas Substitutes from Coal and Oil (Book) | OSTI.GOV." n.d. https://www.osti.gov/biblio/5791766.

Rajkumar, R., and C. Kurinjimalar. 2022. "Food Wastes/Residues: Valuable Source of Energy in Circular Economy." *Handbook of Biofuels*. https://doi.org/10.1016/b978-0-12-822810-4.00007-5.

Saravana Sathiya Prabhahar, R., K. Jeyasubramanian, P. Nagaraj, and A. Sakthivel. 2022. "Catalytic Pyrolysis of Rice Husk with Nickel Oxide Nano Particles: Kinetic Studies, Pyrolytic Products Characterization and Application in Composite Plates." *Biomass Conversion and Biorefinery 2022* 1 (April): 1–18. https://doi.org/10.1007/S13399-022-02703-X.

Senatore, Adolfo, Massimo Cardone, Vittorio Rocco, and Maria Vittoria Prati. 2000. *A Comparative Analysis of Combustion Process in DI Diesel Engine Fueled with Biodiesel and Diesel Fuel*. Warrendale, PA: SAE International.

Singh, Krishna Veer, Hari Om Bansal, and Dheerendra Singh. 2019. "A Comprehensive Review on Hybrid Electric Vehicles: Architectures and Components." *Journal of Modern Transportation* 27 (2): 77–107. https://doi.org/10.1007/s40534-019-0184-3.

Singla, Shelly, Nagaraj P. Shetti, Soumen Basu, Kunal Mondal, and Tejraj M. Aminabhavi. 2022. "Hydrogen Production Technologies – Membrane Based Separation, Storage and Challenges." *Journal of Environmental Management*. https://doi.org/10.1016/j.jenvman.2021.113963.

Sotoudehnia, Farid, Abdulkarim Baba Rabiu, Abdulbaset Alayat, and Armando G. McDonald. 2020. "Characterization of Bio-Oil and Biochar from Pyrolysis of Waste Corrugated Cardboard." *Journal of Analytical and Applied Pyrolysis* 145. https://doi.org/10.1016/j.jaap.2019.104722.

Subramanian, A. S., and S. Ramalingam. 2022. "An Alternative Fuel to CI Engine : Delonix Regia Seed through Biochemical and Solar: Assisted Thermal Cracking Process." *International Journal of Environmental Science and Technology*. https://doi.org/10.1007/s13762-022-04299-1.

Thompson, Tammy M., Carey W. King, David T. Allen, and Michael E. Webber. 2011. "Air Quality Impacts of Plug-in Hybrid Electric Vehicles in Texas: Evaluating Three Battery Charging Scenarios." *Environmental Research Letters* 6 (2). https://doi.org/10.1088/1748-9326/6/2/024004.

Wang, Jie, Mingquan Jiang, Yihong Yao, Yanmei Zhang, and Jianqin Cao. 2009. "Steam Gasification of Coal Char Catalyzed by K2CO3 for Enhanced Production of Hydrogen without Formation of Methane." *Fuel* 88 (9). https://doi.org/10.1016/j.fuel.2008.12.017.

Wang, Yeqin, Zhen Wu, Yuyan Chen, Aoyun Xia, Chang Guo, and Zhongyi Tang. 2018. "Research on Energy Optimization Control Strategy of the Hybrid Electric Vehicle Based on Pontryagin's Minimum Principle." *Computers & Electrical Engineering* 72: 203–13. https://doi.org/10.1016/j.compeleceng.2018.09.018.

Yang, Zixu, Ajay Kumar, and Raymond L. Huhnke. 2015. "Review of Recent Developments to Improve Storage and Transportation Stability of Bio-Oil." *Renewable and Sustainable Energy Reviews* 50: 859–70. https://doi.org/10.1016/j.rser.2015.05.025.

Zhang, Li Hui, Qing Chao Gong, Feng Duan, Chien Song Chyang, and Cheng You Huang. 2020. "Emissions of Gaseous Pollutants, Polychlorinated Dibenzo-p-Dioxins, and Polychlorinated Dibenzo-Furans from Medical Waste Combustion in a Batch Fluidized-Bed Incinerator." *Journal of the Energy Institute* 93 (4): 1428–38. https://doi.org/10.1016/j.joei.2020.01.005.

26 Microwave Pyrolysis of Composite Fuels with Biomass

Dmitrii O. Glushkov, Pavel A. Strizhak, Anatoly S. Shvets, and Ksenia Y. Vershinina

CONTENTS

26.1 Introduction 349
    26.1.1 Use of Biomass as an Energy Source...349
    26.1.2 Joint Pyrolysis of Plant Biomass with Coal and MSW 350
26.2 Materials 351
26.3 Experimental Setup and Methods 352
26.4 Results and Discussion 355
26.5 Conclusion 359
Acknowledgments..359
References..359

26.1 INTRODUCTION

26.1.1 USE OF BIOMASS AS AN ENERGY SOURCE

At present, more than 85% of the world's energy consumption is provided by fossil fuels (coal, oil, gas) (Weiland 2010). According to the forecasts for the world energy development (IEA 2021), annual energy consumption by 2050 will increase three times compared to 2020.

The growing production of fossil hydrocarbons, on which the socioeconomic development of most countries depends, raises serious concerns about the intensive depletion of non-renewable energy sources (Mushtaq, Mat, and Ani 2014). In addition, the extraction of energy from fossil fuels entails negative environmental (greenhouse gas emissions, nitrogen and sulfur oxides, acid rain) and social (increase in the incidence of asthma and heart disease, reduced life expectancy) consequences (Prasad et al. 2019). In this regard, the combined use of coal and biomass as an alternative to traditional fuels can help solve the emerging global problems of mankind in the future.

Today, biomass resources (forest and agricultural waste, municipal solid waste, animal and crop waste, food waste, and industrial wastewater) are estimated at hundreds of billion tons per year due to rapid urbanization and intensive anthropogenic activities (Foong et al. 2020). Such waste is widely distributed and readily available in many parts of the world, has a lower cost than the traditional energy sources, and is considered environment-friendly due to its CO_2 neutrality (Bhuiyan et al. 2018). Energy can be extracted from the biomass waste by direct combustion to produce heat and electricity (Saidur et al. 2011) or by converting it into valuable products (biochar, bio-oil, generator gas) using thermochemical conversion methods (torrefaction, pyrolysis, gasification) (Mendoza Martinez et al. 2021; Sukiran et al. 2017; Cheng et al. 2019). In recent years, one of the main research areas in the field of renewable energy sources has been the development of microwave pyrolysis technologies (Foong et al. 2020; Parvez et al. 2020). This is due to the advantages of fast and efficient heating, higher energy concentration to extract valuable products from biomass (Arshanitsa et al. 2016). Modern research in the field of microwave pyrolysis often considers the joint pyrolysis of plant biomass with another type of raw material (coal, MSW, sludge, etc.) (Wang et al. 2020). The main focus in the research is on the fact that this method not only improves the physicochemical properties of pyrolysis products (the yield of biochar, bio-oil, and generator gas), but also effectively uses the value of biomass waste and recycles it.

26.1.2 JOINT PYROLYSIS OF PLANT BIOMASS WITH COAL AND MSW

Today, the global scientific community recognizes the joint pyrolysis of plant biomass with coal and MSW as a potential solution to the problems associated with the efficient extraction of valuable substances from low-grade coals, as well as the rational use of fossil fuels and biomass waste. The effect of synergy in the joint pyrolysis of coal and biomass is partly studied (P. Yang et al. 2021; Chen et al. 2019; Y. Zhang et al. 2018). It has been proven (Suresh et al. 2021) that biomass, as a solid source of hydrogen, can provide enough of it for hydrogenated low-grade coal pyrolysis, which effectively improves the coal conversion rate into tar and producer gas. Li et al. (Li et al. 2013) suggested that the synergistic effect may be due to secondary pyrolysis reactions in the gas phase. Jones et al. (Jones et al. 2005) observed that the synergistic effect reduces the aromatic compounds content and increases the yield of phenolic compounds in the tar. However, the co-pyrolysis synergistic effect of coal and biomass remains controversial. Existing studies confirm both the positive synergistic effect that can be achieved with the right choice of the biomass type, the biomass-to-coal ratio, pyrolysis temperature, reactors types, etc., and the negative effect with a certain combination of these parameters. For example, in L. Wu et al. (2022), an attempt was made to optimize the experimental conditions for the combined microwave pyrolysis of wheat straw and low-grade coal. The effect of pyrolysis time, microwave power, pulverized coal particle size, and coal/straw ratio on the yield characteristics of pyrolysis products has been studied. The results showed that the temperature rate increase was consistent with resin yield for all conditions except for the coal/straw ratio. Co-pyrolysis of coal and straw increased the yield of tar and generator gas by 13.21% and 12.40%, respectively. At the same time, the tar yield reached a maximum value of 17.20% under the following conditions: coal particle size 0.68–1.00 mm, microwave power 700 W, coal/straw ratio 50/50, pyrolysis time 20 min. An et al. (An, Tahmasebi, and Yu 2017) studied the synergistic effect of the combined pyrolysis of the biomass of Hailar lignite and palm kernel shell under the action of microwave radiation in the temperature range of 400–600 °C. The results showed that the synergistic effect increases the H_2 yield from 34.06 vol% (at 400 °C) to 45.83 vol% (at 600 °C). In this case, the total concentration of H_2 and CO in the generator gas was 64.88 and 83.74 vol%, respectively. This characteristic corresponds to the pyrolysis temperature (the higher the temperature, the more significant the synergistic effect). A significant effect of the pyrolysis temperature was also proved (Y. Zhang et al. 2018) when studying the joint pyrolysis of brown coal and corn straw under the influence of microwave radiation at various components ratios (0, 0.33, 0.50, 0.67, and 1) and pyrolysis temperatures (500 °C, 550 °C, 600 °C). The results showed that increasing the pyrolysis temperature from 500 °C to 600 °C increased the oil yield by 3.2% and the gas yield by 11.6% (based on the average concentrations of these components for all fuel ratios). At the same time, it was noted that most of the synergistic effects on oil yield at various straw/coal ratios were negative. This may be due to several reasons: the co-pyrolysis process complexity, different absorption capacity of materials, different fuel characteristics. Together, this led to various thermal decomposition processes during pyrolysis.

Potential renewable and widespread sources also include municipal solid waste (used tires, plastic, cardboard) (L. Zhang et al. 2022; Niu et al. 2022). Research aimed at studying the synergy of the combined pyrolysis of plant biomass and used tires confirms that the use of both wastes allows achieving a positive effect in the fuel production (oil, coal, and gas). Niu et al. (Niu et al. 2022) studied the effect of biomass types and the ratio of mixture components on the joint pyrolysis of rice husk, wheat straw, and moso bamboo with waste tire. Co-pyrolysis showed positive synergistic effect on CO and CO_2 generation and negative synergistic effect on H_2 and CH_4 content. It is also shown that with increasing waste tire, the char yield was increased but the liquid yield tended to decrease. Waste tire char showed apparently higher BET surface area than biomass char. The adsorption ability of co-pyrolysis char was mainly controlled by the biomass char character. Lahijani et al. (Lahijani et al. 2013) carried out experiments on the joint gasification of biomass and tires in a CO_2 atmosphere. It has been established (Duan et al. 2015) that alkali metals in biomass contribute to the production of activated carbon. Duan et al. (Duan et al. 2015) carried out a joint pyrolysis of microalgae and waste tires in supercritical ethanol and found that the interaction of microalgae and waste tires during pyrolysis improves the quality of the resulting bio-oil.

The study of joint pyrolysis and gasification of plant biomass and plastic is also a promising area for waste disposal (L. Zhang et al. 2022). Xue et al. (Xue et al. 2015) studied the joint pyrolysis of cellulose with polypropylene in various proportions. It has been established that high-quality bio-oil is formed at temperatures above 500–800 °C. Brebu et al. (Brebu et al. 2010) studied the biomass effect and the plastic nature on the product yield and quality of pyrolysis oils and semi-coke. Joint pyrolysis of pine cone and cellulose with synthetic polymers at 500 °C contributes to an increased yield of liquid biofuel compared to the pyrolysis of each considered components separately. The effect of adding different biomass to the fuel composition during co-pyrolysis with low-density polyethylene on the yield and pyrolytic oil characteristics was studied by the authors (J. Yang et al. 2016). It has been established that the synergistic effect, depending on the biomass type, can be both positive and negative. Significant removal of aldehydes, acids, esters, furans, ketones, phenols, and sugars from the final pyrolysis oil was achieved (with a significant increase in alcohol content recorded). Sun et al. (Sun et al. 2022) explored possible ways to achieve synergistic optimization of bio-oil yield and energy recovery efficiency through co-pyrolysis of biomass and plastic in a microwave oven. The results showed that co-pyrolysis of plastic and biomass shortens the reaction time and increases the yield of bio-oil and producer gas. It was noted that an increase in the proportion of plastic in the fuel mixture from 0% to 25%, 50%, and 75% led to an increase in the yield of generator gas by 8.6%, 15.9%, and 36.1%, respectively. Also, an increase in the proportion of plastic in the fuel mixture from 0% to 25% and 50% led to an increase in bio-oil yield by 28.6% and 37.5%, respectively. In addition, a low oxygen content and an increased content of aliphatic hydrocarbons were registered in the resulting bio-oil, which contributed to an increase in the bio-oil energy characteristics.

Thus, the study of the process of joint pyrolysis and gasification of several waste types allows not only their effective disposal, but also generating better products. Information on synergistic effects in the co-pyrolysis of plant biomass and coal or MSW can be crucial for the pyrolysis products final composition. Microwave pyrolysis technology has proven to be a promising approach to converting biomass and other waste types into fuel products (Lam et al. 2019; Suresh et al. 2021). Lam et al. (Lam et al. 2019) studied the vacuum pyrolysis of plastic waste and used cooking oil under the action of microwave radiation. The yield of bio-oil containing light hydrocarbons and having a higher calorific value of 49 MJ/kg (compared to diesel fuel and gasoline) was up to 84 wt%. Luo et al. (J. Luo, Sun, et al. 2021) proved that continuous microwave pyrolysis combined with CO_2 reforming technology has advantages over other pyrolysis methods (batch microwave pyrolysis, traditional electric heating). The yield of biogas, generator gas, and their calorific value reached the maximum values of 71.02 wt%, 85.70 vol%, and 10.87 MJ/Nm^3, respectively.

Thus, the literature analysis led to the conclusion that an integrated approach to the study of the synergistic effect in the combined microwave pyrolysis of biomass with coal and MSW is a promising research area.

The purpose of this research is to study the effect of adding MSW and coal to a plant biomass mixture on the characteristics of the generator gas yield during microwave pyrolysis.

26.2 MATERIALS

The most typical plant biomass types are identified, which are widespread in most regions of the world, namely, sawdust, leaves, and straw (Balat and Ayar 2005). The latter is the most attractive

TABLE 26.1
Fuel Components Characteristics

Name	\multicolumn{5}{c}{Elemental Analysis (wt%)}	\multicolumn{5}{c}{Characteristics}
	C	H	O	N	S	Humidity	Volatiles	Combined Carbon	Ash Content	Calorific Value (MJ/kg)
Sawdust	54.3	5.2	40.0	0.4	–	3.5	80.1	15.1	1.1	18.67
Leaves	49.91	5.92	43.22	0.86	0.09	6.95	76.85	–	6.25	17.05
Straw	39.90	5.75	41.97	0.65	0.13	6.84	–	–	11.59	16.12
Plastic	77.38	15.19	3.62	1.63	0.06	19.00	96.16	1.72	2.12	44.47
Rubber	82.59	8.17	6.14	0.92	2.18	1.15	66.76	28.53	3.56	40
Cardboard	48.6	8.2	39.7	0.58	0.091	17.7	79.5	0.1	2.7	19.5
Hard coal	61.45	4.29	12.20	0.46	0.41	14.34	24.91	53.91	6.85	32.7
Brown coal	78.92	5.75	12.46	1.51	1.36	5.03	34.12	44.54	16.31	23.23

biomass type, since the main components of the gaseous product of microwave pyrolysis are CO, H_2, CO_2, and CH_4 (Zhao et al. 2014; Huang et al. 2013). It was established (Zhao et al. 2014; Huang et al. 2013) that about half of the biomass was converted into H_2 (55–57%), and CO_2 (17–21%) low content was also noted. The authors (Yao et al. 2018) studied slow, fast, and microwave pyrolysis of apple tree leaves, bamboo, cypress, and sycamore trees. It is noted that microwave pyrolysis, compared to slow and fast pyrolysis, shows the highest CO and H_2 content, the highest H_2:CO molar ratio, and also the lowest CO_2 content. In H. Luo et al. (2017), it was noted that sawdust has a high dielectric constant, which means they absorb microwaves. Accordingly, this study identifies the most typical biomass types that are widely distributed in most regions of the world, namely, leaves, straw, and sawdust.

A biomass mixture based on three waste types was used: pine trees sawdust 25%; leaves (birch, poplar) 25%; straw (hay) 50%. Further, typical MSW (plastic, rubber, cardboard) and coals (brown coal of rank B1, Tomsk region, Russia; hard coal of rank T, Kemerovo region, Russia) are added separately to the mixture of plant biomass in a ratio of 50/50. Thus, fuel mixtures were obtained: Mix 1 – biomass mixture 50%, cardboard 50%; Mix 2 – biomass mixture 50%, plastic 50%; Mix 3 – biomass mixture 50%, rubber 50%; Mix 4 – biomass mixture 50%, hard coal 50%; Mix 5 – biomass mixture 50%, lignite 50%. Table 26.1 shows the main fuel components characteristics.

Before experiments, the sawdust was dried for 5 days at room temperature. Then they were crushed using a rotary mill Pulverisette 14 and sieved using laboratory sieving (grain size is 2000 μm). Next, the biomass was soaked for 24 h in water (≈200 mL) to acquire the desired humidity.

26.3 EXPERIMENTAL SETUP AND METHODS

Based on the literature analysis (H. Luo et al. 2017; J. Luo, Ma, et al. 2021; L. Wu et al. 2022; Q. Wu et al. 2020; Yu et al. 2022; Balat and Ayar 2005; Sun et al. 2022; An, Tahmasebi, and Yu 2017), the most promising microwave reactors for the plant biomass pyrolysis, coal, and MSW were identified (Table 26.2).

Based on an analysis of modern experimental facilities (Table 26.2), an experimental setup (Figure 26.1) was designed to solve the problems of this study. Benefits of the experimental setup include simplicity of design and operation, flexibility in conducting experiments, and gas concentrations real-time monitoring. The main elements of the experimental setup are a heating (microwave) oven, which heats fuels, and a gas analyzer Test 1, which implements gas sampling and determination of component concentrations.

An earthen bowl with fuel was placed in the working chamber of the heating oven. Heating of the furnace interior was carried out using microwave waves produced by a magnetron (power from 200 W to 800 W). After the fuel was introduced into the oven, the oven door was hermetically closed,

TABLE 26.2
Overview of Experimental Plants for Microwave Pyrolysis of Plant Biomass with Coal and MSW

Microwave Reactor Characteristics	Process Conditions and Parameters	Fuel	Reference
Microwave oven by Tangshan Microwave Thermal Instrument CO. Ltd., China (maximum power output of 4000 W, a frequency of 2.45 GHz)	Biomass-to-lignite blend ratio of 1:1; microwave receptor (SiC) at the mass ratio of 10:2; temperature at 400, 500, and 600 °C; pyrolysis time is 30 min at a microwave output power of 1000 W	Hailar lignite and palm kernel shell (particle size of less than 1.0 mm)	(An, Tahmasebi, and Yu 2017)
Microwave oven by Hunan Changyi Microwave Technology Co. Ltd. (microwave frequency is 2.45 GHz, microwave power is 500–2900 W, a maximum operating temperature of 1100 °C)	Low-density polyethylene contents of 0%, 25%, 50%, 75%, and 100%; absorbing material: silicon nitride (SiN); temperature 400 °C, 550 °C, 700 °C; the heating rate was 40 °C/min	Cow dung (200 mesh) with low-density polyethylene	(Sun et al. 2022)
Microwave oven NN-SD787S by Panasonic (maximum incident power of 1250 W, frequency of 2450 MHz)	Reaction gas N_2 (500 mL/min); absorber: coke; ratio absorber: biomass 1:5; reaction time 20 min	*Microalgae chlorella* sp.	(Balat and Ayar 2005)
Microwave reactor by Sanle Microwave Technology Co., Ltd. (power intensity 0–3000 W, a frequency 2.45 GHz)	Reaction gas N_2 (3 L/min); microwave power 400–800 W; additives types: K_2CO_3, Na_2CO_3, CuO, Fe_3O_4 (adding ratio 5, 10, 20 mass%)	Wheat straw (particle size 0–0.09 mm)	(Zhao et al. 2014)
Microwave oven by MAX from CEM Corporation (power of 750 W, a frequency of 2450 MHz)	Vacuum degree below 100 mmHg; temperature 480–560 °C; biomass feed 2 g/min; microwave absorber SiC (50 grit)	Sawdust and corn straw, biomass particle size from 0.9 to 1.9 mm	(H. Luo et al. 2017)
The microwave pyrolysis device consisted of a household microwave oven with a hole (ϕ 60 mm) on the side (2.45 GHz, maximal 700 W) and a specially made quartz tube reactor (ϕ 60 mm × 300 mm)	Experiments were carried out by regulating different pyrolysis times, microwave powers, particle sizes of coal, and ratios coal/straw; fuel mass 50 g	Low-rank coal (particle size 0.5–2 mm) and wheat straw (0.5–0.68 mm)	(L. Wu et al. 2022)
MAS-II microwave oven (SINEO) with a frequency of 2450 MHz and power input of 0–800 W	Microwave absorber: SiC with a diameter of 5 mm. Different ratios of HZSM-5 catalyst and metal oxides CoO, NiO, ZrO_2, SrO, CeO_2, CaO (HZSM-5 only, 4:1, 2:1, 1:1, 1:2, and CaO only). Catalytic temperatures 350–550 °C. Catalytic modes (ex situ mixed catalysis, ex situ separate catalysis, and in situ separate catalysis)	Waste vegetable oil (load weight 10 g, flow rate 2 mL/min)	(Q. Wu et al. 2020)
Microwave pyrolysis units (model CY-CP1100C-S, model CY-PY1100C-S) manufactured by Hunan Changyi Microwave Technology Co., Ltd., in China	Model CY-CP1100C-S for continuous microwave pyrolysis. Microwave output power range of 0–4200 W, and a maximum continuous operating temperature of 650 °C.	Air-dried cow manure (50 g)	(J. Luo, Sun, et al. 2021)

(Continued)

TABLE 26.2 *(Continued)*
Overview of Experimental Plants for Microwave Pyrolysis of Plant Biomass with Coal and MSW

Microwave Reactor Characteristics	Process Conditions and Parameters	Fuel	Reference
	Model CY-PY1100C-S for batch feeding condition: microwave output power range of 0–1500 W, and a maximum operating temperature of 1100 °C. N_2 was introduced into the pyrolysis chamber (150 L/h). Power was increased by 400 W every 15 min until the chamber attained the target temperature		
The microwave reactor was purchased from Qingdao MKW Microwave Applied Technology Co., Ltd. (Qingdao, China)	Microwave receptor is SiC (particles 24 mesh, mass 5 g); fuel sample 20 g; N_2 was used as the purge gas to maintain inert atmosphere (flow rate at 1.2 l/min); microwave power is 1000 W; a heating rate of 100 °C/min; temperature is 350–750 °C	Sophora wood and polyvinyl chloride (the particles smaller than 0.105 mm)	(Yu et al. 2022)

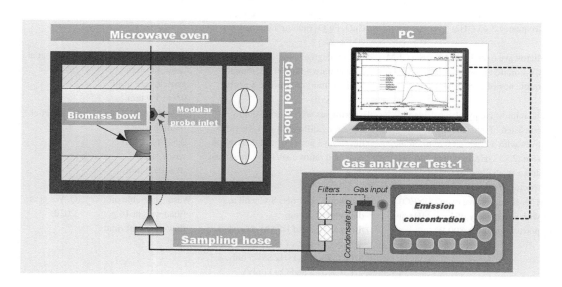

FIGURE 26.1 Schematic of the experimental setup.

the only hole was used for mounting the gas analyzer gas sampling device. To measure the gases concentration, a gas analyzer Test 1 (Boner, Russia) was used, which is a gas analytical system for long-term continuous measurement of the gas mixture components concentrations formed during fuel combustion and gasification. Table 26.3 shows its characteristics.

Gas analyzer Test 1 was connected to a computer. The gas analyzer software broadcast in real time the measured values of the registered gases concentrations, and also archived, integrated, and exported the obtained data into standard text and graphic formats.

TABLE 26.3
Gas Analyzer Test 1 Technical Characteristics (Boner, Russia)

Gas–Air Mixture Component*	Measuring Range	Error	Response Speed
O_2	0–25%	±0.2 vol% (absolute)	≤15 s
H_2 (polarographic sensor)	0–5%	±0.2 vol% (absolute)	≤35 s
CO_2 (optical sensor)	0–30%	±2% (basic percentage)	≤25 s
CH_4 (optical sensor)	0–30%	±5% (relative)	≤25 s
CO	0–40,000 ppm	±5% (relative)	≤35 s
NO	0–1,000 ppm	±5% (relative)	≤35 s
NO_2	0–500 ppm	±7% (relative)	≤45 s
SO_2	0–1,000 ppm	±5% (relative)	≤45 s
H_2S	0–500 ppm	±5% (relative)	≤45 s
HCl	0–2.000 ppm	±5% (relative)	≤45 s

ppm: concentration unit (1 ppm = 0.0001%).
* All sensors without explanation electrochemical.

The experimental procedure included the following stages:

(i) A fuel sample was formed using an analytical balance. The fuel sample weight in each experiment was 15 g. With an increase in the moisture content, the amount of dry biomass decreased.
(ii) The fuel sample was located along the edge of the earthen heat-resistant bowl (substrate) from the magnetron side.
(iii) The bowl was placed in the center of the heating oven, after which the door was hermetically sealed.
(iv) The magnetron and gas analyzer were launched. Fuel sample heating was carried out until gaseous pyrolysis products yield decreased at a power of 700 W. Then the magnetron was turned off and further generator gas release occurred without supplying thermal energy from an external source. During the experiment, the gas analyzer probe carried out the sampling of the formed gaseous substances, which were cleaned in the moisture separator and sent to the gas analyzer to analyze the mixture component composition. Changes in the measured parameters in real time were displayed on the PC monitor. During the experiment, atmospheric air did not enter the cavity of the heating oven. The duration of each experiment averaged 45 min, depending on the material preparation process.
(v) After the experiment, the ash residue was collected and weighed. Before the start of the next experiment, the reactor and all gas paths were purged with compressed atmospheric air. Within one experiment series, at least three experiments were carried out under identical initial conditions.
(vii) Further analysis and data processing were carried out. Experimental results were averaged and random errors were excluded. Calculation of average concentrations was performed using the trapezoidal method (Dorokhov et al. 2021).

26.4 RESULTS AND DISCUSSION

Average and maximum concentrations of the gas mixture components formed during the joint microwave pyrolysis of plant biomass with MSW and coal were obtained (Table 26.4).

TABLE 26.4
Average and Maximum Values of the Producer Gas Components Concentrations

Fuel Sample	\multicolumn{8}{c	}{Concentrations, %}
	CH_4		H_2		CO_2		CO	
	Average	Max	Average	Max	Average	Max	Average	Max
Mix 1	1.7	4.8	1.1	4.2	6.1	11	5.3	10.3
Mix 2	16.3	89.5	1.2	3.7	5.1	12.3	6.5	24.7
Mix 3	7.5	25.1	1.4	3.8	4.4	10.7	6.3	18.7
Mix 4	1.8	8.4	1.8	4.4	1.8	4.4	7.5	27.3
Mix 5	0.3	2.2	0.7	4.2	3.6	12.6	2.8	12.8

It has been found that plastic addition to a plant biomass mixture increases the average concentrations of CH_4 by 89.6% compared to cardboard and by 54% compared to rubber. The hard coal addition increases the average concentrations of the corresponding indicator by 83.3% compared to brown coal. At the same time, the highest CH_4 average concentrations were achieved during mixture pyrolysis with the addition of plastic (16.3%), the lowest – with the addition of brown coal (0.3%). The highest average concentrations of H_2 for all considered additives types range from 0.7% to 1.8%. At the same time, the highest H_2 concentrations were registered for a mixture with the hard coal addition (1.8%), the lowest – for brown coal (0.7%). The average concentrations of the corresponding indicator during mixtures pyrolysis with the MSW addition vary from 1.1% to 1.4%. The addition of rubber as the fuel component of the mixture leads to an increase in the average H_2 concentrations compared to cardboard and plastic by 21.4% and 17.1%, respectively. The average CO_2 concentrations for mixtures with the addition of MSW range from 4.4% to 6.1%. The highest average CO_2 concentrations are characteristic of cardboard, which is 16.4% more than plastic and 27.9% more than rubber. When comparing the corresponding indicators for coals, it was found that during the mixture pyrolysis with the brown coal addition, the average CO_2 values are twice as high as compared to hard coal. The values of average CO concentrations for mixtures with the plastic and rubber are very close, 6.5% and 6.3%, respectively. For a mixture with cardboard, the average CO concentrations are 16–18.5% less. The highest average concentrations of CO were registered for the mixture with hard coal (7.5%), which is 2.6 times more compared to the brown coal addition.

The hard coal addition had the maximum positive effect on the yield of the main gases concentrations: for the mixture under consideration, the lowest average concentrations of CO_2 and the highest average concentrations of H_2 and CO were recorded compared to the rest of the fuel components. Obtained results were compared with other studies. The authors (L. Wu et al. 2022) obtained the following concentrations values of the generator gas main components during the joint microwave pyrolysis of low-grade coal and wheat straw: CO – 24.88 vol%, CH_4 – 14.7 vol%, H_2 – 35.86 vol%, CO_2 – 23.37 vol%. Differences in the compared data can be explained by the fact that many factors affect the yield of pyrolysis products. In our case, the microwave power and the charcoal:straw ratio were the same (700 W and 1:1, respectively). However, the fuel particle size in our experiments exceeded the particle size in the study (L. Wu et al. 2022) by two times. In addition, an important role is played by the pyrolysis time. In this work, it averaged 45 min and did not change. In L. Wu et al. (2022), the reaction time varied from 110 min to 40 min. The fuels component composition and, as a result, their elemental composition also differ.

Among MSW additives, a positive synergistic effect was noted for mixtures with plastic and rubber. With the addition of plastic, the highest average concentrations of CH_4 and CO were achieved. For mixture with rubber, the maximum (out of those presented for MSW) values of average concentrations of H_2 and minimum values of average concentrations of CO_2 are typical. Comparison of the obtained results with data from other studies is ambiguous. In H. Luo et al. (2017), when performing the combined microwave pyrolysis of cow manure and plastic in 50/50 ratio, the following

Microwave Pyrolysis of Composite Fuels with Biomass

concentrations of the gas mixture main components were obtained: H_2 – 9%, CO – 8%, CH_4 – 8.5%, CO_2 – 15.5%. The values of the corresponding components obtained in this study differ by 1.2–2 times. This can be explained by the different component fuels composition, particle size (in our case, the particle size exceeds 25 times), as well as different methods for determining the volume fractions of H_2, CH_4, CO, and CO_2 in gaseous products.

The maximum concentrations of CH_4 (89.5%) and CO (24.7%) were registered for the mixture with plastic (Table 26.4). It is noted that for a mixture with hard coal, the maximum CH_4 values are 3.4 times higher than the corresponding values for a mixture with brown coal. At the same time, the maximum H_2 concentrations for all considered additives are very close and amount to 3.7–4.4%. Maximum CO_2 concentrations for MSW are also close and vary in the range of 10.7–12.3%. The most obvious difference in the values of the corresponding parameters is typical for mixtures with coals. Maximum concentrations of CO_2 for the mixture with brown coal are 2.8 times higher than the corresponding values for the hard coal addition.

Figure 26.2 shows curves in the concentrations of the producer gas individual components (CO, H_2, CO_2) over time.

Based on Figure 26.2, it was found that the start time of the considered gas components emission for all fuel mixtures is the same: for CO it was 20–30 s, H_2 began to release immediately upon heating, the start time for CO_2 emission was 60–65 s. The maximum duration of CO release was registered for mixtures with cardboard, plastic, and brown coal (2000 s). The minimum duration of CO emission is for a mixture with rubber (1200 s). At the same time, for the mixture with cardboard, three peaks of concentration growth were noted; for the remaining mixtures, one growth peak was noted (Figure 26.2a). The maximum duration of H_2 release was registered for mixtures with MSW and coal (2000 s), the minimum for a mixture with brown coal (1400 s). At the same

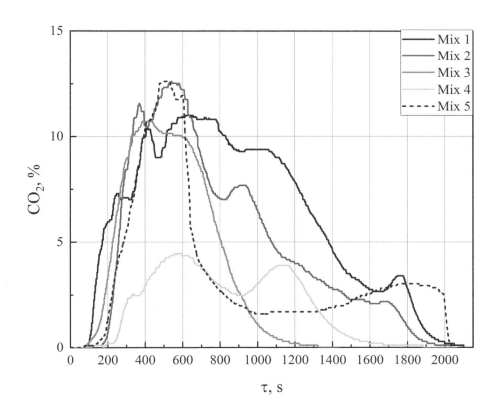

(a)

FIGURE 26.2 Curves of producer gas individual components change: (a) CO; (b) CO_2; (c) H_2.

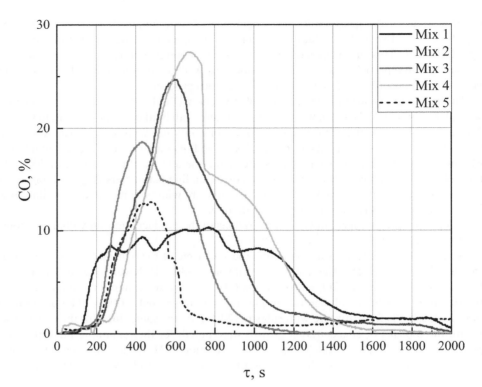

(b)

(c)

FIGURE 26.2 (Continued)

time, for mixtures with cardboard and coal, three peaks of growth in concentrations were noted; for the remaining mixtures, two peaks of growth were noted (Figure 26.2b). The maximum duration of CO_2 release was registered for mixtures with cardboard, plastic, and brown coal; the minimum for a mixture with rubber. At the same time, for a mixture with rubber, one peak of growth in concentrations was registered; for mixtures with coal and brown coal, two peaks of growth; for mixtures with cardboard and plastic, four peaks of growth (Figure 26.2c).

Information on synergistic effects in the co-pyrolysis of plant biomass and coal or MSW can be crucial for the correlation of the pyrolysis products final distribution and composition. The obtained results make it possible to select the additive type to plant biomass in order to obtain generator gas during microwave pyrolysis. At the same time, proceed from the priorities for increasing the yield of one or another gas mixture component.

26.5 CONCLUSION

(i) Joint pyrolysis of plant biomass and coal mixture contributes to an increase in the yield of CO by 63%, H_2 by 61%, CH_4 by 83.3%, and a decrease in the yield of CO_2 by 50% compared to the same indicators for plant biomass and brown coal mixture. The synergistic effect of coal with biomass is associated with the direct interaction of microwave radiation with solid carbon, heating it to a higher temperature compared to biomass particles included in the fuel mixture. The heated coal particles then transfer heat to nearby biomass particles, which quickly reach pyrolysis conditions.

(ii) Among MSW additives, a positive synergistic effect was noted for mixtures with plastic and rubber. Plastic in plant biomass mixture increases average CH_4 concentrations by 89.6% compared to cardboard and by 54% compared to rubber. The presence of rubber in the mixture as a fuel component leads to an increase in the average concentrations of H_2 compared to cardboard and plastic by 21.4% and 17.1%, respectively. Also, for mixtures with plastic and rubber, the average concentrations of CO_2 are lower compared to cardboard by 16.4% and 27.9%, respectively.

(iii) The maximum concentrations of CH_4 (89.5%) and CO (24.7%) are recorded for a mixture with the plastic. In this case, the maximum concentrations of H_2 and CO_2 are 3.7–4.4% and 10.7–12.3%, respectively. For mixtures with coal, the maximum CO_2 concentrations were noted for brown coal, which is 2.8 times higher than the corresponding values for hard coal.

ACKNOWLEDGMENTS

The reported study was funded by the Russian Foundation for Basic Research, National Council of Brazil for Scientific and Technological Development, and Ministry of Science & Technology (Government of India) according to the research project No. 19–53–80019.

REFERENCES

An, Yang, Arash Tahmasebi, and Jianglong Yu. 2017. "Mechanism of Synergy Effect during Microwave Co-Pyrolysis of Biomass and Lignite." *Journal of Analytical and Applied Pyrolysis* 128 (November): 75–82. https://doi.org/10.1016/j.jaap.2017.10.023.

Arshanitsa, Alexandr, Yegor Akishin, Edmund Zile, Tatiana Dizhbite, Valentin Solodovnik, and Galina Telysheva. 2016. "Microwave Treatment Combined with Conventional Heating of Plant Biomass Pellets in a Rotated Reactor as a High Rate Process for Solid Biofuel Manufacture." *Renewable Energy* 91 (June): 386–96. https://doi.org/10.1016/j.renene.2016.01.080.

Balat, Mustafa, and Günhan Ayar. 2005. "Biomass Energy in the World, Use of Biomass and Potential Trends." *Energy Sources* 27 (10): 931–40. https://doi.org/10.1080/00908310490449045.

Bhuiyan, Arafat A., Aaron S. Blicblau, A.K.M. Sadrul Islam, and Jamal Naser. 2018. "A Review on Thermo-Chemical Characteristics of Coal/Biomass Co-Firing in Industrial Furnace." *Journal of the Energy Institute* 91 (1): 1–18. https://doi.org/10.1016/j.joei.2016.10.006.

Brebu, Mihai, Suat Ucar, Cornelia Vasile, and Jale Yanik. 2010. "Co-Pyrolysis of Pine Cone with Synthetic Polymers." *Fuel* 89 (8): 1911–18. https://doi.org/10.1016/j.fuel.2010.01.029.

Chen, Xiye, Li Liu, Linyao Zhang, Yan Zhao, and Penghua Qiu. 2019. "Pyrolysis Characteristics and Kinetics of Coal–Biomass Blends during Co-Pyrolysis." *Energy & Fuels* 33 (2): 1267–78. https://doi.org/10.1021/acs.energyfuels.8b03987.

Cheng, Yoke Wang, Zhan Sheng Lee, Chi Cheng Chong, Maksudur R. Khan, Chin Kui Cheng, Kim Hoong Ng, and Sk Safdar Hossain. 2019. "Hydrogen-Rich Syngas Production via Steam Reforming of Palm Oil Mill Effluent (POME) – A Thermodynamics Analysis." *International Journal of Hydrogen Energy* 44 (37): 20711–24. https://doi.org/10.1016/j.ijhydene.2018.05.119.

Dorokhov, Vadim V., Geniy V. Kuznetsov, Galina S. Nyashina, and Pavel A. Strizhak. 2021. "Composition of a Gas and Ash Mixture Formed during the Pyrolysis and Combustion of Coal-Water Slurries Containing Petrochemicals." *Environmental Pollution* 285 (September): 117390. https://doi.org/10.1016/j.envpol.2021.117390.

Duan, Peigao, Binbin Jin, Yuping Xu, and Feng Wang. 2015. "Co-Pyrolysis of Microalgae and Waste Rubber Tire in Supercritical Ethanol." *Chemical Engineering Journal* 269 (June): 262–71. https://doi.org/10.1016/j.cej.2015.01.108.

Foong, Shin Ying, Rock Keey Liew, Yafeng Yang, Yoke Wang Cheng, Peter Nai Yuh Yek, Wan Adibah Wan Mahari, Xie Yi Lee, et al. 2020. "Valorization of Biomass Waste to Engineered Activated Biochar by Microwave Pyrolysis: Progress, Challenges, and Future Directions." *Chemical Engineering Journal* 389 (June): 124401. https://doi.org/10.1016/j.cej.2020.124401.

Huang, Yu Fong, Pei Te Chiueh, Wen Hui Kuan, and Shang Lien Lo. 2013. "Microwave Pyrolysis of Rice Straw: Products, Mechanism, and Kinetics." *Bioresource Technology* 142 (August): 620–4. https://doi.org/10.1016/j.biortech.2013.05.093.

IEA. 2021. "World Energy Outlook 2021." International Energy Agency (IEA). October 2021. www.iea.org/reports/world-energy-outlook-2021.

Jones, J.M., M. Kubacki, K. Kubica, A.B. Ross, and A. Williams. 2005. "Devolatilisation Characteristics of Coal and Biomass Blends." *Journal of Analytical and Applied Pyrolysis* 74 (1–2): 502–11. https://doi.org/10.1016/j.jaap.2004.11.018.

Lahijani, Pooya, Zainal Alimuddin Zainal, Abdul Rahman Mohamed, and Maedeh Mohammadi. 2013. "Co-Gasification of Tire and Biomass for Enhancement of Tire-Char Reactivity in CO2 Gasification Process." *Bioresource Technology* 138 (June): 124–30. https://doi.org/10.1016/j.biortech.2013.03.179.

Lam, Su Shiung, Wan Adibah Wan Mahari, Yong Sik Ok, Wanxi Peng, Cheng Tung Chong, Nyuk Ling Ma, Howard A. Chase, et al. 2019. "Microwave Vacuum Pyrolysis of Waste Plastic and Used Cooking Oil for Simultaneous Waste Reduction and Sustainable Energy Conversion: Recovery of Cleaner Liquid Fuel and Techno-Economic Analysis." *Renewable and Sustainable Energy Reviews* 115 (November): 109359. https://doi.org/10.1016/j.rser.2019.109359.

Li, Shuaidan, Xueli Chen, Li Wang, Aibin Liu, and Guangsuo Yu. 2013. "Co-Pyrolysis Behaviors of Saw Dust and Shenfu Coal in Drop Tube Furnace and Fixed Bed Reactor." *Bioresource Technology* 148 (November): 24–9. https://doi.org/10.1016/j.biortech.2013.08.126.

Luo, Hu, Liwei Bao, Lingzhao Kong, and Yuhan Sun. 2017. "Low Temperature Microwave-Assisted Pyrolysis of Wood Sawdust for Phenolic Rich Compounds: Kinetics and Dielectric Properties Analysis." *Bioresource Technology* 238 (August): 109–15. https://doi.org/10.1016/j.biortech.2017.04.030.

Luo, Juan, Rui Ma, Jiaman Sun, Guojin Gong, Shichang Sun, and Haowen Li. 2021. "In-Depth Exploration of Mechanism and Energy Balance Characteristics of an Advanced Continuous Microwave Pyrolysis Coupled with Carbon Dioxide Reforming Technology to Generate High-Quality Syngas." *Bioresource Technology* 341 (December): 125863. https://doi.org/10.1016/j.biortech.2021.125863.

Luo, Juan, Shichang Sun, Xing Chen, Junhao Lin, Rui Ma, Rui Zhang, and Lin Fang. 2021. "In-Depth Exploration of the Energy Utilization and Pyrolysis Mechanism of Advanced Continuous Microwave Pyrolysis." *Applied Energy* 292 (June): 116941. https://doi.org/10.1016/j.apenergy.2021.116941.

Mendoza Martinez, Clara Lisseth, Ekaterina Sermyagina, Jussi Saari, Marcia Silva de Jesus, Marcelo Cardoso, Gustavo Matheus de Almeida, and Esa Vakkilainen. 2021. "Hydrothermal Carbonization of Lignocellulosic Agro-Forest Based Biomass Residues." *Biomass and Bioenergy* 147 (April): 106004. https://doi.org/10.1016/j.biombioe.2021.106004.

Mushtaq, Faisal, Ramli Mat, and Farid Nasir Ani. 2014. "A Review on Microwave Assisted Pyrolysis of Coal and Biomass for Fuel Production." *Renewable and Sustainable Energy Reviews* 39 (November): 555–74. https://doi.org/10.1016/J.RSER.2014.07.073.

Niu, Miaomiao, Rongyue Sun, Kuan Ding, Haiming Gu, Xiaobo Cui, Liang Wang, and Jichu Hu. 2022. "Synergistic Effect on Thermal Behavior and Product Characteristics During Co-Pyrolysis of Biomass and Waste Tire: Influence of Biomass Species and Waste Blending Ratios." *Energy* 240 (February): 122808. https://doi.org/10.1016/j.energy.2021.122808.

Parvez, Ashak Mahmud, Muhammad T. Afzal, Peng Jiang, and Tao Wu. 2020. "Microwave-Assisted Biomass Pyrolysis Polygeneration Process Using a Scaled-up Reactor: Product Characterization, Thermodynamic Assessment and Bio-Hydrogen Production." *Biomass and Bioenergy* 139 (August): 105651. https://doi.org/10.1016/j.biombioe.2020.105651.

Prasad, S., V. Venkatramanan, S. Kumar, and K. R. Sheetal. 2019. "Biofuels: A Clean Technology for Environment Management." In *Sustainable Green Technologies for Environmental Management*, 219–40. Springer. https://doi.org/10.1007/978-981-13-2772-8_11.

Saidur, R., E. A. Abdelaziz, A. Demirbas, M. S. Hossain, and S. Mekhilef. 2011. "A Review on Biomass as a Fuel for Boilers." *Renewable and Sustainable Energy Reviews* 15 (5): 2262–89. https://doi.org/10.1016/j.rser.2011.02.015.

Sukiran, Mohamad Azri, Faisal Abnisa, Wan Mohd Ashri Wan Daud, Nasrin Abu Bakar, and Soh Kheang Loh. 2017. "A Review of Torrefaction of Oil Palm Solid Wastes for Biofuel Production." *Energy Conversion and Management* 149 (October): 101–20. https://doi.org/10.1016/j.enconman.2017.07.011.

Sun, Jiaman, Juan Luo, Junhao Lin, Rui Ma, Shichang Sun, Lin Fang, and Haowen Li. 2022. "Study of Co-Pyrolysis Endpoint and Product Conversion of Plastic and Biomass Using Microwave Thermogravimetric Technology." *Energy* 247 (May): 123547. https://doi.org/10.1016/j.energy.2022.123547.

Suresh, Aravind, Alaguabirami Alagusundaram, Ponnusamy Senthil Kumar, Dai-Viet Nguyen Vo, Femina Carolin Christopher, Bharkavi Balaji, Vinatha Viswanathan, and Sibi Sankar. 2021. "Microwave Pyrolysis of Coal, Biomass and Plastic Waste: A Review." *Environmental Chemistry Letters* 19 (5): 3609–29. https://doi.org/10.1007/s10311-021-01245-4.

Wang, Guanyu, Yujie Dai, Haiping Yang, Qingang Xiong, Kaige Wang, Jinsong Zhou, Yunchao Li, and Shurong Wang. 2020. "A Review of Recent Advances in Biomass Pyrolysis." *Energy & Fuels* 34 (12): 15557–78. https://doi.org/10.1021/acs.energyfuels.0c03107.

Weiland, Peter. 2010. "Biogas Production: Current State and Perspectives." *Applied Microbiology and Biotechnology* 85 (4): 849–60. https://doi.org/10.1007/s00253-009-2246-7.

Wu, Lei, Jiao Liu, Pan Xu, Jun Zhou, and Fu Yang. 2022. "Biomass Hydrogen Donor Assisted Microwave Pyrolysis of Low-Rank Pulverized Coal: Optimization, Product Upgrade and Synergistic Mechanism." *Waste Management* 143 (April): 177–85. https://doi.org/10.1016/j.wasman.2022.02.020.

Wu, Qiuhao, Yunpu Wang, Yujie Peng, Linyao Ke, Qi Yang, Lin Jiang, Leilei Dai, et al. 2020. "Microwave-Assisted Pyrolysis of Waste Cooking Oil for Hydrocarbon Bio-Oil Over Metal Oxides and HZSM-5 Catalysts." *Energy Conversion and Management* 220 (September): 113124. https://doi.org/10.1016/j.enconman.2020.113124.

Xue, Yuan, Shuai Zhou, Robert C. Brown, Atul Kelkar, and Xianglan Bai. 2015. "Fast Pyrolysis of Biomass and Waste Plastic in a Fluidized Bed Reactor." *Fuel* 156 (September): 40–6. https://doi.org/10.1016/j.fuel.2015.04.033.

Yang, Jingxuan, Jenny Rizkiana, Wahyu Bambang Widayatno, Surachai Karnjanakom, Malinee Kaewpanha, Xiaogang Hao, Abuliti Abudula, and Guoqing Guan. 2016. "Fast Co-Pyrolysis of Low Density Polyethylene and Biomass Residue for Oil Production." *Energy Conversion and Management* 120 (July): 422–9. https://doi.org/10.1016/j.enconman.2016.05.008.

Yang, Panbo, Shuheng Zhao, Quanguo Zhang, Jianjun Hu, Ronghou Liu, Zhen Huang, and Yulong Gao. 2021. "Synergistic Effect of the Cotton Stalk and High-Ash Coal on Gas Production During Co-Pyrolysis/Gasification." *Bioresource Technology* 336 (September): 125336. https://doi.org/10.1016/j.biortech.2021.125336.

Yao, Benzhen, Tiancun Xiao, Xiangyu Jie, Sergio Gonzalez-Cortes, and Peter P. Edwards. 2018. "H2–Rich Gas Production from Leaves." *Catalysis Today* 317 (November): 43–9. https://doi.org/10.1016/j.cattod.2018.02.048.

Yu, Hejie, Junshen Qu, Yang Liu, Huimin Yun, Xiangtong Li, Chunbao Zhou, Yajie Jin, Changfa Zhang, Jianjun Dai, and Xiaotao Bi. 2022. "Co-Pyrolysis of Biomass and Polyvinyl Chloride Under Microwave Irradiation: Distribution of Chlorine." *Science of the Total Environment* 806 (February): 150903. https://doi.org/10.1016/j.scitotenv.2021.150903.

Zhang, Le, Dingding Yao, To-Hung Tsui, Kai-Chee Loh, Chi-Hwa Wang, Yanjun Dai, and Yen Wah Tong. 2022. "Plastic-Containing Food Waste Conversion to Biomethane, Syngas, and Biochar via Anaerobic Digestion and Gasification: Focusing on Reactor Performance, Microbial Community Analysis, and Energy Balance Assessment." *Journal of Environmental Management* 306 (March): 114471. https://doi.org/10.1016/j.jenvman.2022.114471.

Zhang, Yaning, Liangliang Fan, Shiyu Liu, Nan Zhou, Kuan Ding, Peng Peng, Erik Anderson, et al. 2018. "Microwave-Assisted Co-Pyrolysis of Brown Coal and Corn Stover for Oil Production." *Bioresource Technology* 259 (July): 461–4. https://doi.org/10.1016/j.biortech.2018.03.078.

Zhao, Xiqiang, Wenlong Wang, Hongzhen Liu, Chunyuan Ma, and Zhanlong Song. 2014. "Microwave Pyrolysis of Wheat Straw: Product Distribution and Generation Mechanism." *Bioresource Technology* 158 (April): 278–85. https://doi.org/10.1016/j.biortech.2014.01.094.

27 Combustion and Pyrolysis Characteristics of Composite Fuels with Waste-derived and Low-grade Components

Galina S. Nyashina, Pavel A. Strizhak, and Ksenia Y. Vershinina

CONTENTS

27.1	Introduction	363
27.2	Materials	366
27.3	Experimental Setup and Methods	367
27.4	Results and Discussion	368
27.5	Conclusion	372
	Acknowledgments	372
	References	372

ABBREVIATIONS AND NOMENCLATURE

A^d	ash content of a dry sample
C^{daf}, H^{daf}, N^{daf}, O^{daf}	content of carbon, hydrogen, nitrogen, and oxygen of a dry ash-free sample, %
C^{avg}	average concentration of a substance in flue gas, mg/m³
m	mass, g
$Q^a_{s,V}$	higher heating value, MJ/kg
S^d_t	total sulfur content of a dry sample, %
t	time of gas concentration measurement, s
V^{daf}	volatile content of a dry ash-free sample, %
V_p	capacity of a gas analyzer pump, m³/s
W^a	moisture content, %

27.1 INTRODUCTION

The need to ensure the environmental protection and the continuity of the energy production are the basis for the search for alternative resources that reduce the consumption of fossil fuels. One such source is biomass, an organic material derived from plants, animals, or microorganisms (Gao et al. 2021). Biomass is a widely available resource with high-energy potential (Foong et al. 2020; W. Liu et al. 2020). In thermal power engineering, biomass can be used to produce heat, electricity, and transportation fuels (Collard and Blin 2014). The use of such raw materials can positively affect the economic development of rural areas, increasing the country's internal energy security due to the diversity of existing biomass resources and low levels of greenhouse gas emissions (Bhuiyan et al. 2018). The use of biomass for the production of biomaterials and chemicals (food additives, pharmaceuticals, surfactants, organic solvents, and fertilizers) is also quite common (Shao et al. 2021).

DOI: 10.1201/9781003334415-27

The following main categories of biomass can be distinguished: (1) waste from the woodworking industry (wood chips, sawdust, tree branches); (2) crops grown for energy use (rapeseed, *Jatropha*, eucalyptus, *Miscanthus*, sugarcane); (3) agricultural waste (sugarcane cake, nut shells (coconut, sunflower), corn husks, wheat straw, oil production waste (olive, rapeseed, sunflower waste), palm kernels; (4) municipal solid waste, animal husbandry, and food industry waste. It should be noted that the rapid urbanization and intensive anthropogenic activity lead to the formation of enormous volumes of these components that require systematic disposal. Table 27.1 presents data on the types of biomass utilized in different regions of the world.

TABLE 27.1
Characteristics of Biomass

Biomass Type	C	H	O	N	S	W^a	V^{daf}	C^{daf}	A^d	$Q^a_{s,V}$ (MJ/kg)	Reference
Woody biomass											
Wood pellet	52.3	6.8	40.7	0.16	–	6.7	84.3	15.7	0.8	20.8	(Riaza et al. 2019)
Soft wood	45.34 ± 0.13	5.86 ± 0.04	42.45 ± 0.04	0.58 ± 0.11	0.17 ± 0.07	5.15	–	–	5.60 ± 0.38	18.23 ± 0.13	(Zubkova et al. 2019)
Woody biomass	51.3	6.2	42	0.1	0.021	4.3	83.8	–	0.3	19.36	(Johansson et al. 2018)
Pine sawdust	50.3	6	42.99	0.69	–	6.09 ± 0.3	78.03 ± 0.2	12.16 ± 0.1	2.07 ± 0.03	18.44 ± 09	(Mishra and Mohanty 2018)
Herbaceous and agricultural biomass											
Palm leaves	40.76	5.55	52.14	1.32	0.24	12.03	58.17	15.41	14.4	18.9	(Makkawi et al. 2019)
Corncob	46.6	5.8	47.0	0.4	0.2	–	86.9	11.8	1.3	–	(Yi et al. 2019)
Olive waste	52.8	6.5	39.1	1.6	–	5.9	80.1	19.9	7.6	20.1	(Riaza et al. 2019)
Bagasse	46.4	6,7	45.8	0.7	0.4	–	87.4	9.7	2.9	–	(Yi et al. 2019)
Sugarcane Straw	43.79	5.16	38.90	0.29	–	7.32	74.86	13.27	4.55	17.81	(Pighinelli et al. 2014)
Wheat straw	40.58	4.84	53.84	0.74	0	5.19	64.24	15.60	14.97	–	(Akubo, Nahil, and Williams 2019)
Rice straw	42.66	5.68	37.37	1.03	0.44	1.51	69.09	18.09	11.31	–	(X. Liu et al. 2019)
Cotton stalk	46.8	6.4	46.8	0.3	0.2	8.9	71	16.6	3.5	19.2	(Makkawi et al. 2019)
Rice Husk	37.60	5.26	55.45	1.69	0	8.02	61.43	12.53	18.02	–	(Akubo, Nahil, and Williams 2019)
Husks and shells of nuts											
Sunflower husks	45.82 ± 0.08	6.32 ± 0.02	38.31 ± 0.08	2.61 ± 0.05	0.14 ± 0.02	6.1	–	–	6.81 ± 0.51	19.31 ± 0.13	(Zubkova et al. 2019)
Peanut shell	49.7	5.8	43.7	0.6	0.1	–	84.1	14.5	1.4	–	(Yi et al. 2019)
Palm kernel shell	48.82	5.68	45.08	0.42	–	13.65	75.32	20.81	3.87	14.88 (14.75)	(Maliutina et al. 2017)
Walnut shells	43.41 ± 0.17	5.66 ± 0.06	48.44 ± 0.08	1.98 ± 0.06	0.11 ± 0.03	4.1	–	–	0.41 ± 0.11	16.79 ± 0.08	(Zubkova et al. 2019)
Coconut shell	48.32	5.26	46.14	0.29	0	7.16	68.58	22.00	2.26	–	(Akubo, Nahil, and Williams 2019)
Other											
Cellulose	41.61	5.63	52.64	0.11	0	4.74	84.16	9.85	1.25	–	(Akubo, Nahil, and Williams 2019)
Natural rubber	83.63	11.97	2.71	1.58	0.12	1.71	89.98	4.71	3.60	45	(Ahmad, Abnisa, and Wan Daud 2018)
Brewer's spent grain	42.2	7.2	37.6	3.6	1.1	3.97	83.3	9.51	3.22	21.6	(Borel et al. 2018)
Microalgae	49.6	7.0	25.4	8.2	0.5	10	81	16	9	–	(Chang, Duan, and Xu 2015)

In the past, landfill and open burning were the two most common methods of biomass waste disposal. Despite their simplicity, these methods ultimately contribute to global warming and groundwater pollution (Lam et al. 2016; Omar and Rohani 2015). Thus, there is a need to apply more environment-friendly methods. They include thermochemical conversion methods: incineration, torrefaction, hydrothermal carbonization, gasification, and pyrolysis. Table 27.2 provides a summary of the methods listed.

TABLE 27.2
Characteristics of Thermochemical Methods for Biomass Conversion

Process and Its Description	Product	Applications	Features	Reference
Combustion is an oxidation process at a temperature of 200–600 °C	Thermal and electrical energy	Electricity generation with steam turbines, heat energy generation in wood stoves	Air pollution due to the emission of flue gases containing ash, CO, and NO_x	(Glushkov et al. 2020; Khan et al. 2009; Bhuiyan et al. 2018)
Torrefaction is a heating process without air access, temperature is of 200–300 °C and residence time is of 30–60 min	Torrefied biomass (low grade solid fuel)	Fuel for steam boilers and cooking stoves	Formation of a sticky tar (combustion chamber fouling) that contributes to a decrease in the thermal efficiency of the boiler due to the incomplete decomposition of lignin under moderate temperatures	(Sukiran et al. 2017; Szwaja et al. 2019; Wilk and Magdziarz 2017)
Hydrothermal carbonization is a process that takes place in an aqueous medium at a temperature of 180–260 °C in a closed system at pressures of 1–5 MPa and external heat supply	Gases (1–3%, mainly CO_2), aqueous chemicals (inorganic salts, sugars and organic acids), and solids (hydrochar/biochar/HTC char)	Solid fuel or raw material for pellet production, absorbent for water treatment, soil improvement and carbon sequestration	The method is applied to wet biomass (sewage sludge, food and agricultural waste, algae) and is characterized by faster high heat transfer compared to torrefaction	(Wilk et al. 2020; Mendoza Martinez et al. 2021; B. Zhang et al. 2018)
Gasification is a thermochemical method when biomass components are oxidized by gasifying agents (air, O_2, steam, and CO_2) at the temperature of 800–1600 °C	Generator gas (H_2 and CO) with a small proportion of CO_2 and CH_4	Energy production in gas turbines, internal combustion engines, and boilers	High process temperature, catalyst deactivation (with catalytic approach)	(Chai et al. 2020; Lv et al. 2022)
Pyrolysis is a thermal decomposition in an inert atmosphere at a temperature of 300–1000 °C	Solid biochar, liquid biooil, non-condensable generator gas	Production of chemicals, heat and electricity, catalysts, activated carbon, and high-quality complex fertilizer for the agricultural industry	Less NO_x and SO_x emissions, more value-added products, and less tar formation	(Shao et al. 2021; Foong et al. 2020)

Among the methods listed in Table in 27.2, incineration is the most widespread and relatively inexpensive way of converting biomass into heat and electricity. This method is considered relatively simple and well-established in practical implementation. At the same time, less costs are required when integrating it with various technologies (Bhuiyan et al. 2018). Biomass combustion characteristics depend on a group of factors: fuel mass flow, excess air ratio, combustion method, and temperature in the combustion chamber. Gaseous and solid products of biomass thermal decomposition consist of volatile substances (carbon dioxide, carbon monoxide, light hydrocarbons, hydrogen, and resins), moisture, char, and ash (Abuelnuor et al. 2014). In turn, biomass combustion is complicated by low calorific value, a significant amount of non-combustible pollutants, and high moisture content (Morris et al. 2018). Incomplete combustion of biomass in boilers can lead to the formation of agglomerates and ash and slag deposits on heating surfaces due to lack of oxygen. As a rule, within the framework of known combustion technologies, large volumes of work are carried out to prepare biomass (Sieradzka et al. 2020) before its direct combustion.

Pyrolysis is also widely used among the methods for converting biomass to biofuel. This method is considered promising (Zhou et al. 2020; Shao et al. 2021) due to the efficiency of energy conversion and the ability to use raw materials with different characteristics. Pyrolysis can be an independent technological process, but it is also the main stage of the chemical reaction during the gasification and combustion processes. As a result of pyrolysis, three main products are formed: char, liquid (tar and water), and non-condensable gases (mainly H_2, CO, CO_2, and CH_4) (Morris et al. 2018). Over the past decade, microwave pyrolysis has become widespread in the scientific community dealing with waste management (Foong et al. 2020; Zhou et al. 2020). This is due to the advantages of fast (increased heating rate and reduced reaction time) and efficient (deeper penetration of microwave energy for internal heat generation) heating and higher energy concentration for extracting products from biomass (Arshanitsa et al. 2016). Research on obtaining valuable products from biomass by microwave pyrolysis is carried out by scientists around the world. It has been proven that the yield and quality of products are influenced by numerous factors, such as the type of raw material and its characteristics (Huang, Kuan, and Chang 2018; Li et al. 2021), fuel pretreatment methods (Huang, Kuan, and Chang 2018), operating conditions (temperature, pressure, heating rate, reaction gas) (Mari Selvam and Paramasivan 2021; Lin et al. 2022), and the presence of catalysts and absorbers (R. Zhang et al. 2016). Co-firing of biomass with other types of waste or low-grade fuels (Chen et al. 2019), reactor configuration, and steam condensation regimes (Lenz and Ortwein 2017; Lin et al. 2022) are also significant factors.

Thus, in order to obtain valuable biofuels, it is required to determine an effective approach to biomass conversion, considering the characteristics of each process. Therefore, the purpose of this work is to conduct a comparative analysis of the characteristics (emission concentrations and main components of generator gas) of the processes of direct combustion and pyrolysis (conventional and microwave) of biomass.

27.2 MATERIALS

In this study, three types of biomass and a mixture based on them were considered: sawdust of pine trees; leaves (birch, poplar); straw (hay); mixture 1 (sawdust 50%, straw 25%, leaves 25%); mixture 3 (sawdust 25%, straw 50%, leaves 25%). The presented types of biomass are among the most typical and widespread in most regions of the world (Fernandez et al. 2020; Bunma and Kuchonthara 2018). The use of these components is possible in the presence of accumulated and annually generated volumes of waste in the region. Also, these types of biomass can be quite easily replaced by materials with similar properties. For example, interchangeable components are different types of sawdust (pine, birch, and oak), forest combustible materials (leaves of different types of trees, needles, and grass), and husks of different crops (rice, wheat, and cotton). Table 27.3 provides the main properties of the components used.

TABLE 27.3
Proximate and Ultimate Analysis of the Components

Material	W^a, %	A^d, %	V^{daf}, %	$Q^a_{s,V}$, MJ/kg	C^{daf}, %	H^{daf}, %	N^{daf}, %	S_t^d, %	O^{daf}, %
Leaves	6.95	6.25	76.85	17.05	49.91	5.92	0.86	0.09	43.22
Straw	7.00	2.80	78.50	17.70	50.20	6.36	1.09	0.02	42.33
Sawdust	7.00	1.60	83.40	18.10	52.50	6.58	0.22	0.02	40.68

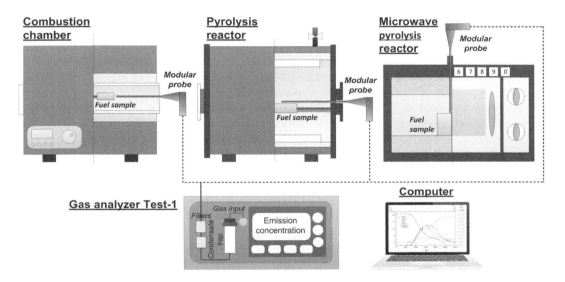

FIGURE 27.1 Schemes of experimental setups for combustion and pyrolysis of biomass and biomass blends.

Preparation of biomass samples consisted of several stages. The components were dried at a temperature of about 20 °C and a humidity of about 7% for 3 days, then they were ground using a Pulverisette 14 rotary mill. After grinding, the biomass was sifted through a sieve with a mesh size of not more than 800 μm. Then, the crushed biomass was distributed in a thin layer over the surface and kept at room conditions for another 3 days to balance with atmospheric moisture. Additional drying of the biomass at high temperatures was not carried out.

27.3 EXPERIMENTAL SETUP AND METHODS

Figure 27.1 presents the schemes of experimental setups used for combustion and pyrolysis of biomass and biomass blends.

Biomass combustion was carried out in an R 50/250/13 muffle tube furnace (Nabertherm GmbH, Lilienthal, Germany) with electric heating (controlled temperature range is 20–1100 °C). To create a high-temperature oxidizing medium, the temperature in the furnace was set to 800 °C. An integrated regulator and an S-type thermocouple (maximum operating temperature is 1350 °C, error ±1 °C) were used to set and control the required heating temperature. The choice of temperature is due to the need for complete combustion of the biomass and the production of a sufficient amount of gas to maintain stable operation of the gas analyzer. After the furnace was heated to a predetermined temperature, the fuel was placed in the central part of the furnace using a coordinate mechanism. In the combustion experiments, the weight of the fuel sample was 0.5 g. The gas sample released during the combustion of biomass was taken by the gas analyzer sampling device. The holes for introducing the fuel sample and the gas analyzer probe were closed with a dense layer of insulation to limit the supply of excess oxygen. The probe was connected to the Test-1 gas analyzer (Bonair-VT, Novosibirsk, Russia), where electrochemical (O_2, NO_x, SO_2) and optical (CO_2, CO, CH_4, H_2) gas sensors were installed. The measurement results were displayed on the computer screen in real time, considering the response speed of the sensors.

Pyrolysis under mixed heat transfer conditions was carried out in a laboratory heating furnace (Figure 27.1) made of stainless steel. To heat the internal space, four ceramic heating panels based on a nichrome heating element were used. The fuel was placed in the furnace at room temperature, after which the chamber was heated to 600 °C (average heating rate was 12 °C/min). The chosen conditions modeled the process of slow pyrolysis and were determined by the design and technical characteristics of the pyrolysis reactor. The input of the fuel sample was carried out through a hole located on the side wall. After placing the fuel in the furnace, the hole was hermetically sealed with a damper, into which the selective device of the gas analyzer was inserted. The mass of the fuel was 3 g. The composition of the flue gases was recorded by the gas analyzer.

Microwave pyrolysis of biomass and their mixture was carried out using a microwave oven (Figure 27.1) with a heater power of 800 W. To determine the composition of the pyrolysis gas, the gas analyzer probe was introduced through a hole in the upper wall of the microwave oven. The fuel sample was placed in a specialized substrate. The weight of the fuel was 6 g. To intensify heat transfer, the biomass was preliminarily wetted with a small amount of moisture. During microwave heating, moisture is involved in heat transfer. It contributes to the heating of the medium in the vicinity of the fuel sample and the transfer of thermal energy directly to the fuel. As a result, the thermal decomposition of biomass is intensified, and the gas yield increased. It has been experimentally established that for the selected biomass mass (6 g), the optimal addition of water is 4 g. After placing the fuel in the microwave oven, the door was hermetically sealed. Next, a magnetron and a gas analyzer were turned on to record the gas composition of microwave pyrolysis products.

After each experiment, a mass of the coke residue was measured. The interior space of all three furnaces (Figure 27.1) was ventilated with atmospheric air to remove residual flue gases. The gas analyzer was purged with compressed atmospheric air to clean the system elements from flue gases and resins.

In each type of the study (combustion or pyrolysis), a set of experiments was carried out, including five to eight repetitions. Then the experimental results were averaged, confidence intervals were determined, and gross and random errors were excluded. The results were averaged by integrating the time trend of the gas concentration. Also, when calculating the average value of a gas concentration, the mass of the fuel was considered since for all three types of experiments (combustion or pyrolysis), a sample mass was different. This fact is due to the features and characteristics of the setups and equipment used. Thus, the mass concentration of gas was calculated by the following formula (by the example of CO_2):

$$CO_2 = C^{avrg} \cdot V_p \cdot t / m$$

where CO_2 is the mass concentration of emissions relative to the initial mass of fuel, mg/g; C^{avrg} is the average concentration of a substance in flue gases, mg/m³; V_p is the capacity of gas analyzer pump, m³/s; t is the measurement time, s; and m is the mass of biomass sample, g.

27.4 RESULTS AND DISCUSSION

Figure 27.2 illustrates the mass concentrations of the main gases produced during direct combustion and pyrolysis of biomass.

As expected, CO_2 was the main gaseous product of biomass combustion. Its concentrations were 54–94% higher than during pyrolysis. Among the types of biomass considered, sawdust combustion

Combustion and Pyrolysis Characteristics of Composite Fuels

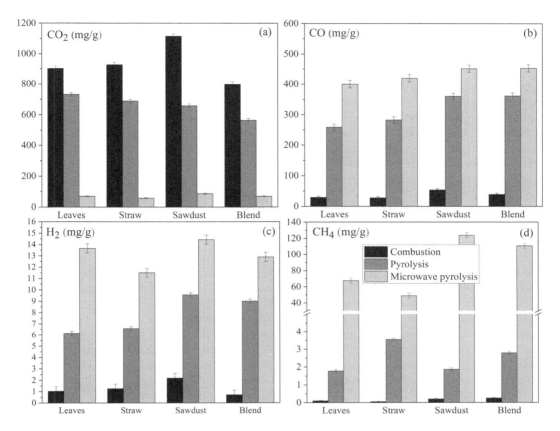

FIGURE 27.2 Average mass concentrations of the main gases formed during the combustion and pyrolysis of various types of biomass.

was characterized by the highest CO_2 emissions. This can be explained by the high content of carbon and volatiles in combination with low ash content of sawdust. The intense CO_2 release may indicate a high degree of the sample burnout and the involvement of a larger amount of carbon in the combustion reactions.

Low concentrations of CO confirm that the main part of the fuel carbon during combustion is transformed into CO_2 (Figure 27.2b). High concentrations of CO in the combustion chamber may indicate incomplete combustion of the fuel, insufficient mixing of fuel with air, a short residence time of particles in the combustion zone, or low temperature. In the experiments performed, the range of CO variation during the combustion of certain types of biomass and a mixture based on it was 30–54 mg/g, which is a rather low value and indicates good biomass burnout characteristics. When burning biomass, low concentrations of H_2 and CH_4 were also recorded, which indicates a rapid and complete burnout of the fuel.

The formation of gaseous pyrolysis products is a consequence of tar cracking, char decomposition, and reactions between particles formed during pyrolysis. CO and CO_2 were the main gaseous products formed during slow pyrolysis. High concentrations of CO_2 and CO in pyrolysis gases are mainly associated with a high degree of deoxygenation that occurs during the decarbonylation and decarboxylation reactions. However, compared to slow pyrolysis, microwave pyrolysis was characterized by a high content of combustible gases (CO, H_2, and CH_4) and a low CO_2 yield. The result obtained can be explained by a number of advantages of this type of thermochemical conversion. Microwave heating has advantages over conventional thermal heating due to the direct effect of microwave radiation on fuel particles, which leads to a rapid supply of heat and subsequent fuel decomposition (Lim et al. 2022). Intensive volumetric heating of biomass is achieved by converting electromagnetic energy into thermal energy. Unlike microwave heating, the heating process in a conventional bed reactor is non-selective, and heat is radiated throughout the chamber volume without concentrating on the fuel. In the case of microwave pyrolysis, the metal wall of the reactor reflects the microwaves and directs them to the fuel, which absorbs them. The pyrolysis heating time is shortened, and heat transfer is faster than with conventional heating. Also, in microwave pyrolysis technologies, specialized microwave absorbers (for example, biochar, activated carbon, silicon carbide, and water) are often used (Ge et al. 2021). These additives are mixed with a fuel in order to additionally accumulate thermal energy on the fuel surface. In this study, water was used as a microwave absorber. Thus, water made it possible to concentrate microwave radiation on the surface of the biomass. In addition, water vapor intensified the biomass decomposition with the release of more H_2 and CO due to the following reaction:

$$C + H_2O \leftrightarrow H_2 + CO \tag{27.1}$$

The distribution of pyrolysis gas products also largely depends on the biomass composition. As a result of thermal decomposition of the main components (cellulose, hemicellulose, and lignin), the main part of the gas mixture is formed. Yang et al. (Yang et al. 2007) concluded that the decomposition of hemicellulose with a higher carboxyl content is accompanied by a high CO_2 yield. During thermal cracking of carbonyls and carboxyls, a significant amount of CO is released from cellulose. Due to the high content of aromatic ring and methoxyl functional groups, lignin pyrolysis produces much more H_2 and CH_4. The content of the main components of the biomass is presented in Table 27.4 (results from Dhyani and Bhaskar (2018) are given).

Table 27.4 shows that the leaves contain significantly more hemicellulose than the other considered biomass. Decomposition of hemicellulose may cause average carbon dioxide concentrations to be higher for leaves. However, since leaves contain less cellulose than other biomass, CO concentrations were lower for them (Figure 27.2b). Pyrolysis of the sawdust resulted in a higher H_2 yield (Figure 27.2c), which may be due to the high proportions of lignin and hydrogen in the sawdust. CH_4 concentration in the case of microwave pyrolysis (Figure 27.2d) of sawdust and a mixture based on them (the sawdust content was 50%) was maximum, which may be due to the active decomposition of lignin under the influence of microwave radiation. In general, the generator gas had the best characteristics when sawdust and mixtures with a high proportion of sawdust (50%) were used for pyrolysis. Compared to other types of biomass, the concentrations of CO, H_2, and CH_4 increased to 13%, 26%, and 63%, respectively.

According to the data obtained, biomass combustion was characterized by the highest emissions of nitrogen and sulfur oxides (Figure 27.3). High temperatures and the presence of a sufficient amount of oxygen in the combustion zone contributed to the oxidation reactions of nitrogenous and sulfurous substances contained in the fuel. The main nitrogen-containing emissions from biomass combustion are NO, HCN, NH_3, and N_2O. Most of the NO (65–75%) is formed at the stage of fuel devolatilization. The combustion of the coke part is of secondary importance in the conversion of

TABLE 27.4
Fiber Analysis (Dhyani and Bhaskar 2018)

Component	Cellulose, %	Hemicellulose, %	Lignin, %
Sawdust	32.63	37.23	22.16
Leaves	15.5	80.5	4
Straw	36.70	34.4	28.90

Combustion and Pyrolysis Characteristics of Composite Fuels

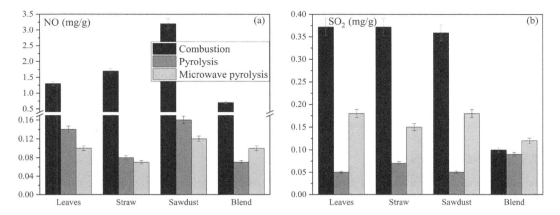

FIGURE 27.3 Average mass concentrations of the main anthropogenic gases released during the combustion and pyrolysis of various types of biomass.

fuel nitrogen, which, as a rule, produces NO. The maximum concentrations of NO were recorded during the combustion of sawdust. NO emissions for this type of biomass were 48% and 58% higher than for straw and leaves. The result obtained is not obvious since the nitrogen content of sawdust is minimal compared to other types of biomass (Table 27.3). As a rule, NH_3 formed during the biomass devolatilization is oxidized to NO according to the following reaction:

$$NH_3 + 5/4 O_2 \rightarrow NO + 3/2 H_2O \quad (27.2)$$

However, a portion of NH_3 (and related radicals, e.g., NH_2) can also react with NO as reducing agents based on the following reaction:

$$2/3 NH_3 + NO \rightarrow 5/6 N_2 + 3/2 H_2O \quad (27.3)$$

There is a hypothesis that the high content of nitrogen in the straw and leaves contributed to the formation of more NH_3 and HCN, which further interacted with NO (according to Equation 27.3) with the formation of free nitrogen. Thus, the conversion of NO to N_2 increased, resulting in a decrease in NO concentrations. Similar dependences were obtained by Winter et al. (Winter, Wartha, and Hofbauer 1999). For Alder wood ($N = 0.26\%$), the conversion to NO was 19.4%, while it was only 6.2% for malt waste ($N = 4.99\%$) (Winter, Wartha, and Hofbauer 1999).

Sufficiently low ash content of sawdust in combination with a high proportion of volatiles stimulate ignition processes and complete burnout of sawdust. This affects the reactivity of the material and the temperatures in the combustion zone. Higher temperatures also intensified the release of NO.

SO_2 emission during the combustion of the considered types of biomass varies in a very narrow range and was approximately 0.37 mg/g. When coal was burned, SO_2 mass concentrations vary in the range from 8 mg/g to 15 mg/g (Rokni et al. 2016; Chakraborty et al. 2008; Nyashina et al. 2022), which is many times higher than emissions for all types of biomass.

A comparison of the results showed that during biomass pyrolysis (slow and microwave), the concentrations of NO and SO_2 were 2–20 times lower than during combustion. This result is largely due to low temperatures and lack of oxygen in the reaction zone. Low emission of nitrogen and sulfur oxides during the pyrolysis is an important advantage of this technology in comparison with direct combustion.

The comparison of the results showed that the use of the fuel mixture made it possible to reduce NO and SO_2 emissions by 10–77% compared to the combustion and pyrolysis of individual components. Such an effect can be explained by synergistic interactions occurring during the interaction of different types of biomass particles. Despite the fact that the characteristics of the components (Table 27.3) are very close to each other, the structure and chemical composition of the samples is of great importance. The internal structure of all components is significantly different. As a consequence, when mixing three different types of biomass, chemical interactions between them are possible, which directly contributed to a decrease in the concentrations of NO and SO_2 in the reaction products. For example, CaO, MgO, and Fe_2O_3 in fuel ash can lead to the formation of an active layer that can catalyze NO reduction and retain sulfur particles in the ash (Tian et al. 2017; Werther et al. 2000; Ohtsuka, Zhiheng, and Furimsky 1997).

27.5 CONCLUSION

The composition of gas products of direct combustion and pyrolysis of biomass has been compared. We established that the direct combustion of biomass produces significantly higher concentrations of anthropogenic gases than pyrolysis.

A comparison of different methods of thermochemical conversion of biomass showed that during pyrolysis (slow and microwave), the concentrations of CO_2 are lower by 54–94%, and NO and SO_2 are many times lower (2–20 times) than during combustion. It has also been found that microwave pyrolysis provides better gas quality than slow pyrolysis. Compared to slow pyrolysis, microwave pyrolysis is characterized by a high content of combustible gases (CO, H_2, and CH_4) and low concentrations of CO_2. The analysis of the obtained results showed that during the microwave pyrolysis of sawdust and mixed biomass, the concentrations of CO, H_2, and CH_4 increased to 63%. The result obtained is due to the interaction of various biomass components and their pyrolysis products.

ACKNOWLEDGMENTS

The reported study was funded by the Russian Foundation for Basic Research, National Council of Brazil for Scientific and Technological Development, and Ministry of Science & Technology (Government of India) according to the research project No. 19–53–80019.

REFERENCES

Abuelnuor, A.A.A., M.A. Wahid, Seyed Ehsan Hosseini, A. Saat, Khalid M. Saqr, Hani H. Sait, and M. Osman. 2014. "Characteristics of Biomass in Flameless Combustion: A Review." *Renewable and Sustainable Energy Reviews* 33 (May): 363–70. https://doi.org/10.1016/j.rser.2014.01.079.

Ahmad, Nabeel, Faisal Abnisa, and Wan Mohd Ashri Wan Daud. 2018. "Liquefaction of Natural Rubber to Liquid Fuels via Hydrous Pyrolysis." *Fuel* 218 (April): 227–35. https://doi.org/10.1016/J.FUEL.2017.12.117.

Akubo, Kaltume, Mohamad Anas Nahil, and Paul T. Williams. 2019. "Pyrolysis-Catalytic Steam Reforming of Agricultural Biomass Wastes and Biomass Components for Production of Hydrogen/Syngas." *Journal of the Energy Institute* 92 (6): 1987–96. https://doi.org/10.1016/j.joei.2018.10.013.

Arshanitsa, Alexandr, Yegor Akishin, Edmund Zile, Tatiana Dizhbite, Valentin Solodovnik, and Galina Telysheva. 2016. "Microwave Treatment Combined with Conventional Heating of Plant Biomass Pellets in a Rotated Reactor as a High Rate Process for Solid Biofuel Manufacture." *Renewable Energy* 91 (June): 386–96. https://doi.org/10.1016/j.renene.2016.01.080.

Bhuiyan, Arafat A., Aaron S. Blicblau, A.K.M. Sadrul Islam, and Jamal Naser. 2018. "A Review on Thermo-Chemical Characteristics of Coal/Biomass Co-Firing in Industrial Furnace." *Journal of the Energy Institute* 91 (1): 1–18. https://doi.org/10.1016/j.joei.2016.10.006.

Borel, Lidja D.M.S., Taísa S. Lira, Jânio A. Ribeiro, Carlos H. Ataíde, and Marcos A.S. Barrozo. 2018. "Pyrolysis of Brewer's Spent Grain: Kinetic Study and Products Identification." *Industrial Crops and Products* 121 (October): 388–95. https://doi.org/10.1016/J.INDCROP.2018.05.051.

Bunma, Teerayut, and Prapan Kuchonthara. 2018. "Synergistic Study Between CaO and MgO Sorbents for Hydrogen Rich Gas Production from the Pyrolysis-Gasification of Sugarcane Leaves." *Process Safety and Environmental Protection* 118 (August): 188–94. https://doi.org/10.1016/j.psep.2018.06.034.

Chai, Yue, Meihong Wang, Ningbo Gao, Yihang Duan, and Jiaqi Li. 2020. "Experimental Study on Pyrolysis/Gasification of Biomass and Plastics for H2 Production under New Dual-Support Catalyst." *Chemical Engineering Journal* 396 (September): 125260. https://doi.org/10.1016/J.CEJ.2020.125260.

Chakraborty, N., I. Mukherjee, A. K. Santra, S. Chowdhury, S. Chakraborty, S. Bhattacharya, A. P. Mitra, and C. Sharma. 2008. "Measurement of CO2, CO, SO2, and NO Emissions from Coal-Based Thermal Power Plants in India." *Atmospheric Environment* 42 (6): 1073–82. https://doi.org/10.1016/j.atmosenv.2007.10.074.

Chang, Zhoufan, Peigao Duan, and Yuping Xu. 2015. "Catalytic Hydropyrolysis of Microalgae: Influence of Operating Variables on the Formation and Composition of Bio-Oil." *Bioresource Technology* 184 (May): 349–54. https://doi.org/10.1016/j.biortech.2014.08.014.

Chen, Xiye, Li Liu, Linyao Zhang, Yan Zhao, and Penghua Qiu. 2019. "Pyrolysis Characteristics and Kinetics of Coal–Biomass Blends during Co-Pyrolysis." *Energy & Fuels* 33 (2): 1267–78. https://doi.org/10.1021/acs.energyfuels.8b03987.

Collard, François Xavier, and Joël Blin. 2014. "A Review on Pyrolysis of Biomass Constituents: Mechanisms and Composition of the Products Obtained from the Conversion of Cellulose, Hemicelluloses and Lignin." *Renewable and Sustainable Energy Reviews* 38 (October): 594–608. https://doi.org/10.1016/J.RSER.2014.06.013.

Dhyani, Vaibhav, and Thallada Bhaskar. 2018. "A Comprehensive Review on the Pyrolysis of Lignocellulosic Biomass." *Renewable Energy* 129 (December): 695–716. https://doi.org/10.1016/J.RENENE.2017.04.035.

Fernandez, Enara, Maider Amutio, Maite Artetxe, Aitor Arregi, Laura Santamaria, Gartzen Lopez, Javier Bilbao, and Martin Olazar. 2020. "Assessment of Product Yields and Catalyst Deactivation in Fixed and Fluidized Bed Reactors in the Steam Reforming of Biomass Pyrolysis Volatiles." *Process Safety and Environmental Protection* 145 (August): 52–62. https://doi.org/10.1016/j.psep.2020.07.039.

Foong, Shin Ying, Rock Keey Liew, Yafeng Yang, Yoke Wang Cheng, Peter Nai Yuh Yek, Wan Adibah Wan Mahari, Xie Yi Lee, et al. 2020. "Valorization of Biomass Waste to Engineered Activated Biochar by Microwave Pyrolysis: Progress, Challenges, and Future Directions." *Chemical Engineering Journal* 389 (June): 124401. https://doi.org/10.1016/j.cej.2020.124401.

Gao, Ningbo, Maciej Śliz, Cui Quan, Artur Bieniek, and Aneta Magdziarz. 2021. "Biomass CO2 Gasification with CaO Looping for Syngas Production in a Fixed-Bed Reactor." *Renewable Energy* 167 (April): 652–61. https://doi.org/10.1016/J.RENENE.2020.11.134.

Ge, Shengbo, Peter Nai Yuh Yek, Yoke Wang Cheng, Changlei Xia, Wan Adibah Wan Mahari, Rock Keey Liew, Wanxi Peng, et al. 2021. "Progress in Microwave Pyrolysis Conversion of Agricultural Waste to Value-Added Biofuels: A Batch to Continuous Approach." *Renewable and Sustainable Energy Reviews* 135 (January): 110148. https://doi.org/10.1016/J.RSER.2020.110148.

Glushkov, Dmitrii, Galina Nyashina, Valeriy Medvedev, and Kseniya Vershinina. 2020. "Relative Environmental, Economic, and Energy Performance Indicators of Fuel Compositions with Biomass." *Applied Sciences* 10 (6): 2092. https://doi.org/10.3390/app10062092.

Huang, Yu Fong, Wen Hui Kuan, and Chun Yuan Chang. 2018. "Effects of Particle Size, Pretreatment, and Catalysis on Microwave Pyrolysis of Corn Stover." *Energy* 143 (January): 696–703. https://doi.org/10.1016/J.ENERGY.2017.11.022.

Johansson, Ann Christine, Linda Sandström, Olov G.W. Öhrman, and Henrik Jilvero. 2018. "Co-Pyrolysis of Woody Biomass and Plastic Waste in Both Analytical and Pilot Scale." *Journal of Analytical and Applied Pyrolysis* 134 (September): 102–13. https://doi.org/10.1016/j.jaap.2018.05.015.

Khan, A. A., W. de Jong, P. J. Jansens, and H. Spliethoff. 2009. "Biomass Combustion in Fluidized Bed Boilers: Potential Problems and Remedies." *Fuel Processing Technology* 90 (1): 21–50. https://doi.org/10.1016/j.fuproc.2008.07.012.

Lam, Su Shiung, Rock Keey Liew, Yee Mun Wong, Elfina Azwar, Ahmad Jusoh, and Rafeah Wahi. 2016. "Activated Carbon for Catalyst Support from Microwave Pyrolysis of Orange Peel." *Waste and Biomass Valorization* 8 (6): 2109–19. https://doi.org/10.1007/S12649-016-9804-X.

Lenz, Volker, and Andreas Ortwein. 2017. "SmartBiomassHeat – Heat from Solid Biofuels as an Integral Part of a Future Energy System Based on Renewables." *Chemical Engineering and Technology* 40 (2): 313–22. https://doi.org/10.1002/CEAT.201600188.

Li, Jian, Huanbo Liu, Liguo Jiao, Beibei Yan, Zhiyu Li, Xiaoshan Dong, and Guanyi Chen. 2021. "Microwave Pyrolysis of Herb Residue for Syngas Production with In-Situ Tar Elimination and Nitrous Oxides Controlling." *Fuel Processing Technology* 221 (October): 106955. https://doi.org/10.1016/J.FUPROC.2021.106955.

Lim, Xin Yi, Peter Nai Yuh Yek, Rock Keey Liew, Meng Choung Chiong, Wan Adibah Wan Mahari, Wanxi Peng, Cheng Tung Chong, et al. 2022. "Engineered Biochar Produced through Microwave Pyrolysis as a Fuel Additive in Biodiesel Combustion." *Fuel* 312 (March): 122839. https://doi.org/10.1016/J.FUEL.2021.122839.

Lin, Junhao, Chongwei Cui, Shichang Sun, Donghua Xu, Rui Ma, Mingliang Wang, Lin Fang, and Biqin Dong. 2022. "Enhanced Combined H2O/CO2 Reforming of Sludge to Produce High-Quality Syngas via Spiral Continuous Microwave Pyrolysis Technology: CO2 Reforming Mechanism and H2/CO Directional Regulation." *Chemical Engineering Journal* 434 (April): 134628. https://doi.org/10.1016/J.CEJ.2022.134628.

Liu, Weiguo, Junming Xu, Xinfeng Xie, Yan Yan, Xiaolu Zhou, and Changhui Peng. 2020. "A New Integrated Framework to Estimate the Climate Change Impacts of Biomass Utilization for Biofuel in Life Cycle Assessment." *Journal of Cleaner Production* 267 (September): 122061. https://doi.org/10.1016/j.jclepro.2020.122061.

Liu, Xiaorui, Zhongyang Luo, Chunjiang Yu, and Guilin Xie. 2019. "Conversion Mechanism of Fuel-N during Pyrolysis of Biomass Wastes." *Fuel* 246 (June): 42–50. https://doi.org/10.1016/j.fuel.2019.02.042.

Lv, Peng, Yonghui Bai, Jiaofei Wang, Xudong Song, Weiguang Su, Guangsuo Yu, and Yuan Ma. 2022. "Investigation into the Interaction of Biomass Waste with Industrial Solid Waste during Co-Pyrolysis and the Synergetic Effect of Its Char Gasification." *Biomass and Bioenergy* 159 (April): 106414. https://doi.org/10.1016/J.BIOMBIOE.2022.106414.

Makkawi, Yassir, Yehya El Sayed, Mubarak Salih, Paul Nancarrow, Scott Banks, and Tony Bridgwater. 2019. "Fast Pyrolysis of Date Palm (Phoenix Dactylifera) Waste in a Bubbling Fluidized Bed Reactor." *Renewable Energy* 143 (December): 719–30. https://doi.org/10.1016/J.RENENE.2019.05.028.

Maliutina, Kristina, Arash Tahmasebi, Jianglong Yu, and Sergey N. Saltykov. 2017. "Comparative Study on Flash Pyrolysis Characteristics of Microalgal and Lignocellulosic Biomass in Entrained-Flow Reactor." *Energy Conversion and Management* 151 (November): 426–38. https://doi.org/10.1016/J.ENCONMAN.2017.09.013.

Mendoza Martinez, Clara Lisseth, Ekaterina Sermyagina, Jussi Saari, Marcia Silva de Jesus, Marcelo Cardoso, Gustavo Matheus de Almeida, and Esa Vakkilainen. 2021. "Hydrothermal Carbonization of Lignocellulosic Agro-Forest Based Biomass Residues." *Biomass and Bioenergy* 147 (April): 106004. https://doi.org/10.1016/j.biombioe.2021.106004.

Mishra, Ranjeet Kumar, and Kaustubha Mohanty. 2018. "Pyrolysis Kinetics and Thermal Behavior of Waste Sawdust Biomass Using Thermogravimetric Analysis." *Bioresource Technology* 251 (March): 63–74. https://doi.org/10.1016/j.biortech.2017.12.029.

Morris, Jonathan D., Syed Sheraz Daood, Stephen Chilton, and William Nimmo. 2018. "Mechanisms and Mitigation of Agglomeration during Fluidized Bed Combustion of Biomass: A Review." *Fuel* 230 (October): 452–73. https://doi.org/10.1016/j.fuel.2018.04.098.

Nyashina, Galina, Vadim Dorokhov, Geniy Kuznetsov, and Pavel Strizhak. 2022. "Emissions from the Combustion of High-Potential Slurry Fuels." *Environmental Science and Pollution Research* 29 (25): 37989–5. https://doi.org/10.1007/s11356-021-17727-5.

Ohtsuka, Yasuo, Wu Zhiheng, and Edward Furimsky. 1997. "Effect of Alkali and Alkaline Earth Metals on Nitrogen Release during Temperature Programmed Pyrolysis of Coal." *Fuel* 76 (14–15): 1361–7. https://doi.org/10.1016/S0016-2361(97)00149-X.

Omar, Hecham, and Sohrab Rohani. 2015. "Treatment of Landfill Waste, Leachate and Landfill Gas: A Review." *Frontiers of Chemical Science and Engineering* 9 (1): 15–32. https://doi.org/10.1007/s11705-015-1501-y.

Pighinelli, Anna L.M.T., Akwasi A. Boateng, Charles A. Mullen, and Yaseen Elkasabi. 2014. "Evaluation of Brazilian Biomasses as Feedstocks for Fuel Production via Fast Pyrolysis." *Energy for Sustainable Development* 21 (August): 42–50. https://doi.org/10.1016/j.esd.2014.05.002.

Riaza, Juan, Patrick Mason, Jenny M. Jones, Jon Gibbins, and Hannah Chalmers. 2019. "High Temperature Volatile Yield and Nitrogen Partitioning During Pyrolysis of Coal and Biomass Fuels." *Fuel* 248 (July): 215–20. https://doi.org/10.1016/j.fuel.2019.03.075.

Rokni, Emad, Aidin Panahi, Xiaohan Ren, and Yiannis A. Levendis. 2016. "Curtailing the Generation of Sulfur Dioxide and Nitrogen Oxide Emissions by Blending and Oxy-Combustion of Coals." *Fuel* 181 (October): 772–84. https://doi.org/10.1016/J.FUEL.2016.05.023.

Mari Selvam, S., and Balasubramanian Paramasivan. 2021. "Evaluation of Influential Factors in Microwave Assisted Pyrolysis of Sugarcane Bagasse for Biochar Production." *Environmental Technology & Innovation* 24 (November): 101939. https://doi.org/10.1016/J.ETI.2021.101939.

Shao, Shanshan, Chengyue Liu, Xianliang Xiang, Xiaohua Li, Huiyan Zhang, Rui Xiao, and Yixi Cai. 2021. "In Situ Catalytic Fast Pyrolysis over CeO2 Catalyst: Impact of Biomass Source, Pyrolysis Temperature and Metal Ion." *Renewable Energy* 177 (November): 1372–81. https://doi.org/10.1016/J.RENENE.2021.06.054.

Sieradzka, Małgorzata, Ningbo Gao, Cui Quan, Agata Mlonka-Mędrala, and Aneta Magdziarz. 2020. "Biomass Thermochemical Conversion via Pyrolysis with Integrated CO2 Capture." *Energies* 13 (5): 1050. https://doi.org/10.3390/en13051050.

Sukiran, Mohamad Azri, Faisal Abnisa, Wan Mohd Ashri Wan Daud, Nasrin Abu Bakar, and Soh Kheang Loh. 2017. "A Review of Torrefaction of Oil Palm Solid Wastes for Biofuel Production." *Energy Conversion and Management* 149 (October): 101–20. https://doi.org/10.1016/j.enconman.2017.07.011.

Szwaja, Stanisław, Aneta Magdziarz, Monika Zajemska, and Anna Poskart. 2019. "A Torrefaction of Sida Hermaphrodita to Improve Fuel Properties. Advanced Analysis of Torrefied Products." *Renewable Energy* 141 (October): 894–902. https://doi.org/10.1016/J.RENENE.2019.04.055.

Tian, Xin, Kun Wang, Haibo Zhao, and Mingze Su. 2017. "Chemical Looping with Oxygen Uncoupling of High-Sulfur Coal Using Copper Ore as Oxygen Carrier." *Proceedings of the Combustion Institute* 36 (3): 3381–8. https://doi.org/10.1016/j.proci.2016.08.056.

Werther, J., M. Saenger, E.-U. Hartge, T. Ogada, and Z. Siagi. 2000. "Combustion of Agricultural Residues." *Progress in Energy and Combustion Science* 26 (1): 1–27. https://doi.org/10.1016/S0360-1285(99)00005-2.

Wilk, Małgorzata, and Aneta Magdziarz. 2017. "Hydrothermal Carbonization, Torrefaction and Slow Pyrolysis of Miscanthus Giganteus." *Energy* 140 (December): 1292–304. https://doi.org/10.1016/J.ENERGY.2017.03.031.

Wilk, Małgorzata, Aneta Magdziarz, Izabela Kalemba-Rec, and Monika Szymańska-Chargot. 2020. "Upgrading of Green Waste into Carbon-Rich Solid Biofuel by Hydrothermal Carbonization: The Effect of Process Parameters on Hydrochar Derived from Acacia." *Energy* 202 (July): 117717. https://doi.org/10.1016/J.ENERGY.2020.117717.

Winter, F., C. Wartha, and H. Hofbauer. 1999. "NO and N2O Formation during the Combustion of Wood, Straw, Malt Waste and Peat." *Bioresource Technology* 70 (1): 39–49. https://doi.org/10.1016/S0960-8524(99)00019-X.

Yang, Haiping, Rong Yan, Hanping Chen, Dong Ho Lee, and Chuguang Zheng. 2007. "Characteristics of Hemicellulose, Cellulose and Lignin Pyrolysis." *Fuel* 86 (12–13): 1781–8. https://doi.org/10.1016/j.fuel.2006.12.013.

Yi, Linlin, Huan Liu, Sihan Li, Meiyong Li, Geyi Wang, Gaozhi Man, and Hong Yao. 2019. "Catalytic Pyrolysis of Biomass Wastes over Org-CaO/Nano-ZSM-5 to Produce Aromatics: Influence of Catalyst Properties." *Bioresource Technology* 294 (December): 122186. https://doi.org/10.1016/j.biortech.2019.122186.

Zhang, Bide, Mohammad Heidari, Bharat Regmi, Shakirudeen Salaudeen, Precious Arku, Mahendra Thimmannagari, and Animesh Dutta. 2018. "Hydrothermal Carbonization of Fruit Wastes: A Promising Technique for Generating Hydrochar." *Energies* 11 (8): 2022. https://doi.org/10.3390/en11082022.

Zhang, Rui, Linling Li, Dongmei Tong, and Changwei Hu. 2016. "Microwave-Enhanced Pyrolysis of Natural Algae from Water Blooms." *Bioresource Technology* 212 (July): 311–17. https://doi.org/10.1016/J.BIORTECH.2016.04.053.

Zhou, Nan, Junwen Zhou, Leilei Dai, Feiqiang Guo, Yunpu Wang, Hui Li, Wenyi Deng, et al. 2020. "Syngas Production from Biomass Pyrolysis in a Continuous Microwave Assisted Pyrolysis System." *Bioresource Technology* 314 (October): 123756. https://doi.org/10.1016/J.BIORTECH.2020.123756.

Zubkova, V., A. Strojwas, M. Bielecki, L. Kieush, and A. Koverya. 2019. "Comparative Study of Pyrolytic Behavior of the Biomass Wastes Originating in the Ukraine and Potential Application of Such Biomass. Part 1. Analysis of the Course of Pyrolysis Process and the Composition of Formed Products." *Fuel* 254 (October): 115688. https://doi.org/10.1016/j.fuel.2019.115688.

28 Analysis of Gaseous Anthropogenic Emissions from Coal and Slurry Fuel Combustion and Pyrolysis

Mark R. Akhmetshin, Galina S. Nyashina, and Pavel A. Strizhak

CONTENTS

28.1 Introduction 377
28.2 Materials 380
28.3 Experimental Setup and Methods 380
28.4 Results and Discussion 381
    28.4.1 Thermogravimetric Analysis 381
    28.4.2 Flue Gas Analysis 383
28.5 Conclusion 385
Acknowledgments 385
References 386

ABBREVIATIONS AND NOMENCLATURE

CWS	Coal-water slurry
A^d	Ash content to a dry basis, %
C^{daf}, H^{daf}, N^{daf}, O^{daf}	Fraction of carbon, hydrogen, nitrogen, and oxygen in coal converted to a dry ash-free state, %
$Q^a_{s,V}$	higher heat value, MJ/kg
R_{max}	maximum weight loss rate, %/min
S_t^d	sulfur content to a dry basis, %
T_b	burnout temperature, °C
T_g	combustion temperature, °C
T_{ign}	ignition temperature, °C
T_{max}	temperature corresponding to the maximum weight loss peak, °C
W^a	moisture content, %
V^{daf}	volatile content to a dry ash-free state, %

28.1 INTRODUCTION

According to statistics (International Energy Agency 2020), about 30% of the world's energy production is produced by burning coal fuels. This process is invariably accompanied by the anthropogenic gas emissions formation. About 55% of sulfur oxide emissions, 25% of nitrogen oxide emissions, and up to 45% of carbon dioxide emissions from the global level come from hard coal combustion. These pollutants directly affect the state of the biosphere and human health. Sulfur and nitrogen oxides, entering the atmosphere, cause a number of cardiovascular and respiratory diseases, and can also contribute to irritation of mucous membranes (Gent et al. 2003; Wang et al. 2021; Munawer 2018). In addition, an increase in the amount of carbon dioxide in the atmosphere creates an excess of greenhouse gases that trap additional heat. This trapping heat causes glaciers to melt and ocean levels to rise.

In this regard, more and more research are directed to the development of environmentally efficient technologies for generating energy using coal. One of the options for solving this problem is the replacement of pulverized coal fuel with coal-water slurries (CWS). Such fuels are a suspended mixture of solid particles and water, the content can vary in the range of 30–60%. Coals of various ranks, coal sludge, and filter cakes (Li et al. 2018; Nyashina, Kuznetsov, and Strizhak 2020; Armesto et al. 2003; Staroń et al. 2019); various types of biomass (straw, sawdust, sewage sludge) (Guo and Zhong 2018; Vershinina, Shlegel, and Strizhak 2019; Olsen et al. 2020); and municipal solid wastes (cardboard, paper) (Xiu et al. 2018) can be used as solid fuel components. Moreover, in order to achieve the required energy, environmental or operational characteristics, various liquid additives, for example, used technical and cooking oils, can be added to the coal-water slurries composition (Gaber et al. 2020; Z. Zhao, Wang, Ge, et al. 2019).

A number of works have shown that such composite fuels can significantly reduce the level of anthropogenic emissions generated in the energy sector (Akhmetshin, Nyashina, and Strizhak 2020; Z. Zhao, Wang, Wu, et al. 2019; Staroń et al. 2019). The combustion of water-containing fuels based on coal and coal processing wastes was characterized by a lower level of anthropogenic emissions compared to pulverized coal fuel (Jianzhong et al. 2014; Dorokhov et al. 2021; Ma et al. 2018). Such suspension fuels have comparable, and sometimes even better energy characteristics, compared to traditional types of boiler fuels (Nyashina, Vershinina, and Strizhak 2020; Jianzhong et al. 2014; Xiu et al. 2018).

For example, in Rokni et al. (2018), gas emissions generated during combustion in a laboratory tube furnace (T_g = 1077 °C) of two coal ranks (high- and low-sulfur bituminous coals), biomass (corn straw and rice husk) and their mixture (50:50) were analyzed. The results showed that mixing both coals with biomass reduced CO_2 emissions by 22–44%, SO_2 by 55–80%, and NO_x by 31–54%. In this case, the calculated combustion efficiency of mixtures group was 4–9% higher than that of coal (Rokni et al. 2018).

The fuels combustion process is significantly influenced by many factors. These include the following: the fuel component composition, the form of its supply, the oxidizing atmosphere composition, the temperature in the combustion chamber, etc.

The characteristics of co-combustion of hard coal and distillery sludge waste in a vertical tube furnace were investigated by Manwatkar et al. (Manwatkar et al. 2021). The experiments were carried out in the combustion chamber at 1000 °C and with 20% excess air. The proportion of distillery sludge waste in co-firing ranged from 2% to 10%. It has been established (Manwatkar et al. 2021) that an increase in the proportion of distillery sludge waste leads to an increase in nitrogen oxide emissions. The increase in NO_x concentrations is associated with an increase in the nitrogen content of the mixed fuel relative to coal. On the other hand, an increase in the proportion of distillery sludge waste leads to a decrease in the sulfur oxides concentration (up to 65%). The reduction in sulfur oxide emissions is caused by a decrease in the fuel calorific value, which in turn leads to incomplete thermal decomposition of fuel sulfur. Also, the addition of distillery sludge waste leads to an increase in CO emissions. This can be caused by incomplete fuel combustion, as well as delayed fuel carbon decomposition, which also contributes to the CO formation.

The composition of the oxidizing atmosphere in the combustion chamber plays a significant role in the characteristics of the fuel combustion process (Riaza et al. 2017; Lei et al. 2020; K. Zhou et al. 2018). Oxygen, nitrogen, and their mixtures with CO_2 and water vapor are most often used as a combustion atmosphere (Yi et al. 2014; Gil et al. 2012; Zou et al. 2021). In Zou et al. (2021), the characteristics of co-combustion of coal and oil sludge were studied by varying the composition of the combustion atmosphere. The lowest fuel ignition temperature (367 °C) was achieved in an atmosphere of 80% O_2 and 20% N_2. With an oxidizing atmosphere of 21% O_2 and 79% N_2, the ignition temperature was 403 °C.

In Dai et al. (2022), the gas composition of the combustion products of coal gangue and semi-coke was studied by varying the O_2/CO_2 ratio in the composition of the combustion atmosphere. An increase in the oxygen proportion in the range of 10–40% led to a decrease in the CO content in the flue gas by up to 5 times, and the fuel burnout degree increased by 1.5 times. NO_x concentrations in the gaseous combustion products increased with an increase in the proportion of oxygen in the combustion atmosphere, which is caused by a greater fuel burnout degree. The lowest concentration of SO_x was registered at 20% O_2 in the oxidizing medium.

Another option aimed at reducing the level of anthropogenic emissions is the thermochemical conversion of coal and coal-water slurries by their pyrolysis. The pyrolysis products are syngas, pyrolysis oil, and combustible char (Chen et al. 2021). Depending on the process conditions and the fuel composition, the obtained solid ratio, liquid and gaseous pyrolysis products can vary in a wide range. The main components of syngas are CO, H_2, and CH_4. Due to the wide variety of possible fuel components and pyrolysis conditions, the qualitative and quantitative composition of the resulting syngas can differ dramatically. The process of fuel thermal decomposition can take place in various combustion atmospheres. The medium used can be air, oxygen, water vapor, CO_2, or a mixture of these gases (Sun et al. 2020; Cao et al. 2019; Saha, Helal Uddin, and Toufiq Reza 2019; Zeng et al. 2016). Due to the high calorific value, syngas can be used as an alternative to natural gas (Janajreh et al. 2021).

Ding et al. (2017) studied the effect of water on the characteristics of pyrolysis and gasification of water-coal suspensions and coals of different ranks. The characteristics of fast pyrolysis (coke residue yield, change in its structure) of fuels with different water contents were studied using a high-frequency furnace at 800–1200 °C in a continuous flow of nitrogen. Additionally, the gasification and pyrolysis characteristics of fuels were studied using a thermogravimetric analyzer (TGA, NETZSCH, STA 449F3) at 950 °C, 1000 °C, and 1200 °C, with a heating rate of 25 °C/min. The rate of nitrogen supply was 80 mL/min and of CO_2 100 mL/min. The results showed that with an increase in the water content in lignite from 2.1% to 13.98%, the coke residue yield during pyrolysis (temperature 800 °C) decreased. With increasing temperature, this trend only intensifies due to the more active interaction of coal with water vapor. The authors also found that the coke residue proportion in the gasification of coal-water suspensions is less than in the coals gasification (Ding et al. 2017).

In Wan et al. (2022), the characteristics of the pyrolysis of coal-water slurry (CWS) and coal-water fuel with the addition of waste lubricating oil – coal-oil-water slurry (COWS) – were studied. The proportion of coal in the composition of CWS varied in the range of 55–65 wt%, water 35–45 wt%, and waste lubricating oil 10–20 wt%. A tubular electric furnace was used as a combustion chamber. The experiments were carried out at 800–1000 °C. As the water proportion in the composition of the suspension fuel increased, the synthesis gas yield increased. The reason for this is the interreaction of water vapor and volatiles during rapid heating, which stimulated the formation of additional gaseous products. At the same time, the content of CO_2 and H_2 in the synthesis gas increased, while the content of CO and CH_4 decreased. The reason for this is the reforming of methane and the water-gas shift reaction, which was more intense with an increase in the proportion of water in the suspension. The addition of waste lubricating oil did not affect the amount of syngas, but its composition changed. The concentrations of H_2 and CH_4 during COWS pyrolysis increased, while the content of CO and CO_2, on the contrary, decreased. With an increase in the pyrolysis temperature, the proportion of water and waste lubricating oil led to a decrease in the reactivity of char (ignition delay time increased).

Based on the literature analysis, it can be concluded that the use of composite fuels is a promising method for the thermochemical fuel carbon conversion. Therefore, it is reasonable to compare the processes of combustion and pyrolysis of coal-water slurry fuel in terms of the kinetics of these processes, as well as the composition of flue gases. Therefore, the purpose of this work is to study the kinetic characteristics of the combustion and pyrolysis processes of pulverized coal and coal-water slurries, as well as the composition of flue gases formed during these processes.

28.2 MATERIALS

As the coal component was used hard coking coal. The results of the technical and elemental analyses of the coal used are presented in Table 28.1. Before use, the coal was dried at 105 °C for 2 h and then ground using a Pulverisette 14 rotary mill to achieve an average particle size of about 140 μm.

To prepare suspension fuel, an AIBOTE ZNCLBS-2500 magnetic stirrer was used. Mixing was carried out for 10 min at a rotor speed of 1500 rpm.

28.3 EXPERIMENTAL SETUP AND METHODS

Figure 28.1 shows a lineup of the experimental setup.

The study consisted of two stages. To study the stages and characteristics of combustion of the studied fuel suspensions, a METTLER-TOLEDO TGA/DSC 3+ TGA analyzer was used. A fuel sample weighing 40 mg was placed in an aluminum crucible and heated from 25 °C to 1100 °C at a rate of 20 °C/min. The combustion medium was atmospheric air (to implement the combustion conditions) or N_2 (to implement pyrolysis conditions) supplied to the combustion chamber at a constant flow rate of 60 mL/min. The systematic error in temperature measurement was ±0.5 °C, and the systematic error in mass measurement was 5 μg. The analyzer was connected to a personal computer and controlled using the STAR^e software. During the experiment, the weight loss of the sample during thermal conversion and the value of the heat flux were recorded.

TABLE 28.1
Proximate and Ultimate Analysis of Coal Used in the Experiments

Component	W^a, %	A^d, %	V^{daf}, %	$Q^a_{s,v}$, MJ/kg	C^{daf}, %	H^{daf}, %	N^{daf}, %	S^d_t, %	O^{daf}, %
Coal	2.05	14.65	27.03	29.76	79.79	4.486	1.84	0.868	12.70

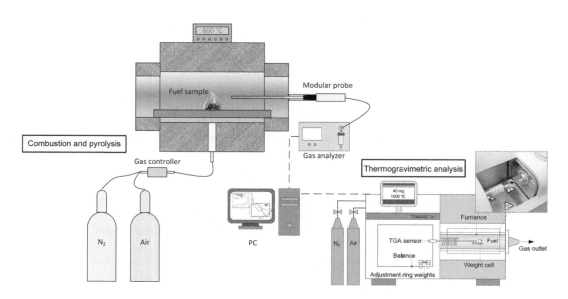

FIGURE 28.1 Scheme of the experimental setup.

Determination of the composition of combustion and pyrolysis gas products was carried out using an electric model combustion chamber made of stainless steel. Four heating elements made of ceramic and nichrome wire were located on all internal surfaces of the combustion chamber, ensuring its uniform heating. To control the temperature, we used a digital controller equipped with a K-type thermocouple, which made it possible to change the power of the heating elements in the range of 100–1000 W. To create a combustion atmosphere, a fitting was provided through which air or nitrogen was supplied. During the experiment, the temperature in the chamber was maintained at 600 °C, because at this temperature, stable coal fuel ignition and combustion is observed, and also, this temperature is quite typical for coal pyrolysis (Wu et al. 2022). To analyze the gas composition, a Test 1 gas analyzer was used. The electrochemical sensors characteristics are following: O_2 (range 0–25%, absolute error ±0.2%), CO (range 0–40,000 ppm, relative error ±5%), SO_2 (range 0–1000 ppm, relative error ±5%), NO (range 0–2000 ppm, relative error ±5%), NO_2 (range 0–500 ppm, relative error ±7%), H_2S (range 0–500 ppm, relative error ±5%), HCl (range 0–2000 ppm, relative error ±5%). The following are the characteristics of optical sensors: CO_2 (range 0–30%, reduced error ±2%), CH_4 (range 0–30%, reduced error ±5%), CO (range 0–30%, reduced error ±5%). Additionally, the gas analyzer is equipped with polarographic sensor – H_2 (range 0–5%, absolute error ±5%). The device includes a modular probe, a condensate collector, and a filtration system for drying and purifying the gas sample. The gas analyzer was connected to a PC on which the Test 1 software was installed. It made it possible to track changes in the concentrations of anthropogenic gases during combustion and pyrolysis in real time. A fuel sample weighing 4 g (for coal) or 8 g (for coal-water slurry) was fed into the combustion chamber on a heat-resistant steel holder. The sample weights were chosen in such a way as to equalize the mass of the combustible component in the fuel sample. The fuel sample weight was controlled using a ViBRA HT 84RCE 2 scale (resolution 10^{-5} g).

28.4 RESULTS AND DISCUSSION

28.4.1 Thermogravimetric Analysis

Figure 28.2 shows TG/DTG profiles obtained by thermal decomposition of coal and coal-water slurry under various gas atmospheres in the combustion chamber. The nature of the curves obtained during the combustion and pyrolysis of fuels differs significantly from each other. Table 28.2 presents the characteristics of the thermal decomposition of the studied fuels under the conditions of combustion and pyrolysis.

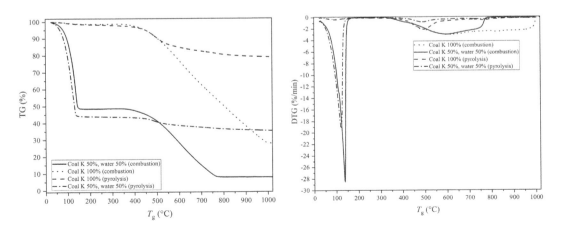

FIGURE 28.2 TG and DTG curves of dry components and mixtures based on them at the heating rate of 20 °C/min.

TABLE 28.2
Combustion and Pyrolysis Characteristic Parameters of Coal and Coal-water Slurry

Fuel	Weight Loss, %	T_{ign}, °C	T_b, °C	R_{max}, %/min	T_{max}, °C
Coal 50%, water 50% (combustion)	92.3	472	796	28.50	137
Coal 100% (combustion)	72.3	468	1004	3.04	603
Coal 100% (pyrolysis)	21.1	–	–	2.12	495
Coal 50%, water 50% (pyrolysis)	64.68	–	–	19.19	117

The process of thermal decomposition of coal when air was supplied to the combustion chamber included two stages. At the first stage (at 25–195 °C), the fuel is dried and the external moisture evaporates. At the second stage (at 350–1004 °C), volatiles are released and the coke residue is burned. This stage was characterized by one broad peak in the mass loss rate on the DTG curve, which is due to the low content of volatiles in the coal composition, as well as a stable carbon structure. The process of release and burnout of volatiles overlapped the combustion of the coke residue. In this regard, it is not possible to separate the stages of degassing and combustion of the coke residue. The maximum weight loss rate (3.04%/min) was recorded at a temperature of 603 °C (Table 28.2). It can be assumed that at this temperature, the combustion of the solid coke residue began and, at the same time, the combustion of volatiles occurred. With a further increase in temperature, the mass loss rate decreased to 2.5%/min and remained practically unchanged until the end of the thermal decomposition process, which indicates the end of the release and volatiles burnout.

The thermal decomposition of a coal-water slurry using air as a combustion atmosphere consisted of three stages. At the first stage, water evaporated from the suspension (25–160 °C); at the second stage, the release and burnout of volatiles occurred (at 359–472 °C); and at the third stage, the decomposition of the carbon residue occurred (at 472–796 °C). The maximum mass loss rate was recorded at 137 °C (Table 28.2), which corresponds to the stage of fuel drying. The nature of the curves at subsequent stages is similar to the combustion of dry coal fuel. As in the case of coal, it is quite difficult to visually separate the stages of combustion of volatiles and solid coke residue. According to the DTG curves, the mass loss rates during the combustion of solid and slurry fuels based on coal are almost the same. For suspension fuel, the maximum mass loss rate at the stage of fuel combustion was 2.98%/min at 602 °C, and during the solid coal fuel combustion, the maximum mass loss rate was 3.04%/min at 603 °C. The complete burnout temperature of the suspension fuel was 796 °C, which is caused by a lower mass of carbon residue in the fuel sample. In addition, the evaporation of water from the fuel suspension leads to the formation of microcracks on the fuel surface, which improves the supply of oxygen to the center of the fuel sample, increasing the speed of its response. The mass loss during combustion of the suspension fuel was 92.3%. Considering the water evaporation from the suspension, the mass loss of the solid part turned out to be higher than that of the solid fuel. This indicates the presence of synergistic effects occurring during the fuel suspensions combustion. In particular, the appearance of microcracks on the surface of the fuel not only intensifies the combustion process, but also ensures its more complete burnout due to the supply of oxygen to the deep layers of the sample.

TG/DTG profiles obtained during the pyrolysis of coal and water-coal suspension in a nitrogen-saturated combustion atmosphere are shown in Figure 28.2. During the pyrolysis process, ignition temperatures (T_{ign}) were not recorded, since this process occurred in the absence of an oxidizing agent. When comparing the mass loss rates obtained during combustion and pyrolysis of the studied fuels, it can be seen that the values of these parameters for both fuels are higher when the combustion conditions are realized. One reason for this is the different properties of the gases used. The temperature of the fuel sample during pyrolysis turned out to be lower than when the fuel was ignited. This is due to the higher heat capacity of nitrogen compared to atmospheric air. The maximum mass loss rate of the fuel suspension at the evaporation stage was 28.50%/min under the combustion conditions and 19.19%/min under the pyrolysis conditions.

The second peak of the mass loss rate is typical for a temperature of 495 °C for hard coal (2.12%/min) and 494 °C (0.79%/min) for coal-water slurry. The reason for this is that half of the weight of the coal-water slurry is water evaporated during the drying stage. Accordingly, at the stage of pyrolysis, a much larger volume of the gas product is released when the solid coal fuel is heated than when the suspension is heated. The mass loss at the pyrolysis stage was 8.81% for the coal-water slurry and 18.98% for the pulverized coal fuel. Based on this, it can be concluded that the form of fuel supply (layer or suspension) does not significantly affect the kinetic characteristics of the pyrolysis process. During the heating of both types of fuels, the end of the pyrolysis process was not registered; therefore, the burnout temperature (T_b) was also not registered.

28.4.2 FLUE GAS ANALYSIS

Figure 28.3 shows the concentrations of anthropogenic gaseous emissions from the combustion of hard coal and coal-water slurry. As expected, despite the equal mass of the coal component, the combustion of coal-water slurry was characterized by a lower level of carbon emissions, sulfur oxide, and nitrogen oxide. Thus, when using coal-water slurry, it is possible to achieve a reduction in CO emissions by 45%, CO_2 by 28%, NO_x by 19%, and SO_2 by 43.5%. This result is explained by a number of physicochemical processes occurring during the suspension fuel combustion (Miller and Bowman 1989; Werther et al. 2000; Armesto et al. 2003).

The formation of NO_x and SO_2 from fuel nitrogen and sulfur occurs as a result of the combustion of particles released along with volatiles and the oxidation of nitrogen and sulfur remaining in the coke part. The distribution of fuel nitrogen and sulfur between the coke and volatile part depends mainly on the structure of the fuel and the combustion temperature. For bituminous coals, the nitrogenous particles emitted with volatiles consist mainly of resinous compounds, which rapidly decompose at high temperatures to hydrogen cyanide (HCN). Nitrogenous volatile substances dominating in concentrations are NH_3 and HCN. Typically, NH_3 can decompose to NH_2 and NH radicals, which can either be oxidized to NO or react with available NO and OH radicals to form N_2. On the contrary, HCN decomposes to NCO with the help of an oxygen radical and CN in the presence of H, OH, O radicals. Further, NCO can react with NO to form N_2O (Miller and Bowman 1989).

The process of water evaporation and the water vapor formation in the combustion chamber directly affects the kinetics of the combustion process. As a result of these processes, the temperature in the combustion zone decreases, which affects the dynamics of the formation of nitrogen and

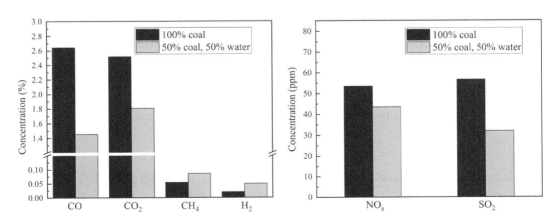

FIGURE 28.3 Anthropogenic emissions from fuel combustion.

sulfur oxides. It was established in Zhao et al. (B. Zhao et al. 2016) that the temperature in the fuel combustion zone directly affects the rate and amount of NO_x and SO_2 formed, and this dependence is non-linear. In addition, the reduction of NO to N_2 in the presence of free H_2 radicals contributed to a decrease in the concentrations of nitrogen oxides (H. Zhou et al. 2019; Xu, Smoot, and Hill 1999):

$$2NO + H_2 \rightarrow 2HNO \qquad (28.1)$$

$$2HNO + H_2 \rightarrow 2H_2O + N_2 \qquad (28.2)$$

The main source of sulfur oxides is the thermal decomposition of pyrite FeS_2. It changes to the more reactive iron sulfide (FeS) and active sulfur at temperatures above 500 °C. SO_2 is the main gaseous product of pyrite oxidation. Active sulfur reacts with oxygen, resulting in the formation of sulfur dioxide.

Also, at temperatures above 500 °C, decomposition reactions of sulfates in coal begin to occur. Ferrous sulfates decompose according to the following reactions at temperatures above 540 °C: $Fe_2(SO_4)_3 \rightarrow Fe_2O_3 + 3SO_3$ (560–700 °C); $2FeSO_4 \rightarrow Fe_2O_3 + 3SO_3$ ($t \geq 620$ °C); $2FeSO_4 \rightarrow Fe_2O_3 + 3SO_3 + SO_2$ (540–680 °C). In the course of these reactions, sulfur oxides SO_2 and SO_3 are formed (Hu et al. 2006).

As in the case of nitrogen oxides, a decrease in the temperature in the combustion zone during the evaporation of water from the composition of the suspension favorably affected the concentrations of sulfur oxides in the composition of the flue gas. The average concentrations of SO_2 during the combustion of coal-water slurry decreased by almost two times (for combustion of coal fuel, the average concentration of SO_2 was 57 ppm, and for combustion of CWS it was 32 ppm).

A significant reduction in emissions from coal-water slurry combustion compared to pulverized coal fuels has been reported for carbon monoxide and carbon dioxide. For example, combustion of CWS instead of pulverized coal reduced CO emissions by 45% and CO_2 emissions by 28%.

The reason for the decrease in CO emissions was the carbon monoxide conversion reaction in the presence of water vapor:

$$CO + H_2O \rightarrow CO_2 + H_2 \qquad (28.3)$$

This effect is confirmed by the increase in the concentration of H_2 in the composition of flue gases during the combustion of CWS, compared with pulverized coal fuel (Figure 28.3).

Due to the decrease in temperature in the combustion zone during the evaporation of water from the suspension, the process of formation of CO_2 was less intense. Therefore, despite the same carbon content in the fuel sample, the average concentrations of carbon dioxide during the combustion of the suspension fuel turned out to be lower than during the combustion of pulverized coal.

Figure 28.4 shows the composition of the pyrolysis gas of pulverized coal and CWS. Methane accounts for the largest share in both cases. The content of CH_4 in the pyrolysis gas was 0.93% and 1.19% for the thermal decomposition of coal and CWS, respectively.

The methane formation during the pyrolysis of the studied fuels occurs in the presence of free H_2 radicals:

$$C + 2H_2 \rightarrow CH_4 \qquad (28.4)$$

In turn, the formation of free hydrogen radicals is realized due to the reaction:

$$C + H_2O \rightarrow CO + H_2 \qquad (28.5)$$

Thus, it can be concluded that the use of a coal-water slurry makes it possible to obtain synthesis gas with a higher content of combustible components compared to pulverized coal fuel. Thus, the proportion of methane in the synthesis gas increased by 29%, and the proportion of hydrogen increased by 65%. The share of CO, at the same time, decreased by 8%. A similar effect was demonstrated in

Gaseous Anthropogenic Emissions from Coal and Slurry Fuel 385

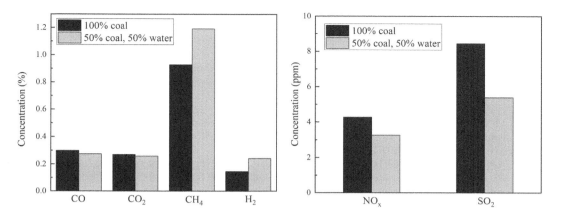

FIGURE 28.4 Anthropogenic emissions during the pyrolysis of the studied fuels.

Wan et al. (2022). The addition of water contributed to an increase in the volume of synthesis gas obtained during the pyrolysis of composite fuels. The reason for this was the interreaction of water vapor with volatile and solid coke residue.

As expected, during the pyrolysis of the studied fuels, extremely low concentrations of sulfur and nitrogen oxides were recorded. The reason for this was the absence of an oxidizing atmosphere. Compared to experiments in the air atmosphere, the concentrations of nitrogen oxides decreased 12.5–13.3 times, and the concentrations of sulfur oxides decreased 5.9–6.7 times.

28.5 CONCLUSION

(i) The conducted experiments have shown that coal combustion is characterized by the maximum emissions of gaseous pollutants. When burning coal-water slurries, the average concentrations of CO are reduced by 45%, CO_2 by 28%, NO_x by 19%, and SO_2 by 43.5%, compared with coal.

(ii) During the pyrolysis of coal-water slurry, synthesis gas is formed with a higher content of combustible components compared to the same process for pulverized coal fuel. The share of CH_4 was higher by 29%, and the share of H_2 by 65%.

(iii) During pyrolysis in an N_2 atmosphere, the average concentrations of nitrogen oxides turned out to be 12.5–13.3 times lower, and the concentrations of sulfur oxides 5.9–6.7 times, compared with the combustion of similar fuels in an air atmosphere.

(iv) It has been established that the burnout completeness of coal-water slurry is higher than that of coal dust. The reason for this is the formation of microcracks on the fuel surface, formed during the moisture evaporation. Microcracks improve the diffusion of oxygen into the bulk of the fuel sample, which ensures a more complete burnout of the carbon residue.

(v) It was found that the form of fuel supply (coal-water slurry or coal dust) does not significantly affect the kinetic characteristics of the thermal decomposition of the coal component.

ACKNOWLEDGMENTS

The reported study was funded by the Russian Foundation for Basic Research, National Council of Brazil for Scientific and Technological Development, and Ministry of Science & Technology (Government of India) according to the research project No. 19–53–80019.

REFERENCES

Akhmetshin, M.R., G.S. Nyashina, and P.A. Strizhak. 2020. "Comparative Analysis of Factors Affecting Differences in the Concentrations of Gaseous Anthropogenic Emissions from Coal and Slurry Fuel Combustion." *Fuel* 270 (June): 117581. https://doi.org/10.1016/j.fuel.2020.117581.

Armesto, L., A. Bahillo, A. Cabanillas, K. Veijonen, J. Otero, A. Plumed, and L. Salvador. 2003. "Co-Combustion of Coal and Olive Oil Industry Residues in Fluidised Bed☆." *Fuel* 82 (8): 993–1000. https://doi.org/10.1016/S0016-2361(02)00397-6.

Cao, Y., Q. Wang, J. Du, and J. Chen. 2019. "Oxygen-Enriched Air Gasification of Biomass Materials for High-Quality Syngas Production." *Energy Conversion and Management* 199 (November): 111628. https://doi.org/10.1016/j.enconman.2019.05.054.

Chen, Zhaohui, Deliang Wang, Hang Yang, Yusheng Zhang, Yunjia Li, Changming Li, Jian Yu, and Shiqiu Gao. 2021. "Novel Application of Red Mud as Disposal Catalyst for Pyrolysis and Gasification of Coal." *Carbon Resources Conversion* 4 (January): 10–18. https://doi.org/10.1016/j.crcon.2021.01.001.

Dai, Ruo Wei, Rui Dong Zhao, Zhi Qi Wang, Jian Guang Qin, Tian Ju Chen, and Jin Hu Wu. 2022. "Study on the Oxy-Fuel Co-Combustion of Coal Gangue and Semicoke and the Pollutants Emission Characteristics." *Journal of Fuel Chemistry and Technology* 50 (2): 152–9. https://doi.org/10.1016/S1872-5813(21)60132-9.

Ding, Lu, Zhenghua Dai, Qinghua Guo, and Guangsuo Yu. 2017. "Effects of In-Situ Interactions Between Steam and Coal on Pyrolysis and Gasification Characteristics of Pulverized Coals and Coal Water Slurry." *Applied Energy* 187 (February): 627–39. https://doi.org/10.1016/j.apenergy.2016.11.086.

Dorokhov, Vadim V., Geniy V. Kuznetsov, Galina S. Nyashina, and Pavel A. Strizhak. 2021. "Composition of a Gas and Ash Mixture Formed during the Pyrolysis and Combustion of Coal-Water Slurries Containing Petrochemicals." *Environmental Pollution* 285 (September): 117390. https://doi.org/10.1016/j.envpol.2021.117390.

Gaber, Christian, Philipp Wachter, Martin Demuth, and Christoph Hochenauer. 2020. "Experimental Investigation and Demonstration of Pilot-Scale Combustion of Oil-Water Emulsions and Coal-Water Slurry with Pronounced Water Contents at Elevated Temperatures with the Use of Pure Oxygen." *Fuel* 282 (December): 118692. https://doi.org/10.1016/j.fuel.2020.118692.

Gent, Janneane F., Elizabeth W. Triche, Theodore R. Holford, Kathleen Belanger, Michael B. Bracken, William S. Beckett, and Brian P. Leaderer. 2003. "Association of Low-Level Ozone and Fine Particles With Respiratory Symptoms in Children With Asthma." *JAMA* 290 (14): 1859–67. https://doi.org/10.1001/JAMA.290.14.1859.

Gil, M. V., J. Riaza, L. A. Lvarez, C. Pevida, J. J. Pis, and F. Rubiera. 2012. "A Study of Oxy-Coal Combustion with Steam Addition and Biomass Blending by Thermogravimetric Analysis." *Journal of Thermal Analysis and Calorimetry* 109 (1): 49–55. https://doi.org/10.1007/s10973-011-1342-y.

Guo, Feihong, and Zhaoping Zhong. 2018. "Co-Combustion of Anthracite Coal and Wood Pellets: Thermodynamic Analysis, Combustion Efficiency, Pollutant Emissions and Ash Slagging." *Environmental Pollution* 239 (August): 21–9. https://doi.org/10.1016/j.envpol.2018.04.004.

Hu, Guilin, Kim Dam-Johansen, Stig Wedel, and Jens Peter Hansen. 2006. "Decomposition and Oxidation of Pyrite." *Progress in Energy and Combustion Science* 32 (3): 295–314. https://doi.org/10.1016/j.pecs.2005.11.004.

International Energy Agency. 2020. *Key World Energy Statistics*. IEA.

Janajreh, Isam, Idowu Adeyemi, Syed Shabbar Raza, and Chaouki Ghenai. 2021. "A Review of Recent Developments and Future Prospects in Gasification Systems and Their Modeling." *Renewable and Sustainable Energy Reviews* 138 (March): 110505. https://doi.org/10.1016/j.rser.2020.110505.

Jianzhong, Liu, Wang Ruikun, Xi Jianfei, Zhou Junhu, and Cen Kefa. 2014. "Pilot-Scale Investigation on Slurrying, Combustion, and Slagging Characteristics of Coal Slurry Fuel Prepared Using Industrial Wasteliquid." *Applied Energy* 115 (February): 309–19. https://doi.org/10.1016/j.apenergy.2013.11.026.

Lei, Kai, Rui Zhang, Buqing Ye, Jin Cao, and Dong Liu. 2020. "Combustion of Single Particles from Sewage Sludge/Pine Sawdust and Sewage Sludge/Bituminous Coal Under Oxy-Fuel Conditions with Steam Addition." *Waste Management* 101 (January): 1–8. https://doi.org/10.1016/J.WASMAN.2019.09.034.

Li, Dan, Daishe Wu, Feigao Xu, Jinhu Lai, and Li Shao. 2018. "Literature Overview of Chinese Research in the Field of Better Coal Utilization." *Journal of Cleaner Production* 185 (June): 959–80. https://doi.org/10.1016/j.jclepro.2018.02.216.

Ma, Xiu Wei, Feng Hai Li, Ming Jie Ma, and Yi Tian Fang. 2018. "Fusion Characteristics of Blended Ash from Changzhi Coal and Biomass." *Ranliao Huaxue Xuebao/Journal of Fuel Chemistry and Technology* 46 (2): 129–37. https://doi.org/10.1016/s1872-5813(18)30007-0.

Manwatkar, Prashik, Lekha Dhote, Ram Avtar Pandey, Anirban Middey, and Sunil Kumar. 2021. "Combustion of Distillery Sludge Mixed with Coal in a Drop Tube Furnace and Emission Characteristics." *Energy* 221 (April): 119871. https://doi.org/10.1016/j.energy.2021.119871.

Miller, James A., and Craig T. Bowman. 1989. "Mechanism and Modeling of Nitrogen Chemistry in Combustion." *Progress in Energy and Combustion Science* 15 (4): 287–338. https://doi.org/10.1016/0360-1285(89)90017-8.

Munawer, Muhammad Ehsan. 2018. "Human Health and Environmental Impacts of Coal Combustion and Post-Combustion Wastes." *Journal of Sustainable Mining* 17 (2): 87–96. https://doi.org/10.1016/J.JSM.2017.12.007.

Nyashina, G.S., G.V. Kuznetsov, and P.A. Strizhak. 2020. "Effects of Plant Additives on the Concentration of Sulfur and Nitrogen Oxides in the Combustion Products of Coal-Water Slurries Containing Petrochemicals." *Environmental Pollution* 258 (March): 113682. https://doi.org/10.1016/j.envpol.2019.113682.

Nyashina, G.S., K. Yu Vershinina, and P.A. Strizhak. 2020. "Impact of Micro-Explosive Atomization of Fuel Droplets on Relative Performance Indicators of Their Combustion." *Fuel Processing Technology* 201 (May): 106334. https://doi.org/10.1016/j.fuproc.2019.106334.

Olsen, Yulia, Jacob Klenø Nøjgaard, Helge Rørdam Olesen, Jørgen Brandt, Torben Sigsgaard, Sara C. Pryor, Travis Ancelet, María del Mar Viana, Xavier Querol, and Ole Hertel. 2020. "Emissions and Source Allocation of Carbonaceous Air Pollutants from Wood Stoves in Developed Countries: A Review." *Atmospheric Pollution Research* 11 (2): 234–51. https://doi.org/10.1016/J.APR.2019.10.007.

Riaza, Juan, Muhammad Ajmi, Jon Gibbins, and Hannah Chalmers. 2017. "Ignition and Combustion of Single Particles of Coal and Biomass under O2/CO2 Atmospheres." *Energy Procedia* 114 (July): 6067–73. https://doi.org/10.1016/J.EGYPRO.2017.03.1743.

Rokni, Emad, Xiaohan Ren, Aidin Panahi, and Yiannis A. Levendis. 2018. "Emissions of SO2, NOx, CO2, and HCl from Co-Firing of Coals with Raw and Torrefied Biomass Fuels." *Fuel* 211 (January): 363–74. https://doi.org/10.1016/j.fuel.2017.09.049.

Saha, Pretom, M. Helal Uddin, and M. Toufiq Reza. 2019. "A Steady-State Equilibrium-Based Carbon Dioxide Gasification Simulation Model for Hydrothermally Carbonized Cow Manure." *Energy Conversion and Management* 191 (July): 12–22. https://doi.org/10.1016/j.enconman.2019.04.012.

Staroń, Anita, Zygmunt Kowalski, Paweł Staroń, and Marcin Banach. 2019. "Studies on CWL with Glycerol for Combustion Process." *Environmental Science and Pollution Research* 26 (3): 2835–44. https://doi.org/10.1007/s11356-018-3814-0.

Sun, Zhao, Zong Chen, Sam Toan, and Zhiqiang Sun. 2020. "Chemical Looping Deoxygenated Gasification: An Implication for Efficient Biomass Utilization with High-Quality Syngas Modulation and CO2 Reduction." *Energy Conversion and Management* 215 (July): 112913. https://doi.org/10.1016/j.enconman.2020.112913.

Vershinina, K. Yu, N. E. Shlegel, and P. A. Strizhak. 2019. "Relative Combustion Efficiency of Composite Fuels Based on of Wood Processing and Oil Production Wastes." *Energy* 169 (February): 18–28. https://doi.org/10.1016/j.energy.2018.12.027.

Wan, Gan, Jie Yu, Xuyang Wang, and Lushi Sun. 2022. "Study on the Pyrolysis Behavior of Coal-Water Slurry and Coal-Oil-Water Slurry." *Journal of the Energy Institute* 100 (February): 10–21. https://doi.org/10.1016/j.joei.2021.10.006.

Wang, Chang'an, Qinqin Feng, Qisen Mao, Chaowei Wang, Guangyu Li, and Defu Che. 2021. "Oxy-Fuel Co-Combustion Performances and Kinetics of Bituminous Coal and Ultra-Low Volatile Carbon-Based Fuels." *International Journal of Energy Research* 45 (2): 1892–907. https://doi.org/10.1002/ER.5871.

Werther, J., M. Saenger, E.-U. Hartge, T. Ogada, and Z. Siagi. 2000. "Combustion of Agricultural Residues." *Progress in Energy and Combustion Science* 26 (1): 1–27. https://doi.org/10.1016/S0360-1285(99)00005-2.

Wu, Chenglin, Yuting Zhuo, Xiuli Xu, Ehsan Farajzadeh, Jinxiao Dou, Jianglong Yu, Yansong Shen, and Zhiqiang Zhang. 2022. "A Combined Experimental and Numerical Study of Coal Briquettes Pyrolysis Using Recycled Gas in an Industrial Scale Pyrolyser." *Powder Technology* 404 (May): 117477. https://doi.org/10.1016/j.powtec.2022.117477.

Xiu, Meng, Svetlana Stevanovic, Md Mostafizur Rahman, Ali Mohammad Pourkhesalian, Lidia Morawska, and Phong K. Thai. 2018. "Emissions of Particulate Matter, Carbon Monoxide and Nitrogen Oxides from the Residential Burning of Waste Paper Briquettes and Other Fuels." *Environmental Research* 167 (November): 536–43. https://doi.org/10.1016/J.ENVRES.2018.08.008.

Xu, H., L. D. Smoot, and S. C. Hill. 1999. "Computational Model for NOx Reduction by Advanced Reburning." *Energy & Fuels* 13 (2): 411–20. https://doi.org/10.1021/ef980090h.

Yi, Baojun, Liqi Zhang, Fang Huang, Zhihui Mao, and Chuguang Zheng. 2014. "Effect of H2O on the Combustion Characteristics of Pulverized Coal in O2/CO2 Atmosphere." *Applied Energy* 132 (November): 349–57. https://doi.org/10.1016/J.APENERGY.2014.07.031.

Zeng, De Wang, Rui Xiao, Zhi Cheng Huang, Ji Min Zeng, and Hui Yan Zhang. 2016. "Continuous Hydrogen Production from Non-Aqueous Phase Bio-Oil via Chemical Looping Redox Cycles." *International Journal of Hydrogen Energy* 41: 6676–84. Elsevier Ltd. https://doi.org/10.1016/j.ijhydene.2016.03.052.

Zhao, Bingtao, Yaxin Su, Dunyu Liu, Hang Zhang, Wang Liu, and Guomin Cui. 2016. "SO2/NOx Emissions and Ash Formation from Algae Biomass Combustion: Process Characteristics and Mechanisms." *Energy* 113 (October): 821–30. https://doi.org/10.1016/j.energy.2016.07.107.

Zhao, Zhenghui, Ruikun Wang, Lichao Ge, Junhong Wu, Qianqian Yin, and Chunbo Wang. 2019. "Energy Utilization of Coal-Coking Wastes via Coal Slurry Preparation: The Characteristics of Slurrying, Combustion, and Pollutant Emission." *Energy* 168 (February): 609–18. https://doi.org/10.1016/j.energy.2018.11.141.

Zhao, Zhenghui, Ruikun Wang, Junhong Wu, Qianqian Yin, and Chunbo Wang. 2019. "Bottom Ash Characteristics and Pollutant Emission During the Co-Combustion of Pulverized Coal with High Mass-Percentage Sewage Sludge." *Energy* 171 (March): 809–18. https://doi.org/10.1016/j.energy.2019.01.082.

Zhou, Hao, Yuan Li, Ning Li, Runchao Qiu, and Kefa Cen. 2019. "Conversions of Fuel-N to NO and N2O During Devolatilization and Char Combustion Stages of a Single Coal Particle Under Oxy-Fuel Fluidized Bed Conditions." *Journal of the Energy Institute* 92 (2): 351–63. https://doi.org/10.1016/j.joei.2018.01.001.

Zhou, Kun, Qizhao Lin, Hongwei Hu, Fupeng Shan, Wei Fu, Po Zhang, Xinhua Wang, and Chengxin Wang. 2018. "Ignition and Combustion Behaviors of Single Coal Slime Particles in CO2/O2 Atmosphere." *Combustion and Flame* 194 (August): 250–63. https://doi.org/10.1016/J.COMBUSTFLAME.2018.05.004.

Zou, Huihuang, Chao Liu, Fatih Evrendilek, Yao He, and Jingyong Liu. 2021. "Evaluation of Reaction Mechanisms and Emissions of Oily Sludge and Coal Co-Combustions in O2/CO2 and O2/N2 Atmospheres." *Renewable Energy* 171 (June): 1327–43. https://doi.org/10.1016/J.RENENE.2021.02.069.

29 Allothermal Approach for Thermochemical Conversion of Coal-enrichment Waste

Roman I. Egorov, Roman I. Taburchinov, Zhenyu Zhao, and Xin Gao

CONTENTS

29.1	Introduction	389
29.2	Materials and Methods	390
29.3	Results and Discussion	392
29.4	Conclusion	397
Acknowledgments		398
References		398

29.1 INTRODUCTION

The permanent decrease of easily accessible fuel deposits imposes limitation on the progress of world economics (Nejat et al. 2015). This fact is independent of any political crises and it makes strong influence on the direction of the scientific investigations in the area of energetics. The very popular way of solution of the increasing deficit is an involvement of different combustible wastes into the industrial use as a low-grade fuel (Yavuz, Küçükbayrak, and Williams 1998). The coal-enrichment waste (CEW) were accumulated in different countries during the last centuries because of the absence of the effective ways of their utilization. This makes the waste deposits a very attractive place for production of the waste-derived composite fuels. Simple estimation of the amount of accumulated CEW gives that Russian Federation has some billion tons of easily accessible waste-derived fuels (British Petroleum 2021). Countries with similar historical scale of the coal mining have comparable potential of deposits of CEW.

The similar great potential has different types of highly available low-grade fuels like peat or lignite. All these fuels have low calorific value due to the presence of the high amount of mineral components (Tabakaev et al. 2017). Numerous publications show that the most appropriate usage of coal-enrichment wastes is in the form of coal-water slurry (Jianzhong et al. 2014). This decreases the fuel reactivity making the storage more suitable together with optimization of the ecological indicators of combustion process (lower air pollution). However, the ignition of the water-filled fuels become much more complicated as well as combustion is not stable enough (Valiullin et al. 2017). Thus, an additional torch of the fuel oil is typically used as supporting heat source for successful combustion of the waste (Osintsev 2012). Such a way of application of the waste-derived fuels as well as low-grade fuels looks not very attractive because of the permanently growing cost of the fuel oil.

The thermochemical conversion (TCC) of waste-derived fuels looks a more effective way because self-sustaining combustion of them does not provide the high-enough temperatures for optimal composition of the gas exhausts (Minchener 2005; Leckner 2015). Traditional techniques of the TCC deal with combustion of the small part of fuel for creation of the heat flow that is used

DOI: 10.1201/9781003334415-29

for gasification of the rest fuel. However, the overall efficiency of such approach is relatively low due to the low calorific value of the fuel.

An application of the allothermal approaches is much more effective (Egorov et al. 2020; Zaitsev et al. 2020). The external heating allows keeping the optimal temperatures and, thus, it gives a very high flexibility according to the fuel type. Therefore, the allothermal approach pave the way for the universal technique of the utilization of CEW and thermochemical conversion of the low-grade fossil fuels.

Various types of the heating techniques (based on different heat exchange mechanisms) can be used for the external heating of the fuel (Egorov, Taburchinov, and Zaitsev 2019). However, the widest range of the heat flows is available with radiative heating by the electromagnetic radiation. The focused flow of the light (Zaitsev et al. 2020) or microwaves (Zhao et al. 2021; Yang et al. 2019) can serve as a very suitable tool for heating the fuel during conversion. It allows different temperature regimes – from typical for self-sustaining combustion to high-temperature regime for extremal thermochemical conversion. An additional feature of such radiative heating is very nice controllability of approaches based on application of the flow of electromagnetic waves.

An estimation of the industrial potential of the allothermal techniques shows that heating by the light flow can be realized by usage of both solar light and light from artificial sources. This combined way appears as a cheapest solution for the long-term perspective. The microwaves can serve as an artificial source instead of light as well.

This chapter shows the overview of the typical regimes available for fuel conversion with heating by the intensive light flow. The fuel-specific features (like temperatures achieved with certain external heat flow intensity or chemical composition of the syngas) were shown in parallel to the process features depending on the heating regime.

29.2 MATERIALS AND METHODS

The allothermal conversion of fuel compositions prepared using CEW and peat was investigated using the peat from the Suhovskoe peat deposit of Tomsk Region of Russian Federation (Tabakaev et al. 2017) and the filter cake of fiery coal from the Kemerovo region of Russian Federation (Egorov et al. 2020). The preparation of water slurries was done immediately before the fuel conversion experiments as well as preparation of oil–peat compositions. The non-refined rapeseed oil was used for preparation of peat–oil composition (Valiullin, Egorov, and Strizhak 2017).

There were three types of fuel samples: the waste-derived coal-water slurry that consists of 60 wt% of filter cake and 40 wt% of water; the peat–water slurry with 30 wt% moisture; and the peat–oil slurry with 60 wt% of the rapeseed oil. These samples allow showing the effect of the fuel properties onto the possible conversion regime at different intensities of external heating. The results of the TGA analysis of these samples is presented in Figure 29.1. The measurements were done by Mettler Toledo TGA3 DSC (Mettler, Switzerland) with heating rate about 10 °C/min and with air flow about 2 mL/min.

One can see that essential thermal decomposition of these fuel mixtures begins after 350 °C. The maximal decomposition rates occur at 386 °C for peat–oil mixture, 633 °C for CEW, and 300 °C and 780 °C for water slurry of peat. In general, the water slurry of both CEW and peat have much lower reactivity than peat–oil mixture. All these fuel show quite similar levels of heat production (see Figure 29.1c).

The experimental investigations of the allothermal conversion of the presented fuel compositions were done using the approach presented in Figure 29.1d. The focused light of halogenic lamp was heating the flat layer of the fuel. The power of the light flow was up to 25 W according to the electrical chosen current (the full power of the lamp is about 600 W). Light was focused by parabolic mirror into the small spot on the sample surface (R ~0.75 mm). This allows the light flow intensity up to 1400 W/cm². The fuel sample size was a bit bigger than light spot for simplification of the optical

Allothermal Thermochemical Conversion of Coal Waste

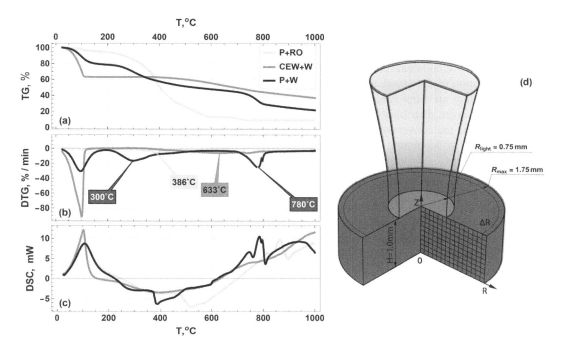

FIGURE 29.1 The results of the TGA analysis of the fuel samples (a–c) and the scheme of the external heating of the fuel (d).

alignment. The fuel was placed into the cube chamber with optical window on the top side (V ~3 cm^3 with special cavity on the bottom surface for accumulation of liquid products of conversion).

The light absorption coefficient of the peat and CEW with high enough moisture is about 0.5–0.6 for a whole visible range of wavelengths. This allows an easy estimation of the effective heat flow that makes an immediate fuel heating.

The temperature of the fuel surface was measured by Thermal Imaging System Optris Pi 1M (Optris GmbH, Germany). The chemical composition of the produced gases was analyzed by gas-analyzer Test-1 (Bonner, Russian Federation) that can real-time measure the volume concentrations of carbon oxides, hydrogen, methane, oxides of nitrogen, and sulfur dioxide. All the measurements were repeated for five times and the averaged data is presented further.

The simple simulation of the allothermal gasification of low-grade fuels (with moisture corresponding to the water content of slurry with CEW or peat) was done using the model presented in Egorov and Taburchinov (2021). This model based on solution of the 2D transient heat transfer equation for cylindrical fuel layer takes into account the external radiative heating as well as the internal heating due to the chemical reactions heat. The model includes four main reactions typically used for description of the gasification process (Minchener 2005):

$$C + O_2 \rightarrow CO_2 + Q_{ox_1} \tag{29.1}$$

$$2C + O_2 \rightarrow 2CO + Q_{ox_2} \tag{29.2}$$

$$CO_2 + C \rightarrow 2CO - Q_R \tag{29.3}$$

$$C + H_2O \rightarrow CO + H_2 - Q_H \tag{29.4}$$

Equations 29.1 and 29.2 are exothermal with oxidation heat of 9.3 and 4.3 MJ/kg, respectively. Otherwise, the Equations 29.3 and 29.4 are endothermal with reaction heat of 3.92 and 4.68 MJ/kg, respectively. The simulation estimates water evaporation and filtration of the gases through the fuel layer. The radiative and convective heat losses were taken in account too.

An application of the simulation allows observing some important features of the fuel conversion process (that are impossible to measure directly) like spatial distribution of the temperature inside the fuel layer, distribution of the reaction performances, and spatial distributions of carbon and water during the conversion process. The simulation results were compared with experimental results and show good enough correlation.

29.3 RESULTS AND DISCUSSION

According to the TGA data (in low oxygen regime), the typical temperature that allows the conversion of water slurry of the peat is more than 300 °C. It corresponds to the fast pyrolysis of the peat in the low-oxygen environment or smoldering of it when there is much enough of oxygen. The maximal heat production occurs at about 400 °C and this fact reflects the possibility of the self-sustaining combustion even of the wet peat. Therefore, it is evident that the thermochemical conversion of the wet peat does not require too high intensities of the external heating.

However, the mixture of peat and plant oil shows a bit another behavior. The maximal reactivity of the mixture is observed at 380–390 °C whatever the maximal heat production is at more than 500 °C. This shows that stable self-sustaining combustion of such fuel compositions is quite problematic. Volatile components of the rapeseed oil yield out effectively removing the heat from the reaction area and decreasing the temperature growth. The external heating principally allows the successful jump of the temperature into the range where the fuel decomposition produces much enough heat for stable conversion.

The water slurry of CEW has maximal reactivity at 633 °C, but in this case, the peak of heat production occurs at lower temperatures. This means that such fuel can burn in self-sustaining regime if the temperature is enough for high-performance decomposition.

Figure 29.2 shows the dependence of the temperature on lit fuel surface on the light flow intensity.

The presented dependences show that wet peat achieve the high enough temperature with relatively weak external heating (~800 °C at 50–60 W/cm²). An easy estimation with Stefan–Boltzmann's law shows that hot surface with temperature about 800 °C produces the heat radiation with intensity about 8–8.5 W/cm². The typical furnace at such temperature gives the total heat flow incoming into the fuel droplet at about 50 W/cm². Combustion of the water-filled fuel is not very stable under such conditions. However, the dry peat can burn good enough. The visible light flow with intensity about 60 W/cm² corresponds to the heat flow of 30–40 W/cm² and the fuel irradiation by the light with

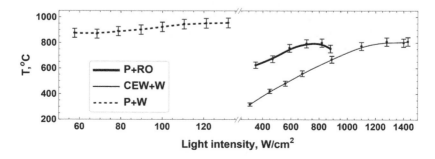

FIGURE 29.2 The dependence of the temperature of fuel surface under the irradiation on the intensity of light flow.

intensity near 100 W/cm² is very close to the furnace heating. As one can see from Figure 29.2, such irradiation allows temperature above 900 °C that is enough for effective fuel conversion.

The mixture of peat and rapeseed oil requires much higher intensities to achieve temperatures that allows effective conversion. According to Figure 29.2, it should be more than 700 W/cm². In this case, the temperature of the fuel layer quickly jumps above the maximal heat production threshold and the conversion will go with maximal involvement of the chemical reaction heat.

The water slurry of CEW demonstrates the fast increase of surface temperatures when the intensity of incoming light flow is growing in range of 400–1200 W/cm². Contrary to the relatively weak dependences of the temperatures versus light intensity for other fuel compositions, the heat conductivity of CEW makes it very sensitive to the performance of external heating.

Heating with much higher light intensities will not increase temperature too much due to the fast increase of the radiative heat losses. The layer of wet peat achieves thermal equilibrium when light flow intensity is about 120–150 W/cm². The layer of peat–oil mixture achieves it at 800–900 W/cm². The water slurry of used CEW achieves the thermal equilibrium when incoming light flow is more than 1200 W/cm². All mentioned mean that radiative heat losses of unit area of fuel layer become greater than light-induced heating of the fuel when incoming heat flow is about 60–90, 400–550, and 600–700 W/cm² correspondingly for wet peat, peat–oil mixture, and water slurry of CEW.

On the one hand, the temperature of the sample depends on its thermal properties. The heat capacity and heat conductivity of dry peat are about 1.7 kJ/(kg °C) and 0.06 W/(m °C), respectively, thus, the wet peat have them at about 2.4 kJ/(m °C) and 0.22 W/(m °C) at low temperatures. The heat capacity and heat conductivity of peat–oil mixture are about 2 kJ/(kg °C) and 0.137 W/(m °C), respectively. Similarly, for CEW slurry, the heat capacity and heat conductivity are more than 2.6 kJ/(m °C) and 0.39 W/(m °C), respectively. One can see that high values of mentioned parameters for CEW slurry determine the sensitivity of fuel temperature to the level of external heating. The light intensity should be high enough for compensation of the heat dissipation over the sample volume. Otherwise, the temperature will vary slowly.

On the other hand, the chemical reaction heat contributes to the fuel heating too. The gross calorific value of CEW slurry is about 10.4 MJ/kg. The same value for peat–oil mixture is about 37 MJ/kg and wet peat has near 11 MJ/kg. However, the intensive evaporation of the volatile matter from the peat–oil mixture defines the very high level of heat losses during the heating. This move such fuel closer to CEW than to peat and decrease the effect of the high calorific value.

Despite the differences of thermal properties, conversion processes of all these fuels are allothermal. This means that process decay when the external heating switches off the inner heat production is not enough for the self-sustaining combustion in low-oxygen atmosphere that is typical for thermal conversion processes. Figure 29.2 shows that addition of the plant oil moves the peat very close to low-grade coals relating to required parameters of conversion process.

The spatial distributions of the process parameters inside the fuel layer obtained from the simulation data show that allothermal process has one feature similar for all mentioned fuels. The high temperature area is mostly placed inside the illuminated area with very small intrusion into the outside zone (the temperature goes down below 500 °C when radial coordinate is 20% bigger than light-spot size). As a result, the initial performance of chemical reactions has a radial distribution that gradually falls around the center of the lit area. Following changes of the concentration of water, carbon and carbon dioxide modulate the reactions performance. The temporal evolution of typical distribution of the combustible non-bonded carbon is shown in Figure 29.3. It shows four states corresponding to thermochemical conversion of CEW after 1 s, 4 s, 20 s, and 50 s of the heating when the intensity of light flow is about 800 W/cm². The shape of these distributions correlate well with corresponding spatial temperature patterns.

As one can see, during the first 20 seconds of fuel conversion, the cavern inside the carbon distribution occurs. The front of the reaction has a very high gradient of the carbon concentration due to the high enough temperature (up to 900–1000 °C) at the external surface on the top of fuel layer.

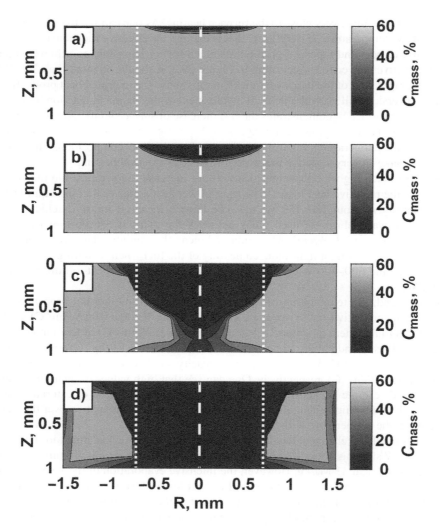

FIGURE 29.3 The simulated distribution of the carbon inside the CEW layer corresponding to heating time about 1 s (a), 4 s (b), 20 s (c), and 50 s (d). White dotted lines at $R = \pm 0.7$ mm correspond to the borders of lit area. The coordinates are normalized to $R_{max} = 0.75$ mm and $Z_{max} = 1$ mm.

The temperature quickly decreases going deeper into the fuel layer. After 1 min of heating, the temperature at the bottom surface of layer achieves more than 600 °C. The focused light supports oxidation of carbon inside the lit area making a hole in its spatial distribution.

After the half a 40 s of the heating, some of the carbon outside the lit area is burned too. However, the oxidation here goes at relatively low temperature and the domination reaction here is (1) whenever the central part of the lit area produces mostly CO. This is result of the domination of reaction (2) at high temperature together with contribution of Boudouard reaction (3).

After 20 s of fuel heating, the carbon distribution is burned out at the central axis of the light flow (Figure 29.3c). However, the temperature on the front of combustion is not so high this moment due to the presence of the ash layer above together with essential decrease of the oxygen filtration. As result, the shape of cavern become much narrower with appearance of the essential gradients of carbon concentration. The oxygen income from the bottom surface initiates the counter-flow of fuel combustion when temperature achieves high enough values.

Further heating of the sample that does not contain combustibles on the central axis leads to enlargement of the cavern inside the spatial distribution of the carbon with certain decrease of the average temperature. It continues up to the moment when the incident light flow still makes a heating of combustibles. The temperature of the fuel layer quickly decays when the diameter of empty cavern inside the spatial distribution of carbon exceeds the size of the focal spot of the light flow.

Finally, one can see, that amount of the burned carbon outside the lit area is about 10% of the volume covered by the irradiated spot. The chemical reactions here go with domination of the CO_2 production.

Switching from CEW to the peat–water composition (Egorov and Taburchinov 2021) moves us to the qualitatively similar results. The external heating determines the spatial distribution of the reaction zone where the high-temperature fuel conversion occurs. The high-temperature oxidation consumes an oxygen so fast that local deficit of the oxidant is always present blocking the self-sustaining combustion as well as the strong production of the CO_2. The process quickly decays with radial coordinate. Such localization of the fuel combustion is suitable because it allows the predictable propagation of the reaction-volume. Therefore, one can estimate that closing the light flow leads to fast stopping of the gasification process. The typical process temperatures vary for different fuels with respect to the intensity of the incident light. However, the domination of the external heat flow on the chemical reaction heat makes all the process patterns qualitatively similar.

Introducing the rapeseed oil to the peat with the corresponding set of the strong heat dissipating processes makes the fuel conversion more dependent on the external heating. As a result, the localization of the volume with high performance of the chemical reactions becomes more compact under lit area of the fuel surface.

Figure 29.4 shows the normalized dependences of the gas production on the incident light intensity. One can see that dependences of concentrations of particular gases on light intensity are different. During the conversion of the wet peat (blue curves), the production of CO_2, CH_4, and NO_x demonstrate the presence of evident intensity threshold (~100 W/cm²) after which production of these gases accelerated. This threshold clearly corresponds to temperature threshold (~900 °C) that defines the moment when chemical reactions achieve the new level of performance. Concentrations of all rest gases (CO, H_2, SO_2) increase gradually with light (heating) intensity. The presence of the threshold reflects initiation of the fuel burning. In the case of wet peat, it switches from slow smoldering to much faster combustion mode.

4292 ppm, 24,536 ppm, 22,411 ppm for CO (produced from P + RO, CEW + W, and P + W correspondingly)
21,167 ppm, 0 ppm, 5600 ppm for CO_2
30,148 ppm, 990 ppm, 520 ppm for CH_4
0 ppm, 139 ppm, 125 ppm for SO_2
1993 ppm, 1360 ppm, 6570 ppm for H_2
2 ppm, 50 ppm, 201 ppm for NO_x

Conversion of the mixture of peat and rapeseed oil always shows the gradually increasing concentrations of all gases with light intensity. This means that there are not any qualitative changes of the process performance inside the described range of the light intensities. Pyrolysis of the oil together with evaporation of the volatile liquids produce very big amount of volatiles per time unit and the produced gases make cooling of the fuel layer. Together with this, the strong flow of outgoing gases decreases the oxygen incomes preventing the fuel oxidation. Therefore, most of the reactions that can produce an additional heat are suppressed even when external heating is so strong (800–900 W/cm²).

Conversion of the CEW slurry shows presence of the light intensity threshold (~800 W/cm²) for CO, H_2, and NO_x that reflects the ignition of fuel. However, it is unstable and combustion stops in couple of seconds after lightening is off. Production of methane and sulfur dioxide has mostly pyrolytic origin.

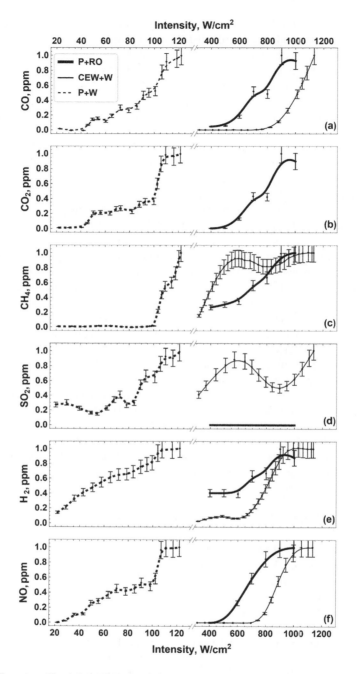

FIGURE 29.4 The normalized dependences of the average gas production on the incoming light intensity. The normalization factors are as follows:

Comparing the presented in Figure 29.4 dependences with previously obtained similar dependencies describing the trivial fuel combustion, one can see, that application of the external heating makes the trends much different. The domination of the CO_2 is absent even for wet peat which was converted at lower intensity of external heating. This fact is clear because higher than usual temperatures lead to stronger production of CO as well as to increasing of the CO_2 reduction when pure carbon is present (during the conversion of CEW slurry). The additional effect of the overheating is acceleration of the oxidation reactions and the yielding gases decrease the incoming of the oxygen in the hot fuel surface (without essential decrease of its temperature). As result, the process mostly goes with a deficit of the oxidant. The residual water oxidize carbon too, making an additional channel of CO production. However, the main benefits of the low-oxygen state take the pyrolysis process. Finally, the fuels with very high amount of macromolecular components (P + RO) can be converted to the mixture of gases containing many combustibles (CH_4, CO, and H_2). The very fast consumption of the oxygen at high temperatures switch the combustion of such fuels to domination of pyrolysis under the strong external heating.

The generalization of presented data allows classifying the fuel properties according to their influence on the thermochemical conversion. Chemical composition of the fuel determines the efficiency of its self-heating at high temperature. When the self-heating is weak, the process totally depends on the external heating. This is typically observed for most of the wastes. In this case, the thermal properties define the maximal level of temperature available for conversion of such composition. The fuels with low heat conductivity potentially achieve higher temperatures with lower heating intensity. This means that optimally fuel have to be maximally crushed. The CEW containing high amount of mineral matter has high enough heat conductivity and thus requires higher heating intensity. The evident recommendation is preparation of the fuel composition of CEW with other components that have a low heat conductivity (like low-grade fossil combustibles).

An application of the electromagnetic radiation (light or microwaves) for external heating of the fuel has a great potential due to the possibility of the extremely high flow focusing with relatively small sizes of supporting infrastructure. The principal limitation of this approach is caused by the thermodynamic fundamentals that require the increase of the radiative heat losses with temperature growth. Certain temperature losses overtake any heat incoming and further increase of the incident light intensity becomes useless. However, choosing the shell material that is transparent for light (or microwaves) and absorbing the long waves of heat radiation, one can achieve higher saturation temperature.

The principal domination of the external heating on the distributed chemical reactions heat makes the process of fuel conversion much more predictable and driven in comparison with self-sustaining processes. This fact allows making different configurations of the reactor adapting it to better recuperation of the residual heat. The presented results show the features of fuel conversion at normal pressure, but the focused flow of electromagnetic waves is also appropriate for the pressurized conversion techniques.

29.4 CONCLUSION

The flexible approach of the allothermal fuel conversion can be realized with usage of the focused flow of electromagnetic radiation as a heating source. It allows getting temperature that is much higher than occurring at self-sustaining combustion of waste or low-grade fuels. The intensity of radiation flow can be adjusted for particular fuel, making the optimization of the conversion process.

The conversion of water slurry of coal-enrichment waste effectively starts when light intensity is higher than ignition threshold (800 W/cm^2) as well as conversion of wet peat (100 W/cm^2). However, the conversion of mixtures with high ratio of liquid macromolecular components gradually accelerates with heating intensity increase (without any threshold effects).

The external heating allows effective conversion of the compositions whose maximum of reactivity occurs at temperature that does not allow high enough heat production. The auto-thermal conversion of such fuels is almost impossible. The domination of the external heating on the chemical reactions heat ultimately determines that conversion process mostly goes inside the lit area of the fuel layer.

The composition of produced gases essentially differs from that obtained at self-sustained combustion. The allothermal approach of CEW conversion allows domination of production of the combustible gases on the non-combustible even if the process goes at normal pressure.

ACKNOWLEDGMENTS

The research was performed in the frame of the project supported by Russian Foundation for Basic Research (project No. 19–53–80019).

REFERENCES

British Petroleum. 2021. *Statistical Review of World Energy 2021*. London: British Petroleum. www.bp.com/content/dam/bp/business-sites/en/global/corporate/pdfs/energy-economics/statistical-review/bp-stats-review-2021-full-report.pdf.

Egorov, Roman I., and Roman I. Taburchinov. 2021. "The Numerical Study of Allothermal Gasification of the Peat by the Focused Light Flow." *Applied Thermal Engineering* 195 (August): 117253. https://doi.org/10.1016/J.APPLTHERMALENG.2021.117253.

Egorov, Roman I., Roman I. Taburchinov, and Alexandr S. Zaitsev. 2019. "Efficiency of Different Heat Exchange Mechanisms for Ignition of Coal–Water Compositions." *Energy & Fuels* 33 (8): 7830–4. https://doi.org/10.1021/acs.energyfuels.9b01360.

Egorov, Roman I., Alexander S. Zaitsev, Hong Li, Xin Gao, and Pavel A. Strizhak. 2020. "Intensity Dependent Features of the Light-Induced Gasification of the Waste-Derived Coal-Water Compositions." *Renewable Energy* 146 (February): 1667–75. https://doi.org/10.1016/J.RENENE.2019.07.146.

Jianzhong, Liu, Wang Ruikun, Xi Jianfei, Zhou Junhu, and Cen Kefa. 2014. "Pilot-Scale Investigation on Slurrying, Combustion, and Slagging Characteristics of Coal Slurry Fuel Prepared Using Industrial Wasteliquid." *Applied Energy* 115 (February): 309–19. https://doi.org/10.1016/j.apenergy.2013.11.026.

Leckner, Bo. 2015. "Process Aspects in Combustion and Gasification Waste-to-Energy (WtE) Units." *Waste Management* 37 (March): 13–25. https://doi.org/10.1016/J.WASMAN.2014.04.019.

Minchener, Andrew J. 2005. "Coal Gasification for Advanced Power Generation." *Fuel* 84 (17): 2222–35. https://doi.org/10.1016/J.FUEL.2005.08.035.

Nejat, Payam, Fatemeh Jomehzadeh, Mohammad Mahdi Taheri, Mohammad Gohari, and Muhd Zaimi Muhd. 2015. "A Global Review of Energy Consumption, CO2 Emissions and Policy in the Residential Sector (with an Overview of the Top Ten CO2 Emitting Countries)." *Renewable and Sustainable Energy Reviews* 43: 843–62. https://doi.org/10.1016/J.RSER.2014.11.066.

Osintsev, K. V. 2012. "Studying Flame Combustion of Coal-Water Slurries in the Furnaces of Power-Generating Boilers." *Thermal Engineering* 59 (6): 439–45. https://doi.org/10.1134/S0040601512060079.

Tabakaev, Roman, Ivan Shanenkov, Alexander Kazakov, and Alexander Zavorin. 2017. "Thermal Processing of Biomass into High-Calorific Solid Composite Fuel." *Journal of Analytical and Applied Pyrolysis* 124 (March): 94–102. https://doi.org/10.1016/j.jaap.2017.02.016.

Valiullin, T.R., Roman I. Egorov, and Pavel A. Strizhak. 2017. "Perspectives of the Use of Rapeseed Oil for the Doping of Waste-Based Industrial Fuel." *Energy and Fuels* 31 (9): 10116–20. https://doi.org/10.1021/acs.energyfuels.7b01598.

Valiullin, T.R., K. Yu. Vershinina, D. O. Glushkov, and S. A. Shevyrev. 2017. "Droplet Ignition of Coal–Water Slurries Prepared from Typical Coal- and Oil-Processing Wastes." *Coke and Chemistry* 60 (5): 211–18. https://doi.org/10.3103/S1068364X17050076.

Yang, Ziqi, Yuanqing Wu, Zisheng Zhang, Hong Li, Xingang Li, Roman I. Egorov, Pavel A. Strizhak, and Xin Gao. 2019. "Recent Advances in Co-Thermochemical Conversions of Biomass with Fossil Fuels Focusing on the Synergistic Effects." *Renewable and Sustainable Energy Reviews* 103 (April): 384–98. https://doi.org/10.1016/j.rser.2018.12.047.

Yavuz, Reha, Sadriye Küçükbayrak, and Alan Williams. 1998. "Combustion Characteristics of Lignite-Water Slurries." *Fuel* 77 (11): 1229–35. https://doi.org/10.1016/S0016-2361(98)00017-9.

Zaitsev, Alexandr S., Roman I. Taburchinov, Irina P. Ozerova, Amaro O. Pereira, and Roman I. Egorov. 2020. "Allothermal Gasification of Peat and Lignite by a Focused Light Flow." *Applied Sciences* 10 (8): 2640. https://doi.org/10.3390/app10082640.

Zhao, Zhenyu, Salah Mohammed Abdullah Abdo, Xiaojun Wang, Hong Li, Xingang Li, and Xin Gao. 2021. "Process Intensification on Co-Pyrolysis of Polyethylene Terephthalate Wastes and Biomass via Microwave Energy: Synergetic Effect and Roles of Microwave Susceptor." *Journal of Analytical and Applied Pyrolysis* 158 (September): 105239. https://doi.org/10.1016/J.JAAP.2021.105239.

30 Membrane Technology in Circular Economy

Current Status and Future Projections

Lukka Thuyavan Yogarathinam, Ahmad Fauzi Ismail, Pei Sean Goh, and Anatharaman Narayanan

CONTENTS

30.1 Introduction 399
30.2 Overview of Pressure-driven Membrane Separation 400
30.3 Membrane Technology for Circular Economy 401
    30.3.1 Membrane Technology for Bio-products Circular Economy 401
    30.3.2 Membrane Technology for Water Circular Economy 404
30.4 Summary 406
References 408

30.1 INTRODUCTION

Water is the basis for humankind and the availability of fresh natural water on earth is only 0.5%. The remaining 97% and 2.5% of water is saline and trapped in the polar ice caps and the atmosphere, respectively (Gude and Nirmalakhandan 2009). Rapid industrialization, urbanization, and climate change are all providing severe challenges in obtaining freshwater in the present era. Additionally, natural disasters and other anthropogenic resources such as agricultural wastewater and toxic pollutants from industrial effluents, as well as home wastewater, have an impact on water quality (Lawrencia et al. 2022). According to the United Nations Environment Program (UNEP), two-thirds of the world population will confront water scarcity by 2025 (UNEP 2007). To overcome the water scarcity, seawater desalination and wastewater reclamation are the sustainable and economical approach to deliver clean potable water. Besides, circular economy of water has been proposed recently to recover the water (product) and value-added products from the wastewater treatment plant (waste). The three main sources of wastewater are municipal, agricultural, and industrial, which constitute various valued-added products such as nutrients, heavy metals, biomolecules (proteins, polyphenols, and lipids), and organic and inorganic compounds (Mehta et al. 2021). Importantly, biomolecule fractionation and water recovery are the most challenging in wastewater management. The circular economy scheme of waste management would aid in achieving the safe drinking water and water safety (Sustainable Development Goal [SGD 6]), secondary source for industrial application (process, design, and manufacturing), and natural resources regeneration (Zarei 2020).

In this regard, membrane technology is a widely deployed technique for the production of potable water and fractionation of value-added bio-products from wastewater.

The sustainable membrane–based techniques for various separation applications are reverse osmosis (RO), nanofiltration (NF), forward osmosis (FO), electrodialysis (ED), and membrane distillation

(MD). Additional membrane processes for wastewater reclamation include ultrafiltration (UF), microfiltration (MF), and membrane bioreactor (MBR). Membrane technology was desired over other conventional techniques due to the unique advantages of high selectivity (trade-off between water permeability and solute rejection), no phase-change operation, low-energy consumption, economic, ease of process intensification and system design, low floor space, and low carbon footprint (Goh, Wong, and Ismail 2022). However, membrane fouling is the common unavoidable phenomenon, which limits the membrane performance and longevity, demanding regular cleaning and maintenance.

In wastewater filtration, fouling occur in the form of organic fouling, inorganic fouling (scaling), biofouling, and colloidal fouling. Fouling rate differs in each membrane separation process due to a variety of factors such as driving force, pore size, and mode of operation. Fouling is either blocking or adsorption of solutes on membrane surface and pore substructures. Concentration polarization is accumulation of solute molecules or reversible cake layer formation on the membrane surface. Fouling and concentration polarization are common in the membrane separation process (Yogarathinam et al. 2022). Fouling relies on the composition of wastewater, membrane surface properties, and filtration process conditions. To overcome this limitation, membrane modification and integrative membrane system are widely adopted prominent strategies to minimize the fouling and deliverance of product recovery under energy-efficient condition. Hence, this chapter aimed to discuss the recent innovation of pressure-driven membrane processes (MF, UF, NF, and RO) on bioproducts and water circular economy from wastewater. It includes the value-added products such as polyphenols, proteins, lipids, and process waters.

30.2 OVERVIEW OF PRESSURE-DRIVEN MEMBRANE SEPARATION

Membrane is a thin selective barrier that controls the solute movement from feed solution based on driving forces such as pressure, temperature, electric current, and concentration gradient. Figure 30.1 shows the classification of pressure-driven membranes (Yang et al. 2019). MF, UF, NF, and RO are the four types of pressure-driven membrane separation and classified based on pore size. Size exclusion is the dominant mechanism in pressure-driven membrane processes and the electrostatic repulsion phenomena also contribute in NF and RO processes. MF and UF membranes have an asymmetric structure and their pore sizes range from 100 nm to 10 μm and 2–100 nm, respectively. In wastewater reclamation, MF and UF membranes are widely used to separate the macromolecular solutes such as oil, colloids, microorganism, proteins, natural organic material, microplastic, and polymers. The common commercial polymeric UF and MF membranes are polyacrylonitrile (PAN), polysulfone (PSf), polyethersulfone (PES), polyvinylidene fluoride (PVDF), and cellulose acetate (CA). NF and RO membranes have a thin film composite (TFC) structure and their pore sizes range from 1 to 2 nm and from 0.1 to 1 nm, respectively. NF and RO membranes are widely used for desalination applications and specifically in removing the heavy metals, dyes, organic acids, micropollutants, divalent salts, and monovalent salts (Lee, Elam, and Darling 2016). Membranes with hydrophilic properties, as well as superior chemical and mechanical stability, are desirable for circular economy. The most effective strategies to controlling fouling in the membrane separation process are to optimize operational parameters and modification of membranes. Hydrophobic polymeric membranes are frequently modified with hydrophilic polymers and biopolymers using coating, grafting, mixing, and self-assembly techniques. Nanomaterials have spurred interest in membrane technology due to their usage in the production of mixed matrix membranes (MMMs) and thin film nanocomposite (TFN) membranes to control membrane fouling and improve separation of desired product (Yin and Deng 2015).

Nanotechnology is the science and development of nanoscale systems, where nanoscale materials are organic or inorganic particles with sizes 1–100 nm. Based on their structural arrangement, nanomaterials are classified as zero-dimensional (0D), one-dimensional (1D), two-dimensional (2D), and three-dimensional (3D). Nanoclusters, nanocrystals, nanotubes, dendrimers, quantum dots, and fullerenes are the most frequent forms of nanomaterials. MMMs membranes are nanocomposite UF

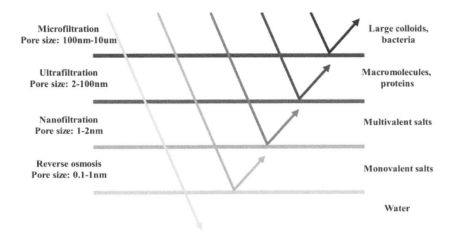

FIGURE 30.1 Classification of pressure-driven membrane (Yang et al. 2019).

and MF membranes made up of nanoparticles and organic polymers. TFN membranes using nanomaterial-based RO and NF membranes have the nanomaterial embedded in a selected thin layer of polyamide/polypiperazine. The key advantages of nanomaterial modification are increased water sorption capacity and hydrophilicity, as well as selective pollution removal and improved stability. The commonly using nanomaterials are nanometal oxides, nanoclays, carbon nanomaterial, metal organic frameworks (MOF), covalent organic frameworks (COFs), and hybrid nanocomposites. Ceramic membranes have the advantage of being able to perform at high temperatures while also being resistant to liquids and gases. Ceramics are made up of an inorganic asymmetric structure with either a dense or a skin porous layer.

30.3 MEMBRANE TECHNOLOGY FOR CIRCULAR ECONOMY

Under the circular economy concept, the latest development in membrane technology provides a sustainable solution for waste minimization and resource recovery in industrial applications.

30.3.1 MEMBRANE TECHNOLOGY FOR BIO-PRODUCTS CIRCULAR ECONOMY

Agro-processing wastes are abundant in bioactive chemicals and can be used as a source of biofuels, biogas, and nutritious substances. This section presents the recent development in membrane technology for the recovery of polyphenols, proteins, and lipids. Table 30.1 present the recent developments in MF, UF, NF, and RO processes for the recovery of bio-products from wastewater. Giacobbo et al. reported the integrated UF-NF system for the recovery of polyphenols and polysaccharides from racking wine lees (Giacobbo, Bernardes, and de Pinho 2017). UF membrane of 1 and 10 kDa were used to separate the polyphenols from polysaccharides in wine lees. NF membrane of 150–300 kDa displayed almost complete retention of total phenolic compounds and anthocyanins. Ochando-Pulido and Martínez-Férez compared the micro- and ultracentrifuge membranes for the recovery of polyphenols from Olive Mill Wastewater (Ochando-Pulido and Martínez-Férez 2017). The MWCO of chosen membranes are 3, 10, 30, 50, 100 kDa and 0.45 μm. A maximum of 57.1% total phenolic compound rejection was achieved in 3-kDa membrane. Uyttebroek et al. recovered the total phenolic compounds from apple pomace extracts using NF membrane (Uyttebroek et al. 2018). NF membrane of 150–300 Da (NFX, Synder Filtration [USA]) showed effective in organic solvent filtration as compared to other commercial NF membranes. NFX membranes exhibited

TABLE 30.1
Membrane Technology for the Recovery of Value-added Bio-products from Wastewater

Products	Source	Membrane Technology	Mechanism	Inferences	Reference
Polyphenols and polysaccharides	Wine lees	Integrated UF–NF system	Sieving	Integrated UF-NF system displayed a 92% and 99% rejection of total phenolic compounds and anthocyanins, respectively	(Giacobbo, Bernardes, and de Pinho 2017)
Protein	Potato processing water	Integrated approach of centrifugation–paper filtration–MF–UF	Sieving	Integrated UF (PES 10 kDa) process with conventional pretreatment (centrifugation, paper filtration, and MF) enhanced the protein concentration up to 3.5 than initial concentration. The antifouling tendency also improved up to 72% as compared to the conventional pretreatments	(Dabestani, Arcot, and Chen 2017)
Lycopene	Tomato waste	UF	Sieving	Tubular membrane of 200 kDa UF was studied for concentration and neutralization of lycopene from tomato waste under constant volume filtration. The total lycopene carotenoid improved significantly from 399 mg per 100 g to 1037.2 mg per 100 g dry matter for the neutralized samples	(Phinney et al. 2017)
Protein hydrolysate	Tuna wastes	Integrated UF–NF system	Sieving	Cascade membrane system also showed slight higher product yield than ideal system. Besides, cascade UF–NF system was effective for the water recovery and reclamation	(Abejón et al. 2018)
Whey protein	Cheese whey effluent	Metal oxide–embedded PES MMMs	Sieving	PES-TiO$_2$, PES-ZrO$_2$, and PES-ZnO MMMs were attempted to recover the whey protein from cheese whey effluent. A 2.0 wt% PES-ZrO$_2$ MMMs showed a maximum total whey protein rejection of 96% and concentrated up to 43.9 mg/L	(Yogarathinam et al. 2018)
Proteins	Starch wastewater	Integrated UF–NF system	Sieving	Integrated UF (2000 Da)–NF (200–1000 Da) hollow fiber membrane showed the protein recovery of 85.62 and 92.1%, respectively. Fouling was observed in both UF and NF module. Integrated UF–NF system is also a sustainable technology for the protein recovery and desalting of proteins	(Li et al. 2020)
Proteins	Sardine cooking effluent	Integrated UF–NF–RO system	Sieving	Integrated UF–NF showed effective system for the pretreatment of sardine cooking effluent. The hydrolyzed RO concentrate displayed higher antioxidant and antimicrobial activity	(Ghalamara et al. 2022)
Biodiesel	Dairy, paper and pulp, and biomass gasification	Hybrid bioreactor and MF system	Biodegradation and sieving	Integrated bioreactor-coupled ceramic MF system achieved a maximum COD rejection, and toxicity removal of 92.7% and 97.71% was attained in dairy wastewater. *R. opacus* yield a total lipid concentration in the range of 53–67% for industrial wastewaters	(Goswami et al. 2019)
Short-chain fatty acids	Waste sludge	Hybrid bioreactor and MF system	Biodegradation and sieving	Anaerobic fermentation-combined MF membrane module showed low-cost method for the recovery of short-chain fatty acids from waste sludge. The combination of membrane module and biocarriers enhanced the production of acetic, propionic, isobutyric, *n*-butyric, isovaleric, and *n*-valeric acid	(Q. Zhang et al. 2022)

97–98% rejection of phenolic compounds and the lower rejection of 92%, 78%, and 87% was also noticed for quinic acid, catechin, and epicatechin. According to the techno-economic impact evaluation, the price of electricity and the use of ethanol had a profound influence in a pilot-scale analysis. Sinichi et al. attempted the recovery of phenolic compounds from yellow mustard protein isolates wastewater using TFC polyamide NF (150–300 Da) membrane (Sinichi, Siañez, and Diosady 2019). UF and diafiltration method was used to permeate the total polyphenol from the protein isolates. In phenol recovery, TFC NF membrane showed the pH of feed solution had significant impact in control of phenolic retention due to membrane surface charge and phenolic chemical structures. Rodrigues et al. attempted to recover the phenolic compounds from winery wastewater using PES 10 kDa UF membrane and chelating agent as esterquat (Rodrigues, Gando-Ferreira, and Quina 2020). Lozada evaluated the recovery of bioactive antioxidant compounds from olive oil washing wastewater using bench-scale tubular 0.8 μm Al_2O_3 MF membrane (Lozada et al. 2022). The influence of transmembrane pressure (TMP), flow rate, temperature and pH on membrane flux, and total antioxidant rejection was also analyzed. A maximum of 51.4% of total antioxidants was achieved under the experimental condition of TMP, temperature, flow rate, and pH of 2.5 bar, 25°C, 3 L/min, and 7.63, respectively.

Dabestani et al. studied the recovery of protein from potato processing water using UF PES 10 kDa membrane (Dabestani, Arcot, and Chen 2017). The impact of various pretreatment methods on producing a high protein recovery was also evaluated and the chosen pretreatment methods are sedimentation, centrifugation, paper filtration, and MF. The protein concentration increased up to 3.5 times for the pretreatment methods combination of centrifugation, paper filtration, and MF followed by UF. The pretreatment methods also reduced the fouling by the elimination of secondary products and the effective cleaning solution was found to be NaOH solution. Zhang et al. attempted the protein recovery and reclamation of alfalfa wastewater using rotating disk UF membrane module (W. Zhang et al. 2017). Figure 30.2a shows the schematic of protein recovery from alfalfa wastewater. Rotating disk UF membrane module (Figure 30.2b) was mainly used to minimize the fouling and the influence of different MWCO membranes ranging from 5 kDa to 100 kDa was also analyzed. UF membrane of 30 kDa membrane under 55 °C displayed a superior permeability and crude protein concentration of 9.36 $L/m^2/h/bar$ and 23.35 g/L, respectively. The scheme also proposed as permeated water for agricultural irrigation and concentrated protein for fodder for animals. Abejón et al. designed the cascade UF–NF system for the recovery of protein hydrolysates from tuna wastes (Abejón et al. 2018). Cascade UF–NF system model displayed a highest product purity of 49.5% and the integrated system also showed effective approach in control of clogging, water management, and cost-optimization. Pezeshk et al. studied the different MWCO regenerated cellulose UF membrane (3, 10, and 30 kDa) for the recovery of protein hydrolysates from yellowfin tuna viscera (Pezeshk et al. 2019). The protein fractions from low MWCO PES 3 kDa displayed antibacterial and antioxidant activities. Li et al. integrated the self-fabricated hollow fiber UF (20,000 Da) and TFC NF (200–1000 Da) membranes for the recovery of protein from starch wastewater (Li et al. 2020). The water permeability of hollow fiber UF and NF membranes are 58.4 and 4.1 $L/m^2/h/bar$, respectively. UF and NF membranes displayed rejection of 85.62% of high-molecular-weight protein and 92.1% of low-molecular-weight protein, respectively.

Zhang et al. recovered the short-chain fatty acids from waste sludge using aerobic bioreactor–coupled MF membrane module (Q. Zhang et al. 2022). To improve recovery, 0.1-μm PVDF membrane and low-density polyethylene were used as membrane material and biocarriers, respectively. The combination of membrane module and biocarriers enhanced the concentration of short-chain fatty acids up to 3500 mg/L. Biocarriers also aided to decrease the membrane fouling and MF membrane aided to retain the organic matter and bacteria. Goswami et al. attempted the biodiesel production from the integrated bioreactor–coupled ceramic MF system using *Rhodococcus opacus* from the dairy, biomass gasification, and paper and pulp wastewater (Goswami et al. 2019). Integrated bioreactor–coupled ceramic MF system achieved a chemical oxygen demand (COD) rejection of 92.7%, 87.6%, and 88.2% for the dairy, paper and pulp, and biomass gasification wastewater, respectively.

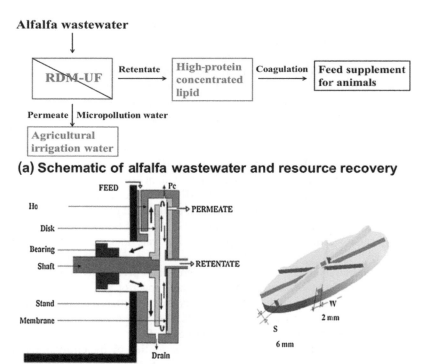

FIGURE 30.2 Schematic of (a) alfalfa wastewater reclamation and (b) rotating disk membrane module (W. Zhang et al. 2017).

The toxicity removal percentage of 97.71%, 94.48%, and 92.86% were also observed for the dairy, paper, and pulp and biomass gasification wastewater, respectively. The higher concentration of total intracellular lipids of *R. opacus* was 67.36% for the dairy wastewater.

30.3.2 Membrane Technology for Water Circular Economy

Membrane technology offered a promising technique in the reuse of saline water for various purposes which include process water and potable water. The sources of saline water are municipal wastewater and industrial effluents such as oily wastewater, heavy metal polluted water, and textile wastewater. The waste streams constitute organic micropollutant, microorganism, and inorganic heavy metals and ions. MF, UF, and NF membranes were used as a pretreatment stage in the wastewater reclamation process prior to the RO process. Mahmodi et al. coated the ZnO layer on α-alumina MF membrane using atomic layer deposition (ALD) for the produced water reclamation under vacuum filtration (Mahmodi et al. 2022). Figure 30.3a and b shows the schematic of ZnO layer ALD on α-alumina membrane and vacuum-assisted PW filtration. The influence of coating cycles (0–120) and thickness (0–14 nm) of ZnO layer deposition on α-alumina membrane performance was also studied. The pore size of α-alumina MF membrane decreased from 200 to 184.7 nm for the 120-cycle ALD–based ZnO layer–coated α-alumina membrane. A maximum TOC rejection of 99% with a flux of 185.7 L/m^2/h was achieved in 120-cycle ZnO layer–coated α-alumina membrane for the concentration of 4500 mg/L. Treated water standards also meet the Environmental Protection Agency (EPA) guidelines for livestock and irrigation watering.

Manouchehri and Kargari reported the reclamation of laundry wastewater using 0.22-μm hydrophilic cellulose ester membrane (Manouchehri and Kargari 2017). The impact of process conditions such as TMP (0.2–1.5 bar) and flow rate (30 t to 80 L/h) on membrane flux and fouling analysis was also evaluated. The optimum condition was found to be 0.5 bar and 80 L/h. In domestic laundry water filtration, cellulose ester membrane exhibited 98.4% and 95.8% for turbidity and total suspended solids, respectively. The treated water would also be used for toilets and irrigation. Manni et al. developed a low-cost magnesite ceramic MF membrane for the treatment of textile wastewater (Manni et al. 2020). Flat disk ceramic membrane was fabricated using uniaxial pressing and sintering methods. The effect of sintering temperature ranging from 900 °C to 1200 °C was studied corresponding to membrane formation. The porous structure with no defect was observed in magnesite MF membrane fabricated at 1100 °C. A 1.12-μm magnesite membrane showed the superior water permeability and turbidity rejection of 922 L/h/m^2/bar and 99.9%, respectively. Ozbey-Unal et al. studied the integrated chemical treatment-MF-RO system for the treatment of industrial zone wastewater (Ozbey-Unal et al. 2020). The performance of three commercial polyamide RO membranes (BW30 [Dow- Filmtech], HP [TriSep], and LE [Dow- Filmtech]) and different modes of integrated approach of chemical treatment, MF, and RO were also evaluated. Among the RO membranes, BW30 membrane displayed 80% recovery of water within 260 min under lab-scale. In the pilot-scale operation, the chemical dosing and pH control have significant influence in control of membrane fouling in integrated membrane process. The organic and inorganic compounds rejection efficiency in RO membranes ranged from 87% to 95.5.%. The permeate water quality has good alignment with standards for the usage in industrial cooling and boiler water systems. An et al. attempted the treatment of semiconductor wastewater using hybrid membrane system of ceramic MF-advanced oxidation process (AOP) with RO (An et al. 2022). Commercial Al_2O_3/SiO_2-ZrO_2 MF membrane (mean pore size: 50 nm) coupled AOP system displayed the turbidity rejection of 98% for the real-time semiconductor wastewater and was used as a pretreatment for RO system. The performance of three commercial polyamide RO membranes such as ESPA2-LD (NittoDenko, Japan), RE4040-BE (CSM, Toray Advanced Materials Korea Inc.), and TMG-10D (Toray, Japan) were also analyzed. ESPA2-LD membrane showed a nominal permeability of 29 L/m^2/h/bar with better salt rejection of ~95% and TOC rejection for the real-time semiconductor wastewater.

Huang et al. attempted the treatment of cold-rolling emulsion wastewater treatment using commercial 20-kDa $ZrO_2/α$-Al_2O_3 ceramic UF membrane (Huang et al. 2022). A maximum of 99.88% for oil and 98.41% for COD rejection was observed. The parameter optimization and membrane cleaning (alkaline and acid solution washing) method indicated the sustainable technique for the prolonged UF membrane performance. Such techniques also improve the process economy and waste minimization. Yogarathinam et al. studied the binary metal oxide titania-zirconia (TiZr) and titania-zinc oxide (TiZn) incorporated sulfonated poly(ether ether ketone) (SPEEK)/PES MMMs for the pretreatment of seawater desalination (Lukka Thuyavan et al. 2021). In longer duration filtration, 0.5 wt% TiZr/SPEEK/PES MMMs showed a steady-state flux of 88 L/m^2/h with a water recovery of 47%. Binary metal oxide–incorporated SPEEK/PES MMMs also showed antifouling tendency against natural organic matter fouling and biofouling resistant against *E. coli*. Jarma et al. evaluated the reclamation of geothermal wastewater using mini-pilot-scale spiral wound NF and RO process (Jarma et al. 2022). The influence of brine to feed ratio was studied to evaluate the fouling and membrane performance analysis. The brine management study revealed that the geothermal reinjection fluid can be used as a feed solution for the recovery of water for secondary applications. At constant TMP condition, the water recovery of 60% was achieved within 4 h. Özgün et al. compared the different pretreatment methods to the NF-RO system for the real-time textile wastewater reclamation (Özgün et al. 2022). The chosen pretreatment methods are MF–UF, ozonation, ultraviolet (UV) irradiation, and titanium dioxide (TiO_2) and zeolite adsorption. A 0.05-μm PVDF membrane demonstrated a superior pretreatment technique for the COD and color removal for real-time textile wastewater. The MF-RO hybrid system showed the maximum COD, conductivity, and color rejection of 97.5%, 93.6% and 99.3%, respectively.

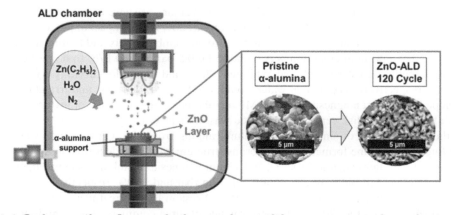

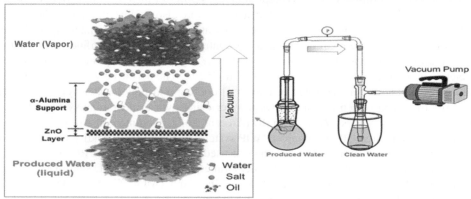

FIGURE 30.3 Schematic of (a) ALD coating and (b) vacuum-assisted filtration (Mahmodi et al. 2022).

Recently, TFN membranes showed increased water permeability without offsetting for solute rejection in RO desalination. The recent advancement of NF and RO TFN membranes with different types of functional membranes are listed in Table 30.2. As shown in Table 30.2, emerging RO membranes have shown to be a feasible and futuristic membrane for desalination.

30.4 SUMMARY

Membrane technology has proven to be an important separation technique in a variety of industrial applications for the recovery of value-added products from waste resources. This chapter described the recent innovations in the recovery of bio-products and water from wastewater using pressure-driven membrane separation process. In the bio-product circular economy, MF and UF membranes are widely used, especially to reject large molecular weight proteins and lipids from agro-food industry wastewater. The recovery of polyphenols from wastewater using NF and integrated membrane technology has been demonstrated to be a viable method for recovering smaller organic compounds. RO has been deployed as a promising and sustainable method in production of potable water from wastewater and saline resources. MF and UF membranes are also practiced in combination with conventional treatment methods to produce process water from industrial wastewater. Fouling is a critical issue in membrane separation–based bio-product and water circular economy.

TABLE 30.2
TFN NF and RO Membrane for Desalination Application

Nanomaterials	Membrane Technology	Modification	Inferences	Reference
Polydopamine (PDA)-modified MoS$_2$ (MoS$_2$ @ PDA) nanosheets	NF	Interfacial polymerization	A 0.01 wt% MoS$_2$ @PDA nanosheets–incorporated TFN NF membrane exhibited superior water permeability and Na$_2$SO$_4$ rejection of 6.7 L/m^2/h/bar and 98.2%, respectively. PDA improved the molecular interaction between MoS$_2$ and polyamide layer	(Xie et al. 2022)
Polyethyleneimine (PEI)-assembled ZIF-8	NF	Interfacial polymerization	PEI-assembled ZIF-8 TFN NF membrane was studied for the removal of pharmaceuticals and personal care products (PCPs) from aqueous solution. PEI-ZIF-8 TFN membrane held the maximum water permeability of 18.55 L/m^2/h/bar with Na$_2$SO$_4$ rejection of 93.4%. The selected PCPs are methylparaben, propylparaben, bisphenol A, atenolol, norfloxacin, and ofloxacin. The higher rejection of more than 80% was observed for norfloxacin and ofloxacin. The rejection decreased in both TFC and TFN membranes due to the hydrophobic characteristics of PCPs	(Guo et al. 2022)
Titanium dioxide/carbon dots (TiO$_2$/CDs)	RO	Interfacial polymerization	Photocatalyst TiO$_2$/CDs nanocomposite–incorporated polyamide TFN membrane displayed a maximum permeability of 1.39 L/m^2/h/bar with a NaCl rejection of 99.2%. TiO$_2$/CDs nanocomposite also improved the chlorine resistance and antifouling tendency in the membrane	(Vatanpour et al. 2022)
Titania nanosheets (TNS) and polyethyleneimine (PEI)	RO	Interfacial polymerization and layer by layer (LbL)	Alternate layers of titania nanosheets (TNS) and polyethyleneimine (PEI) were formed on polyamide TFC membrane using spraying-assisted layer-by-layer assembly technique. The single-layer spray-coated PEI–TNS membrane displayed highest water permeability and NaCl rejection of 1.39 L/m^2/h/bar and 97%, respectively. The surface hydrophilic and antifouling property was also improved in spray-coated techniques	(Ahmad et al. 2022)

Novel nanocomposite MMMs and TFN membranes have recently risen to prominence in the wastewater and desalination sectors for their potential to deliver solute rejection without flux balance. Integrative membrane systems have shown to be successful in reducing fouling and environmental pollution. The other emerging membrane technologies such as forward osmosis (FO), membrane distillation (MD), and electrodialysis (ED), in combination with such pressure-driven integrative membrane process, could provide a sustainable and energy-efficient waste management strategy. The other emerging membrane technique such as forward osmosis (FO), membrane distillation (MD), and electrodialysis (ED) with integrative membrane process could also be a sustainable and energy-efficient technique in waste management. Overall, this chapter would be beneficial to industrialists and scholars interested in using pressure-driven membrane processes in the bio-product and water circular economies.

REFERENCES

Abejón, R., M. P. Belleville, J. Sanchez-Marcano, A. Garea, and A. Irabien. 2018. "Optimal Design of Industrial Scale Continuous Process for Fractionation by Membrane Technologies of Protein Hydrolysate Derived from Fish Wastes." *Separation and Purification Technology* 197. Elsevier: 137–46. doi:10.1016/j.seppur.2017.12.057.

Ahmad, Nor Akalili, Pei Sean Goh, Kar Chun Wong, Stanley Chinedu Mamah, Ahmad Fauzi Ismail, and Abdul Karim Zulhairun. 2022. "Accelerated Spraying-Assisted Layer by Layer Assembly of Polyethyleneimine/Titania Nanosheet on Thin Film Composite Membrane for Reverse Osmosis Desalination." *Desalination* 529 (October 2021). Elsevier: 115645. doi:10.1016/j.desal.2022.115645.

An, Sun-A., Jonghun Lee, Jeonghoo Sim, Cheol-Gyu Park, Jin-San Lee, Hojung Rho, Kwang-Duck Park, Han-Seung Kim, and Yun Chul Woo. 2022. "Evaluation of the Advanced Oxidation Process Integrated with Microfiltration for Reverse Osmosis to Treat Semiconductor Wastewater." *Process Safety and Environmental Protection* 162. Elsevier: 1057–66. doi:10.1016/j.psep.2022.05.010.

Dabestani, Shirin, Jayashree Arcot, and Vicki Chen. 2017. "Protein Recovery from Potato Processing Water: Pre-Treatment and Membrane Fouling Minimization." *Journal of Food Engineering* 195. Elsevier: 85–96. doi:10.1016/j.jfoodeng.2016.09.013.

Ghalamara, Soudabeh, Ezequiel R. Coscueta, Sara Silva, Carla Brazinha, Carlos D. Pereira, and Manuela E. Pintado. 2022. "Integrated Ultrafiltration, Nanofiltration, and Reverse Osmosis Pilot Process to Produce Bioactive Protein/Peptide Fractions from Sardine Cooking Effluent." *Journal of Environmental Management* 317. Elsevier: 115344. doi:10.1016/j.jenvman.2022.115344.

Giacobbo, Alexandre, Andréa Moura Bernardes, and Maria Norberta de Pinho. 2017. "Sequential Pressure-Driven Membrane Operations to Recover and Fractionate Polyphenols and Polysaccharides from Second Racking Wine Lees." *Separation and Purification Technology* 173. Elsevier: 49–54. doi:10.1016/j.seppur.2016.09.007.

Goh, P. S., K. C. Wong, and A. F. Ismail. 2022. "Membrane Technology: A Versatile Tool for Saline Wastewater Treatment and Resource Recovery." *Desalination* 521. Elsevier: 115377. doi:10.1016/j.desal.2021.115377.

Goswami, Lalit, R. Vinoth Kumar, Kannan Pakshirajan, and G. Pugazhenthi. 2019. "A Novel Integrated Biodegradation – Microfiltration System for Sustainable Wastewater Treatment and Energy Recovery." *Journal of Hazardous Materials* 365. Elsevier: 707–15. doi:10.1016/j.jhazmat.2018.11.029.

Gude, Veera Gnaneswar, and Nagamany Nirmalakhandan. 2009. "Desalination at Low Temperatures and Low Pressures." *Desalination* 244 (1–3). Elsevier: 239–47. doi:10.1016/j.desal.2008.06.005.

Guo, Zhiqiang, Hongli Wang, Liang Wang, Bin Zhao, Yiran Qian, and Hongwei Zhang. 2022. "Polyamide Thin-Film Nanocomposite Membrane Containing Star-Shaped ZIF-8 with Enhanced Water Permeance and PPCPs Removal." *Separation and Purification Technology* 292 (January). Elsevier: 120886. doi:10.1016/j.seppur.2022.120886.

Huang, Yanjun, Heng Liu, Yi Wang, Guangsen Song, and Lei Zhang. 2022. "Industrial Application of Ceramic Ultrafiltration Membrane in Cold-Rolling Emulsion Wastewater Treatment." *Separation and Purification Technology* 289 (February). Elsevier: 120724. doi:10.1016/j.seppur.2022.120724.

Jarma, Yakubu A., Aslı Karaoğlu, Alper Baba Islam Rashad Ahmed Senan, and Nalan Kabay. 2022. "Brine Minimization in Desalination of the Geothermal Reinjection Fluid by Pressure-Driven Membrane Separation Processes." *Desalination* 535. Elsevier: 115840. doi:10.1016/j.desal.2022.115840.

Lawrencia, Dora, Lay Hong Chuah, Phatchani Srikhumsuk, and Phaik Eong Poh. 2022. "Biodegradation Factors and Kinetic Studies of Point-of-Use Water Treatment Membrane in Soil." *Process Safety and Environmental Protection* 161. Elsevier: 392–408. doi:10.1016/j.psep.2022.03.053.

Lee, Anna, Jeffrey W. Elam, and Seth B. Darling. 2016. "Membrane Materials for Water Purification: Design, Development, and Application." *Environmental Science: Water Research and Technology* 2 (1). Royal Society of Chemistry: 17–42. doi:10.1039/c5ew00159e.

Li, Hongbin, Xianhua Zeng, Wenying Shi, Haixia Zhang, Shoufa Huang, Rong Zhou, and Xiaohong Qin. 2020. "Recovery and Purification of Potato Proteins from Potato Starch Wastewater by Hollow Fiber Separation Membrane Integrated Process." *Innovative Food Science and Emerging Technologies* 63. Elsevier: 102380. doi:10.1016/j.ifset.2020.102380.

Lozada, Gabriela Soledad Live, Ana Isabel García López, Antonio Martínez-Férez, and Javier M. Ochando-Pulido. 2022. "Boundary Flux Modelling of Ceramic Tubular Microfiltration Towards Fouling Control and Performance Maximization for Olive-Oil Washing Wastewater Treatment and Revalorization." *Journal of Environmental Chemical Engineering* 10 (2): 107323. doi:10.1016/j.jece.2022.107323.

Lukka Thuyavan, Y., G. Arthanareeswaran, A. F. Ismail, P. S. Goh, M. V. Shankar, B. C. Ng, R. Sathish Kumar, and K. Venkatesh. 2021. "Binary Metal Oxides Incorporated Polyethersulfone Ultrafiltration Mixed Matrix Membranes for the Pretreatment of Seawater Desalination." *Journal of Applied Polymer Science* 138. Elsevier: e49883. doi:10.1002/app.49883.

Mahmodi, Ghader, Anil Ronte, Shailesh Dangwal, Phadindra Wagle, Elena Echeverria, Bratin Sengupta, Vahid Vatanpour, David N. McIlroy, Joshua D. Ramsey, and Seok Jhin Kim. 2022. "Improving Antifouling Property of Alumina Microfiltration Membranes by Using Atomic Layer Deposition Technique for Produced Water Treatment." *Desalination* 523. Elsevier: 115400. doi:10.1016/j.desal.2021.115400.

Manni, A., B. Achiou, A. Karim, A. Harrati, C. Sadik, M. Ouammou, S. Alami Younssi, and A. El Bouari. 2020. "New Low-Cost Ceramic Microfiltration Membrane Made from Natural Magnesite for Industrial Wastewater Treatment." *Journal of Environmental Chemical Engineering* 8 (4). Elsevier: 103906. doi:10.1016/j.jece.2020.103906.

Manouchehri, Massoumeh, and Ali Kargari. 2017. "Water Recovery from Laundry Wastewater by the Cross Flow Microfiltration Process: A Strategy for Water Recycling in Residential Buildings." *Journal of Cleaner Production* 168. Elsevier: 227–38. doi:10.1016/j.jclepro.2017.08.211.

Mehta, Nidhi, Kinjal J. Shah, Yu I. Lin, Yongjun Sun, and Shu Yuan Pan. 2021. "Advances in Circular Bioeconomy Technologies: From Agricultural Wastewater to Value-Added Resources." *Environments – MDPI* 8: 20. doi:10.3390/environments8030020.

Ochando-Pulido, Javier M., and Antonio Martínez-Férez. 2017. "About the Recovery of the Phenolic Fraction from Olive Mill Wastewater by Micro and Ultracentrifugation Membranes." *Chemical Engineering Transactions* 60: 271–6. doi:10.3303/CET1760046.

Ozbey-Unal, Bahar, Philip Isaac Omwene, Meltem Yagcioglu, Çigdem Balcik-Canbolat, Ahmet Karagunduz, Bulent Keskinler, and Nadir Dizge. 2020. "Treatment of Organized Industrial Zone Wastewater by Microfiltration/Reverse Osmosis Membrane Process for Water Recovery: From Lab to Pilot Scale." *Journal of Water Process Engineering* 38 (August). Elsevier: 101646. doi:10.1016/j.jwpe.2020.101646.

Özgün, H., H. Sakar, M. Ağtaş, and I. Koyuncu. 2022. "Investigation of Pre-Treatment Techniques to Improve Membrane Performance in Real Textile Wastewater Treatment." *International Journal of Environmental Science and Technology*. Springer. doi:10.1007/s13762-022-04034-w.

Pezeshk, Samaneh, Seyed Mahdi Ojagh, Masoud Rezaei, and Bahareh Shabanpour. 2019. "Fractionation of Protein Hydrolysates of Fish Waste Using Membrane Ultrafiltration: Investigation of Antibacterial and Antioxidant Activities." *Probiotics and Antimicrobial Proteins* 11 (3). Probiotics and Antimicrobial Proteins: 1015–22. doi:10.1007/s12602-018-9483-y.

Phinney, David M., John C. Frelka, Jessica L. Cooperstone, Steven J. Schwartz, and Dennis R. Heldman. 2017. "Effect of Solvent Addition Sequence on Lycopene Extraction Efficiency from Membrane Neutralized Caustic Peeled Tomato Waste." *Food Chemistry* 215: 354–61. doi:10.1016/j.foodchem.2016.07.178.

Rodrigues, Rafaela P., Licínio M. Gando-Ferreira, and Margarida J. Quina. 2020. "Micellar Enhanced Ultrafiltration for the Valorization of Phenolic Compounds and Polysaccharides from Winery Wastewaters." *Journal of Water Process Engineering* 38. Elsevier: 101565. doi:10.1016/j.jwpe.2020.101565.

Sinichi, Sayeh, Ana Victoria Legorreta Siañez, and Levente L. Diosady. 2019. "Recovery of Phenolic Compounds from the By-Products of Yellow Mustard Protein Isolation." *Food Research International* 115. Elsevier: 460–6. doi:10.1016/j.foodres.2018.10.047.

UNEP. 2007. *Global Environment Outlook: GEO 4 Environment for Development*. United Nations Environment Programme.

Uyttebroek, Maarten, Pieter Vandezande, Miet Van Dael, Sam Vloemans, Bart Noten, Bas Bongers, Wim Porto-Carrero, Maria Muñiz Unamunzaga, Metin Bulut, and Bert Lemmens. 2018. "Concentration of Phenolic Compounds from Apple Pomace Extracts by Nanofiltration at Lab and Pilot Scale with a Techno-Economic Assessment." *Journal of Food Process Engineering* 41 (1). Elsevier: e12629. doi:10.1111/jfpe.12629.

Vatanpour, Vahid, Shadi Paziresh, Seyed Ali Naziri Mehrabani, Solmaz Feizpoor, Aziz Habibi-Yangjeh, and Ismail Koyuncu. 2022. "TiO2/CDs Modified Thin-Film Nanocomposite Polyamide Membrane for Simultaneous Enhancement of Antifouling and Chlorine-Resistance Performance." *Desalination* 525 (December 2021). Elsevier B.V.: 115506. doi:10.1016/j.desal.2021.115506.

Xie, Fei, Wen-Xuan Li, Xin-Yu Gong, Dovletjan Taymazov, Han-Zhuo Ding, Hao Zhang, Xiao-Hua Ma, and Zhen-Liang Xu. 2022. "MoS2 @PDA Thin-Film Nanocomposite Nanofiltration Membrane for Simultaneously Improved Permeability and Selectivity." *Journal of Environmental Chemical Engineering* 10 (3). Elsevier: 107697. doi:10.1016/j.jece.2022.107697.

Yang, Zi, Yi Zhou, Zhiyuan Feng, Xiaobo Rui, Tong Zhang, and Zhien Zhang. 2019. "A Review on Reverse Osmosis and Nanofiltration Membranes for Water Purification." *Polymers* 11. Elsevier: 1252. doi:10.3390/polym11081252.

Yin, Jun, and Baolin Deng. 2015. "Polymer-Matrix Nanocomposite Membranes for Water Treatment." *Journal of Membrane Science* 479. Elsevier: 256–75. doi:10.1016/j.memsci.2014.11.019.

Yogarathinam, Lukka Thuyavan, Arthanareeswaran Gangasalam, Ahmad Fauzi Ismail, Sivasamy Arumugam, and Anantharaman Narayanan. 2018. "Concentration of Whey Protein from Cheese Whey Effluent Using Ultrafiltration by Combination of Hydrophilic Metal Oxides and Hydrophobic Polymer." *Journal of Chemical Technology and Biotechnology* 93 (9). Elsevier: 2576–91. doi:10.1002/jctb.5611.

Yogarathinam, Lukka Thuyavan, Kirubakaran Velswamy, Arthanareeswaran Gangasalam, Ahmad Fauzi Ismail, Pei Sean Goh, Anantharaman Narayanan, and Mohd Sohaimi Abdullah. 2022. "Performance Evaluation of Whey Flux in Dead-End and Cross-Flow Modes via Convolutional Neural Networks." *Journal of Environmental Management* 301 (January 2021). Elsevier: 113872. doi:10.1016/j.jenvman.2021.113872.

Zarei, Mohanna. 2020. "Wastewater Resources Management for Energy Recovery from Circular Economy Perspective." *Water-Energy Nexus* 3. KeAi Communications Co.: 170–85. doi:10.1016/j.wen.2020.11.001.

Zhang, Qianqian, Linyu Wu, Jianghao Huang, Yuetong Qu, Yu Pan, Li Liu, and Hongtao Zhu. 2022. "Recovering Short-Chain Fatty Acids from Waste Sludge via Biocarriers and Microfiltration Enhanced Anaerobic Fermentation." *Resources, Conservation and Recycling* 182 (April). Elsevier: 106342. doi:10.1016/j.resconrec.2022.106342.

Zhang, Wenxiang, Luhui Ding, Nabil Grimi, Michel Y. Jaffrin, and Bing Tang. 2017. "A Rotating Disk Ultrafiltration Process for Recycling Alfalfa Wastewater." *Separation and Purification Technology* 188. Elsevier: 476–84. doi:10.1016/j.seppur.2017.07.037.

Index

2D CFD model, 247
3D CFD model, 247

A

ablative reactor, 62
accelerating model, 214–215
activated carbon, 2, 6, 9–13
activated charcoal, 7
adsorbent, 2, 6–12, 14–18, 20, 22
adsorption, 15, 17, 19, 22, 23, 25, 26, 28, 29
agricultural waste, 8, 11, 12, 290
air-cathode, 110
algae, 103–106, 108–115, see also microalgae
algal fuel cell, 105–107
algal microbial fuel cell, 104–107
allothermal conversion, 390
anaerobic digestion, 54, 56
analysis, 280–281
analytic hierarchy process, 302–303, 305
anode, 104, 108–113, 115, 116
    for BES, 17
    material, 17
    performance, 16
    properties, 16
anoxic, 104, 109–112
Appiko Movement, 181
application of nanoparticles (NPs) and nanocomposites (NCs), 31–33, 35
aromatization, 94, 95
Arrhenius equation, 207, 208
artificial neural network approach, 248
ASPEN plus modeling, 247
atomic layer deposition, 404
Auger reactor, 62, 195

B

bacterio-algal fuel cell, 105, 106
baffled microbial fuel cell, 81
bagasse, 28
batteries, 12, 18, 19, 21, 22, 24, 25
bed material, 251, 253, 254
benthic microbial fuel cell, 80
BES
    challenges/outlook, 21
    future scope, 22
    parameters, 15
bibliographical, 105–106
biocatalyst, 264–270
    synthesis, 267
biocathode, 105, 112
bio-char, 8, 17, 190
biodegradable, 234, 236, 237
biodiesel, 222, 261–267, 288, 297–298, 300, 302–307
    consumption, 261
    production, 261–267, 298
    synthesis, 261, 262–265
biodigester, 292

biodigestors, 312
bioelectricity, 104–105, 108–110, 114–115
bioelectrochemical system, 15, 105, see also BES
bioenergy, 297–300, 302
    potential, 297
biofilm, 111
    enrichment factors, 16
biofuel, 104, 113, 297–302, 336–338
    technology, 295
biogas, 288
biological treatment, 54
biomass, 94–97, 100, 104, 108–110, 112, 114, 191, 224, 226, 276, 277, 298, 349–359, 363–372
    particle size, 251–254
    waste derived anode, 18
bio-oil, 2, 6–9, 12, 96–99, 101, 129, 190, 337, 341
bio-products, 401, 402
bioresources, 103
Brazil, 297, 299, 302–303, 307
bubbling fluidized bed combustors, 60
buffers, 108, 112, 116

C

calcium oxide, 224
capital expenditure, 281
carbonaceous material, 99
carbon anode, 17
carbon dioxide, 107–109
carbon footprint, 104, 109
carbon nanotube, 73
carbon sequestration, 109
carcinogenic, 16, 17
case study, 297–298, 302, 307
catalyst concentration, 263–265, 268
catalyst, 41, 46, 47, 251, 253, 254, 261–265, 268–269
catalytic pyrolysis, 94–95
cathode, 108, 110–111, 113–114, 116
cellulose, 11, 12, 13
CFD – DEM model, 247
CFD modeling approach, 246, 259
challenges, 4, 7, 15, 25
char, 95, 129
characteristics, 53–54, 61
characterization, 261–263, 267–268
charcoal, 7
chemical kinetics, 206–208
Chipko Movement, 179
circular economy, 400, 401
circulating bed reactor, 62
circulating fluidized bed, 60
Citespace, 105
classification, 122–123
coal, 349–359, 377–385
coal-enrichment waste, 389–397
coal-water slurry, 379–385, 390–397
Coastal Regulation Zone, 184
co-gasification, 336
coke, 96–97, 100–101

combined heat and power, 319
combined kinetic and equilibrium approach, 246
combustible gases, 369
combustion, 363–372, 377–385
concentration, 366–372
condenser, 195
conducting polymer, 74
contamination, 10, 25
conventional biodiesel production, 261–264
conventional methods, 1, 6, 17, 25
conversion, 1, 4, 10, 11, 15, 25
cooking oil, 265, 268
co-pyrolysis mechanism, 191, 192
cost analysis, 81
cost-benefit index, 293
cryogenic separation, 280

D

deactivation, 97–98, 100
decelerating models, 214, 215
disease, 234
dissolved oxygen, 108, 110
double chambered, 109–110
double chamber microbial fuel cell, 71, 76
dual fuel system, 288
dumping, 234, 236–237
dyes, 17, 24, 26, 27, 28

E

edible oil, 263–264
electric infrared incinerators, 61
electric vehicle, 342
electrode materials, 72
electrode reactions, 70
electrogenic, 108, 113
electrolysis, 113, 283
electrolytic coagulation, 113
electrolytic flocculation, 113
electrolytic flotation, 113
electron acceptor, 104, 110
electron transfer in BES, 16
electronic, 1–8, 11, 12, 14, 15, 17, 25
emissions, 377–385
energy consumption, 261, 267
energy crops, 298, 302–303, 305–307
energy demand, 261
energy scenario, 261, 262
engine, 2, 7, 9, 12
Environment Protection Act, 184
environment, 2, 4–8, 10, 17, 25, 26
environmental, 233, 236, 237, 239
    movements, 177–178
    sustainability, 167
enzymatic hydrolysis, 56
equilibrium modeling, 243
equivalence ratio, 252–254
eutrophication, 104, 108, 114–115
e-waste, 2–12, 14–17, 25

F

fast pyrolysis, 61–62
fatty acid, 263, 265–266
feedstock, 222
    selection, 261–262
fixed bed reactor, 61
flash pyrolysis, 61
flat plate microbial fuel cell, 81
Flexible Diesel-Natural Gas System, 288
fluidized bed incinerators, 60
fluidized bed reactor, 62, 197
fly ash, 13, 16
focused light, 390
footprint, 109
fossil fuels, 221
free fatty acids, 224
FTIR, 264–267
fuel cell, 342–344
fuel mixture, 351

G

gas analyzer, 352–355, 367–368
gasification, 62, 121, 123, 238, 243, 274–277, 312
gasifier types, 63
gasifying agent, 251, 253, 254
gas mixture, 354–359
generation, 233, 234, 235, 236, 237, 239
graphitic anode, 17
greenhouse gases, 1, 339, 344
green hydrogen, 273, 277, 281, 283
green synthesis of NPs/NCs by using agro-waste materials, 33, 35
gross-calorific value, 393

H

hazardous, 2, 4–6, 8–10, 15, 25
    waste management, 323
health, 2, 4, 6, 7, 16, 18, 25
heating rate, 95
heterogeneous catalyst, 224, 225, 263, 265, 268
HHV, 3, 8
high-density polyethylene, 323
homogeneous catalyst, 223, 263–265
hybrid electric vehicle, 342
hybrid vehicles, 342
hydrocarbon, 221
hydrogen demand, 272
hydrogen production, 273, 276–277, 281, 283
hydrometallurgical, 2, 9, 14, 15, 25

I

incinerating, 12, 16, 25, 41, 42, 46, 47, 54, 57, 60–61, 63, 312
India, 235, 237
industrial waste, 6, 9
industrialization, 3, 17, 25
influence of gasifier parameters, 250
influencing factors, 125–128
innovations, 54–59, 60–63

Index

intensity threshold, 395
interfacial polymerization, 407

K

kinetic modeling, 246, 256

L

landfill, 237
landfill depletion, 287
land filling, 2, 7, 25, 26, 41, 46
leaching, 10, 13, 16, 25
LHV, 3, 6
life cycle assessment, 298–303, 324
life cycle cost, 298–299, 301–302
life cycle sustainability assessment, 298–303, 305–306
light intensity, 104, 107–109
linear curve fitting, 209, 211–213, 216
lubricant oil plastic containers, 323

M

machine learning approaches, 248
maximal heat production, 392
maximum concentration, 355–359
mechanical biologic treatment, 312
mechanical recycling, 41, 42, 46
membrane, 108–109, 111–112, 116, 280
membrane technology for biodiesel production, 227
mesoporous silica, 99
metabolic pathway, 108, 110
metal anode, 17
metal oxides, 98
microalgae, 262–263, 267
microbial fuel cell, 104–105, 106, 107, 109, 111, 113
microcrack, 382
micropollutant, 7
microwave, 226
    -assisted biodiesel production, 226, 265
    -assisted pyrolysis, 62
    -assisted transesterification, 265
    pyrolysis, 349–359, 366–372
    reactor, 199
mixed matrix membranes, 400
mixture of peat and rapeseed oil, 393–395
model fitting approach, 213, 215
model free approach, 209, 210–211
moisture content, 252–254
molar ratio, 263–265
multiple criteria decision-making, 299–300, 302
multiple hearth furnaces, 60
municipal, 233, 234–236, 239
municipal solid waste, 279, 311, 335

N

Nafion, 109, 111, 112
nanoadsorbents, 1, 2, 6, 9, 10, 16, 19, 21
nanobiotechnology, 31
nanocatalyst, 264, 270
nanocellulose, 15, 25, 28, 30
nanofiltration, 399

nanokaolin, 15
nanomaterials, 1, 11, 12, 14, 15, 17, 25
    and nanoparticles, 29, 30, 31
nanotechnology, 17, 25, 27, 28
Narmada Bachao Andolan, 181
non-biodegradable, 15
non-condensable gas, 192
non-edible oil, 262–263
non-governmental organizations, 171
non-stoichiometric approach, 245

O

oil extraction, 263
operating conditions, 63
optimization, 264
organic waste, 205
output, 105
oxygen reduction reaction, 73

P

palm oil, 297–298, 302–307
petrochemical, 7
pharmaceutical, 18, 19
photobioreactor, 108, 111
photoinhibition, 109
photosynthesis, 104, 108–109
photosynthetic algal microbial fuel cell, 104, 106
photosynthetic microbial fuel cell, 105, 109, 111
plasma pyrolysis reactor, 199
plastic, 191
pollution, 2, 5, 8, 17, 25, 26, 233, 236, 238, 239
posttreatments, 227
potash, 2–3, 6, 9–12
power density, 104–105, 108–109, 110, 112
power to cost ratio, 81
pressure, 250–251, 253, 254
pressure-driven membrane processes, 400
pressure swing adsorption, 279
process, 7, 9, 10, 14, 15, 17, 18, 25
proton exchange membrane, 75
    fuel cell, 344
purification, 10, 25
pyrolysis, 1–2, 6–7, 9–12, 25, 41, 43, 45–49, 51, 54, 57–58, 60–63, 121, 122, 205, 216, 233, 236, 239, 337, 363–372, 379–385
    reactor, 61
pyro-metallurgical, 2, 14–16, 25

R

reaction rate, 206, 207
reaction time, 264–265, 267
recovery, 2, 9–12, 14, 15, 17, 25, 234, 236, 237, 239
recyclable, 234
recycling, 2, 4, 6–11, 14, 15, 17, 20, 25
renewable, 103
resource depletion, 41
reusability, 225
reverse logistics, 325
reverse osmosis, 399
rice husk, 8, 10, 11, 15, 17, 19–21

rotary kiln reactor, 195
rotating disk UF membrane module, 403

S

sanitation, 172, 174
scientometric, 105
seaweed, 14
self-help groups, 170
SEM, 264–265, 269
semi kinetic model, 246
separation, 7, 9, 10, 13, 14, 15, 25
sigmoidal mode, 214, 215
The Silent Valley Movement, 180
single chambered, 108, 110, 111
single chamber microbial fuel cell, 71, 76
slag, 8, 13
slow pyrolysis, 61, 122, 124
social-environmental impacts, 299
social life cycle assessment, 298–299, 302
solar gasification, 1–5, 9, 11–12
solid acid catalyst, 263
solid state reaction, 214, 215
solid waste, 287
sources of plastic wastes, 41, 44
soybean, 297–298, 302–307
spent batteries, 6, 12, 25
stacks, 78
steam gasification, 123, 275
steam-injected gasification, 276
stirrer speed, 263
stoichiometric approach, 244–245
support vector machine approach, 248
surface modification/functionalization, 19
sustainability, 297–303, 305–307
    dimensions, 298
    indicators, 300
    of palm oil, 303
    of soybean, 303
sustainable, 104–105, 109, 116, 233, 239
sustainable BES, 21
sustainable development goals, 167
sustainable waste management strategies, 323
synergistic effect, 191
syngas, 129–130

T

tar, 129
technique, 15, 17, 25, 236, 238, 239
techno-economic, 1–2, 11–12, 280–281

TEM, 263–268
temperature, 250, 253, 254, 263–265, 267
thermal degradation, 205, 206
thermochemical power plant, 315
thermochemical recycling, 41, 46
thermochemical treatment, 57–59, 63
thermogravimetric analysis, 206
total kinetic model, 246
toxic, 235
transesterification, 265–266
transition metals, 4, 25
transition metal oxides, 72
triple bottom line, 298, 300, 303
tubular reactor, 198

U

ultrafiltration, 400
ultrasonic, 56
ultrasonic-assisted biodiesel production, 226, 267–269
ultrasonic-assisted transesterification, 266
up flow microbial fuel cell, 81
urban waste, 291

V

value-added products, 105, 113, 114

W

waste, 233–239
    management, 3, 8, 11, 12, 17, 26, 185
    plastic oil, 41, 43, 49, 50, 51
    plastic production, 41, 43
    shells, 224
    to value, 205, 217
waste-to-energy technologies, 311
wastewater, 2, 4, 6–8, 12, 13, 22
water, 1–19, 21, 22, 399
    pollution, 3, 4

X

XRD, 264, 266

Z

zeolite, 11, 13, 19, 96–98
zero carbon, 105, 109